ESO ASTROPHYSICS SYMPOSIA
European Southern Observatory

Series Editor: Philippe Crane

Springer-Verlag Berlin Heidelberg GmbH

Peter A. Shaver (Ed.)

Science with Large Millimetre Arrays

Proceedings of the ESO-IRAM-NFRA-Onsala
Workshop, Held at Garching, Germany
11–13 December 1995

 Springer

Volume Editor

Peter A. Shaver
European Southern Observatory
Karl-Schwarzschild-Strasse 2
D-85748 Garching, Germany
email: pshaver@eso.org

Series Editor

Philippe Crane
European Southern Observatory
Karl-Schwarzschild-Strasse 2
D-85748 Garching, Germany

Cataloging-in-Publication data applied for

Science with large millimetre arrays : proceedings of the ESO
IRAM NFRA Onsala workshop, held at Garching, Germany, 11
- 13 December 1995 / Peter A. Shaver (ed.). - Berlin ;
Heidelberg ; New York ; Barcelona ; Budapest ; Hong Kong ;
London ; Milan ; Paris ; Santa Clara ; Singapore ; Tokyo :
Springer, 1996
(ESO astrophysics symposia)

NE: Shaver, Peter A. [Hrsg.]; European Southern Observatory

ISBN 978-3-662-22395-6   ISBN 978-3-540-69999-6 (eBook)
DOI 10.1007/978-3-540-69999-6

Typesetting: Camera ready by authors
Cover design: Springer-Verlag Design & Production
SPIN: 10517782   55/3142-543210 - Printed on acid-free paper

Preface

The next major step in millimetre astronomy, and one of the highest-priority items in radio astronomy today, is a large millimetre array with a collecting area of up to 10 000 m². A project of this scale will almost certainly require international collaboration, at least within Europe, and possibly with other major partners elsewhere. In order to establish a focal point for this project within Europe, a study has been undertaken by the Institut de Radio Astronomie Millimétrique (IRAM), the European Southern Observatory (ESO), The Onsala Space Observatory (OSO), and The Netherlands Foundation for Research in Astronomy (NFRA). In the context of this project, a workshop attended by some 100 participants was held at ESO Garching on December 11–13, 1995 to discuss the scientific advances such an array will make possible.

Throughout the three days of the workshop the strong enthusiasm for the concept of a large millimetre array in the southern hemisphere (the Large Southern Array, or LSA) was obvious, and it became clear that such a facility would have a profound impact on almost all areas of observational astrophysics. It was particularly clear that, since their main science drivers (cosmology, and the origins of galaxies, stars and planets) are the same, and their angular resolutions and sensitivities similar, the LSA and the VLT would strongly complement each other. The LSA was repeatedly described as the "millimetre equivalent of the HST and the VLT". It is hoped that this enthusiasm is fully conveyed in these proceedings, and that these activities will ultimately lead to the realization of the LSA itself.

It is a pleasure to acknowledge the input from the Scientific Organizing Committee: P. Andreani (Padova), R. Booth (Onsala), L. Bronfman (Santiago), D. Downes (IRAM), P. Encrenaz (Paris), M. Grewing (IRAM), S. Guilloteau (IRAM), R. Hills (Cambridge), F. Israel (Leiden), L.-A. Nyman (SEST), E. van Dishoeck (Leiden), F. Viallefond (Paris), J. Whiteoak (Culgoora), and T. Wilson (Bonn). I would also like to thank the participants for helping to make this workshop such a success, the ESO fellows and students for their assistance during the meeting, Pamela Bristow for her help with these proceedings, and especially Christina Stoffer for her indispensable role in the organization of all aspects of the workshop.

European Southern Observatory   Peter Shaver
Garching, June 1996

Contents

(b) Galactic and Solar System Studies

3. A Large Millimetre Array in the Southern Hemisphere

4. Workshop Summary

List of Participants

Name	Institution
ANDREANI, Paola	Università di Padova, Dip. di Astronomia andreani@pdmida.pd.astro.it
BAARS, Jacob	MPI für Radioastronomie, Bonn jbaars@mpifr-bonn.mpg.de
BÅÅTH, Lars	Halmstad University, Centre for Imaging Technologies Lars.Baath@cbd.hh.se
BACHILLER, Rafael	Observatorio Astronomico Nacional (IGN) Bachiller@cay.es
BAUDRY, Alain	Observatoire de Bordeaux baudry@observ.u-bordeaux.fr
BECKWITH, Steven	MPI für Astronomie, Heidelberg svwb@mpia-hd.mpg.de
BERGERON, Jacqueline	ESO, Garching jbergero@eso.org
BJÖRNSSON, Claes-Ingvar	Stockholm Observatory bjornsson@astro.su.se
BLAIN, Andrew	University of Cambridge, Cavendish Laboratory awb@mrao.cam.ac.uk
BOOTH, Roy	Onsala Space Observatory roy@oso.chalmers.se
BREMER, Michael	IRAM, Grenoble bremer@iram.fr
BRONFMAN, Leonardo	Universidad de Chile, Dep. de Astronomia leo@calan.das.uchile.cl
BROWN, Robert	NRAO, Charlottesville rbrown@nrao.edu
BUTCHER, Harvey	NFRA, Dwingeloo hbutcher@nfra.nl
CARRASCO, Luis	INAOE/UNAM, Puebla, Mexico carrasco@tonali.inaoep.mx
CHIN, Yi-nan	Radioastronomisches Institut der Universität Bonn einmann@astro.uni-bonn.de
CHINI, Rolf	MPI für Radioastronomie, Bonn rchini@mpifr-bonn.mpg.de

CLEMENTS, David — ESO, Garching
dclement@eso.org

COMBES, Françoise — Observatoire de Paris – DEMIRM
bottaro@obspm.fr

COX, Pierre — Obs. de Marseille
cox@obmara.cnrs-mrs.fr

CROVISIER, Jacques — Observatoire de Meudon
crovisie@mesioa.obspm.fr

CRISTIANI, Stefano — Università di Padova,
Dip. di Astronomia
cristiani@astrpd.pd.astro.it

DE ZOTTI, Gianfranco — Osservatorio Astronomico di Padova
dezotti@astrpd.pd.astro.it

DEWDNEY, Peter — DRAO/NRC, Penticton
ped@drao.nrc.ca

DOWNES, Dennis — IRAM, Grenoble
downes@iram.fr

DUTREY, Anne — IRAM, Grenoble
dutrey@iram.fr

EGAMI, Eiichi — MPI für extraterrestrische Physik,
Garching
egami@mpe-garching.mpg.de

ENCRENAZ, Pierre — Observatoire de Paris – DEMIRM
pencrenaz@mesiob.obspm.fr

FELDT, Markus — Max-Planck-Gesellschaft AG, Jena
mfeldt@astro.uni-jena.de

FELLI, Marcello — Osservatorio Astrofisico di Arcetri
mfelli@arcetri.astro.it

FRANSSON, Claes — Stockholm Observatory
claes@astro.su.se

GENSHEIMER, Paul — MPI für Radioastronomie, Bonn
paulg@fs1.mpifr-bonn.mpg.de

GENZEL, Reinhard — MPI für extraterrestrische Physik,
Garching
genzel@mpe-garching.mpg.de

GIACCONI, Riccardo — ESO, Garching
dg@eso.org

GREVE, Albert — IRAM, Grenoble
greve@iram.fr

GREWING, Michael — IRAM, Grenoble
grewing@iram.fr

GROENEWEGEN, Martin — MPI für Astrophysik, Garching
groen@mpa-garching.mpg.de

GUÉLIN, Michel — IRAM, Grenoble
guelin@iram.fr

GUILLOTEAU, Stéphane — IRAM, Grenoble
guillote@iram.fr

HALL, Peter — Australia Telescope National Facility, Parkes
phall@atnf.csiro.au

HARJU, Jorma — University of Helsinki, Observatory
jorma.harju@helsinki.fi

HILLS, Richard — MRAO, Cavendish Lab., Cambridge
richard@mrao.cam.ac.uk

ISHIZUKI, Sumio — Nobeyama Radio Observatory, National Astronomical Observatory
ishizuki@nro.nao.ac.jp

ISRAEL, Frank — Leiden Observatory
israel@strw.leidenuniv.nl

KRAMER, Carsten — IRAM, Granada
kramer@iram.es

KRICHBAUM, Thomas — MPI für Radioastronomie, Bonn
p459kri@mpifr-bonn.mpg.de

KUAN, Yi-Jehng — Institute of Astronomy & Astrophysics, Taipei (Taiwan)
kuan@biaa3.biaa.sinica.edu.tw

LAMB, James — IRAM, Grenoble
lamb@iram.fr

LAZAREFF, Bernard — IRAM, Grenoble
guelin@iram.fr

LEQUEUX, James — Observatoire de Paris – DEMIRM
lequeux@mesioa.obspm.fr

LONGAIR, Malcolm — Cavendish Laboratory, University of Cambridge
msl@mrao.cam.ac.uk

LUCAS, Robert — IRAM, Grenoble
lucas@iram.fr

MARTEN, André — Observatoire de Paris-Meudon, DESPA
marten@obspm.fr

MARTIN-PINTADO, Jesus — Centro Astronómico de Yebes, Guadalajara
martin@cay.es

MATAGNE, Jean — ST-ECF, Garching
jmatagne@eso.org

MAUERSBERGER, Rainer — University of Arizona, Steward Observatory
mauers@as.arizona.edu

MENTEN, Karl — Harvard Smithsonian Center for Astrophysics
menten@cfa.harvard.edu

MEZGER, Peter — MPI für Radioastronomie, Bonn
—

MONNET, Guy — ESO, Garching
gmonnet@eso.org

NATTA, Antonella — Osservatorio Astrofisico di Arcetri
natta@arcetri.astro.it

NÜRNBERGER, Dieter — Universität Würzburg, Astronomisches Institut
nurnberg@astro.uni-wuerzburg.de

NYMAN, Lars-Åke — ESO/SEST, La Silla
lnyman@eso.org

OLOFSSON, Hans — Stockholm Observatory
hans@astro.su.se

OMONT, Alain — Institut d'Astrophysique, Paris
omont@iap.fr

OSTERLOH, Martin — Max-Planck-Gesellschaft AG, Jena
osterloh@fred.astro.uni-jena.de

PALLAVICINI, Roberto — Osservatorio Astronomico di Palermo
pallavic@oapa.astropa.unipa.it

PLATHNER, Dietmar — IRAM, Grenoble
plathner@iram.fr

PRADERIE, Françoise — C.N.R.S. – DRI, Paris
francoise.praderie@cnrs-dir.fr

PRIETO, Almudena — MPI für extraterrestrische Physik, Garching
alm@rosat.mpe-garching.mpg.de

RENZINI, Alvio — ESO, Garching
arenzini@eso.org

RIGOPOULOU, Dimitra — MPI für extraterrestrische Physik, Garching
dar@mpe.mpe-garching.mpg.de

ROBSON, Ian — Joint Astronomy Centre, Hilo
eir@jach.hawaii.edu

ROWAN-ROBINSON, Michael — Imperial College, London
m.rrobinson@ic.ac.uk

SEIRADAKIS, John-Hugh — University of Thessaloniki, Department of Physics
jhs@astro.auth.gr

SHAVER, Peter — ESO, Garching
pshaver@eso.org

STAGUHN, Johannes — Universität Köln,
I. Physikalisches Institut
staguhn@ph1.uni-koeln.de

TACCONI, Linda — MPI für extraterrestrische Physik,
Garching
linda@mpe-garching.mpg.de

THUM, Clemens — IRAM, Grenoble
thum@iram.fr

TOFANI, Gianni — Osservatorio Astrofisico di Arcetri
tofani@arcetri.astro.it

ULRICH-DEMOULIN, Marie-Helene — ESO, Garching
mhulrich@eso.org

VAN ARDENNE, Arnold — NFRA, Dwingeloo
ardenne@nfra.nl

VAN DER WERF, Paul — Leiden Observatory
pvdwerf@strw.leidenuniv.nl

VIALLEFOND, François — Observatoire de Paris,
Radioastronomie Millimétrique, Paris
viallefond@obspm.fr

WAGNER, Stefan — Landessternwarte Heidelberg
swagner@mail.lsw.uni-heidelberg.de

WALL, William — INAOE, Puebla, Mexico
wwall@tonali.inaoep.mx

WALMSLEY, Malcolm — Universität Köln,
I. Physikalisches Institut
walmsley@apollo.ph1.uni-koeln.de

WENDKER, Heinz — Hamburger Sternwarte
hjwendker@hs.uni-hamburg.de

WHITE, Simon — MPI für Astrophysik, Garching
swhite@mpa-garching.mpg.de

WHITEOAK, John — Australia Telescope National Facility,
Narrabri
jwhiteoa@atnf.csiro.au

WIELEBINSKI, Richard — MPI für Radioastronomie, Bonn
p022rwi@mpifr-bonn.mpg.de

WIKLIND, Tommy — Onsala Space Observatory
tommy@oso.chalmers.se

WILD, Wolfgang — IRAM, Granada
wild@iram.es

WILSON, Tom — MPI für Radioastronomie, Bonn
p073twi@mpifr-bonn.mpg.de

WOLTJER, Lodewijk — Observatoire de Haute-Provence
—

ZENSUS, Anton — NRAO, Charlottesville
azensus@nrao.edu

Part 1

Introduction

Millimetre Astronomy in the 21st Century

Malcolm S. Longair

Cavendish Laboratory, University of Cambridge, Cambridge CB3 0HE, England

Abstract. The Large Southern Array (LSA) will open up new ways of tackling many of the most important problems of the astrophysics of the 21st century. It is shown that high resolution, high sensitivity observations in the millimetre waveband open up new ways of tackling the problems of the origin of stars and of the origin and evolution of galaxies, in particular, the rate at which their stellar populations were formed as a function of cosmic epoch. The large negative K-corrections for star-forming galaxies at large redshifts plus the effects of evolutionary changes with cosmic epoch suggest that the counts of objects at redshifts $z \gtrsim 1$ should be very steep and the LSA should easily reach flux densities at which this steep count can be observed. If the counts are as steep as expected, strong gravitational lensing is likely to be important for sources on the steep region of the source counts.

1 Introduction

The scientific objectives of the Large Southern Array (LSA) are simply summarised by listing the topics highlighted in the volume *The Large Southern Array* prepared by the ESO-IRAM-NFRA-Onsala consortium.

1. Galaxies and quasars at high redshift.
2. Radio galaxies and quasars at redshifts $0.1 < z < 2$.
3. Molecular clouds in our Galaxy and nearby galaxies.
4. Accretion discs and outflow jets of protostars.
5. Protoplanetary discs around young stellar objects.
6. Circumstellar envelopes and expelled shells of stars of all types.
7. Objects within our own solar system.

These constitute areas of current research of the highest scientific importance and the case for studying them using the techniques of millimetre and submillimetre astronomy is outstanding. The science is predominantly the *Astrophysics of the Cold Universe*. I would argue, however, that, although the topics appear to be somewhat disparate, they are in fact all part of one story.

I will also argue that they are part of a multi-disciplinary, multi-wavelength approach to the great problems of astronomy. To obtain the complete picture, we also have to look outside the millimetre/sub-millimetre wavebands. For example, the remarkable observation of jets from young stellar objects such as HH30 by the *Hubble Space Telescope* (*HST*) is precisely the type of observation which is complementary to those made in the millimetre wavebands. Likewise, the protoplanetary discs about low-mass stars in the Orion Nebula observed by O'Dell and his colleagues are the bread and butter of millimetre astronomy but now replicated in the optical waveband (O'Dell 1996). These are two obvious examples but similar cases can be found in all the topics in the above list. A key point is that the images obtained by the LSA will be of similar angular resolution to those obtained by the *HST* and this is becoming the state-of-the-art for all the complementary wavebands.

An important aspect of the multi-wavelength approach is complementarity with the plans in the USA for the Millimetre Array (MMA) and the Japanese Large Millimetre and Submillimetre Array (LMSA). These projects have as their primary goal the scientific exploitation of the very highest submillimetre windows available from high dry sites, whereas the principal goal of the LSA is the exploitation of the wavelength interval 1 to 3 mm with large collecting area.

In one way or another, many aspects of the science of these projects is related to the rôle of the interstellar gas and star formation in different guises. For example,

1. The origin of planetary systems and of our Solar System is an integral part of the process of star formation.
2. The recycling of processed material to the interstellar medium provides the molecules and dust for the next generation of new stars.
3. The evolution of galaxies depends crucially upon understanding in some detail the processes of star formation.
4. The understanding of the origin of galaxies depends upon working out the sequence of star formation activity which leads to the structures of galaxies as we know them today
5. There may be the need to form baryonic dark matter in the form of population III stars and various types of inert lumps of matter.

The traditional story is that the essential ingredients of regions of star formation are present in spiral and irregular galaxies but one of the more remarkable developments, which was described by Garth Illingworth at the Paris HST-2 conference, was the fact that there is certainly dust, and presumably molecules in various S0 and elliptical galaxies as well (Illingworth 1996). These results have come from careful subtraction of the average light profiles of galaxies to reveal regions of patchy obscuration which would be difficult to detect on saturated optical photographs of the inner regions of galaxies. Among the examples he presented were the cD galaxy NGC 6166, which contains a prominent dust lane threading its way through the multiple nuclei of this system; the elliptical galaxy IC 1459 which shows patchy obscuration towards the nucleus and the S0 galaxy NGC 6861, which to all appearances is a standard spiral galaxy. There must be significant amounts of dust and molecules in these systems.

In the same spirit, it is notable that a number of active galaxies appear to have significant masses of gas which are ultimately responsible for powering the active galactic nuclei. For example, in the cases of the active galaxies M87 and NGC 4261, *HST* observations have been made of ionised gas close to their nuclei and these have enabled the masses of the central black holes to be determined. In both of these examples, there is evidence for widespread gas throughout the galaxies, despite the fact that they are classified as giant ellipticals. There must be molecules and cold gas present as well in these systems. This is certainly the case for the galaxy NGC 4258 in which H_2O megamasers have been observed within 10 milliarcsec of the nucleus and which have enabled the highest mass densities of any active galactic nucleus to be determined (Miyoshi *et al.* 1995). There is also the increasingly convincing picture of dusty tori in the nuclei of active galaxies which are required by unification schemes for active galaxies. These must also contain molecules and cool gas.

The reason for drawing attention to these observations is that it appears that dust and molecules are everywhere and not just in spiral and irregular galaxies. Millimetre and submillimetre observations are thus important for all classes of galaxy.

2 Observations of Distant Galaxies

I will review some of the observations presented at the Paris HST-2 meeting, since I believe that they are of the greatest importance for millimetre astronomy and cosmology. First of all, it is important to appreciate exactly what it is that is observed when optical HST observations extend to redshifts $z \geq 0.5$. Giavalisco *et al.* (1996) described computations of what galaxies look like at large redshifts by using Astro-1 Shuttle observations of nearby galaxies in ultraviolet wavebands at 152 nm and 249 nm. Their computations make the important point that, if galaxies are to be classified successfully at redshifts $z > 1$ by the HST, they must either have greater surface brightnesses or luminosities than the typical galaxies which we observe nearby. We will find that there are good observational reasons why this is, in fact, the case. It is also clear that it is possible to make excellent progress in the morphological classification of galaxies out to redshifts of about 0.5, and greater for the more luminous galaxies, and this has been accomplished in a number of recent studies.

The morphologies of faint galaxies have been studied by Abraham *et al.* (1996) and Driver *et al.* (1995, 1996). Different approaches are taken to assigning morphological types to the galaxies. These procedures are of considerable importance in understanding the excess counts of faint blue galaxies observed in deep galaxy surveys. Figure 1 shows the counts of galaxies of different morphological types from the *HST* Medium Deep Survey. The solid lines show the expected counts of uniform world models with $q_0 = \frac{1}{2}$ and it can be observed that the E/S0 and spiral samples follow closely these expectations. The important result is that the excess blue galaxies are associated with galaxies classified as Irregulars/Mergers. A similar result has been found by Driver *et al.* (1995) in their analysis of a single ultra-deep WFPC2 deep field which was observed for 5.7 hours in each of the V (F606W) and I (F814W) wavebands.

There has been considerable debate about the nature of the excess faint blue galaxy population. It had been expected that there might well be more blue galaxies at faint apparent magnitudes because even passively evolving models of galaxies suggest that the old stellar population should be brighter in the past

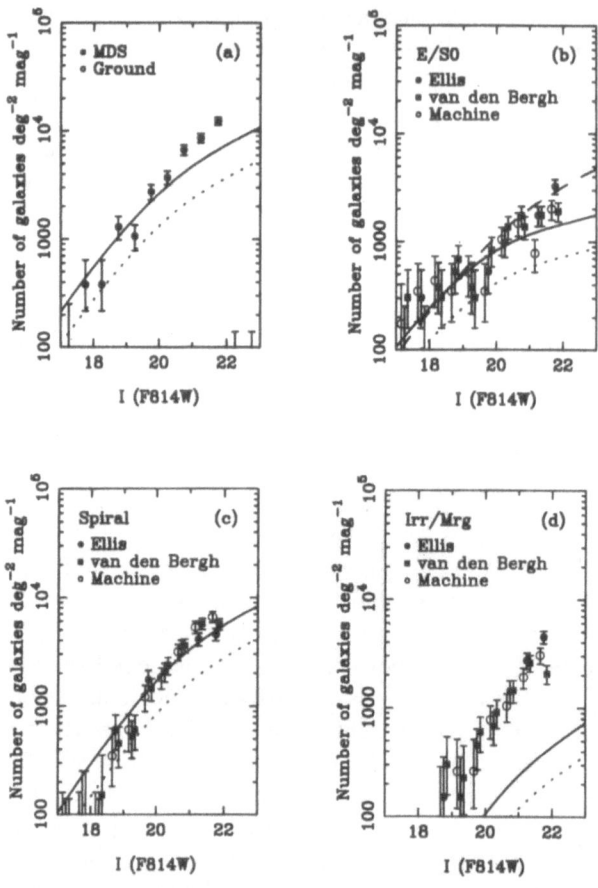

Fig. 1. The counts of faint galaxies of different morphological types. The counts as determined by van den Bergh (vdB), Ellis (RSE) and according to automatic classification procedures for the same sample of galaxies are indicated by different symbols. The solid lines show the expected galaxy counts for a uniform world model with no evolution. The pronounced excess of irregular/merging galaxies is apparent (Abraham *et al.* 1996).

and there should be more star formation activity. One of the surprises of the first redshift surveys, which extended to apparent magnitudes at which the blue excess was observed, was that the mean redshift of the galaxies at the faintest magnitudes did not increase any more rapidly than would have been expected if the galaxy population had remained unchanged with cosmic epoch (Glazebrook *et al.* 1995). This result was interpreted as indicating that the excess blue galaxies were associated with a population of dwarf galaxies. There must, however, be more to the story than this.

It is now becoming possible to study fainter samples of galaxies and I give three examples which indicate what is now becoming the state of the art. Schade reported the most recent results of the Canada-France Redshift Survey (Le Fèvre *et al.* 1995). This magnitude-limited complete survey contains 943 objects with magnitudes $17.5 \leq I_{AB} \leq 22.5$. The objects for which redshifts have been measured extend to redshifts $z = 1.2$. *HST* images have been secured for 32 randomly selected galaxies with redshifts $z > 0.4$ and these display the normal range of morphological types. There are however important differences. The mean rest frame surface brightnesses of the late type galaxies are about 1.2 magnitudes greater than those of nearby galaxies. Some degree of peculiarity/asymmetry is observed in 30% of the objects and 13% show clear signs of mergers or interactions. There are compact blue components in 30% of the galaxies and these occur predominantly in the peculiar systems, but a few of them are also present in normal systems. According to Schade *et al.* (1995), these galaxies are predominantly the cause of the excess of faint blue galaxies and the numbers are consistent with those found in the morphological surveys described above.

The second example concerns the recent results of Cowie *et al.* (1995) on the redshift distribution of faint galaxies. The galaxies were selected from very deep surveys in small areas of sky and consisted of all the objects in these areas which satisfied the magnitude selection criteria $K < 20$, $I < 22.5$ and $B < 24.5$. Spectra for these galaxies have been obtained with the Keck 10-m telescope and the programme has been remarkably successful in discovering large redshift galaxies. There were 367 objects which satisfied the selection criteria and, among them, 91 are already known to be galaxies with redshifts $z > 0.7$ and 40 with redshifts $z > 1$. The reason for their success is immediately apparent from the typical spectra of galaxies in their sample (Figure 2). It is apparent that the large redshift galaxies have strong emission lines, characteristic of regions of star formation. According to their analysis, these are luminous galaxies with absolute B magnitudes, $M_B \sim -21$. The luminosities of the large redshift galaxies in the [OII] line are typically at least an order of magnitude greater than those of a reference sample at redshifts $z < 0.7$. They interpret these galaxies as star-forming galaxies at redshifts $z \sim 1$ and the rates at which the stars are being formed can account for about 5 to 20% of their present stellar populations.

Deep *HST* images have been obtained for 9 of the large redshift galaxies in the I waveband. As they note, the galaxies have 'strikingly unusual morphologies, often consisting of chains or structures of compact blobs, suggesting that they are generally not dominated by uniformly distributed star formation'. These results suggest that, among the faint blue galaxies, there is a population of distant star forming galaxies in addition to objects at smaller redshifts.

Similar preliminary results were reported by Koo for a deep survey undertaken with the Keck 10-m Telescope in the *HST* Groth strip (Koo 1996). Their sample extends to $I = 24$ and in this area they already have redshifts for 35 galaxies with redshifts in the range $0.3 < z < 1.2$, with a mean redshift of about 0.8, significantly greater than that of the Canada-France Redshift Survey. Again

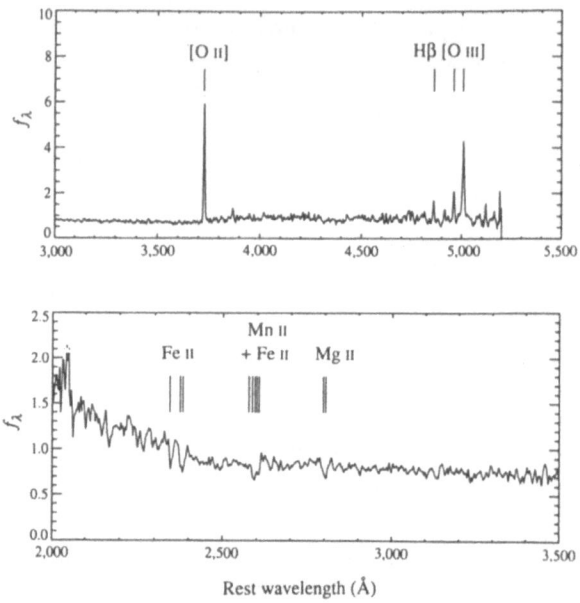

Fig. 2. The average rest-wavelength spectra of the objects in the deep survey field SSA13. It is apparent that they are strong emission line objects with only weak absorption lines in the ultraviolet region of the spectrum (Cowie *et al.* 1995).

the sample contains a large number of unusual galaxies including objects with multiple knots and the types of chain galaxy reported by Cowie *et al.* (1995). In addition, they have found 9 red galaxies which are as red as elliptical galaxies are at the present epoch. The inference is that these galaxies have already had time to form old stellar populations and so must have undergone their last burst of major star formation at redshifts $z > 2$.

Two examples of the study of galaxies at very large redshifts were described. The first of these results from the study by Windhorst and Keel (1995) of a young 'elliptical' radio galaxy at a redshift $z = 2.390$. Pasarelle described new observations of the field of this radio galaxy which, by great good fortune, lies at such a redshift that the Lyman-α line is redshifted into the narrow F410W filter. They find evidence for 18 Lyman-α objects at this redshift, all of them with luminosities between about 0.1 and 1 L^*. All of these objects seem to be compact and again the sum of their brightness distributions seems to follow the $r^{1/4}$ law. They suggest that this is evidence for the early formation of the bulges of galaxies.

The largest redshift systems which have been identified as young star-forming galaxies have been discovered by searching for the redshifted Lyman limit by multicolour photometry. The technique is similar to that described by Lilly and Cowie (1987) and refined for the detection of 'Lyman-limit galaxies' by Steidel and Hamilton (1992, 1993). The predicted spectrum of a starburst galaxy is illustrated in Figure 5(a) in which it can be seen that, as the starburst ages, the spectrum remains of the same characteristic form, namely, it is flat from the Lyman limit at 91.2 nm to longer wavelengths with an abrupt cutoff at $\lambda < 91.2$ nm. At a redshift $z = 3$, the Lyman limit is shifted to 400 nm and so the characteristic signature of these objects is that roughly equal intensities are observed in the G and R_S wavebands but the intensity in the ultraviolet waveband is very low as illustrated in Figure 5(b). The story began with the successful attempt to identify the large redshift absoption systems present in the background quasar QSO 0000-262, which has an emission redshift $z = 4.11$ (Steidel and Hamilton 1992, 1993). In this field, Macchetto et al. (1993) identified a 'Lyman-α radio quiet galaxy' at a redshift $z = 3.428$. Searches in four other QSO fields are described by Steidel et al. (1995).

Macchetto and Giavalisco (1995) and Steidel et al. (1996) have described further observations of these fields. Macchetto and Giavalisco (1995) described the application of this multicolour technique to the field containing the galaxy at redshift $z = 3.428$ and several objects with the signature of star-forming galaxies were found. At the Paris meeting, Giavalisco described the exciting result that spectroscopy with the Keck 10-m telescope has confirmed that these objects are indeed galaxies at redshift $z \sim 3.2$. These galaxies have been imaged by the WFPC2 and, when the images of the galaxies are summed, they are found to follow the standard de Vaucouleurs $r^{1/4}$ dependence of surface brightness upon radius of elliptical galaxies.

Steidel et al. (1996) have obtained the spectra of 24 candidate star-forming galaxies selected in both the quasar fields and in random regions of sky and have had great success in measuring redshifts for these with the Keck 10-m telescope. 17 of the objects have redshifts in the interval $3.01 \leq z \leq 3.43$. They find the important result that the comoving space density of these star-forming galaxies in the redshift interval $3 \leq z \leq 3.5$ is about half that of luminous galaxies with $L \geq L^*$ at the present epoch. The inferred velocity dispersions within the galaxies suggest that they are indeed massive galaxies. The star formation rates correspond to about about $8.5 h_{50}^{-2}$ $M_\odot$ yr^{-1}, similar to the star formation rates per galaxy found by Cowie et al. (1995). Steidel et al. infer that they have discovered the formation of the spheroidal components of the progenitors of massive galaxies — massive galaxy formation was certainly well underway by a redshift of 3.

3 Millimetre Astronomy and Galaxy Formation and Evolution

In general terms, the evidence presented above suggests that star formation activity has taken place in galaxies throughout the redshift interval $0.3 \lesssim z \lesssim 3$ and probably at larger redshifts as well. It is certain that a considerable fraction of the galaxies we observe today formed a significant fraction of their stellar populations at redshifts greater than 2. The star-formation activity associated with the formation of these populations should be readily observable with the LSA. Prototypes of the type of activity to be found in star-forming galaxies are

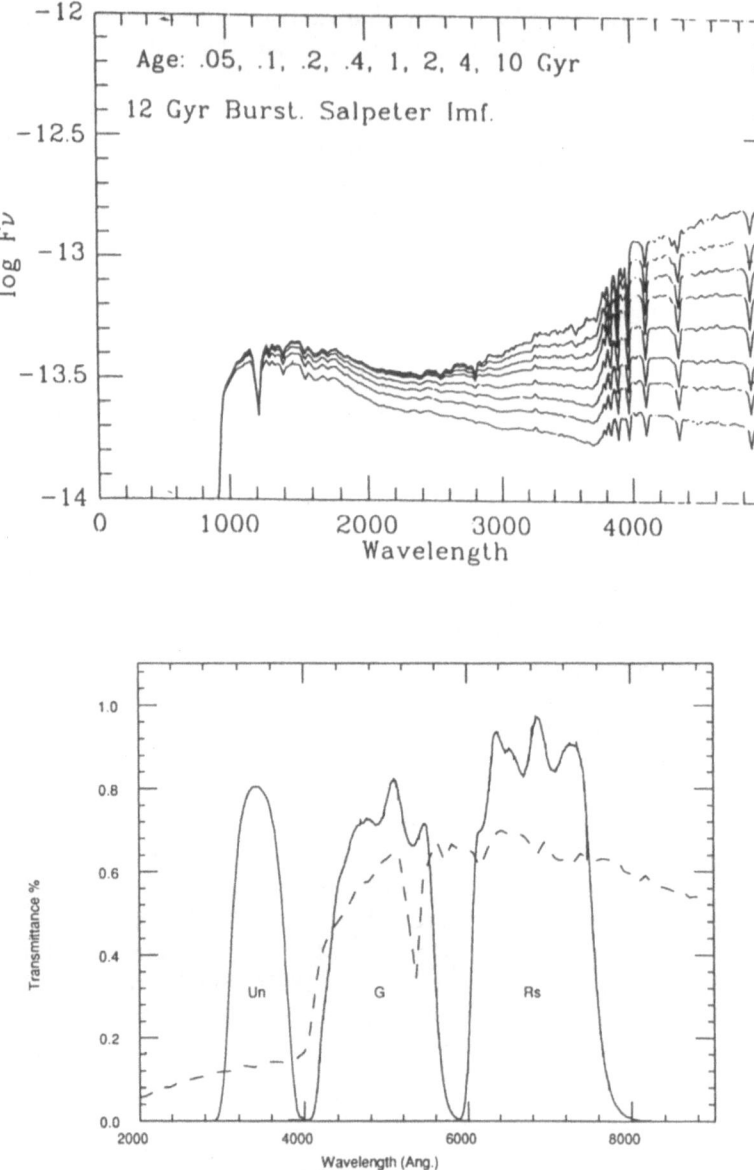

Fig. 3. (a). The spectrum of a starburst of duration 12 Gyr as observed at different ages (White 1989, from computations by G. Bruzual). (b) Illustrating how three colour photometry in the U_n, G and R_s wavebands can isolate star-forming galaxies at large redshifts. The dashed line shows the spectrum of a star-forming galaxy at a redshift $z = 3$ (Macchetto and Giavalisco 1995).

provided by the non-Seyfert Markarian galaxies, the ultraviolet and far-infrared properties of which have been studied by Mazzarella and Balzano (1986) and by Weedman (1988). They show that the majority of the star-forming galaxies are stronger emitters in the far-infrared waveband as compared with the optical and ultraviolet wavebands. Thus, wherever there is a significant amount of star formation taking place in galaxies, the sources must also be intense far-infrared emitters and hence modelling what is likely to be observed in the millimetre and submillimetre wavebands is of the greatest interest (Blain and Longair 1993 a, b; 1996).

This story is by now well-known. Because of the very large negative K-corrections in the millimetre and submillimetre waveband for the typical dust emission spectrum associated with the regions of star formation, the millimetre flux density-redshift relation is more or less flat from redshifts of about 1 to 10, depending sensitively upon the observing frequency and the temperature of the dust. Andrew Blain and I have presented a number of models of the expected counts of star-forming galaxies in the millimetre and submillimetre waveband and they are all characterised by very steep source counts as soon as flux densities are reached at which the flux density-redshift relation becomes flat.

In Figure 4, typical source counts at a wavelength of 1 mm which result from these computations are shown. The dotted lines show the expected source counts if the local luminosity function of IRAS galaxies is adopted and different assumptions made about the strength of the cosmological evolution of their properties. The lowest counts assume that there is no evolution of the comoving luminosity function, while the highest count assumed the form of evolution observed for radio galaxies, X-ray sources and quasars, $L(z) = L(0)(1 + z)^3$ at redshifts $0 < z < 2$, $L(z) = $ constant, for $2 < z < 5$. This form of evolution would exceed the limits to the background radiation at 0.5 mm as observed by $COBE$. The intermediate dotted line shows an evolutionary model which would be consistent with the background spectrum: $L(z) = L(0)(1 + z)^{1.2}$ for $0 < z < 5$; $L(z) = 0$, for $z > 5$. The solid line shows the count expected in a hierarchical clustering model, based upon the Press-Schechter picture of galaxy formation (Blain and Longair 1993b). It can be seen that the source count is very steep indeed. The reason for this is that, in the Press-Schechter, the bulk of the luminous galaxies are close to the turn-over luminosity L^* and it turns out that the flux densities of these galaxies at millimetre wavelengths are almost independent of redshift as the galaxies increase in mass with decreasing redshift.

The key point is that very steep source counts are expected in the millimetre waveband and that, according to our computations, the best survey wavelength for these studies, taking into account the spectrum of the sources, the field of view of the telescope and the expected sensitivity of the receivers is about 850 μm. We have estimated that it should be possible to reach flux densities at which the steep rise in the source counts can be detected by long integrations with the SCUBA array detector on the JCMT (Blain and Longair 1996), and it should certainly be possible to observe these very distant star-forming galaxies with the LSA. With the high angular resolution available, revealing details of

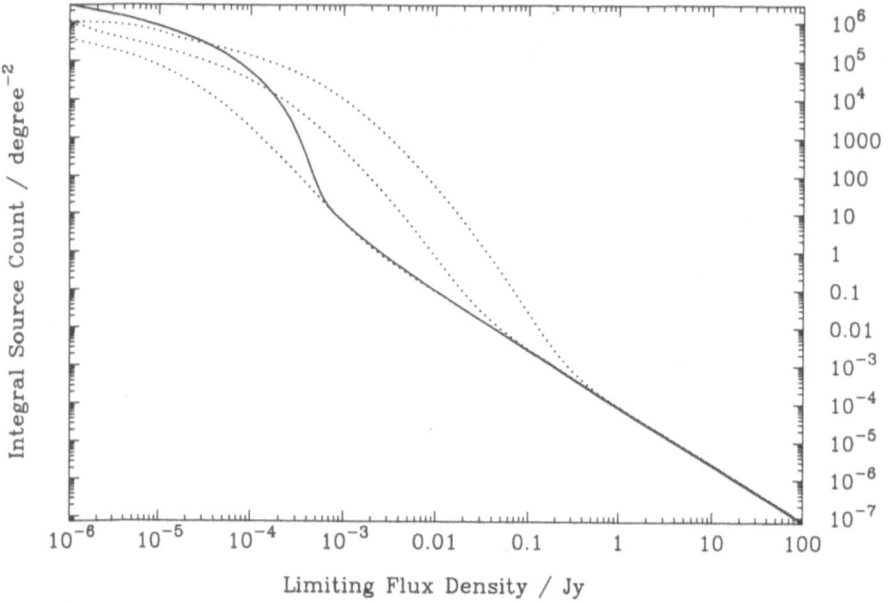

Fig. 4. Examples of counts of galaxies at 1 mm wavelength for evolving IRAS galaxies (dotted lines) and for a hierarchical clustering model (solid line). In increasing numbers of source, the lowest IRAS count assumes no evolution of the comoving luminosity function of IRAS galaxies, the highest count assumes $L(z) = L(0)(1 + z)^3$ at redshifts $0 < z < 2$, $L(z) = \text{constant}$, for $2 < z < 5$ and the intermediate counts assume $L(z) = L(0)(1 + z)^{1.2}$ for $0 < z < 5$; $L(z) = 0$, for $z > 5$. In the hierarchical clustering model, the scheme is based up the Press-Schechter formalism (Blain and Longair 1993(b), 1996).

the star-formation process in these distant galaxies should be quite feasible.

It is important to emphasise that these techniques provide a very powerful tool indeed for investigating galaxies in the process of formation at very large redshifts. Accompanying the strong dust emission, there will be the molecular, atomic and ionic lines which will enable the redshifts of the young galaxies, or protogalaxies, to be determined. In principle, galaxies could be observed out to redshifts of about 10 or 20 and thus provide a glimpse of what must have been taking place during the 'dark ages' between redshifts of about 4 and 1000, when so much of the key activity in galaxy formation took place. The evolution of the large scale distribution of galaxies through these epochs can be investigated directly. These points are taken up in more detail by Andrew Blain in his contributed papers.

There is another unique feature of the steep counts expected in the millimetre wavebands. If the counts of objects are steep enough and the population is at a large redshift, both of which conditions apply to millimetre observations of very distant galaxies, it is expected that a significant fraction of the detected objects will be strong gravitational lenses. Generally, the counts of objects are too shallow for this to be a large effect, but this is not necessarily the case in the millimetre waveband. The tail of strong lensing results has a flux density distribution which falls off as A^{-3}, where A is the amplification of the image (see, for example, Peacock 1982). Thus, if the integral source count is steeper than S^{-4}, the amplified sources can make a very significant contribution to the counts. From these considerations, it is apparent that the very steep source counts in Figure 4 provide the best opportunity for observing strongly-lensed objects. These important ideas are illustrated most simply in Figure 5 and are dealt with in more detail by Andrew Blain in a contributed paper.

It is apparent that there is the greatest probability of finding high amplification sources in the hierarchical model because of the very steep source count. Indeed, in certain flux density intervals, the counts may well be dominated by strongly-lensed objects (Figure 5a). The effect is smaller in the *IRAS* models, but still much more significant than is expected in, say, the optical or radio wavebands. Figure 5b illustrates how the presence of the strong gravitational lensing can give rise to a high redshift tail to the predicted redshift distribution in particular flux density intervals. It would be of the greatest importance to determine the fraction of gravitationally-lensed images as a function of flux density, since this could provide a unique tool for cosmology and for studies of the very large redshift Universe. The high angular resolution of the LSA would enable the structures of these strongly lenses objects to be distiguished.

4   Conclusions

I have only had space to describe briefly a few of the exciting types of science which can be undertaken with the LSA. I am firmly convinced that it is magnificent project which will open up completely new ways of studying the origin of stars and the origin and evolution of galaxies. In many ways, it is the millimetre counterpart of the *Hubble Space Telescope*. I am also convinced that the interactions of observations with the LSA with observations made in other wavebands, particularly the optical, infrared and ultraviolet wavebands, adds enormously to the strength of the case for the LSA. Without the complementary information from observations in other wavebands with similar angular resolution, the picture will remain incomplete. In these endeavours, the millimetre waveband has a crucial and unique rôle to play.

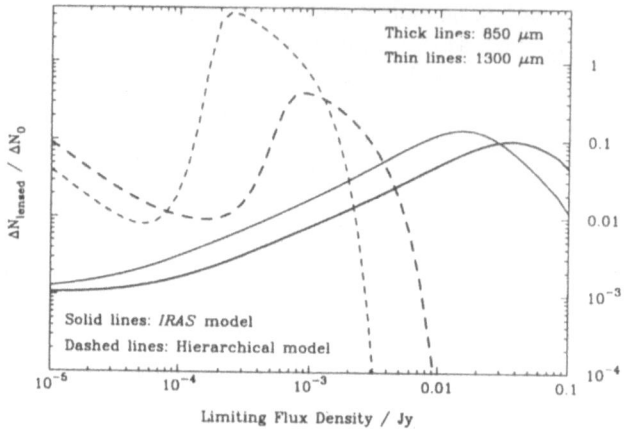

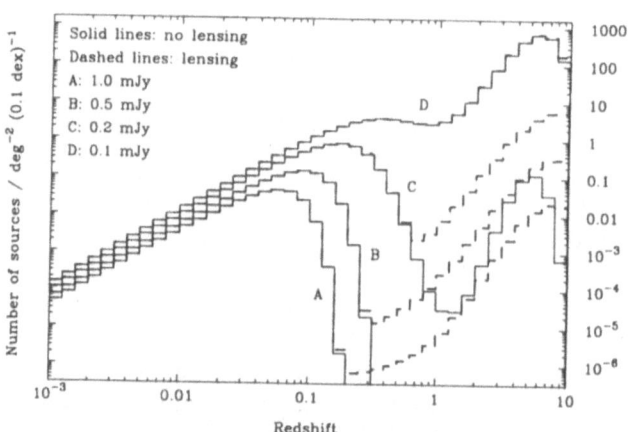

Fig. 5. (a) The ratio of the differential source counts of high-magnification lensed sources to the total differential source counts predicted at 850 and 1300 μm in the IRAS-based and hierarchical models. (b) The redshift distributions predicted for sources with flux densities above four different thresholds at 1.3 mm in the hierarchical model. The solid lines show the expected redshift distribution without lensing and the dashed lines the contribution of lensed sources (Blain 1996).

References

Abraham, R.G., van den Bergh, S., Glazebrook, K., Ellis, R.S., Santiago, B.X., Surma, P. and Griffiths, R.E. (1996). MNRAS, (in press).

Blain, A.W. (1996). MNRAS, (in preparation).

Blain, A.W. and Longair, M.S. (1993a). MNRAS, **264**, 509.

Blain, A.W. and Longair, M.S. (1993b). MNRAS, **265**, L21.

Blain, A.W. and Longair, M.S. (1996). MNRAS, (in press).

Cowie, L.L. (1988). In *The Post-recombination Universe* (eds. N. Kaiser and A.N. Lasenby), 1. Dordrecht: Kluwer Academic Publishers.

Cowie, L.L., Hu, E.M. and Songaila, A. (1995). Nat, **337**, 603.

Driver, S.P., Windhorst, R.A. and Griffiths, S.P. (1995). ApJ, **453**, 48.

Driver, S.P., Windhorst, R.A., Windhorst, A., Ostrander, E.J., Keel, W.C., Griffiths, R.E. and Ratnatunga, K.U. (1995). ApJ, **449**, L23.

Giavalisco, M, Livio, M., Bohlin, R.C., Macchetto, F.D. and Stecher, T.P. (1996). (preprint)

Glazebrook, K., Ellis, R., Colless, M., Broadhurst, T., Allington-Smith, J. and Tanvir, N. (1995). MNRAS, **273**, 157.

Illingworth, G. (1996). In *HST2*, Proceedings of the Second European Conference on the Hubble Space Telescope, (eds. P. Benvenuti, D. Macchetto and E. Schreier), (in press).

Koo, D. (1996). In *HST2*, Proceedings of the Second European Conference on the Hubble Space Telescope, (eds. P. Benvenuti, D. Macchetto and E. Schreier), (in press).

Le Fèvre, O., Crampton, D., Lilly, S.J., Hammer, F. and Tresse, L. (1995). ApJ, **455**, 60..

Lilly, S.J. and Cowie, L.L. (1987). In *Infrared Astronomy with Arrays*, (eds. C.G. Wynn-Williams and E.E. Becklin), 473. Honolulu: Institute for Astronomy, University of Hawaii Publications.

Macchetto, D and Giavalisco, M. (1995). *ESO Messenger*, September 1995, p. 14..

Macchetto, D., Liparo, S., Giavalisco, M., Turnshek, D.A. and Sparks, W.B. (1993). ApJ, **404**, 511.

Mazzarella, J.M. and Balzano, V.A. (1986). ApJS, **62**, 751.

Miyoshi, M., Moran, J., Herrnstein, J., Greenhill, L., Nakai, N., Diamond, P. and Inoue, M. (1995). Nat, **373**, 127.

O'Dell, C.R. (1996). In *HST2*, Proceedings of the Second European Conference on the Hubble Space Telescope, (eds. P. Benvenuti, D. Macchetto and E. Schreier), (in press).

Schade, D., Lilly, S.J., Crampton, D., Hammer, F., Le F'evre, O. and Tresse, L. (1995). ApJ, **451**, L1.

Steidel, C.C., Giavalisco, M., Pettini, M., Dickinson, M. and Adelberger, K.L. (1996). AJ, (in press)

Steidel, C.C. and Hamilton, D. (1992). AJ, **104**, 941.

Steidel, C.C. and Hamilton, D. (1993). AJ, **105**, 2017.

Weedman, D. (1988). Astro. Letts. and Comms. **27**, 117.

White, S.M. (1989). In *The Epoch of Galaxy Formation* (eds. C.S. Frenk, R.S. Ellis, T. Shanks, A.F. Heavens and J.A. Peacock), 1. Dordrecht: Kluwer Academic Publishers.

Windhorst, R.A. and Keel, W.C. (1996). ApJ, (in press).

A $10000\,\mathrm{m}^2$ Southern MM Array

D. Downes

Institut de Radio Astronomie Millimétrique, Grenoble, France

Abstract. To advance beyond the impressive achievements of current mm astronomy, we should build a **Large Southern millimeter Array** with a collecting area of $10000\,\mathrm{m}^2$. This new array should also have a resolution of $0.1''$ at $3\,\mathrm{mm}$, and at least 1000 simultaneous baselines. The area and resolution would be 10 times better than available so far, and the large number of baselines would give vastly improved image quality. Scientifically, such an array would open new directions in the study of all types of mm radiation — molecular lines, thermal emission of dust, line and continuum emission of ionized gas, and synchrotron emission of relativistic particles — from galaxies and quasars at high redshift in the early universe, molecular clouds in our Galaxy and nearby galaxies, accretion disks and outflow jets of protostars, protoplanetary disks around young stars, envelopes of stars of all types, and objects in our solar system. This new mm array should be on a site above $3000\,\mathrm{m}$ that allows $10\,\mathrm{km}$ baselines. We favor the southern hemisphere because there are many mm/sub-mm quality sites on flat land at high altitudes, and because the southern sky contains the central part of the Milky Way, the Magellanic Clouds, and many spectacular galaxies. Such a mm array would complement the European investment in southern hemisphere optical astronomy for the ESO VLT. Design goals should be reliability, robustness, and simplicity. Combinations that would give a $10000\,\mathrm{m}^2$ area would be $50 \times 16\,\mathrm{m}$ or $100 \times 11\,\mathrm{m}$ dishes.

1 Introduction

The millimeter and sub-mm bands are unique in astronomy in having more than 1000 radio spectral lines of interstellar and circumstellar molecules as well as broad-band emission from cool dust in space. They are the only bands in the electromagnetic spectrum where we can detect cool dust and molecules far away in young, high redshift galaxies in the early universe, and nearby in low-temperature cocoons of protostars in the Milky Way. They are the only bands that give us details about young stars with bipolar jets or protoplanetary disks and old stars with outflowing carbon and oxygen-rich envelopes. Because of this science, mm astronomy has become a world-wide effort, with at 18 mm/sub-mm telescopes presently operating in North and South America, Asia, and Europe. Many of the best telescopes are European built, and include the IRAM $30\,\mathrm{m}$ telescope in Spain, the IRAM $5 \times 15\,\mathrm{m}$ array in France, the $15\,\mathrm{m}$ JCMT on Hawaii, the $15\,\mathrm{m}$ SEST in Chile, and the $10\,\mathrm{m}$ HHT in Arizona. With such instruments, mm astronomy is now providing detailed studies of many types of astronomical objects. Its rich harvest of results amply justifies the construction of a more powerful, next generation instrument. The detections of CO at high redshifts in the objects 10214+4724, H1413+117, and BR 1202−07 mean that mm astronomy will become a significant investigative tool in cosmology, provided

Table 1. 10-Sigma Sensitivity at 230 GHz in 1^{hr} with $T_{sys} = 100\,\mathrm{K}$ and Area $= 10^4\,\mathrm{m}^2$.

max. baseline	230 GHz beam	—— Galactic —— resolution at 10 kpc	$10\,\sigma$ at δv = 2 km/s	— Extragalactic — resolution at 10 Mpc	$10\,\sigma$ at δv = 20 km/s	Continuum $10\,\sigma$ at $\delta\nu$ = 2 GHz
0.3 km	$1''$	$5.0\ 10^{-2}$ pc	0.1 K	50 pc	0.04 K	0.2 mJy
1.0 km	$0.3''$	$1.5\ 10^{-2}$ pc	1 K	15 pc	0.4 K	0.2 mJy
3.0 km	$0.1''$	$5.0\ 10^{-3}$ pc	10 K	5 pc	4 K	0.2 mJy
10. km	$0.03''$	$1.5\ 10^{-3}$ pc	100 K	1.5 pc	40 K	0.2 mJy

we can detect many more high-z objects in reasonable times and with high spatial resolution. Maps of the star IRC+10216 in rare molecules like MgNC and NaCl show the power of mm interferometry to probe the chemistry of circumstellar envelopes. New studies of the star GG Tauri in ^{13}CO show the ability of mm interferometry to derive the kinematics of disks around pre-Main Sequence stars. Maps of galaxies like M51 show the potential of mm astronomy to map molecules in nearby galaxies. In fact, nearly all objects studied in mm astronomy have weak, fine structure. To further resolve this structure, we need longer baselines, which are only usable with more sensitivity. This can only be reached with an array of large collecting area. Arguments for a large mm array have been reviewed recently by Booth (1994), Ishiguro (1994), Downes (1994), and, in an earlier context, in NRAO's MMA Proposal (1990). A mm array for the 21st century should have a substantially greater collecting area than existing mm telescopes. For comparison, the IRAM 30 m telescope has a collecting area of $700\,\mathrm{m}^2$; with 6 antennas, the IRAM interferometer will have $1000\,\mathrm{m}^2$. The Nobeyama 45 m telescope has an area of $1590\,\mathrm{m}^2$, and the planned Mexico–UMass 50 m dish will have $1960\,\mathrm{m}^2$. A feasible goal for a 21st century array is a collecting area of $10000\,\mathrm{m}^2$. The resolution, now $\sim 1''$ with existing mm interferometers, should also be improved by a factor of 10, which implies baselines of 5 to 10 km. The interrelated goals of collecting area, sensitivity, and resolution could be attained with an array of $50 \times 16\,\mathrm{m}$ antennas, or $100 \times 11\,\mathrm{m}$ antennas.

The brightness temperature sensitivity of a large array varies as

$$\Delta T_b \propto \frac{\lambda^{2.5}\, T_{sys}\, B_{max}^2}{n\, D^2\, (t\, \Delta V)^{0.5}} \tag{1}$$

where λ is wavelength, T_{sys} is system temperature, B_{max} is maximum baseline, n is number of dishes, D is their diameter, t is integration time, and ΔV is velocity resolution. Expected sensitivities of the LSA are summarized in Table 1. There is very interesting science to be done with sub-arcsecond beams, but because maximum usable baseline is related to collecting area, this science can only become accessible if a new large mm array has a collecting area of $\sim 10000\,\mathrm{m}^2$.

2 Astrophysical Goals

Because it is the domain of radiation processes of *cold* matter, mm astronomy is well-suited to the study of the origin of galaxies and the origin of stars and planets. Scientifically, a next generation array should provide a quantum jump in mm astronomy and be able to open new directions in astrophysics such as the study of protogalaxies, protostars, and protoplanetary disks.

Early Universe Studies: The discovery of CO lines and dust continuum emission in the $z = 2.3$ galaxy IRAS F10214+4724 dramatically opened up the distant universe to mm astronomy. Since then, dust emission, CO lines, and C I lines have been detected in the gravitationally lensed Cloverleaf quasar H1413+117 at $z = 2.5$, and several molecular lines have been detected at $z = 0.25$, 0.7, and 0.9 in absorption against distant background radio sources. Furthermore, mm/sub-mm thermal emission from dust has been detected in quasars with redshifts as high as 4.7. The mm emission from dust may be one of the best tracers for finding primeval galaxies at $z \geq 5$. If high-mass starbursts injected large amounts of dust into the disks of young galaxies, the resulting far IR emission will be detectable at high z (the increasing distance is compensated by increased flux as the far IR bump is redshifted into the mm bands). Millimeter studies can potentially determine the redshift range in which most of the early star formation and dust injection occured. This information will greatly improve our understanding of the timescales of galaxy and structure formation in the universe. The epoch of peak quasar activity was at $z = 2$ to 3.5 (e.g., Rees 1994). This rise and fall of the quasar population spans about 2 billion years, or about 50 generations of quasars, and was obviously a fundamental epoch in the evolution of galaxies. **The study of this epoch of galaxy development is one of the main goals of a new millimeter array, and it is one of the main reasons to have a huge collecting area. This aspect — very large collecting area for objects at $z = 2$ to 5 — is how this proposal differs from previous proposals to build new arrays or extend existing mm arrays.**

These studies can be done best in the *millimeter* region. A advantage of the mm region is the ability to detect thermal emission from dust, which is too weak to detect at cm or meter wavelengths. Another advantage is that for lines with the same brightness temperatures and velocity linewidths, the line power varies as $\nu^2 T_b \Delta\nu$, and hence as ν^3. A CO(3–2) line redshifted to 100 GHz emits $3 \ 10^7$ times more power than an H I line shifted to 400 MHz. Even if the H I line could someday be detected at $z > 2$, it is unfortunately redshifted to the meter band where there is a high level of radio noise from our Galaxy, perturbation of array phases by the ionosphere, and much man-made radio interference. Another advantage of the mm bands is that most molecules have a ladder of spectral lines. If a redshift is too high for a line to stay in the mm region, there is a good chance the next transition up the ladder will be shifted into it. At high z, the submm spectrum is redshifted into the mm region, so one can do submm astronomy at mm-quality sites. This new field of study of high z objects means that mm astronomy can play an important role in studies of the early universe, – but only if we have an instrument with large collecting area.

Molecular Clouds in Nearer Galaxies and in our Galaxy: Galactic disks, with their spiral arms and Giant Molecular Clouds (GMCs), are the birthplaces of stars. Millimeter line and continuum observations provide clues to how stars form and how they affect the interstellar medium and the chemical evolution of galactic disks. The mm lines of CO and its isotopic species trace both high and low density gas, HCN and CS trace high density gas, and HCO^+ traces ionization. These mm data can yield the physical properties, locations, and distributions of star-forming GMCs across galactic disks, and will be a valuable complement to high-resolution ground-based IR observations. For such mm studies a large new mm array is indispensible.

Galaxies at 30 to 200 Mpc will be prime targets for a large mm array. The array could image of dust and gas emission in galaxies at 200 Mpc with 100 pc resolution. An array with $10000\,m^2$ of collecting area could image an enormous number of galaxies. Particularly interesting would be maps of gas and dust in a large sample of irregular galaxies, The nearest merger galaxies are at 40 to 100 Mpc; with a resolution of 0.1 Mpc, it would be possible to determine how merger-induced starbursts progress spatially in the gas.

Galaxies to 30 Mpc would be targets for which a new large array would give us the same linear resolution at 30 Mpc that we now have for the Local Group. A resolution of $0.1''$ will permit detailed imaging of GMCs in nearby galaxies. With a resolution of $0.1''$, clumps within GMCs could be identified to distances of 10 Mpc — a region containing hundreds of galaxies of greatly varying type — enabling us to derive sizes, masses, temperatures, densities of star-forming clouds. With the same resolution, the statistics of the distribution of GMCs inside galaxies could be derived to distances $> 100\,Mpc$. In many nearby galaxies, a $30''$ field of view is sufficient for studies of spiral arms and GMCs. To image whole galaxies (typical sizes $3'$ to $15'$), a mosaic is needed; this is already a familiar technique with mm interferometers.

Galactic Nuclear Regions: A new large array will allow us to determine the masses and kinematics of optically obscured galactic nuclei with a resolution of a few parsecs and image the distributions of a variety of molecules and isotopic species. In nearby galaxies, a resolution of $1''$ to $2''$ gives a detailed look on a scale of 15 to 30 pc. This is sufficient to resolve the clumps in the central concentrations. A next-generation array with ten times more sensitivity than the IRAM interferometer could extend such studies to all the known Seyfert galaxies, and give high resolution maps of the condensations in innermost parts of the circumnuclear disks. A new large mm array will be able to study not only the gas but also the dust obscuring the galaxy nuclei. High resolution and the sensitivity of a $10000\,m^2$ collecting area will allow us not only to map the dust in the nearer galaxies, but also to detect the dust in IR luminous galaxy nuclei to several Gpc. A large mm array with a $0.1''$ beam could reveal the structure, masses, and kinematics of circumnuclear material with a precision equal to, or surpassing that of the Hubble Space Telescope.

Local Group Galaxies are now being studied with a resolution of $1''$, which shows clumps within GMCs and yields the clumps' mass distribution, kinematics, dynamics, and relation to ongoing star formation. With a resolution of 0.1″, the clumps themselves could be mapped, and the properties of the birthplaces of stars in many low-metallicity Local Group galaxies could be compared directly with those of stars in the Milky Way. This would be highly interesting, as stellar properties and formation are closely tied to metallicity.

Magellanic Cloud Studies would also benefit from a new mm array. The beams of current southern mm dishes cannot resolve compact molecular sources in the Clouds. What we need are beams of 0.1″ to a few arcsec. At the Magellanic Clouds' distance, this would give the same linear resolution as in CO studies of the Milky Way with beams of several arcsec to a few arcmin.

The Galactic Center would also be a favored region of study for a next-generation mm array. Of particular interest would be the gas dynamics of the 1-pc circumnuclear disk around the galactic center source Sgr A*, the numerous star-forming centers in the cloud Sgr B2, the environs of the various black-hole candidates detected by gamma-ray satellites, the interactions of the extensive magnetic field structures with molecular clouds in the arcs near Sgr A, and the hundreds of molecular clouds in the bar at the center of the galaxy. Because of the southerly declination, the study by mm interferometers of this more extensive region around the galactic center has not even begun.

Molecular Clouds in the Central $l = \pm 30°$ **of the Milky Way** make up most of the molecular clouds in our Galaxy. There is already a wealth of information about these molecular clouds from low-resolution, single-dish studies in various lines, from maser source surveys, and from near and far IR observations. What has been lacking so far is extensive, high-resolution study with mm interferometers. Because of the great number of dense cores in these molecular clouds, and the number of spectral lines that can be mapped, this is a program which will keep a large mm array busy for decades.

Chemistry of Star Forming Regions: A large new mm array will revolutionize studies of star formation. It will allow us to better trace the chemistry of star-forming regions, to detect and map protoplanetary disks, and to study the kinematics and dynamics of bipolar flows from young stellar objects. Molecular line profiles can indicate the infall, rotation, and outflows accompanying the birth of stars. Intensity ratios of lines from different levels of the same molecule can tell us the density and temperature in the circumstellar gas. Molecular abundances are diagnostics of the evolutionary stage of a young stellar object. In the protostellar core, the chemistry is dominated by ion–molecule reactions driven by cosmic ray ionization. In the subsequent contraction, the density becomes so high that molecules condense onto dust grains, where abundances can be modified by grain surface reactions and photochemical processes in ices, to yield strongly hydrogenated molecules like H_2O, NH_3 and CH_3OH. Once the young star has formed, radiation heats the gas and evaporates the icy grain mantles. Shocks in the outflows can also return the mantles' refractory elements like silicon to the gas phase. These molecules subsequently drive a rich chemistry yielding complex organic molecules like CH_3OCH_3 and CH_3CN, until the normal ion-molecule chemistry resumes after 10^5 yr.

Protoplanetary Disks: With a resolution of $0.1''$ at 3 mm and $0.05''$ at 1.3 mm, a Large Southern Array could resolve 5 AU at a distance of 100 pc, the distance of the nearest star-forming regions and many low-mass pre-main-sequence stars. This will greatly advance the study of (< 100 AU-sized) protoplanetary disks, and will allow us to study the properties of disks around high-mass stars, which has up to now been impossible due to lack of resolution. It will be exciting to extend current research to the $0.1''$ scale and to weak lines in protoplanetary disks around young massive stars like MWC 349 as well as to stars of lower mass, and to study the chemistry of protosolar nebulae.

Outflow Jets from Young Stellar Objects: While some bipolar flows can be studied with existing arrays, the protostellar accretion disks will become accessible for detailed study only with a large new array. Recent detections of the thermal emission of dust in the disks around HL Tau and L1151 IRS 5 show beams $< 1''$ are needed to resolve the protostellar accretion disks, and obviously $0.1''$ is needed to derive functions like temperature vs. radius.

Dynamics and Chemistry of Circumstellar Envelopes: A large mm array will make fundamental contributions to our understanding of the dynamics and chemistry of the envelopes of evolved, oxygen-rich and carbon-rich stars. A large mm array would give us the ability to study these objects anywhere in the Galaxy with better linear resolution than we now have for objects within 500 pc of the Sun. Since stars are concentrated in the Milky Way's inner disk, a southern array would permit study of most of the large circumstellar envelopes in the Galaxy.

Solar System Research: A new large array could map CO in Mars and Venus and give high-resolution data on wind, temperature, CO distribution, and atmospheric dynamics. Observations of HDO at 226 GHz would allow mapping of water vapour on Mars and Venus. A large new mm array would permit searches for spatial variations of CO and HCN on Neptune, which would give clues to their origins in Neptune's troposphere or stratosphere. An array with $0.1''$ resolution could detect SO_2 in atmosphere around the volcanic centers of Jupiter's moon Io. A large mm array would be able to detect and map complex nitriles like C_2H_3CN and HC_5N on Titan, and open up the study of Titan's organic chemistry. It could also search for CO on Pluto and Neptune's moon Triton, the detection of which would be important for cosmogony.

Comets : The chemistry of comets is relevant to our understanding of the evolution of pre-planetary disks. Future ground-based mm observations will complement the planned Rosetta space probe, which may sample a single comet, while a new, powerful mm array could investigate the abundance of molecules and their isotopes in many comets. A large mm array with $0.1''$ resolution would be able to map the cometary molecules in the coma, and thus gain access, via line shapes and Doppler shifts, to the dynamics of the coma. An array with an area of $10000 \, m^2$ could search for lower-abundance molecules, radicals, and ions in comets. Since the intensities of cometary spectral lines can vary on short time scales, a large new array would be able to rapidly image some molecular lines.

3   Characteristics of a Next-Generation MM Array

To make a major impact in astrophysics, the interferometer must have:

• **A capability to open up new fields of science.** An obvious domain is cosmology and the epoch of galaxy formation. A powerful array such as proposed here will also revolutionize research on protoplanetary disks around nearby stars, the formation of protostars, the structure and chemistry of circumstellar envelopes, and molecular gas in nearby galaxies.

• **For research on star formation and evolution, a capability to see across the Galaxy** all the objects — outflows, young stars, old stars, molecular clouds — that we now study in detail at 1 kpc distance.

• **For research on galaxy evolution, a capability to see across the universe**, i.e., enough sensitivity and resolution to detect dust and gas in galaxies at $z > 2$. These criteria imply a factor of 10 improvement over present telescopes. This is feasible with current technology, and can be achieved with:

• **An angular resolution of 0.1″ at a wavelength of 3 mm.** An important requirement for early universe studies is angular resolution. To study objects at $z > 2$, we need 0.1″ resolution at 3 mm and 0.05″ at 1.3 mm. This implies baselines of 5 to 10 km. For constant system temperature, the noise in Jy is constant as baselines increase, but the noise in brightness temperature increases as baseline squared (eq.1). This noise can rapidly dominate the signal, which is limited to a few tens of K by the emission physics. Thus we cannot use long baselines (10 km) unless we have plenty of sensitivity. Most mm astronomy is spectroscopy, with bandwidths governed by the linewidths, not the receivers. Furthermore, in some of the mm windows, receiver performance is close to the atmospheric noise limits. *So for much of the spectral line research, the only way to increase sensitivity is to increase the collecting area.* In particular, we can only reach angular resolutions of $\leq 0.1″$ for thermal lines if we increase the collecting area by an order of magnitude over current values.

• **A collecting area of 10000 m².** The 1000 m² area of present mm telescopes allows us to detect some high-z galaxies, but only gas-rich and gravitationally-lensed galaxies like IRAS F10214+4724 and H1413+117 at $z = 2.5$. What is needed for further advances at high z is the next factor of 10 step in collecting area that usually goes with progress in telescope building. In optical astronomy, these steps went from 1 m mirrors to 3.5—5 m mirrors to 10 m mirrors (or 16 m equivalent with the VLT). The same progress by factors of ten in collecting area was also made in radio astronomy, successively in the meter, decimeter, and short cm ranges (10 m dishes to 30 m dishes to 100 m and 300 m dishes). In mm astronomy we've gone from the 10 m class to the 30–45 m class, but must now take the next step — another factor of ten in collecting area. This means a mm telescope with the same collecting area as the Effelsberg 100 m telescope or the VLA, about 10000 m². An appropriate mm telescope for early universe studies should have a collecting area of 10000 m², equivalent to ten times the IRAM interferometer, and achievable with 50 × 16 m dishes or 100 × 11 m dishes. In *number* of antennas, such an array may be compared with the VLA (27 × 25 m dishes), the GMRT in India (30 × 45 m dishes), or the Culgoora radioheliograph

(96×13 m dishes in a 3 km ring plus 48 other aerials in an inner 2.8 km ring (Wild 1967). The requirements of a factor of ten increase in collecting area and a site with enough room to exploit the sensitivity and give 0.1″ resolution imply:

• **Many antennas:** $10000 \, \mathrm{m}^2 = 50 \times 16$ m dishes, or 100×11 m dishes.

• **A large number of baselines** (≥ 1000) for high-quality, fast images.

• **A site with 10×10 km** of flat terrain above an altitude of 3000 m in a dry climate, preferably near a city or pre-existing infrastructure.

Field of View: If the antennas are 16 m dishes, their field of view at 3 mm will be about 50″, which is sometimes felt to be too small for the type of interferometry done in the 1980's and 1990's. These considerations will be less important in 2010 to 2040, when scientific interest will have shifted to objects at greater distances or more compact than those studied today. Relevant points will be:

• For studies of faint, distant, and compact objects for which we want 0.1″ beams, **calibration** will be crucial. This includes the per-baseline sensitivity on calibrator sources, the number of usable calibrator sources on the sky, their separation from the source being observed, and the number of objects on which self-calibration can be used. Calibration is easier with large (16 m) dishes, and for a 21st-century array, may outweigh considerations about field of view.

• High z source sizes will be a few arcsec or less. The fields of view provided by 16 m dishes will be more than adequate.

• The ratio of field of view to beamsize will be greatly improved over present day mm arrays. The field of view of 16 m dishes will be more than adequate for most projects. With the long baselines made possible by a big collecting area, the ratio of field of view to resolution (baseline to dish size) will be the same as in cm astronomy: a 0.1″ beam in a 55″ field of view means 300,000 beam areas per field of view, not the present 750 beam areas with 2″ beams.

• Mosaicing is now a standard technique of mm arrays and ongoing improvements to the software have made wide-field mapping rather routine.

• If needed, field of view could also be increased by use of multi-beam receivers. In the electronics, there might be a choice between simultaneous multi-frequencies at a single position, or simultaneous multi-fields at a single frequency.

A Zero–Order Cost Estimate: For a collecting area of $10000 \, \mathrm{m}^2$, we adopt a model array of 50×16m dishes and assume costs can be scaled to those of current IRAM dishes. The 1995 price of 4 M Ecu per IRAM dish includes the cost of a built-in transporter for each antenna; it would be lower if the dishes were fixed or moved by an independent transporter. Based on the experience with the 6 antennas built by IRAM, we expect a 25% cost reduction due to a simplified design, and another 20% reduction for a large production run like 50×16 m. To benefit from such economies of scale, we must obtain approval for the whole project, rather than partial funding for the array at widely-spaced epochs. We also extrapolated from IRAM receiver costs and the per-baseline cost of correlators and IFs. We estimated costs of control and assembly buildings on site and a headquarters in a city, to obtain the following investment costs (in M Ecu): 150 for antennas, 75 for electronics, 18 for infrastructure, 2.5 for management, 24.5 for contingency, all for a total of 270 M Ecu (1995). Experience suggests a new large mm array will need a staff of 150. The laboratory for computer, receiver and electronic engineers should be within 2 hours' drive of the site. There should be $\geq$ 20 people on-site on work days. If the array were in northern Chile, most of the staff's families might live in Santiago, and personnel might fly in for weekly shifts at the observatory. A staff of 150 means an operating budget of $\geq$ 30 M Ecu/yr (1995), depending on the site. In sum, the project will approach the cost of the VLT or a scientific satellite. The investment cost is 0.1 Space Telescope, and the running cost will be less than 5%/yr of the Space Telescope running costs.

A Southern Site: Where is the best site for a next generation mm array? The case for the southern hemisphere rests on both the possible sites and the science. There are many good, flat, mm-quality sites at high altitude, and there are many interesting mm sources in the southern sky. Advances in mm interferometry will require 10 $\times$ 10 km flat terrain at an altitude over 3000 m in a dry climate, preferably near a city or pre-existing infrastructure. Some of the world's best mm sites are in the Atacama desert in Northern Chile where there are many dry, large, flat areas at altitudes between 3000 and 5000 m. At the higher elevations, 22 potential sub-mm sites were found by Raffin & Kusunoki (1992). These sites are extremely dry, with precipitable H_2O around 1 mm most of the time. Further sites are being tested by teams from Nobeyama, Onsala/SEST, and the University of Chile, and other sites at altitudes $>$ 5000 m are being tested by the NRAO. A scientific reason for building a large mm array in the southern hemisphere is that most Galactic objects of interest for mm astronomy are in the south. These include the Galactic nucleus and the central part of the Milky Way, which contains most of the Galactic star forming regions, proto-planetary discs, circumstellar envelopes, and planetary nebulae. The southern sky also has some spectacular galaxies that are prototypes for their classes, and the Magellanic Clouds, of special interest because of their clues to chemical evolution and their proximity to our Galaxy. With optical observatories in Chile, Australia, and South Africa, and with the Australia Telescope and the SEST, there are good astronomical resources in the southern hemisphere to guide and complement research with a large mm array. These resources will become more powerful, as the European Southern Observatory is building its Very Large Telescope (VLT), and both the U.S.–U.K.–Canada–Brazil–Argentina–Chile GEMINI consortium and the Carnegie Institution will have new 6 to 8 m-class telescopes in Chile. A large mm array would greatly enhance the value of these new investments in southern hemisphere optical astronomy.

4 Design Study

Many tasks of the design phase could be done by a consortium of existing institutes. The study could use the IRAM array to test new correlators, IF systems, and one or more prototype antennas. One of the most difficult tasks will be the IF processing and phase tracking system, which defines interfaces for receivers and correlators, and hence requires excellent coordination. If the site is in South

America, it may be useful to get help during the study phase from the European Southern Observatory, and some sharing of existing ESO infrastructure, transport, etc., in the initial period.

Antennas: To minimize repairs and maintenance on a spread-out, remote site, a design goal should be reliability of all components. The dishes should be simple and robust, and able to keep their mm-quality surface without any active corrections. The goal should be to have strong antennas which can survive in the outdoors, without astrodomes, and without active optics. The design study should consider a low-maintenance, fault-tolerant antenna with a minimum of serviceable parts; possibly no receiver cabin; receivers in a compact, plug-in, cryogenic module instead; new antenna mounts; simpler designs; a robust reflector, possibly without an enclosed back structure; a subreflector support that is stable under wind gusts and thermal changes; a minimum number of parts to assemble; and improved pointing accuracy and pointing stability. Because of the small beamwidths at mm wavelengths, strong efforts should be made to improve pointing and focussing accuracy. Thermal effects dominate the precision of pointing and focussing. The study should be aimed at keeping pointing precision and focus stability even when part of the antenna structure is in sunlight, and part in shadow. The dishes should be heavy enough to keep their pointing accuracy in winds up to 20 m/s. **The telescope pointing deserves strong engineering efforts.** It may be desireable, in the late 1990's, to try out one or more prototypes of improved antennas with the IRAM array before starting a production run for a next-generation mm array. These prototypes might then be re-integrated into the new instrument when it is built.

Receivers should be designed to be tunerless, maintainence-free, have a minimum of moving parts, and require no helium filling. Operation on a remote, spread-out site will require upgrades in current receivers to maximize robustness and reliability and minimize maintenance. The receiver box must hold ≥ 6 different mixers, as observing in several frequency bands simultaneously would give higher scientific throughput and better capability to correct for atmospheric phase variations. The design study must evaluate the impact on sensitivity of simultaneous frequency operation, given the additional optics required and the difference in the primary beams. Closed-cycle cryo-generators for the receivers must be maintenance-free for long periods, and possibly installable in modular form. The modular concept must be evaluated relative to other designs optimized for system noise. Work on higher-temperature superconducting junctions would be of great interest, especially if these components make cryo-generators cheaper and easier to maintain. Testing of prototypes is needed to see whether SIS receivers can be replaced by HEMT receivers, at least at 3 mm.

IF stages need further development to cover large bandwidths, i.e., several GHz for simultaneous observing at different frequencies. Larger IF bandwidths than those used at present are needed to increase spectral-line observing efficiency and continuum sensitivity. Bandwidth requirements are particularly demanding for extragalactic observing at 1.3 mm. Detailed work is needed on optical fiber transport of the IF signals and the L.O. phase reference. Cost comparisons are needed for transport of digital vs. analog signals by optical fibers. Study and prototype testing is needed for development of integrated IF filters and low-cost LO systems. Phase stability of the electronics is a challenge. Over 1 to 100 second intervals, the electrical length variations over 10 km baselines must be kept to less than $10\,\mu$m, or equivalently the timing precision must be better than 0.04 picoseconds per 1 – 100 sec interval for operation at 300 GHz. Even more stringent accuracy is needed if sub-mm operation is planned.

Correlators must be developed to cover broad bandwidths and to work for long periods without chip failures. Engineers should consider dedicated broad and narrow-band correlators, possibly of different types. A broad band correlator might cover 2 to 8 GHz at 5 to 10 MHz resolution for continuum and extra-galactic work, and a narrow band correlator might have ten windows each 10 to 160 MHz wide, at 50 kHz to 1 MHz resolution, for lines in Galactic sources or in nearer, strong-line galaxies. Ultra-fast samplers must be developed and tested to evaluate the technological possibilities and reliability of ultra-fast chips vs. many units of slower chips. Work must be done on the control of timing for mm array electronics over several kilometers. Is it possible to make the phase switching that separates the sidebands an option, depending on whether the dominant noise is instrumental or atmospheric?

Software needs a large effort, to develop the ability to handle multi-frequency spectral-line operation on $\geq$ 1000 baselines. Data cubes will be much larger than those produced by mm arrays so far, or by the VLA. Because of the long base-lines, the array will operate partly in a multi-speckle regime, and new software strategies may be needed to overcome atmospheric fluctuations.

Phase Correction: A critical development will be a method to correct interferometer phases for atmospheric fluctuations. This development will have implications for site selection, frequency coverage, computing needs, mode of operation, baseline lengths, and array configurations. At cm wavelengths, the signal-to-noise ratio permits many VLA programs to use self-calibration and correct for instrumental and atmospheric phase errors. A similar procedure is used in the modeling and global fringe fitting in VLBI. To do this, the signal to noise ratio must be > 3 in integration times of a few seconds. However, most sources of interest for mm interferometry are weak, thermal sources, requiring coherent integration times of many hours ! If there are no corrections to the phase fluctuations, the atmospheric seeing limit for mm interferometers is about 0.5″, independent of frequency. Hence the maximum usable baseline, without phase corrections, would be about 450 m at 3 mm, to within a factor of two. One way to correct the phase is to use the variations of sky brightness temperature, which are proportional to the water vapor variations within a primary beam. This method requires receivers stable to one part in 10^5. Current IRAM results show a radiative correction method does indeed work in clear sky conditions. Whatever the method finally adopted, large dishes, e.g., 16 m dishes instead of 8 m dishes, are important to increase the signal-to-noise ratio per baseline, and to maximize the number of sources usuable for self-calibration. *A workable method of phase correction will be essential before a large new mm array can be built.*

We intend to refine this method with the IRAM array in the coming years.

Interferometer Configuration: The antennas might be in a ring or in a Gaussian random pattern, so natural weighting would yield a gaussian beam with low sidelobes. Some ideas on configurations to optimize the u, v coverage are given by Cornwell (1987, 1988) and Keto (1992). Servicing might be done with an elevator platform on a truck. This solution must be weighed against the ease of assembling antennas in a hall, and bringing them back for maintenance and upgrades. However, for $\geq$ 50 dishes, it may be better to build antennas in the open air. It would be feasible to construct 10 dishes per year. One could imagine two work crews, with one crane, with each team finishing a dish every two months. A separate team would do cabling and equip the dishes with a receiver package. With such an assembly, the configuration needs careful thought so the new array can start working progressively. Since construction will extend over several years, it would be good to start doing science as soon as 10 dishes are on site, and then exploit the full instrument several years later when all antennas are operating. The first 10 dishes would operate with a maximum baseline $\leq$ 1 km. The next 10 dishes could then be used of baselines twice as long as the first group, and so on, up to the final group of 10 antennas allowing the longest baselines. In this schedule, there might be 30 antennas within a core $\sim$ 1 km in diameter. This 1-km core might contain half the total collecting area and form a sensitive compact array for initial detections without serious phase problems. The best telescopes, with the best panels from a large production run, might all be put in the inner compact group for observing at 345 GHz or in the sub-mm windows, while the outer antennas might be mainly used for 3 and 1.3 mm. While construction of the complete array might take 5 years, some observing might be able to start after the first year of assembly. When could such a mm array enter into full operation? For large facilities like the VLA, IRAM, and the VLT, the time from initial designs to initial operation was 10 to 20 years, So if serious planning starts in 1996, a large new mm array could be in a productive phase around 2010.

5 Conclusions

• The LSA, a Large Southern millimeter Array with a collecting area of $10000\,\mathrm{m}^2$, ten times greater than available at present, would open new directions in astrophysics and greatly enhance the rich harvest of results already obtained from this unique part of the electromagnetic spectrum.

• The southern hemisphere is of special interest for a new mm array because of the excellent mm sites, the richness of the southern sky for Milky Way objects, and the powerful new optical telescopes being built in Chile.

• The mm array could consist of $50 \times 16\,\mathrm{m}$ dishes or $100 \times 11\,\mathrm{m}$ dishes with pointing accuracy $1''$ r.m.s., above 3000 m altitude on a site allowing baselines of 5 to 10 km. With the larger dishes, The global surface accuracy might be $50\,\mu\mathrm{m}$ r.m.s. with the larger dishes or $25\,\mu\mathrm{m}$ with the smaller dishes. The goal is to synthesize a $0.1''$ beam at 3 mm.

• Receivers should cover, in first priority, **the 3 mm and 1.3 mm windows**. These are likely to be the two windows where most of the science will be done. It would also be desireable to equip the antennas in the other mm windows (2 mm, 7 mm, and 0.8 mm), and depending on the site chosen and the dish surface accuracy, possibly the sub-mm windows as well.

• The main requirements could be met with existing technology, but to optimize this technology and minimize operating costs, design studies are needed:

• For the antennas: low production costs; robust, open-air antennas that can be assembled easily, and that need minimal maintenance. For antennas up to 16 m diameter, operating in the mm ranges, **the surface accuracy is not the problem. The problem is the pointing accuracy.** The antenna structures must be thermally stable, giving good pointing and focus stability, even when part of the structure is shadowed.

• For the receivers: low production costs, robust modules, broad bandwidths, no tuning, no helium filling, with simultaneous multi-frequency operation.

• For the IF system: broad-band IFs, fiber optic transport of analog or digital IF signals and local oscillator phase reference, analog or digital delay lines; excellent instrumental phase stability.

• For the correlators: broad-bandwidths; low chip failure rates; ultra-fast samplers, evaluate ultra-fast chips vs. many units with slower chips; evaluate the need for separate correlators for continuum and spectroscopy.

• Investment and operating costs would be ∼ 270 M Ecu (1995) and 30 M Ecu/yr, respectively. A staff of ∼ 150 persons would be needed. On the likely sites, staff may have to be flown in weekly from a large city.

• Incorporation of one or more prototype antennas in the IRAM array would be an important intermediate step before starting large-scale industrial production for a new large mm array. This would allow testing of new antenna and receiver technology, help to develop methods to correct for atmospheric phase fluctuations, and would be scientifically rewarding as well.

Finally, we emphasize how this proposal differs from other programs to extend the existing mm arrays of BIMA, OVRO, NRO, or IRAM, or to build new arrays such as the Smithsonian sub-mm array, the MMA proposed by NRAO, or the large mm/sub-mm array proposed by Nobeyama: **The main point is the huge collecting area.** The aim is to achieve the maximum feasible sensitivity and resolution, and to concentrate on compact mm sources, rather than extended ones. Two scientific domains in which this telescope will excel will be detection of gas and dust in galaxies at high redshifts and the detection of accretion disks and proto-planetary disks around young stellar objects in our Galaxy.

The second point is the wavelength range. We put our highest priority on a **millimeter** interferometer, that will operate mainly at 3 mm and 1.3 mm. While it is difficult to predict discoveries in advance, the relative richness of the different mm windows in molecular lines and the seeing limits on usable baselines both suggest that the scientific productivity will be greatest in the 3 mm and 1.3 mm windows. Although the inner parts of the instrument might have the best antennas, allowing operation to 0.8 mm or even shorter in some seasons, the main goal would be to have access to the rich spectrum of molecular emission in the millimeter ranges. Hence, in contrast to other proposed new arrays, the goal of the Large Southern Array is to attain the collecting area and resolving power needed to explore the universe in the very productive millimeter bands.

References

Booth, R.S. 1994, in Astronomy with Millimeter and Submillimeter Wave Interferometry, ed. M. Ishiguro & W.J. Welch, San Francisco: Astron. Soc. Pacific., 413

Cornwell, T.J., 1987, MMA Memo No. 38. Socorro: NRAO

Cornwell, T.J., 1988, IEEE Trans. Ant. Prop., AP-36, 1165

Downes, D., 1994, in Frontiers of Space and Ground-Based Astronomy, ed. W. Wamsteker, M. Longair, Y. Kondo, Dordrecht: Kluwer, 133

Ishiguro, M. 1994, in Astronomy with Millimeter and Submillimeter Wave Interferometry, ed. M. Ishiguro, W.J. Welch, San Francisco: Astron. Soc. Pacific, 405

Keto, E., 1992, in Design Study for the SAO Submm Array, Chap. III.

NRAO, 1990, Proposal for a Millimeter Array, Charlottesville, Va.

Raffin, P., Kusunoki, A. 1992, SAO Submillimeter Array Tech. Memo 59, Cambridge, Mass.: Smithsonian Astrophysical Observatory

Rees, M., 1994, Quart. Journ. Roy. Astron. Soc. 35, 391

Wild, P. 1967, Proc. IREE Australia, 28, No. 9

Part 2

Science with
Large Millimetre Arrays

The nature and observability of protogalaxies

Simon D.M. White

Max-Planck-Institut für Astrophysik, Karl-Schwarzschild-Straße 1,
D-85740 Garching bei München, Germany

Abstract. I discuss recent theoretical work on the formation and evolution of galaxies paying particular attention to the ability of current models to make detailed comparisons with observations of the galaxy population both nearby and at high redshift. These models suggest that much (perhaps most) star formation in the universe took place in objects that are already detected in deep galaxy samples. In addition, they predict that systems with large star formation rates are unlikely to be much more abundant in the past than they are at present. Recent data show that the star formation rate in the nearby universe is, in fact, a substantial fraction of that required to make all the stars seen in galaxies, and that the observed abundance of objects forming stars at rates in excess of $10 M_\odot/yr$ is approximately the same at redshifts of 1.25 and 3.25 as it is at $z = 0$. Both the epoch of galaxy formation and "typical" protogalaxies may already have been observed but not recognised. Thermal emission from dust in such protogalaxies could be detected by a large millimeter array, and molecular line observations could explore the dynamical state of the gas in the more massive systems.

1 Introduction

The traditional approach to interpreting observational data on the faint galaxy population is grounded firmly in the work which Beatrice Tinsley carried out in the 1970's (see Tinsley 1980). One starts from a characterisation of the local galaxy population in terms of the abundance of objects as a function of luminosity and Hubble type. One assumes that star formation always occurs with an Initial Mass Function (IMF) similar to that inferred from observations of the disk population in the Solar Neighborhood. One picks a simple parameterized model for the star-formation history of a galaxy specified, for example, by the redshift at which star formation starts, the fraction of the stars form in an initial burst, and the characteristic timescale with which formation of the remaining stars decays exponentially in time. These parameters are then adjusted separately for each Hubble type so that the synthesised stellar population at redshift zero has colours which agree with those observed for nearby galaxies. These simple models then allow one to predict the luminosity and colour which a present day galaxy would have if it were seen, for example, at $z = 1$.

This simple scheme can make full use of the information we have about nearby galaxies and of our knowledge of stellar evolution. In addition, it is easily extended to include, for example, a separate treatment of the bulges and disks of spiral galaxies, or a phenomenological treatment of starbursts. It allows one to use the counts, colours and redshift distributions of faint galaxies, to look for possible constraints on cosmological parameters and on the "epoch of galaxy formation". Furthermore, the assumption of a well defined collapse epoch when a burst forms a significant fraction of the final stellar population has been traditional since the earliest discussions of galaxy formation (e.g. Partridge and Peebles 1967) and has virtue of predicting that the collapsing protogalaxy should be very bright and so, perhaps, easily observable. The failure to find such objects has convinced many observers that galaxy formation must occur at high redshift. An interesting variant of the traditional approach is presented by Gronwall and Koo (1995) who show that it can still explain much of the recent data.

With the refurbishment of the HST and the advent of spectroscopy on 10 meter telescopes, this kind of approach is no longer adequate. Samples of field galaxies can now be studied to redshift 3 and beyond, and the new morphological and spectral data show that these objects cannot simply be considered as present-day galaxies whose star-formation is at a less advanced stage but whose structure is otherwise unaltered. The distant objects are often disturbed, are in most cases relatively small, and typically show evidence for substantial ongoing star formation (Cowie et al 1995; Steidel et al 1996; Giavalisco et al 1996, Abraham et al 1996). It seems likely that an understanding of their relation to nearby systems will only be obtained through more detailed consideration of how different types of galaxies were assembled and made their stars.

Theoretical understanding of the formation and evolution of galaxies has improved greatly over the past two decades. Surveys of the spatial and kinematic distributions of galaxies, together with simulations which demonstrate how gravity can produce these distributions, have led to a much clearer picture of the likely context for galaxy formation than was available in the 1970's. This picture can be further explored at high redshift using quasar absorption lines (e.g. Cen et al 1994; Katz et al 1996). In addition, studies of the stellar populations within galaxies and of interacting galaxies have highlighted the role of galaxy transformation processes like starbursts and mergers. As I now discuss, hierarchical clustering theory can be used to a build a phenomenology of galaxy formation which is physically based, relatively simple, tuned to agree with detailed simulation results where these are available, and designed to allow direct comparison with observation.

2   The current structure formation paradigm

A well-developed "standard" picture for the formation of structure has been adopted as a working hypothesis by most cosmologists. Its main elements are the following.

• Most of the matter in the universe is in some dark, nonbaryonic form. It must be dark because all attempts to observe radiation from the dominant component in galaxy clusters or in the outer halos of galaxies have failed. It must be nonbaryonic because the mass content of the universe inferred from dynamical and gravitational lensing measurements exceeds the baryonic content required if the observed abundances of light elements is to be consistent with the theory of Big Bang nucleosynthesis.

• Baryonic matter is present in the amount predicted by Big Bang nucleosynthesis. This is few percent of the closure density, and is significantly larger than the amount observed directly in the form of stars or intergalactic gas.

• Deviations from uniformity were small in the early universe and were generated at very early times, probably by quantum effects during inflation. This process imposed no characteristic scale relevant to the formation of galaxies or larger structures, and produced a gaussian field of linear density fluctuations. An alternative, structure generation at late times by topological defects such as cosmic strings or textures, has attracted less attention because its consequences are much harder to calculate.

• Structure grows through gravitational instability. Radiation pressure on the gas and dispersive motions in the dark matter can have significant effects at early times, but later evolution is driven entirely by the gravity of the dark matter until the collapse of the dark halos of galaxies.

• Galaxies form at late times by the dissipative settling of gas to the centres of the potential wells provided by dark matter halos. This explains the observed segregation between luminous and dark matter in galaxies as well as the origin of the spin of galactic disks.

At cosmology conferences this framework is usually taken for granted; arguments tend to centre on the parameter values which define specific implementations of it ($H_0, \Omega, \Omega_b, \Lambda$, dark matter type, etc.). For non-specialists, however, it can seem more relevant to question whether the universe really is made primarily of some entirely new form of matter, and whether all the structure we see really arose from quantum zero-point fluctuations when the universe was $\sim 10^{-35}$s old. As a scheme for forming galaxies and larger structures, this hierarchical picture was outlined by White and Rees (1978) and has been developed in recent years both by numerical simulations (e.g. Katz et al 1992; Navarro et al 1995) and by detailed semi-analytic treatments of the relevant physical processes.

3   Hierarchical galaxy formation

The formation of galaxies is expected to occur in a very similar way in most currently popular versions of the paradigm just discussed. (These are usually designated by acronyms, for example, SCDM, CHDM, τCDM, OCDM, ΛCDM, TCDM, PIB, texture+HDM; I will not here discuss the relative merits of these possibilities.) The dominant processes shaping the evolution of the galaxy population and the structure and morphology of galaxies are the following.

• The dark matter clusters hierarchically from a gaussian field of initial density fluctuations. Small objects form first and merge together to make larger ones. A well-developed analytic theory for the dynamics and statistics of this process and also for the structure of the resulting objects has been tested in considerable detail against numerical simulations (Lacey & Cole 1993, 1994; Cole & Lacey 1996; Mo & White 1996; Navarro et al 1996).

• Gas cools and collects at the centres of dark halos to form cold rotationally supported disks. As first shown by Fall and Efstathiou (1980), the observed angular momenta of spiral galaxy disks can be produced by tidal torques at early times *only* if the disks formed within an extended massive halo in this way.

• Star formation occurs: (a) in quiescent disks; (b) in bursts during galaxy collisions and mergers; (c) during the initial collapse of a galaxy/halo system. In the present universe most star formation occurs in mode (a), but mode (b) is also significant. In a hierarchical model there is little distinction between collapse and merging, and so between modes (b) and (c).

• Winds from massive stars and shocks from supernovae may reheat cold gas in galaxy disks and may affect the accretion of new gas from the surrounding halo. Such feedback effects appear necessary to limit the efficiency of star formation and to ensure that sufficient gas remains at late times to form the large galaxies which contain most observed stars. Direct evidence for strong feedback associated with vigorous star formation is seen in the "superwinds" generated by some starburst galaxies (Heckman et al 1990).

• Ellipticals form by the merger of disk/bulge systems made primarily of stars. Gas may condense to form a new disk around such an elliptical and so transform it back into the bulge of a spiral. The first process is observed directly in the nearby universe, although there is controversy over the fraction of ellipticals produced by it. Direct evidence for the second can, perhaps, be found in the discovery of a nearby spiral with a counter-rotating bulge (Prada et al 1996).

• When dark halos merge the galaxies within them remain distinct. Thus a massive "cluster" dark halo can contain many galaxies, and the halo of a spiral galaxy may include a number of small satellites. Galaxy mergers occur when dynamical friction brings the orbit of a subsidiary galaxy near the centre of its dark halo where it can encounter the dominant galaxy which resides there. Thus the Milky Way will accrete the Magellanic Clouds, and the cD galaxies at cluster centres can grow by swallowing other cluster galaxies as well as by accreting more gas from surrounding cooling flows.

All of the above processes seem certain to play a role in shaping the observed galaxy population. The difficulty lies in specifying when they should occur and what their relative importance should be. In addition one needs methods to calculate their effect. The analytic description of hierarchical clustering referred to above can be extended to generate Monte Carlo realisations of the full merging history of a present-day dark halo (Kauffmann & White 1993). Within such a merger tree it is possible use simplified analytic descriptions of the relevant physical processes to follow the formation, evolution and interaction of all the galaxies which end up in the final halo. By considering many halos with the appropriate mass distribution, it is then possible to reconstruct the galaxy population in a representative region of the universe.

This programme was first carried out by Kauffmann et al (1993) who showed that if parameters are tuned so that a dark halo with circular velocity of 220 km/s typically contains a system resembling the Milky Way, then hierarchical models reproduce many of the observed regularities of the local galaxy population. For example, the Tully-Fisher relation, the bulge-to-disk ratios of spirals, the morphology-environment and gas fraction-environment relations, the colour-morphology relation, the rich cluster luminosity function and luminosity-morphology relation, all these are well reproduced. The most serious discrepancy affects the field galaxy luminosity function which has the correct number of bright galaxies but too many faint ones. Models which come close to fitting the nearby galaxy population can also be consistent (with no further adjustment) with the available counts and redshift distributions for faint galaxies (Kauffmann et al 1994). An independent implementation of this programme by Cole et al (1994) differs substantially in many details but comes to similar conclusions. In Heyl et al (1995) these authors also studied how the success of such schemes depends on the particular cosmology in which they are implemented.

Such attempts to build physically based models differ fundamentally from the traditional Tinsley approach. Both schemes use population synthesis models to predict the photometric properties of galaxies, but the resemblance stops there. The traditional approach treats each galaxy independently of all others. The galaxy population at high redshift can be identified one-to-one with the nearby population whose properties are adopted from observation rather than derived from a model. The epoch of galaxy formation is a parameter of the arbitrary analytic form chosen to describe star formation histories, and is adjusted to fit faint galaxy data. In contrast, galaxies form and transform continually in the hierarchical models. There is no simple relationship between the galaxies at $z = 1$ and those at $z = 0$. Some old ellipticals have grown new disks, some old spirals have merged to make new ellipticals, and a significant amount of the matter in nearby galaxies was in *no* galaxy at $z = 1$. The parameters which are adjusted describe the efficiencies of uncertain physical processes (for example, star formation, feedback and dynamical friction) and only indirectly influence the formation history of galaxies. The present properties of the galaxy population are not built in *a priori*, but are predictions of the model; as a result they usually show some discrepancies with observation.

The advantage of the hierarchical models is that because they offer a complete, albeit schematic, description of the history of the galaxy population, they can be used to address a very broad range of questions. For example, with Monte Carlo realisations of the full formation history of cluster galaxies one can study the conditions necessary for an observable Butcher-Oemler effect, for the elliptical colour-luminosity relation to be as tight as observed, and for this relation to evolve as observed. One can also check whether the star formation histories of "field" ellipticals (and thus their colours) are expected to differ systematically from those of cluster ellipticals. These questions were addressed by Kauffmann (1995, 1996) who found that the observed strength of the Butcher-Oemler effect appears to require a high density universe, that such a universe can be consistent with the tight colour-luminosity relation of cluster ellipticals, and that field ellipticals are predicted to be younger (and so bluer) than cluster ellipticals.

Figure 1 shows the average formation history predicted for elliptical galaxies in rich clusters for a CDM universe with $\Omega = 1$ and $H_0 = 50$ km/s/Mpc. About 55% of the stars in the ellipticals formed more than 10 Gyr ago and almost none less than 4 Gyrs ago. On the other hand, the typical elliptical had its last major merger about 7 Gyrs ago, so that most of the stars were formed well before the observed galaxies were assembled. From this figure we can infer that the stars in, say, a $10^{11} M_\odot$ elliptical galaxy were forming rapidly around 11 Gyr ago, corresponding to $z = 2.5$. At this time the star formation rate was of order $0.2 \times 10^{11} M_\odot / 1\text{Gyr} = 20 M_\odot / \text{yr}$. However, the galaxy was in several pieces, so the star formation rate in the largest piece was $\sim 10 M_\odot / \text{yr}$. This is an order of magnitude smaller than the rate inferred in traditional models where a bright elliptical forms in a single burst of duration 1 Gyr or less.

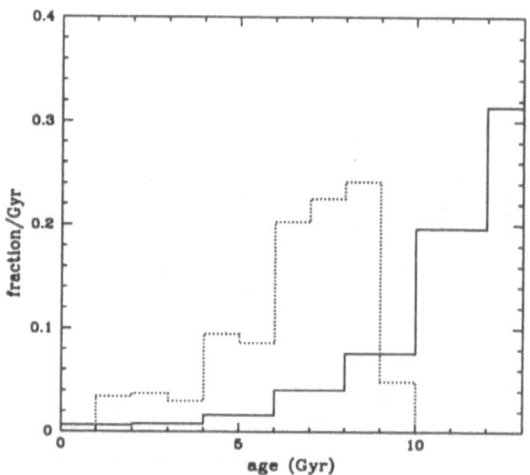

Fig. 1. The solid histogram is the stellar age distribution for elliptical galaxies in rich galaxy clusters according to a standard CDM model with $\Omega = 1$ and $H_0 = 50$ km/s/Mpc. The dotted histogram gives the time of the last major merger for this same population of galaxies. For field ellipticals both histograms would be shifted to lower ages. For field galaxies in general the stellar age distribution becomes much flatter. (Adapted from fig. 3 of Kauffmann 1996).

Hierarchical models automatically specify luminosities and morphologies for all galaxies at all redshifts, and so it is easy to predict galaxy counts as a function of morphology as well as of colour and apparent magnitude. In the Hubble Deep Field galaxies can be classified to much fainter limits than was previously possible. Baugh et al (1996) show that the counts presented by Abraham et al (1996) are in good agreement with their previously published hierarchical models – the rapidly rising count of of irregular and star-forming systems is reproduced quite naturally. The models also predict redshift distributions as a function of morphology, and it should soon be possible to test these predictions directly. Hierarchical models can also predict the spatial distribution of galaxies as a function of luminosity and type. A first analysis has been presented by Kauffmann et al (1996) who show that the observed difference in clustering between spirals and ellipticals and between bright galaxies and dwarfs is easily reproduced. However, in the rest of this contribution I concentrate on the star formation history of the galaxy population since it is this which relates most directly to millimeter observations. The case of cluster ellipticals, discussed above, illustrates the major point; star formation occurs at lower redshifts, in smaller objects, and at lower rates than in traditional models.

4   The abundance of star-forming galaxies

The best observational measure of the star-formation rate in nearby galaxies is generally thought to be Hα luminosity. A recent survey by Gallego et al (1995) has determined the first reliable luminosity function for the local universe based purely on Hα selection. These authors find that the galaxies which provide most of the nearby star-formation differ from those which dominate the optical luminosity density. The star-forming galaxies are typically late-type systems with luminosities well below L_* and with star formation occurring predominantly in a compact nuclear region. The Hα luminosities can be converted to star-formation rates assuming a standard solar neighborhood IMF, and the data then give the local abundance of galaxies as a function of star-formation rate. This distribution is shown in cumulative form in Figure 2. By integrating over all luminosities Gallego et al estimate a total star-formation rate of 0.013 $M_\odot$/yr/Mpc^3 where here and below $H_0 = 50$ km/s/Mpc.

It is interesting to compare this number with the total mass density of stars in the local universe. The APM redshift sample of Loveday et al (1992) covers the largest volume surveyed to date and leads to an estimated luminosity density of $6.5 \times 10^7 L_\odot$/Mpc^3 in the B band. According to Efstathiou et al (1988) 40% of the B light comes from E/S0 galaxies and the rest from spirals. The detailed kinematic study of van der Marel (1991) shows that the mean M/L of early type galaxies is 3.2 in B after averaging over an appropriate luminosity function. For spirals Persic and Salucci (1992) estimate $M/L_B \approx 1.0$ after subtracting the contribution of dark halos to the estimated dynamical masses. Putting all these numbers together gives a total star density of $1.2 \times 10^8 M_\odot$/Mpc^3. Thus the local star formation rate observed by Gallego et al appears enough to make all the stars observed by Loveday et al in about 10 Gyrs. The overall star formation rate need never have been higher in the past than it is today.

In fact there is mounting evidence that the luminosity density inferred from the APM sample is too low by about a factor of 2, perhaps because of photometric problems (Ellis et al 1996; Bertin & Dennefeld 1996). In this case the average past star formation rate would need to be twice the currently observed value. This seems consistent with the detection of strongly enhanced star formation in intrinsically *faint* galaxies at redshifts beyond 0.3 (Ellis et al 1996; Lilly et al 1995). However, these same samples show little evidence for an enhancement of the abundance of rapidly star-forming systems, and in fact, as I now show, there appears to be no such enhancement out to redshifts beyond 3.

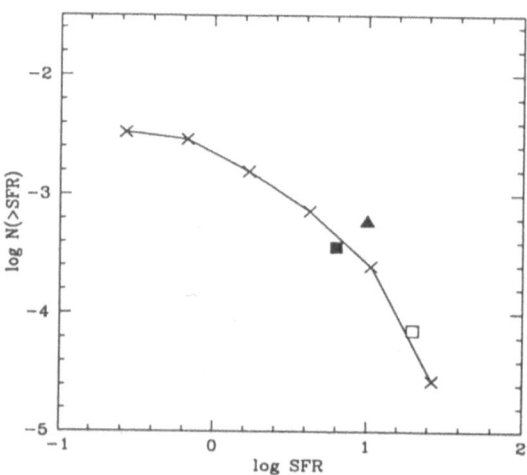

Fig. 2. The crosses linked by a solid line show the abundance (in Mpc^{-3}) of nearby galaxies with star-formation rates exceeding a given value (in $M_\odot/yr$) plotted against that value. The data are taken from Gallego et al (1995). The triangle is the abundance estimated at $z \sim 1.25$ from the data of Cowie et al (1995) while the squares are similar estimates at $z \sim 3.25$ from the data of Steidel et al (1996). Filled symbols assume $q_0 = 0.5$ while the open square assumes $q_0 = 0.05$.

Cowie et al (1995) present redshift and emission line data for an almost complete sample of galaxies with $B < 24.5$. Both population modelling of the UV continuum and the emission line data in their figure 3 suggest that their sample should be complete for objects with unobscured star-formation rates greater than 10 $M_\odot/yr$ over the redshift range $1.0 < z < 1.5$. They find an average of 0.7 such objects per square arcmin, leading to a mean comoving abundance of $6.4 \times 10^{-4} Mpc^{-3}$ assuming $q_0 = 0.5$. The same exercise can also be carried out at higher redshift. Steidel et al (1996) argue that their Lyman limit imaging technique finds all galaxies in the redshift range $3 < z < 3.5$ with $R > 25$. This band corresponds to the far UV continuum in the galaxy rest frame, and they estimate $R > 25$ to correspond to an unobscured star-formation rate exceeding 6.3 $M_\odot/yr$. They find 0.4 such objects per square arcmin giving a comoving volume density of $3.6 \times 10^{-4} Mpc^{-3}$. I plot both abundances on top of the local data in figure 2. Remarkably, the abundance of systems forming stars faster than 10 $M_\odot/yr$ is similar at $z = 3.25$, at $z = 1.25$ and at $z = 0$. If one instead adopts $q_0 = 0.05$ the star-formation rate for the Steidel et al galaxies increases by a factor of 3 but their abundance drops by a factor of 5. As figure 2 shows the inferred abundance is still similar to that seen locally.

It thus appears that although global star-formation rates were higher in the past, the enhancement resulted purely from an increase in the number of systems forming stars at modest rates. A new population of unobscured massive star-bursts is not seen at redshifts less than 3.5. It should, however, be noted that the star-formation rates are estimated in different ways for each of the three samples in figure 2; hence systematic shifts in the relative calibration might change the impression given by this plot.

5 Millimeter observations of young galaxies

In hierarchical models most star formation occurs relatively late. (The equivalent of Figure 1 but for all the stars in the universe shows a much more gentle decline to the present.) In addition there is no period when bulges and ellipticals form rapidly as single units and so are very bright. Instead star formation in the past occurred in lower mass and more gas-rich objects than today but at rates comparable to those in nearby systems. It seems unlikely that extreme objects like the superluminous IRAS starburst sources were ever much more abundant than they are today. The observational data in figure 2 support this conclusion but it is important to note that they refer to unobscured star-formation. Most of the radiation in the strongest nearby starbursts is absorbed by dust and reradiated in the far infrared. In addition, the fraction of the energy emitted by dust seems to be greater in systems which are more dynamically disturbed. Both observational and theoretical arguments suggest that a larger fraction of star-forming objects are irregular at high redshift, so the fraction of the energy radiated in the FIR may be much greater than it is locally.

The amount of heated molecular gas and the amount of hot dust are the key parameters which determine the observability of distant galaxies by a millimeter array. A further important parameter is the size of the emitting region. Recent studies of distant galaxies find that the UV-emitting regions are almost always small with typical angular sizes well below one arcsec, corresponding to physical sizes of a kiloparsec or two. The physically-based models I have discussed make significantly more pessimistic predictions than *ad hoc* models which typically extrapolate the strong evolution detected in nearby samples of IRAS galaxies in order to predict the FIR emission at $1 < z < 5$ (e.g. Rowan-Robinson, this volume). Objects with star-formation rates of 30 $M_\odot$/yr are found in both the Cowie et al and Steidel et al samples. Since these surveys covered very small areas it is likely that objects will be found forming stars at 100 $M_\odot$/yr over the full redshift range $z < 4$; at least some of these objects may emit most of their radiation in the FIR and would then be only a few times less luminous than nearby ultraluminous IR galaxies (e.g. Clements et al 1996). Scaling from such systems shows that an order of magnitude increase in sensitivity of millimeter interferometers would allow line observations of young galaxies at $z > 1$ and detections of the dust emission from the most luminous systems at high redshift ($z = 4$ to 5). The detection of millimeter lines will be particularly important because it will give information about the dynamical state of the gas during the early stages of galaxy assembly. An angular resolution approaching 0.1 arsec will be needed to map the emitting gas in most systems.

References

Abraham, R.G., et al. (1996): MNRAS **279**, L47
Baugh, C.M., Cole, S., Frenk, C.S. (1996): MNRAS, submitted
Bertin, E., Dennefeld, M. (1996): A&A, in press
Cen, R., Miralda-Escude, J., Ostriker, J.P., Rauch, M. (1994): ApJ **437**, L9
Clements, D.L. et al (1996): MNRAS **279**, 459
Cole, S. et al. (1994): MNRAS **271**, 781
Cole, S., Lacey, C.G. (1996): MNRAS, in press
Cowie, L.L., Hu, E.M., Songaila, A. (1995): Nature **377**, 603
Efstathiou, G., Ellis, R.S., Peterson, B.A. (1988): MNRAS **232**, 431
Ellis, R.S. et al. (1996): MNRAS **280**, 235
Fall, S.M., Efstathiou, G. (1980): MNRAS **193**, 189
Gallego, J., Zamorano, J., Aragon-Salamanca, A., Rego, M., (1995): ApJ **455**, L1
Giavalisco, M., Steidel, C.C., Macchetto, F.D. (1996): ApJ, in press
Gronwall, C., Koo, D. (1995): ApJ **440**, L1
Heckman, T., Armus, L., Miley, G.L. (1990): ApJS **74**, 833
Heyl, J.S, Cole, S., Frenk, C.S., Navarro, J.F. (1995): MNRAS **274**, 755
Lilly, S.J., LeFevre, O., Hammer, F., Crampton, F. (1996): ApJ, in press
Loveday, J., Peterson, B.A., Efstathiou, G., Maddox, S.J. (1992): ApJ **390**, 338
Katz, N., Hernquist, L., Weinberg, D.H. (1992): ApJ **399**, L109
Katz, N., Weinberg, D.H., Hernquist, L., Miralda-Escude, J. (1996): ApJ, in press
Kauffmann, G. (1995): MNRAS **274**, 161
Kauffmann, G. (1996): MNRAS, in press
Kauffmann, G., Nusser, A., Steinmetz, M. (1996): MNRAS, in press
Kauffmann, G., Guiderdoni, B., White, S.D.M. (1994): MNRAS **267**, 981
Kauffmann, G., White, S.D.M. (1993): MNRAS **261**, 921
Kauffmann, G., White, S.D.M., Guiderdoni, B. (1993): MNRAS **264**, 201
Lacey, C.G., Cole, S. (1993): MNRAS **262**, 627
Lacey, C.G., Cole, S. (1994): MNRAS **271**, 676
Mo, H.J., White, S.D.M. (1996): MNRAS, in press
Navarro, J.F., Frenk, C.S., White, S.D.M. (1995): MNRAS **275**, 56
Navarro, J.F., Frenk, C.S., White, S.D.M. (1996): MNRAS, in press
Partridge, R.B., Peebles, P.J.E. (1967): ApJ **147**, 868
Persic, M., Salucci, P. (1992): MNRAS **258**, 14P
Prada, F., Guttierez, C.M., Peletier, R.F., McKeith, C.D. (1996): ApJ **463**, L9
Steidel, C.C., Giavalisco, M., Pettini, M., Dickinson, M., Adelberger, K.L., (1996): ApJL, in press
Tinsley, B.M. (1980): Fund. Cosm. Phys. **5**, 287
van der Marel, R.P. (1991): MNRAS **253**, 710
White, S.D.M., Rees, M.J. (1978): MNRAS **183**, 341

Observability of Early Evolutionary Phases of Galaxies at mm Wavelengths

Gianfranco De Zotti[1], Paola Mazzei[1], Alberto Franceschini[2], and Luigi Danese[3]

[1] Osservatorio Astronomico, Vicolo dell'Osservatorio 5, I-35122 Padova, Italy
[2] Dipartimento di Astronomia dell'Università di Padova, Vicolo dell'Osservatorio 5, I-35122 Padova, Italy
[3] International School for Advanced Studies (SISSA/ISAS), Via Beirut 2-4, I-34013 Trieste, Italy

Abstract. Several lines of evidence and theoretical arguments suggest that a large fraction of starlight is absorbed by interstellar dust and re-radiated at far-IR wavelengths, particularly during early evolutionary phases of early type galaxies, which may even, under some circumstances, experience an optically thick phase. Therefore far-IR to mm observations are crucial to understand the galaxy evolution. The strong K-correction makes surveys at mm wavelengths ideally suited for studying high-z galaxies. The broad redshift range covered by mm surveys at sub-mJy flux limits offers a good chance for gaining important information also on the geometry of the Universe.

1 Introduction

The IRAS survey has clearly demonstrated the crucial role of dust in shaping the spectral energy distribution (SED) of galaxies. A direct comparison of their local luminosity functions in the optical and in the far-IR shows that, locally, about 30% of starlight is reprocessed by dust.

It is very likely that this fraction was higher in the past, when the star formation activity was more intense and also early type galaxies possessed a plentyful interstellar medium, which might have been metal enriched on a very short timescale, the lifetime of the first generation of massive stars.

Thus, the observational study of evolution of dust emission in galaxies is crucial to understand the evolution of galaxies themselves.

2 Luminosity Evolution, Dust Absorption and Far-IR/mm Emission

A useful scheme for modelling the evolution of galaxies relies on the consideration of two extreme patterns (Sandage 1986): on one side, disk galaxies, characterized by dissipational collapse, with slow gas depletion, i.e. star formation rate (SFR) never much higher than today; on the other side, spheroidal galaxies thought to have used up most of their gas to form stars in a time short compared with the collapse time, i.e. with a spectacularly large initial SFR. The evolution of galaxies of different Hubble types can then be modelled as a suitable combination of the two basic components.

This scheme is certainly oversimplified in many respects; in particular, it does not allow for important facts such as merging, interactions, star formation induced by nuclear activity, and so on. Still, it may be useful to sketch out some of the chief features.

Our approach (Mazzei et al. 1992, 1994) exploits chemical and photometric evolution models of stellar populations, complemented by allowing for the effect of dust. Simple assumptions are adopted for the dust component, namely: the dust to gas ratio is proportional to the metallicity (i.e. a constant fraction of metals is locked up in dust grains); stars and dust are well mixed; the "standard" grain model (Mathis et al. 1977; Draine and Lee 1984), including a power law grain size distribution, holds at any time.

If indeed the metallicity and the star formation rate in *galactic disks* did not vary much throughout their lifetime, their SED too remained essentially unchanged, except for optical colours being somewhat bluer and the dust being somewhat warmer during the early phases (Mazzei et al. 1992).

On the contrary, dramatic far-IR evolution is expected for early-type galaxies due to the fast (exponential with a timescale of a few Gyr) decrease of the SFR with increasing galactic age. Their bolometric luminosity increases by a substantial factor with decreasing galactic age, T (a factor $\simeq 10$ from $T = 15\,\mathrm{Gyr}$ to $T = 2\,\mathrm{Gyr}$). Moreover, the far-IR to optical luminosity ratio increases from local values $\lesssim 10^{-2}$ (Mazzei and De Zotti 1994a) to $\simeq 1$ or even $\gg 1$ at early times. On the whole, the far-IR luminosity of early type galaxies may have increased by about three orders of magnitude (if the effect of merging may be ignored), a luminosity evolution rate more extreme than even that quoted for optically selected quasars.

Under the above assumptions for dust properties, a key parameter in determining the evolution of the SED of early type galaxies, is the gas consumption rate: in the case of a fast conversion of gas into stars, the far-IR emission is never dominant; but if the gas depletion is slower, the galaxy may experience a prolonged opaque phase, with most of the luminosity emitted in the far-IR (Mazzei and De Zotti 1996).

In any case, *a substantial dust emission is expected during early evolutionary phases of all galaxies.*

2.1 Are primeval galaxies heavily obscured?

As noted above, under some circumstances, galaxies which are in the processes of transforming into stars a large fraction of their mass in a relatively short time (the conventional definition of primeval galaxies) may be heavily obscured by dust. We are unable to work out an a priori estimate of how frequently this may occur.

However, as shown by Franceschini et al. (1994), under the assumption that, during the phases of intense star formation, most of the optical radiation was absorbed by dust and reradiated in the far-IR, a consistent picture obtains in the framework of simple luminosity evolution models. In particular, we may account for: the remarkable lack of high redshift galaxies in optically selected samples down to $B \simeq 24$ (Colless et al. 1993; Cowie et al. 1991); the failure to detect Lyα emission in searches for primeval galaxies (e.g., Thompson et al. 1995); the deep 60 μm IRAS counts and, exploiting the far-IR/radio correlation for galaxies, most of the observed sub-mJy flattening of radio counts over a couple of decades in flux.

Also, contrary to recent claims (e.g. Thompson et al. 1995), predictions of models entailing strongly obscured primeval galaxies are not in conflict, but may even be supported by COBE data on the far-IR to mm background (cf. Fig. 1).

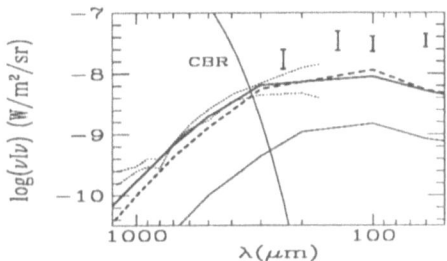

Fig. 1. Contributions to the far-IR to mm background intensity predicted by two different models for galaxy evolution, compared with limits (Hauser 1995) and estimates (dotted lines, Puget et al. 1996) based on COBE data. The lower thin curve is a minimal contribution from normal galaxies, estimated assuming no evolution up to $z = 1$. The dashed line is a *merging* model, in which most galaxies form stars and build up at $z \simeq 1$. The thick continuous line is an evolution model at constant galaxy mass function (for more details on both models see Franceschini et al. 1995).

2.2 Direct Evidences of Large Amounts of Dust in High-z Sources

The observed spectral energy distribution of the ultraluminous object IRAS F10214+4724 (Rowan-Robinson et al. 1991; Lawrence et al. 1993), at $z \simeq 2.3$ is remarkably well fitted by a model for young (age $\lesssim 1$ Gyr) spheroidal galaxies with strong dust extinction (Mazzei and De Zotti 1994b).

On the other hand, this source hosts an active nucleus which may well be the main energy source, as strongly suggested by the evidences of gravitational lensing (Eisenhardt et al. 1996). Indeed, Granato et al. (1996) obtain a good fit of its SED with the dusty torus model that fits the SEDs of both broad and narrow line AGNs in the framework of the unified model (Granato & Danese 1994). The diameter of the far-IR emitting dusty torus is, in their model, of $\simeq 2$ kpc (for $H_0 = 50$), corresponding to an angular size of $0''.3$.

Dust masses $\simeq 10^8$–10^9 m$_\odot$ (suggesting gas masses of between 10^{10} and 10^{12} m$_\odot$) are indicated by mm/sub-mm detections (Dunlop et al. 1994; Chini

& Krügel 1994; Ivison 1995) of the high z radio galaxies 4C41.17 ($z = 3.8$), 53W002 ($z = 2.39$), and 8C1435+635 ($z = 4.26$); such dust masses are 1–2 orders of magnitude higher than found for nearby radio galaxies (Knapp et al. 1990; Knapp & Patten 1991).

Furthermore, evidences of vast reservoirs of dust at high z are provided by mm/sub-mm detections of a number of distant radio quiet QSOs (Andreani et al. 1993; Ivison 1995, and references therein); several high-z radio loud QSOs were also detected, but the observed mm fluxes could be accounted for by synchrotron emission.

As in the case of IRAS F10214+4724, the relative importance of the nucleus and of a possible gigantic burst of star formation in heating the dust is still unclear. A better determination of the effective dust temperature, that may soon be provided by ISO observations, will certainly be extremely helpful: dust temperatures exceeding $\simeq 60$ K are probably difficult to explain with a starbust model. But for a firm assessment of the problem, we will probably need the sensitivity and the angular resolution of the large mm array which would allow an observational characterization of the structure and luminosity of both the torus and the starburst.

3 Flux vs Redshift at mm Wavelengths

The spectral energy distribution (SED) of galaxies during the early evolutionary phases, characterized by intense star formation activity, is likely to be qualitatively similar to that of the nearby starburst galaxy M 82, shown in Figure 2.

Above a few mm, the luminosity is dominated by radio emission. The study of this emission and of its evolution is an interesting problem *per se* since a better understanding of the relative contributions of synchrotron and free-free processes at these wavelengths would shed light on the properties of the magnetic field on one side and of HII regions on the other.

In the range $100\,\mu$m–1 mm, dust emission dominates and the continuum spectrum of star forming galaxies can be described by a power law of the form $L_\nu \propto \nu^\alpha$ with $\alpha \simeq 3.5$ (Franceschini and Andreani 1995; Chini et al. 1995), although the far-IR to optical luminosity ratio in the case of normal late type galaxies is substantially smaller than for M 82, consistent with the correlation between far-IR emission and star formation rate.

Then, as we observe at mm wavelengths galaxies at larger and larger redshifts, the K-correction works to rapidly increase the flux, as we move to higher and higher frequencies along a steeply increasing spectrum (Franceschini et al. 1991; Blain and Longair 1993). For steep enough spectral indices, the K-correction eventually overcomes the effect of increasing distance, so that, above some redshift z_m, the flux actually increases with distance.

In the absence of any luminosity evolution, if $\alpha = 3.5$ we have $z_m = 2$, 1.83, 1.60, 1.29, and 0.96 for $\Omega = 0$, 0.03, 0.1, 0.3, and 1, respectively (De Zotti et al. 1996).

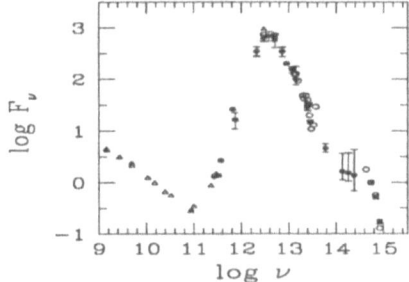

Fig. 2. Observed continuum spectrum of M 82 from UV to radio wavelengths. Data are from Hughes et al. (1989) and references therein, Cohen and Volk (1989), Huang et al (1994) and references therein, Kennicut (1992).

The minimum, however, is shallow, so that it can hardly be exploited to determine the density parameter. On the other hand, the weak dependence of flux on distance implies a rather uniform distribution of sources over a broad redshift range (up to $z \simeq 10$, if galaxies were already present) thus offering a good chance of exploring the geometry of the universe, particularly once the photometric evolution of galaxies will be reasonably well understood.

4   Predictions for Deep Millimeter Surveys

Estimates have been attempted by Franceschini et al. (1991), Blain & Longair (1993, 1996), Mazzei et al. (1996), based on different evolution models.

Such estimates, however, are a very delicate exercise because they require large extrapolations with very limited observational constraints. The nearest counts on the far-IR side, the IRAS counts at $60\,\mu m$, span a limited range of flux and are rather uncertain at the faint end. On one hand, the redshift survey by Ashby et al. (1996) of the deep IRAS field at the North ecliptic pole (Hacking & Houck 1987) has discovered that the counts may be significantly above average because of the presence of a large supercluster at $z = 0.088$. At the other hand, Gregorich et al. (1995), from a study of a set of deep IRAS fields covering a total area about three times larger than that of Hacking and Houck (1987) report faint counts about a factor of two higher (note, however, that the completeness limit adopted by Gregorich et al. (1995), $\simeq 50\,mJy$, is only 2.5 times higher than the estimated rms confusion noise; there is thus a serious danger that counts are overestimated at the faint end because of source confusion).

Also, there is a considerable spread in the observed $(1.3\,mm/60\,\mu m)$ flux ratios of galaxies (cf. Fig. 3), implying a correspondingly large uncertainty in the estimated local luminosity function of galaxies at mm wavelenghts; this uncertainty is strongly amplified by the extreme steepness of the counts.

From a theoretical point of view, there is a great deal of uncertainty on the physical processes governing galaxy formation and evolution. At one extreme

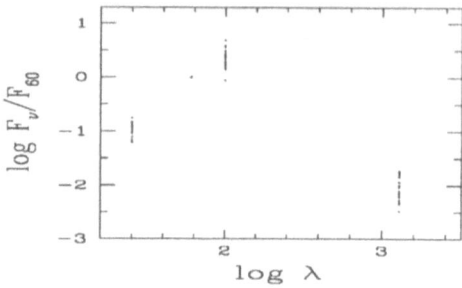

Fig. 3. SEDs of galaxies observed at mm wavelengths by Chini et al. (1995), normalized to 60 μm fluxes.

there are models assuming that the comoving density of galaxies remained essentially constant after their formation, while they evolved in luminosity due to the ageing of stellar populations and the birth of new generations of stars (pure luminosity evolution). At the other extreme, according to some hierarchical clustering models, large galaxies are formed by coalescence of large numbers of smaller objects. The observed properties of both disk and spheroidal galaxies imply that extensive merging cannot have occurred in the last several billion years (cf. e.g. Franceschini et al. 1994); it must, therefore, have occurred at significant redshifts.

The faint end of expected counts may be strongly different in the two cases. Of course, models advocating extensive merging at $z \gtrsim$ 1–2 imply a sharp enhancement of the counts at faint flux densities (Blain and Longair 1993, 1996). The flux density at which such enhancement begins is model dependent and is sensitive to several unknown quantities such as the merging history, the star formation rate and the initial mass function during the merging process, the dust temperature. On the other hand, a sub-$L_\star$ galaxy with a dust temperature of 60 K, typical of galaxies with very intense star formation, at $z \gtrsim$ 1 has a flux density $< 10\,\mu$Jy at 1.3 mm.

All in all, the actual physical processes governing the formation and the early evolution of galaxies may be very complex and may depend on an impressive number of unknown or poorly known parameters: spectrum of primordial density perturbations, hydrodynamic processes in the primordial gas, merging rate, star formation rate, initial mass function, galactic winds, infall, interactions, dust properties, and so on. The observational contraints are still very poor. Hence, a direct observational study of these phases is essential.

If, as argued above, metal enrichment and condensation of metals into dust grains occurs very quickly in primeval galaxies, far-IR to mm observations will play a crucial role in this field. ISO and SCUBA offer excellent prospects; their data will certainly help very much to discriminate between different scenarios. On the other hand, the much better sensitivity of the large mm array is required to test the possibility of substantial merging at $z >$ 1–2, as expected, in particular, in the framework of cold dark matter cosmologies (Kauffmann et al. 1993).

5   Conclusions

The emission from interstellar dust, which locally comprises $\simeq 30\%$ of the global bolometric luminosity of galaxies, is very likely to have been significantly larger during earlier evolutionary phases, when the (metal enriched) ISM was more abundant. Therefore observations of the dust emission are crucial to understand the galaxy evolution.

In the case of early type galaxies, very poor of dust and gas at present, the evolution in the far-IR/mm bands could have been very spectacular, to the point that during the first 1–2 Gyr of their lifetime, most of their starlight could have been reprocessed by dust; this corresponds to an increase of the far-IR luminosity by more than three orders of magnitude.

A related issue is the primary power source of ultraluminous IRAS sources (and in particular IRAS F10214+4724) and of the very large far-IR/mm emission from some high-z radiogalaxies. From the far-IR luminosity and the available constraints on dust temperature it is concluded that the dust distribution in the most luminous sources has a size of at least several hundred parsecs ($H_0 = 50$). The planned large mm array might resolve these sources even if the far-IR emission comes from a dusty torus surrounding an active nucleus.

The very steep increase of spectra of galaxies with increasing frequency in the range few-mm to $\simeq 100\,\mu$m makes surveys at mm wavelengths exceptionally well suited for investigating the evolution of galaxies up to $z \simeq 10$. The very weak dependence of flux on distance for $z > 1$ implies a very uniform coverage of the full redshift range over which galaxies presumably exist; once the physics of the evolution is understood, this gives a good chance of investigating also the geometry of the universe.

It may be debated whether the large mm array is the appropriate instrument for surveys aimed at investigating the early evolutionary phases of galaxies. Indeed SCUBA is expected to be capable of following the evolution of dust emission from bright galaxies up to very high redshifts. On the other hand, substantially better sensitivity and spatial resolution than achievable with SCUBA may be necessary to investigate e.g. the process of extensive merging of small dusty clumps at $z \gtrsim 1$–2. Other methods, such as the study of absorption lines in the spectra of high-z quasars, may be heavily biased if the clumps are very dusty.

On the other hand, for such surveys a not too small field of view at $\simeq 1\,$mm is essential.

Work supported in part by ASI.

References

Andreani, P., La Franca, F., Cristiani, S. (1993): MNRAS **261**, L35

Ashby, M.L.N., Hacking, P.B., Houck, J.R., Soifer, B.T., Weisstein, E.W. (1996): ApJ **456**, 428

Blain, A.W., Longair, M.S. (1993): MNRAS **264**, 509

Blain, A.W., Longair, M.S. (1996): MNRAS in press

Chini, R., Krügel, E. (1994): A&A **288**, L33

Chini, R., Krügel, E., Lemke, R., Ward-Thompson, D. (1995): A&A **295**, 317

Cohen, M., Volk, K. (1989): AJ **98**, 1563

Colless M., Ellis R., Taylor K., Hook R. (1993): MNRAS **261**, 19

Cowie, L.L., Songaila, A., Hu, E.M. (1991): Nat **354**, 460

De Zotti, G., Franceschini, A., Mazzei, P., Toffolatti, L., Danese, L. (1996): Ap. Lett. & Comm. in press

Draine, B.T., Lee, H.M. (1984): ApJ **285**, 89

Dunlop, J.S., Hughes, D.H., Rawlings, S., Eales, S.A., Ward, M.J. (1994): Nat **370**, 347

Eisenhardt, P., Soifer, B.T., Armus L., Hogg, D., Neugebauer, G., Werner, M.: (1996), ApJ, in press

Franceschini, A., Andreani, P. (1995): ApJ **440**, L5

Franceschini, A., Granato, G.L., Mazzei, P., Danese, L., De Zotti, G. (1995): Proc. COBE Workshop on "Unveiling the Cosmic IR Background", College Park, MD

Franceschini, A., Mazzei, P., De Zotti, G., Danese, L. (1994): ApJ **427**, 140

Granato, G.L., Danese, L. (1994): MNRAS **268**, 235

Granato, G.L., Danese, L., Franceschini, A. (1996): ApJ in press

Gregorich, D.T., Neugebauer, G., Soifer, B.T., Gunn, J.E., Herter, T.L. (1995): AJ **110**, 259

Hacking, P., Houck, J.R. (1987): ApJS **63**, 311

Hauser, M. (1995): Proc. COBE Workshop on "Unveiling the Cosmic IR Background", College Park, MD

Huang, Z.P., Thuan, T.X., Chevalier, R.A., Condon, J.J., Yin, Q.F. (1994): ApJ **424**, 114

Hughes, D.H., Gear, W.K., Robson, E.I. (1989): MNRAS **244**, 759

Kauffmann, G., White, S.D.M., Guiderdoni, B. (1993): MNRAS **264**, 201

Ivison, R.J. (1995): MNRAS **275**, L33

Kennicutt, R.C. Jr. (1992): ApJ **388**, 310

Knapp, G.R., Bies, W.E., van Gorkom, J.H. (1990): AJ **99**, 476

Knapp, G.R., Patten, B.M. (1991): AJ **101**, 1609

Lawrence, A., et al. (1993): MNRAS **260**, 28

Mathis, J.S., Rumpl, W., Nordsieck, K.H. (1977): ApJ **217**, 425

Mazzei, P., De Zotti, G. (1994a): MNRAS **266**, L5

Mazzei, P., De Zotti, G. (1994b): ApJ **426**, 97

Mazzei, P., De Zotti, G. (1996): MNRAS, in press

Mazzei, P., De Zotti, G., Xu, C. (1994): ApJ **422**, 81

Mazzei, P., Lonsdale, C., Chokshi, A. (1996): in preparation

Mazzei, P., Xu, C., De Zotti, G. (1992): ApJ **256**, 45

Puget, J.L., Abergel, A., Bernard, J.-P., Boulanger, F., Burton, W.B., Désert, F.-X., Hartmann, D. (1996): A&A, in press

Rowan-Robinson, M., et al. (1991): Nat **351**, 719

Sandage, A. (1986): A&A **161**, 89

Thompson, D., Djorgovski, S.G. (1995): AJ **110**, 982

Thompson, D., Djorgovski, S., Trauger, J. (1995): AJ **110**, 963

Studying High Redshift Starburst Galaxies with a Large (Sub)millimetre Array

Paul P. van der Werf and Frank P. Israel

Leiden Observatory, P.O. Box 9513, NL–2300 RA Leiden, The Netherlands

Abstract. We review the current observational status of observations of high redshift galaxies in the (sub)millimetre continuum, in molecular lines and in low-excitation fine-structure lines, and discuss the significance and implications of the results obtained so far. The prospects for studying high-z galaxies with future large (sub)mm interferometric facilities are discussed. The thermal dust emission from luminous starburst galaxies is shown to be detectable throughout the universe, while the Milky Way may be detected out to $z \lesssim 5$. Sub(mm) surveys are shown to provide powerful tools for tracing the evolution of starforming galaxies. We discuss the optimum strategy for CO observations of high redshift galaxies. We finally show that starburst galaxies will be detectable in low-excitation fine structure lines throughout the $z \lesssim 10$ universe. The very luminous [C II] line gives access to Milky Way type galaxies at $4 \lesssim z \lesssim 10$ for $q_0 = 0.5$, or to ten times more luminous objects for $q_0 = 0.1$.

1 Introduction

The first observable evolutionary stage of a galaxy is expected to be the phase of the buildup of the initial stellar population. The redshift at which this process takes place is not known, and probably varies with galaxy type. For instance, some low-metallicity starburst dwarf galaxies may be undergoing their first major starburst and are in this sense forming at the present epoch. The disks of spiral galaxies probably undergo a continuous buildup over an extended redshift interval. In contrast, bulges and ellipticals, which have little recent star formation, must have formed the bulk of their stellar populations at an earlier epoch. While the redshift interval at which spheroid formation takes place is a subject of debate (e.g., Peebles 1989), it is clear that a substantial burst of star formation must be involved: in order to build up a stellar population of $10^{11} \, M_\odot$ by $z = 3$, a star formation rate (SFR) $\dot{M}_* \gtrsim 10^2 \, M_\odot \, \mathrm{yr}^{-1}$ at $z > 3$ is required. In addition, if large galaxies are assembled by the successive merging of smaller galaxies, such as in hierarchical clustering models, numerous merger-induced starbursts are expected to occur. Numerical models of "bottom-up" structure formation in a cold dark matter (CDM) universe (e.g., White & Frenk 1991) predict a maximum merger rate at $1 < z < 3$.

While the first stars must have formed in a dust-free environment, a starburst is expected to rapidly enrich the ambient interstellar material, producing solar metallicities in a few times 10^8 years (Matteucci & Padovani 1993). As a result, only during the first few 10^8 years the burst will take place in a dust-free interstellar medium (ISM), and most of the time the initial starburst will be optically obscured and emit more than 90% of its energy in the far-infrared (FIR) wavelength range (Mazzei et al. 1994). This effect may account for the lack of success of searches for high-z starburst galaxies in Lyα emission surveys (e.g., De Propris et al. 1993; Parkes et al. 1994; Thompson et al. 1995; Thompson & Djorgovski 1995, and references therein), since the resonantly scattered Lyα line will be destroyed even if only a small amount of dust is present.

Such dusty, gas-rich, optically obscured high-z starburst galaxies are ideally studied in the FIR/submillimetre continuum, in emission lines of polar molecules (principally CO), and in low-excitation fine-structure lines. These diagnostics all lie in the (sub)millimetre region. First detections of high-z objects in all of these tracers have been obtained in recent years. In this review we summarize the current observational situation in this field and examine the possibilities and optimum observing strategies for finding and studying dusty high-z starburst galaxies with a large (sub)millimetre array. As a specific application, we discuss the observability of these objects as a function of redshift with the proposed Large Southern Array (LSA). Throughout this paper we adopt a value $H_0 = 75\,\mathrm{km\,s^{-1}\,Mpc^{-1}}$ for the Hubble constant.

2 Thermal emission from dust in high redshift galaxies

A dusty starburst galaxy emits more than 90% of its luminosity in the FIR wavelength region. The thermal dust emission obeys the relation $S_\nu \propto Q_\nu B_\nu(T_d)$, where $B_\nu(T_d)$ is the Planck function at the dust temperature T_d, and Q_ν is the dust emissivity, which in the FIR-submm region obeys the relation $Q_\nu \propto \nu^k$, with k increasing with wavelength from 1 to 2 (Draine & Lee 1984). Therefore, in the long wavelength Rayleigh-Jeans regime, the thermal dust emission falls off as $S_\nu \propto \nu^{-4}$. Hence for dusty high-z objects, where the peak of the thermal dust emission is shifted into the (sub)mm regime, a very large, negative K-correction applies. As result, the (sub)mm regime is the *ideal* spectral region for studying dusty high-z starburst galaxies.

Thermal dust emission has been detected in the (sub)mm region now in more than 20 high-z objects (e.g., Isaak et al. 1994; McMahon et al. 1994; Dunlop et al. 1994; Ivison 1995). Implied FIR luminosities L_{FIR} for these *hyperluminous infrared galaxies* range from $10^{13}\,L_\odot$ to several times $10^{14}\,L_\odot$. If this luminosity is powered by star formation, the corresponding SFR follows from

$$\frac{\dot{M}_*}{\mathrm{M_\odot\,yr^{-1}}} = A \times 10^{-10}\,\frac{L_{\mathrm{FIR}}}{L_\odot}, \tag{1}$$

where the constant A is of order unity and depends on the details of the initial mass function (IMF). Thus for the hyperluminous infrared galaxies the SFRs exceed $10^3\,\mathrm{M_\odot\,yr^{-1}}$, sufficient for building up an entire stellar population in less than 10^8 yrs, and suggesting that these objects are indeed undergoing their *initial* starburst. Modeling of the spectral energy distribution (SED) with a dust-producing initial starburst for three of these objects yields ages in the range from 0.05 to 2 Gyr (Mazzei & De Zotti 1996). Some caution is required in converting the observed FIR luminosities directly into SFRs, since an unknown fraction of L_{FIR} is provided by the AGNs embedded in these objects. However, even if only 10% of the observed L_{FIR} is powered by star formation, SFRs of 10^2 to a few times $10^3\,M_\odot\,yr^{-1}$ are implied. Given the large dust masses, $M_d \sim 10^8 - 10^9\,M_\odot$ (see Hughes (1996) for a detailed discussion), implied by the observed SEDs, it would in fact be very surprising if these objects were not forming stars at prodigious rates. Furthermore, as argued by Rowan-Robinson (1996), the observed SEDs cannot be reproduced with only a dust-embedded AGN, but *require* the presence of a starburst providing a substantial part of the observed FIR luminosity.

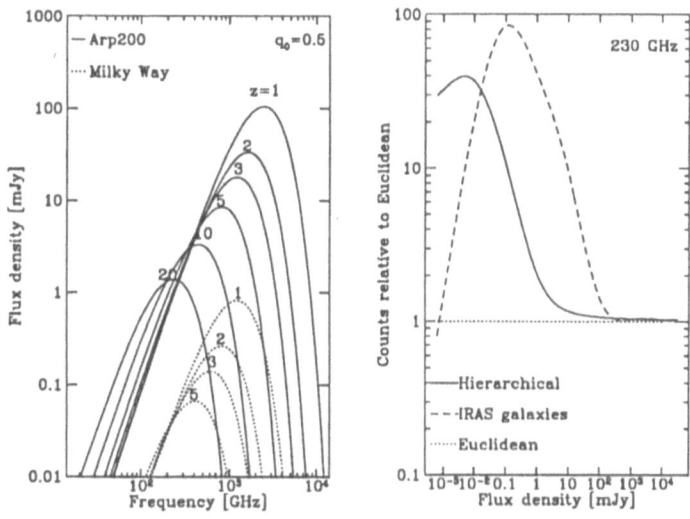

Fig. 1. *Left panel:* Continuum flux of Arp 220 and the Milky Way as a function of redshift for $q_0 = 0.5$. *Right panel:* Source count predictions at 230 GHz relative to Euclidean counts for two evolution models: one model based on IRAS galaxies evolving in luminosity as $(1 + z)^3$ for $z < 2$ and with no evolution at higher z, and one based on a hierarchical clustering model with starbursts occurring during mergers.

The extraordinary possibilities in this field offered by a large (sub)millimetre wave array are illustrated in Fig. 1 (left panel). The large negative K-correction makes a starburst galaxy with $L_{FIR} = 10^{11}\,M_\odot$ (10% of the L_{FIR} of Arp 220) detectable as a point source with the LSA at the 5σ level in 4 hours out to $z \gtrsim 20$, i.e, *throughout the universe*. Point source detection of a galaxy a few times more luminous than the Milky Way (which has $L_{FIR} \approx 10^{10}\,L_\odot$) in the same integration time and at the same S/N ratio would be possible out to $z \sim$ 5. For imaging, sensitivity limits will depend on the details of the brightness temperature distribution on the sky (and of the positions of the antennas in the array), but a crude estimate is given by dividing the total flux into ten independent beams, so that the detection limit for imaging is about a factor of ten higher than that for point source detection. The angular size distance D_A is (for a universe with cosmological constant $\Lambda = 0$) given by

$$D_A = \frac{c}{H_0 \, q_0^2 \, (1+z)^2} \left(z q_0 + (q_0 - 1) \left[\sqrt{2 q_0 z + 1} - 1 \right] \right). \qquad (2)$$

Therefore, 1 kpc corresponds to a minimum angular size of $\sim 0\!''\!.3$ between $z = 1$ and $z = 3$ for $q_0 = 0.5$ and to $\sim 0\!''\!.2$ at $z > 1$ for $q_0 = 0.1$. Hence, for sub-kpc spatial resolution at these distances, $0\!''\!.1$ angular resolution is required, corresponding to baselines of 3 km.

As shown in the left panel of Fig. 1, the detectability of a high-z starburst galaxy at 230 GHz remains almost constant, or may even increase somewhat with redshift for $z > 1$. As a result, source counts at (sub)mm wavelengths will be strongly non-Euclidean, and (sub)mm-selected samples will contain a very high proportion of high-z objects, which makes surveys very attractive. As shown in Fig. 1 (right panel), the counts are extremely sensitive to the luminosity and density evolution of the FIR galaxy population and to the SED of the high-z objects, and hence will provide exceedingly powerful diagnostics of these properties of the high-z population. In order to utilize these diagnostics, the source counts must be established to at least the flux levels where they begin to strongly invert, i.e., increase over the level of purely Euclidean counts. Source count predictions in the (sub)mm have been presented by Blain & Longair (1993) and indicate that, depending on the SED and evolutionary properties of the high-z starburst population, counts will begin to invert strongly at 230 GHz flux densities of 0.1 to 10 mJy. The corresponding predicted source densities at 0.1 mJy are typically a few tens per arcmin^2 (Blain & Longair 1993). The LSA should, with a primary beam of 0.1 arcmin^2 at 230 GHz (for 15 m array elements) thus typically have a few 0.1 mJy sources in every target field, and detect these sources at the 5σ level in 1 hour of integration time. The uncertainty in the predicted source densities is at least an order of magnitude, but it is clear from these estimates that even if the more pessimistic source counts are adopted, the LSA will be very well suited for cosmology with 230 GHz source counts. As shown by Blain & Longair (1993), a further powerful diagnostic is provided by the redshift distribution of a (sub)mm-selected sample. Hence a programme to construct a sufficiently large and faint 230 GHz-selected sample with spectroscopic redshifts would be extremely interesting, and here the LSA and VLT would form an exceedingly powerful combination.

It should be noted that, due to the ν^{-4} dependence of the flux densities of starburst galaxies in this spectral region, the capabilities of the LSA outlined above would be greatly enhanced if it would operate down to submm (850 μm) wavelengths. Source flux densities rise by a factor of five from 1300 to 850 μm, as do (Euclidean) count rates *per primary beam* to a fixed flux density limit. At an excellent submm site such as in northern Chile, and with an array of high 850 μm beam-efficiency dishes, a five times larger sample can be obtained in the same integration time, enhancing the capabilities in of the LSA in this important field by almost an order of magnitude.

3 CO emission from high redshift starburst galaxies

Detections of CO emission from high-z objects have so far only been obtained in two gravitationally lensed systems: IRAS F10214+4724 at $z = 2.286$ (Brown & Vanden Bout 1991; Solomon et al. 1992b) and the $z = 2.546$ "Cloverleaf quasar" H1413+117 (Barvainis et al. 1994). However, even after accounting for gravitational lensing the molecular gas content of these two objects exceeds that of the most gas-rich objects in the local universe, which shows that galaxies with very large masses of dense, enriched molecular gas exist at high redshifts. Searches in other (non-lensed) high redshift objects have, despite painstaking efforts (e.g., Wiklind & Combes 1994; Van Ojik et al. 1996) yielded only upper limits, which however do not rule out molecular masses such as found in IRAS F10214+4724 and H1413+117, after gravitational lensing in these two objects has been taken into account. With current instrumentation, CO in such objects can only be detected if amplified by a fortuitous foreground gravitational lens. However, future large collecting area facilities will be able to detect the *unlensed* objects.

It is a fortunate coincidence that the brightest rotational lines of CO are distributed such that a particular (sub)millimetre wavelength band always contains at least one CO transition *regardless* of the redshift of the emitting galaxy. However, determining which transitions are likely to yield a detection depends on the (expected) intrinsic line ratios of the target. In order to compare the strengths of different transitions of CO, it is convenient to express line luminosities on the L' scale, which can be easily related to observable quantities, as follows

$$L'_{\text{line}} = \int\int T_{\text{b}}\, dv\, d\Omega\, D_{\text{A}}^2 \approx T_{\text{b}}\, \Delta v\, \Omega_{\text{s}}\, D_{\text{A}}^2. \tag{3}$$

Here T_{b} is the *intrinsic* Rayleigh-Jeans brightness temperature in the line and Ω_{s} is the source solid angle. This equation shows that lines with equal intrinsic brightness temperatures T_{b} have equal L' luminosities, as long as source dimensions and linewidth do not change between transitions. Since T_{b} is constant for opaque, thermalized lines (often true for the lower CO transitions), L' will also be constant under these conditions. Note that the line luminosities L_{CO} in the usual units of erg s^{-1} still *do* increase with increasing rotational quantum number under these conditions, as can be seen from the equation relating L and L' luminosities:

$$L_{\text{line}} = L'_{\text{line}} \frac{8\pi k \nu_0^3}{c^3}, \tag{4}$$

where ν_0 is the *rest* frequency of the transition.

A compilation of observed L' luminosities of the lowest CO lines of a number of well-studied galaxies (Israel & Van der Werf 1996) shows that in starburst galaxies, where the average temperature is 50 K or higher, levels up to $J = 6$ (116 K above the ground state) are still relatively easily excited. Due to a fairly high average density, the levels stay thermalized up to $J = 4$, and L' begins to drop significantly only if the upper level $J \geq 7$. In quiescent spiral galaxies such as the Milky Way however, where average densities and temperatures are lower, the gas becomes subthermally excited at much lower rotational levels and L'

drops already by a factor 2 from the 1→0 to the 2→1 line. Gas-rich spirals such as M51 or NGC 6946 are intermediate between these two cases, having constant L' up to the 3→2 line. It is significant that the line ratios in IRAS F10214+4724 (Solomon et al. 1992a) and H1413+117 (Barvainis 1996) agree with those observed in low-z starburst galaxies, indicating similar conditions. In order to estimate the observability of similar objects at high redshift, we assume the same intrinsic line ratios as in the corresponding low-redshift galaxies. Only for Milky Way type objects, where even the low-J levels are subthermally excited, additional radiative excitation by the cosmic microwave background has to be taken into account (e.g., Solomon et al. 1992a). Radiative transfer calculations based on a Large Velocity Gradient (LVG) model show that for low densities a cosmic microwave background $T_{CMB} \sim 10$ K (valid for $z \sim 2$) significantly enhances the populations of the $J = 2 - 4$ levels over the low-redshift ($T_{CMB} = 2.7$ K) values, with a corresponding increase in L' luminosity in these lines.

For point source detection, we express the expected signal in terms of the flux density integrated over the line:

$$S_{line} \Delta v = \frac{2k\nu_0^2}{c^2} L'_{line} \frac{1+z}{D_L^2}, \tag{5}$$

where $D_L = (1 + z)^2 D_A$ is the luminosity distance. Inserting numbers, this equation reduces to

$$\frac{S_{line} \Delta v}{Jy\, km\, s^{-1}} = 3.1 \times 10^{-8} \frac{L'_{line}}{K\, km\, s^{-1}\, pc^2} \left(\frac{\nu_0}{GHz}\right)^2 \left(\frac{D_L}{Mpc}\right)^{-2} (1+z). \tag{6}$$

Based on this equation and the line ratios described above, we now calculate the expected signal for starburst galaxies, gas-rich spirals, and Milky Way type galaxies. To fix absolute line strengths, we use the prototypical ultraluminous infrared galaxy Arp 220 as a template starburst galaxy, and M51 as a template CO-rich spiral. Predictions for fainter or brighter objects can be derived by scaling with respect to the relevant luminosity (see Israel & Van der Werf 1996 for a compilation).

The expected linestrengths as a function of redshift are plotted in Fig. 2, for the 3 mm and 1.3 mm atmospheric windows, and for $q_0 = 0.5$ (the somewhat lower flux predictions for $q_0 = 0.1$ can be found in Israel & Van der Werf (1996)). The benefit of having a number of lines available is seen most clearly in the lefthand panel of Fig. 1, which shows the remarkable result that a starburst galaxy detectable at $z \sim 1$ can *in the same integration time* be detected out to $z \sim 5$, and, in principle, with only a few times more integration time out to much higher redshifts. For Milky Way type galaxies the situation is somewhat less favourable because significant flux is lost by going to lines higher than $J = 4 \rightarrow 3$. These results can easily be understood by noting that as long as the levels involved are thermalized, the flux that is to be detected follows a blackbody curve at the excitation temperature of the gas, and thus increases as ν_0^2 in the Rayleigh-Jeans regime. This increase partly compensates for the larger distance, until such high lines are reached that T_b begins to drop significantly. This effect

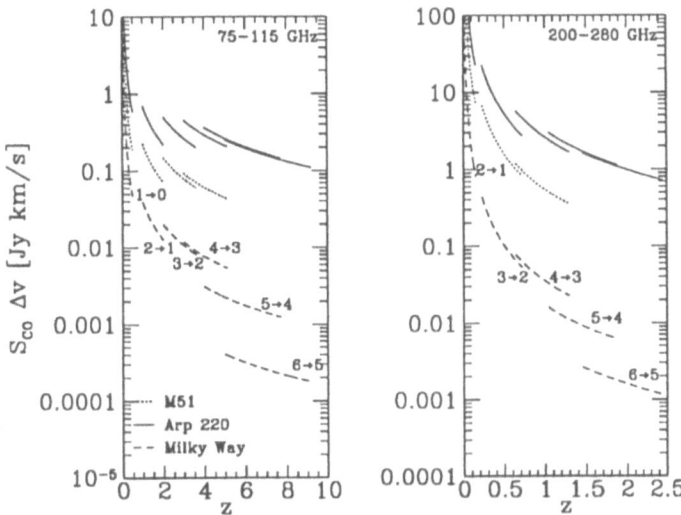

Fig. 2. Integrated strengths of CO lines as a function of redshift for various types of galaxies, in the 3 mm and 1.3 mm windows for $q_0 = 0.5$

is entirely analogous to the advantageous K-correction that operates in (sub)mm continuum studies of high redshift objects, except that for the CO lines the fluxes increase proportional to ν_0^2, while for the continuum emission the flux densities increase as ν_0^4. Therefore the considerations related to the K-correction that hold for detection estimates for dust emission from high-z starburst galaxies, hold to a lesser extent also for thermalized CO emission. Specifically, just like dust emission from luminous starburst galaxies will be detectable throughout the universe in the (sub)mm band with future instrumentation, CO emission from such objects will be detectable out to $z \sim 5$.

We place our *point source* detection limit at the 5σ level for a velocity resolution of $30\,\mathrm{km\,s^{-1}}$, sufficient to obtain line profiles, allowing applications such as CO Tully-Fisher studies. This detection limit corresponds to line integrals of about $0.08\,\mathrm{Jy\,km\,s^{-1}}$ in the 3 mm and 1.3 mm bands respectively for a $300\,\mathrm{km\,s^{-1}}$ wide line in a 12 hour synthesis with the LSA. Inspection of Fig. 2 now reveals the possibilies and optimum observing strategies, which we summarize below.

1. For *starburst galaxies* at $z \lesssim 2.5$ the highest S/N ratio is obtained in the 1.3 mm band, where M51-type galaxies or starburst galaxies 10 times fainter than Arp 220 can be *imaged* out to $z \lesssim 1$, while Arp 220-like objects can be imaged out to $z \sim 2.5$. Starburst galaxies 10 times fainter than Arp 220 and M51-type galaxies will be detectable as point sources out to $z \sim 2.5$ in the 1.3 mm band.

2. For $z > 2.5$ the 1.3 mm band is less favourable, since only fairly high rotational lines, which may not be efficiently excited, are available. Therefore, at $z > 2.5$ the 3 mm band is the better choice, and starburst galaxies with

Arp 220 luminosities will be detectable throughout the $z < 10$ universe in this band for $q_0 = 0.5$ (out to $z \sim 6$ for $q_0 = 0.1$).

3. For Milky Way type galaxies it is necessary to observe low-J lines, and as a result the 3 mm band is favoured over the 1.3 mm band. Point-source detection of Milky Way type galaxies is possible in the 3 mm band out to $z \lesssim 1$ for $q_0 = 0.5$. Out to $z \sim 1$, fairly low lines are still available in the 1.3 mm band as well, and for these redshifts the 3 mm and 1.3 mm bands yield about the same S/N ratio. Imaging is restricted to about $z < 0.25$, for both the 3 mm and the 1.3 mm band.

4. Given a choice of transition, for any redshift the higher transition should be more easily detectable up to $J = 5$ for starburst galaxies and up to $J = 4$ for Milky Way type galaxies.

Since the initial starburst in galaxies forming the bulk of their stellar population may well resemble Arp 220 and similar objects in luminosity and observable properties (Van der Werf 1996), these considerations show the immense potential of future large millimetre wave arrays for observing early stages of galactic evolution. As another application, we point out the possibility offered by the LSA to extend Tully-Fisher studies to much larger distances than is possible with atomic hydrogen, allowing the accurate measurement of perturbations to the general Hubble flow out to very high redshift.

4 Detectability of low-excitation fine-structure lines

Low-excitation atomic and ionic fine-structure lines lie in the short submm and FIR region, and shift into the (sub)mm regime only at high redshifts. However, these lines are expected to be very luminous and detectable out to extremely large distances. With its excitation potential of 11.3 eV (i.e., below the Lyman limit), upper level temperature of 91 K and critical density in molecular gas of $5 \times 10^3 \, \mathrm{cm}^{-3}$, the [C II] is the principal cooling transition of a neutral, dense, warm ISM. The lowest excitation lines of ionized gas are the [N II] lines at 122 and 204 μm with an excitation potential of 14.3 eV (just above the Lyman limit). We adopt the following line luminosities in units of L_{FIR}, based on COBE data for the Milky Way (Wright et al. 1991): [N II] 122 μm: 0.08%; [C II] 158 μm: 0.5%; [N II] 205 μm: 0.05%; [C I] 370 μm: 0.003%; [C I] 609 μm: 0.002%. While for other galaxies these ratios will be somewhat dependent on physical parameters (densities, far-UV radiation fields, abundances), luminosities of the [C II] and [N II] lines in low-z objects are observed to scale linearly with L_{FIR} (Stacey et al. 1991; Petuchowski et al. 1994). Note that in a low-metallicity starburst some of these lines can become even more important as coolants. For instance, in the Large Magellanic Cloud (LMC) the [C II] line contains $\sim 1\%$ of the total luminosity, while this fraction rises to $\sim 3\%$ in individual star forming complexes in the LMC (Israel et al. 1996). Thus during the metal-poor first stage of an "initial starburst", the [C II] 158 μm line may carry an even higher fraction of the luminosity than in the Milky Way.

So far, the only confirmed high-z detection of one of these lines has been obtained in the gravitationally lensed "Cloverleaf quasar", where Barvainis (1996) detected the [C I] 609 μm line. Searches for [C II] 158 μm (Isaak et al. 1994) and [N II] 205 μm (Ivison & Harrison 1996) have so far not been successful, although for QSO observations this may be due to the lack of accurate redshifts, which are based on very wide lines, strongly distorted by the Lyα forest.

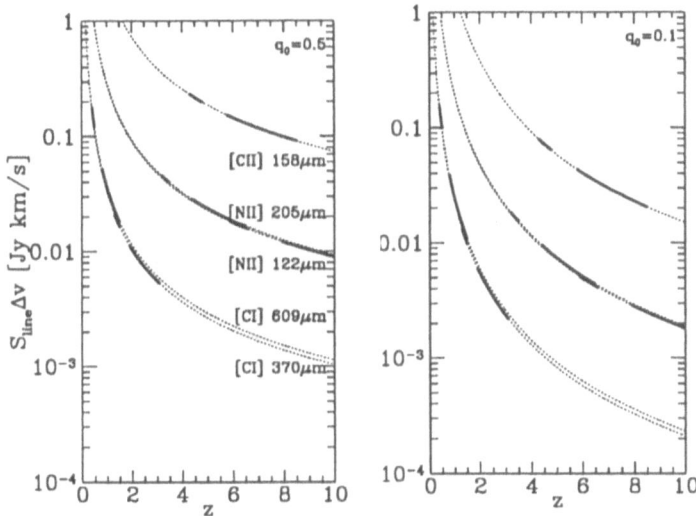

Fig. 3. Integrated strengths of luminous low-excitation fine-structure lines of the Milky Way ($L_{\mathrm{FIR}} \approx 10^{10}\, L_\odot$) as a function of redshift. Drawn lines indicate redshift ranges where the relevant line is located in the $330 - 360\,$GHz (lowest redshifts) or $200 - 280\,$GHz (highest redshifts) atmospheric window. Note that for Arp 220-like objects flux predictions are about 150 times higher.

Flux predictions for these lines are presented in Fig. 3. Placing again the detection limit at $0.08\,\mathrm{Jy\,km\,s^{-1}}$, this figure shows that for $q_0 = 0.5$, the Milky Way would easily be detected at $z = 4.2 - 4.8$ in the [C II] 158 μm line in the $350\,$GHz window. Starburst galaxies with $L_{\mathrm{FIR}} \gtrsim 10^{11} L_\odot$ will be detectable out to $z \sim 3$ in the [C I] lines. At very high redshifts ($4 \lesssim z \lesssim 10$), the $200 - 280\,$GHz window gives access to starburst galaxies in the [N II] lines and to Milky Way type galaxies in the [C II] line, provided that suitable targets with accurately determined redshifts are available. Although at these very high redshifts detection rates will be significantly lower for $q_0 = 0.1$, the possibilities offered by these very bright lines are extremely exciting.

Acknowledgements. The research of Van der Werf has been made possible by a fellowship of the Royal Netherlands Academy of Arts and Sciences.

References

Barvainis, R., Tacconi, L., Antonucci, R., Alloin, D., & Coleman, P. 1994, Nat, 371, 586

Barvainis, R., 1996, CO, C I and (possibly) HCN in the Cloverleaf quasar. In: Bremer, M.N., Van der Werf, P.P., Röttgering, H.J.A., & Carilli, C.L. (eds.), Cold gas at high redshift, Kluwer, Dordrecht

Blain, A.W., & Longair, M.S. 1993, MNRAS, 264, 509

Brown, R.L., & Vanden Bout, P.A. 1991, AJ, 102, 1956

De Propris, R., Pritchet, C.J., Hartwick, F.D.A., & Hickson, P. 1993, AJ, 105, 1243

Draine, B.T., & Lee, H.M. 1984, ApJ, 285, 89

Dunlop, J.S., Hughes, D.H., Rawlings, S., Eales, S.A., & Ward, M.J. 1994, Nat, 370, 347

Hughes, D.H., 1996, Thermal emission from dust in high-z galaxies. In: Bremer, M.N., Van der Werf, P.P., Röttgering, H.J.A., & Carilli, C.L. (eds.), Cold gas at high redshift, Kluwer, Dordrecht

Isaak, K.G., McMahon, R.G., Hills, R.E., & Withington, S. 1994, MNRAS, 269, L28

Israel, F.P., & Van der Werf, P.P., 1996, Considerations for detecting CO in high redshift galaxies. In: Bremer, M.N., Van der Werf, P.P., Röttgering, H.J.A., & Carilli, C.L. (eds.), Cold gas at high redshift, Kluwer, Dordrecht

Israel, F.P., Maloney, P.R., Geis, N., Herrmann, F., Madden, S.C., Poglitsch, A., & Stacey, G.J., 1996, ApJ, in press

Ivison, R.J., & Harrison, A.P., 1996, MNRAS, in press

Ivison, R.J. 1995, MNRAS, 275, L33

Matteucci, F., & Padovani, P. 1993, ApJ, 419, 485

Mazzei, P., & De Zotti, G., 1996, ApJ, in press

Mazzei, P., De Zotti, G., & Xu, C. 1994, ApJ, 422, 81

McMahon, R.G., Omont, A., Bergeron, J., Kreysa, E., & Haslam, C.G.T. 1994, MNRAS, 267, L9

Parkes, I.M., Collins, C.A., & Joseph, R.D. 1994, MNRAS, 266, 983

Peebles, P.J.E., 1989, Galaxy formation: high redshift or low? In: Frenk, C.S. (ed.), The epoch of galaxy formation, Kluwer, Dordrecht, p. 1

Petuchowski, S.J., Bennett, C.L., Haas, M.R., Erickson, E.F., Lord, S.D., Rubin, R.H., Colgan, S.W.J., & Hollenbach, D.J. 1994, ApJ, 427, L17

Rowan-Robinson, M., 1996, The evolution of the far-infrared galaxy population. In: Bremer, M.N., Van der Werf, P.P., Röttgering, H.J.A., & Carilli, C.L. (eds.), Cold gas at high redshift, Kluwer, Dordrecht, p. 61

Solomon, P.M., Downes, D., & Radford, S.J.E. 1992a, ApJ, 398, L29

Solomon, P.M., Radford, S.J.E., & Downes, D. 1992b, Nat, 356, 318

Stacey, G.J., Geis, N., Genzel, R., Lugten, J.B., Poglitsch, A., Sternberg, A., & Townes, C.H. 1991, ApJ, 373, 423

Thompson, D., & Djorgovski, S.G. 1995, AJ, 110, 982

Thompson, D., Djorgovski, S., & Trauger, J. 1995, AJ, 110, 963

Van der Werf, P.P., 1996, Ultraluminous infrared galaxies: dissipation in forming spheroidal systems. In: Bremer, M.N., Van der Werf, P.P., Röttgering, H.J.A., & Carilli, C.L. (eds.), Cold gas at high redshift, Kluwer, Dordrecht, p. 37

Van Ojik, R., et al., 1996, submitted to A&A

White, S.D.M., & Frenk, C.S. 1991, ApJ, 379, 52

Wiklind, T., & Combes, F. 1994, A&A, 288, L41

Wright, E.L., et al. 1991, ApJ, 381, 200

Deep Surveys with a Large Submillimetre Array

Michael Rowan-Robinson and Chris Pearson

Astrophysics Group, Blackett Laboratory, Imperial College, London

Abstract. We review the evidence for evolution of the far infrared galaxy population and present a new model for source counts and background radiation from radio to X-ray wavelengths. The model is used to predict the numbers of sources expected in a number of proposed far infrared and submillimeter surveys, including the proposed Large Southern Array.

1 Far Infrared Surveys

The IRAS mission demonstrated the power of far infrared surveys for cosmology. The IRAS Point Source Catalog (PSC) at 60 μm was used by several groups to study the large-scale galaxy distribution out to 150 h^{-1} Mpc and to determine the cosmological density parameter Ω_o. Redshift surveys of samples from the IRAS Faint Source Survey (FSS) have shown that the infrared galaxy population is undergoing strong evolution (Oliver et al 1995) and have led to the discovery of a population of hyperluminous infrared galaxies (Rowan-Robinson et al 1991, Rowan-Robinson and McMahon 1996, Rowan-Robinson 1996 and references therein), further examples of which have been found through submillimetre observations of high redshift radio-galaxies and quasars. Fig 1 shows the distribution of starburst luminosity with redshift for the 25 hyperluminous infrared galaxies ($L_{ir} > 10^{13} L_o$) known to date.

The cosmological results from IRAS have inspired several survey proposals with the ISO satellite. Surveys on selected small areas will be carried out in Guaranteed Time by both the CAM and PHOT teams (Franceschini et al 1996) and a 200 μm Slew Survey will be carried out during slews of the telescope, which is expected to cover about 20 % of the sky and reach a limiting sensitivity of about 1 Jy. In addition a consortium of 19 European Institutes has been awarded 215 hours of Open Time to carry out a survey of about 20 sq deg of sky at 15 and 90 μm, the ELAIS (European Large Area ISO Survey) survey. The predicted sensitivities are 15 mJy at 90 μm and 1.7 mJy at 15 μm.

In order to predict the source-counts and redshift distribution expected in such surveys, we need to determine the luminosity function and rate of evolution of the infrared galaxy population.

2 Direct Estimates of the Rate of Evolution of IRAS Galaxies

The evolution of the 60 μm luminosity function can be characterized by:

$$\eta(L, z) = f(z)\eta_o(L, L^*(z))$$

where, for density evolution we take $f(z) = (1 + z)^P, z < z_f$,
and for luminosity evolution, we take $L^*(z) = (1 + z)^Q, z < z^*$,
$= (1 + z^*)^Q, z^* < z < z_f$,
$= 0, z > z_f$.

The most direct evidence for evolution comes from large redshift surveys. Saunders et al (1990) found that for the QDOT 1-in-6 sparsely sampled IRAS galaxy redshift survey to $S(60) = 0.6$ Jy:

$$Q = 3.1 \pm 1.0$$

$$P = 6.7 \pm 2.3.$$

After correction for redshift errors, for the effects of Malmquist bias, and for non-linearity in the IRAS PSC flux-scale, Oliver et al (1995) found that this should be revised to:

$$P = 4.2 \pm 2.3.$$

From the shallower but larger 1.2 Jy survey, Fisher et al (1992) found no clear evidence for evolution: $P = 2 \pm 3$ but the uncertainty is such that the QDOT values are not inconsistent with this.

The much deeper IRAS FSS galaxy redshift survey of Oliver et al (1995), which consists of 1400 galaxies in 700 sq deg with $S(60)$ 0.2 Jy, after correction for Malmquist bias, and with a K-correction based on a 2-component (starburst+cirrus) fit to the 100/60 μm colours, yields strong evidence for evolution, with average values for the evolution rate from several methods:

$$P = 5.6 \pm 1.6$$

$$Q = 3.3 \pm 0.8$$

3   Evidence for Evolution from 60 μm Counts

Several groups have found that the 60 m source-counts can only be understood if the IRAS galaxy population is subject to strong evolution (Hacking and Houck 1987, Hacking et al 1987, 1989, Danese et al 1987, Lonsdale and Hacking 1989, Lonsdale et al 1990, Hacking and Soifer 1991, Oliver et al 1992). In Fig 1 of Oliver et al (1992), for example, observations from the IRAS PSC, the IRAS FSS and the pointed observations of Hacking and Houck (1987), are compared with predictions for the case of no evolution, and for strong density and luminosity evolution. The no evolution case is clearly inconsistent with the observations, while either density or luminosity evolution give consistent results. Counts with ISO will allow the latter two models to be distinguished.

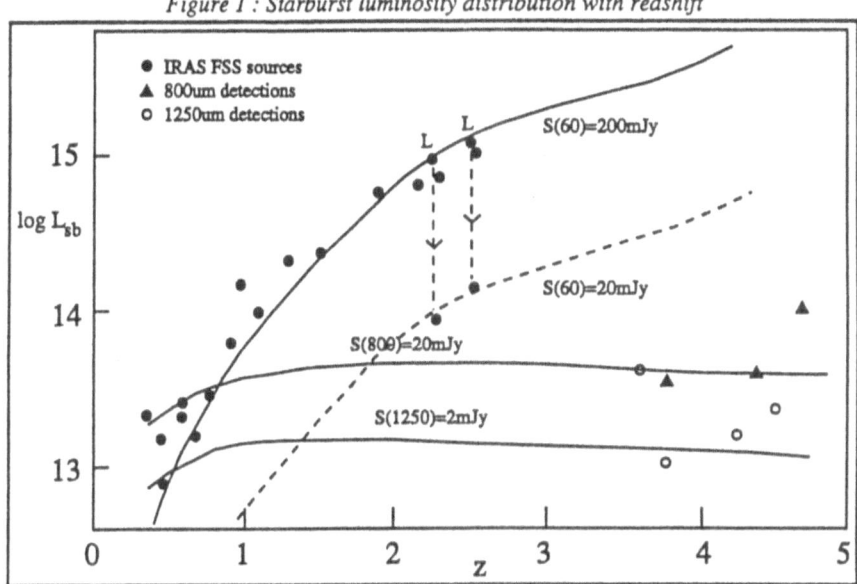

Figure 1 : Starburst luminosity distribution with redshift

Figure 2 : N-z distribution at 90μm over 25 sq. deg. to a flux limit of S=15mJy

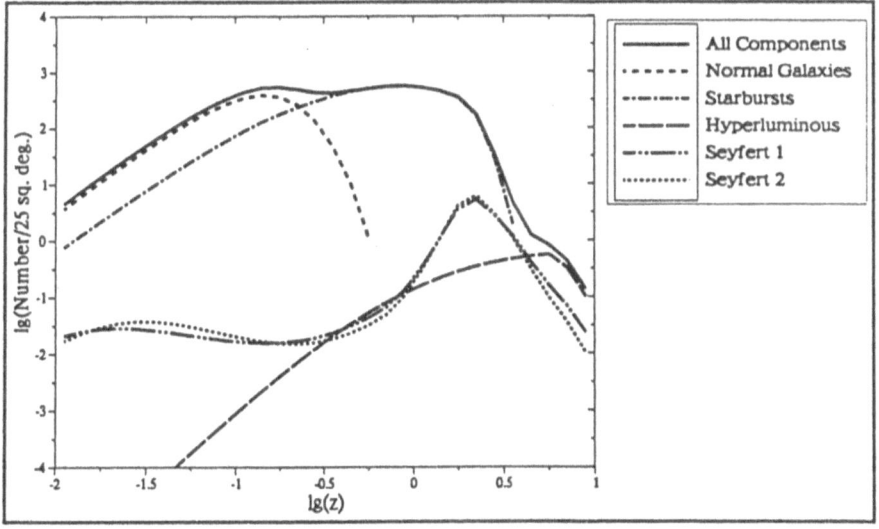

Figure 3 : Normalized differential counts at 60μm

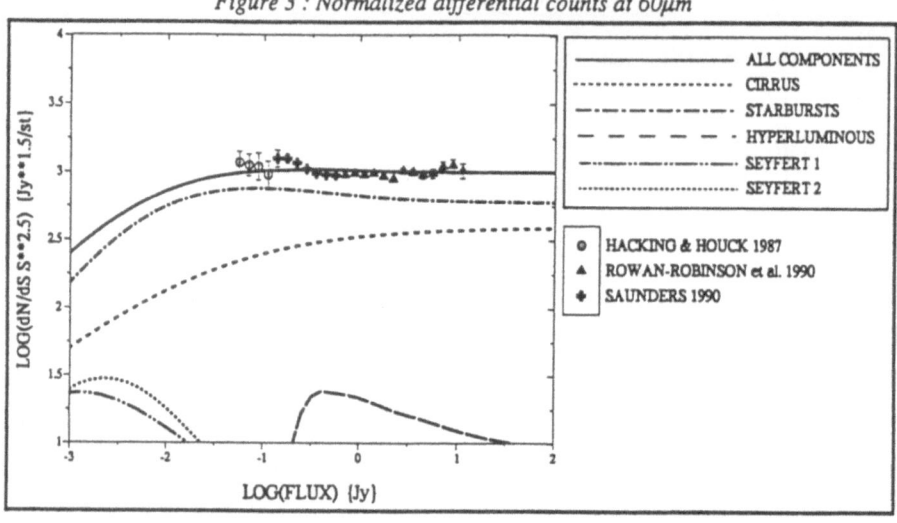

Figure 4 : Differential counts of sub-millijansky sources at 1.4GHz

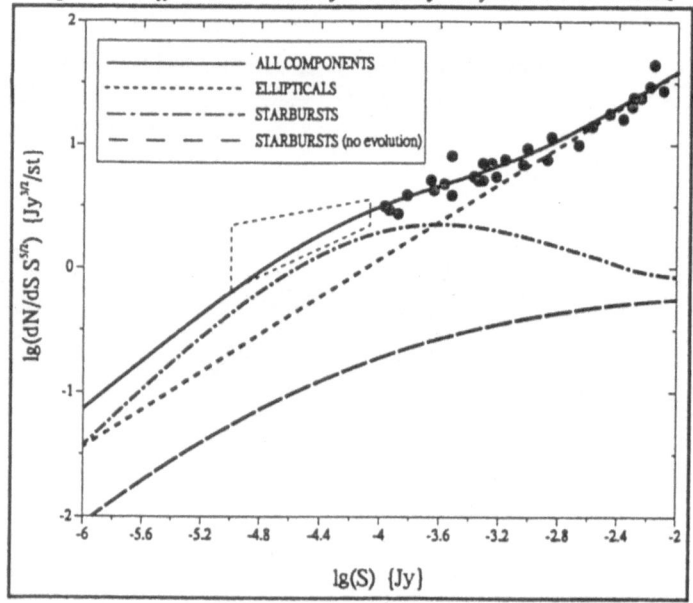

4   Sub-mJy Radio-sources as Starburst Galaxies

The bright radio source-counts ($S(1.4GHz) >> 1mJy$) are due to radio-galaxies and quasars undergoing strong evolution, which is approximately of the form of luminosity evolution (eg Condon 1984, Dunlop and Peacock 1990), though some models involve a small amount of density evolution also. Below 1 mJy at 1.4 GHz the slope of the counts steepens again and there is evidence of a new population of blue radio-emitting galaxies (Mitchell and Condon 1985, Windhorst et al 1987, Thuan and Condon 1987, Franceschini et al 1988, Condon 1989, Lonsdale and Harmon 1991).

Benn et al (1993) have carried out spectroscopy of a sample of 112 idensitifications of sources with $S(1.4GHz) > 0.1mJy$ and shown that below 1 mJy most of the sources are starburst galaxies very similar to those seen by IRAS. The luminosity function for these galaxies agrees well with that at 60 μm, shifted by the radio-fir relation

$$S(60\mu m) = 90S(1.4GHz).$$

Fits to the sub-mJy source-counts then show that strong evolution is required in this population, with luminosity evolution strongly favoured over density evolution (Rowan-Robinson et al 1993a).

5   A New Model for Source Counts and Background Radiation from Radio to X-ray Wavelengths

A new model for extragalactic source counts proposed by Pearson and Rowan-Robinson (1996) defined in the infra-red consisting of a normal spiral + cirrus component defined by cool $100/60\mu m$ colours, a starburst/IR ultraluminous galaxy component defined by warm $100/60\mu m$ colours, hyperluminous IR galaxies ($L_{60} > 10^{12.5}L_\odot$), a 3-30 μm Seyfert/QSO + dust torii component and an elliptical component (negligible in far IR) has been used to predict source counts from radio to X-ray wavelengths ($H_o = 50$, $\Omega = 1$).

K-corrections are found using model source spectra. For the normal galaxies the cirrus model of Rowan-Robinson (1992) is used in the IR, while the model Sbc data of Coleman et al. (1980) and Yoshii and Takahara (1988) is used for the near IR/optical spectrum assuming that 30% of the optical light is re-radiated in the IR. For the starburst/ultra/hyperluminous components, the model spectra of Rowan-Robinson and Efstathiou (1993) are used in the far-IR while at near-IR/optical wavelengths, for $\nu < \nu_B$, the Sab spectrum of Coleman et al. (1980) is used and for $\nu > \nu_B$ a HII galaxy spectrum (Mk36, Neugebauer et al. (1976)) is used. In modelling the optical spectra of starburst galaxies it is assumed that 95% of the optical light is re-radiated in the far-IR. For the Seyfert galaxies the torus model of Rowan-Robinson (1995) is used and the elliptical galaxies are modelled using the spectral energy distributions of Yoshii and Takahara (1988) and Bertola et al. (1982)

The normal spirals and starburst/ultra/hyperluminous populations are respectively modelled using the cool and warm 60 μm galaxy luminosity functions of Saunders et al. (1990). For optical counts an assumption of $L_{IR}/L_B \approx 30\%$ is assumed for the normal spiral galaxies to shift the luminosity function to the B-band. The Seyfert/QSOs use the 12μm parameters of Rush et al. (1993) who represented their sample with the 2 power law luminosity function of Lawrence et al. (1986). the elliptical galaxies use a B-band Schecter function to model their contribution to the B and K band counts.

Pure luminosity evolution is assumed for the starburst, ultra/hyperluminous and AGN components and is of the form used by Boyle et al. (1988) found from fits to optical QSO data. An evolution rate of $(1+z)^Q$ $\{z < 2\}$, $(1+z_*)^Q$ $\{z > 2\}$ is assumed where $Q = 3.1$, $z_* = 2$ and the formation redshift, $z_f = 5$. This form of evolution also fits the data for radio counts of radio galaxies and QSOs, sub-mJy counts of starburst galaxies at 1.4GHz and far-IR counts of IRAS galaxies. The physical motivation behind such evolution would be galaxy-galaxy interaction driven star formation and feeding of black holes in galactic nuclei.

6 Submillimetre Surveys

Submillimetre surveys have been proposed for the new SCUBA bolometer array on the JCMT and for ESA's Fourth Cornerstone Mission FIRST. It is clearly of interest to compare the performance of these with what could be achieved with large ground-based millimetre arrays like LSA and MMA.

Rowan-Robinson, Pearson, Mobasher, Griffin, Gear, Dunlop, Longair and Blain are collaborating to propose a large survey at 450 and 850 μm with SCUBA on the JCMT. We estimate that with 120 hours of observing time we could cover 1 sq deg to a sensitivity of 11 mJy or 10 sq deg to a sensitivity of 36 mJy at 850 μm, in either case detecting 30-100 sources.

For FIRST it has been proposed that 10% of the observing time over 2 years could be allocated to a survey at 200, 400 and 800 μm. A sensitivity of about 10 mJy could be achieved at 200 and 400 μm for a 10 sq deg survey, and some 6000 galaxies should be detected, a high proportion of them at $z > 1$.

Finally, for LSA/MMA, if 10% of the observing time were allocated to a survey over the first 3 years of operation, 7 sq deg could be surveyed to a sensitivity of 0.9 mJy at 1 mm (5-σ), and some 18000 galaxies would be expected to be detected, again most at high redshift.

The performance of the different proposed far infrared and submillimetre surveys is compared in Table 2.

In conclusion, the proposed large millimetre arrays could have great interest for cosmology, because of their capacity to detect large numbers of high redshift galaxies. However such surveys would be complementary to those of ISO and FIRST, because without multi-wavelength information, nothing would be known of the bolometric power of the galaxies detected, their dust mass, star formation rate etc.

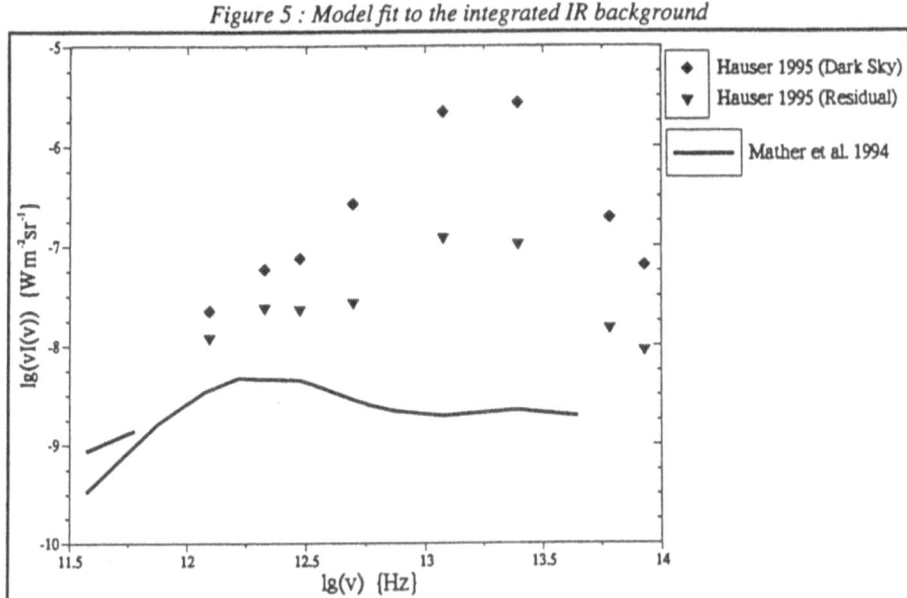

Figure 5 : Model fit to the integrated IR background

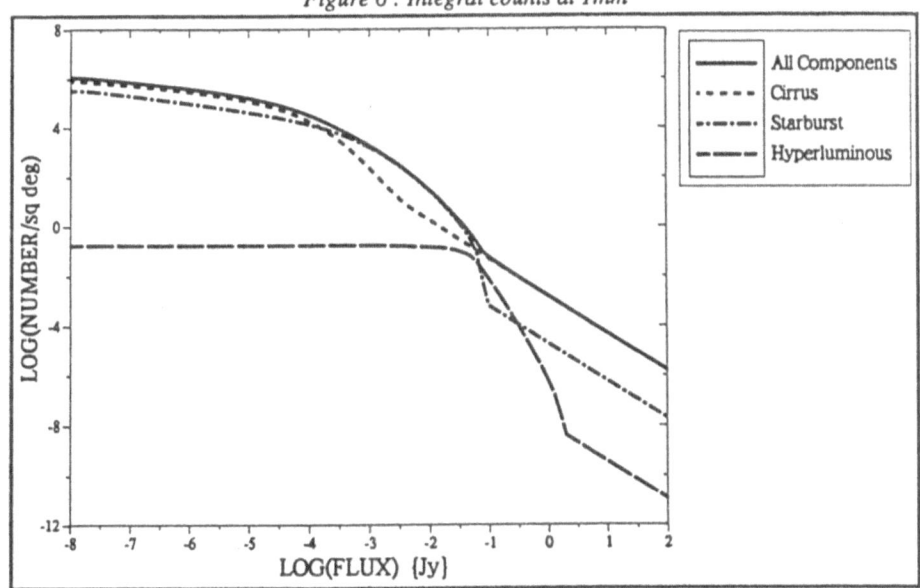

Figure 6 : Integral counts at 1mm

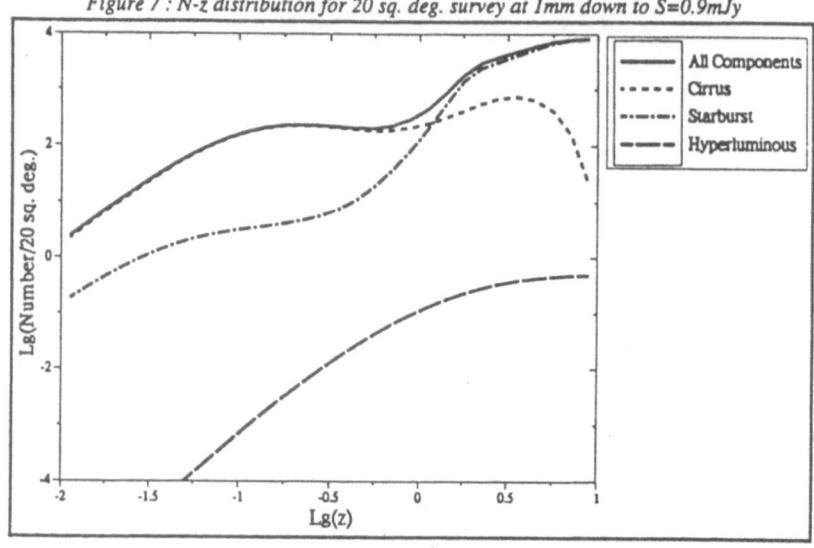

Figure 7 : N-z distribution for 20 sq. deg. survey at 1mm down to S=0.9mJy

Table 1 : Number of sources expected for ISO survey at 90μm and 12μm covering 25 sq. deg.

λ (μm)	lgν	S (mJy)	lgS	ALL COMPONENTS N	N(z>1)	%(z>1)	NORMAL (CIRRUS) N	N(z>1)	%(z>1)	STARBURST N	N(z>1)	%(z>1)
90	12.5214	15	-1.8239	7362	1548	21	2323	0	0	5003	1513	30
15	13.2996	1.7	-2.7696	3298	592	18	1526	0	0	1120	438	39

λ (μm)	lgν	S (mJy)	lgS	HYPERLUMINOUS N	N(z>1)	%(z>1)	SEYFERT 1 N	N(z>1)	%(z>1)	SEYFERT 2 N	N(z>1)	%(z>1)
90	12.5214	15	-1.8239	3.75	3.41	91	15.6	15.0	96	16.8	16.2	96
15	13.2996	0.9	-3.05	2.77	2.44	88	294	65	22	355	87	25

Table 2 : Comparison of FIR and sub-mm Surveys

Survey	Date	λ	S_{lim} (mJy)	Area	No. sources	Main type	Positional accuracy
IRAS-FSS	1987	60μm	200	whole sky	60000	spirals/starburst	30"
VLA	1995	20cm	2	80% sky	10^6	AGN	1"
ISO-ELAIS	1997	90μm	15	20□ (215hrs)	7000	starburst	10"
ISO-slew	1997	200μm	1000	8000□	5000	spirals	30"
SCUBA	1993	850μm	13	10□ (900hrs)	700	starburst/protogalaxies	3"
FIRST	2007	400μm	10	10□ (1800hrs)	6000	starburst/protogalaxies	7"
LSA	2010	1mm	0.9	7□ (2700hrs)	18000	starburst/protogalaxies	0.1"

References

Benn C.R., Rowan-Robinson M., McMahon R.G., Broadhurst T.J., Lawrence A., 1993, MNRAS, **263**, 98

Bertola F., Capaccioli M. & Oke J.B., 1982, ApJ, **254**, 494-499

Boyle B.J., Shanks T. & Peterson B.A., 1988, MNRAS, **235**, 935-948

Coleman G.D., Wu C. & Weedman D.W., 1980, ApJS, **43**, 393-416

Condon J.J., 1989, ApJ, **228**, 13

Danese L., de Zotti G., Franceshini A., Toffolatti L., 1987, ApJ, **318**, L15

Fisher K.B., Strauss M.A., Davis M., Yahil A., Huchra JP., 1992, ApJ, **389**, 188

Franceshini A., Cesarsky C., Rowan-Robinson M., 1996, Memoria della Societa Astronomica Italiana (in press)

Franceshini A., Danese L., de Zotti G., Xu C., 1988, MNRAS, **233**, 175

Hacking P.B., Houck J.R., 1987, ApJS, **63**, 311

Hacking P.B., Soifer B.T., 1991, ApJ, **367**, L49

Hacking P.B., Condon J.J., Houck J.R., 1987, ApJ, **316**, L15

Lawrence A., Walker D., Rowan-Robinson M., Leech K.J. & Penston M.V., 1986, MNRAS, **219**, 687-701

Lonsdale C.J., Hacking P.B., 1989, ApJ, **338**, 712

Lonsdale C.J., Hacking P.B., Conrow T.P., Rowan-Robinson M., 1990, ApJ, **358**, 60

Lonsdale C.J., Harmon R.T., 1991, Adv.Space.Res., **11**, 255

Mitchell K.J., Condon J.J., 1985, AJ, **90**, 1957

Neugebauer G., Becklin E.E., Oke J.B., & Searle L., 1976, ApJ, **205**, 29-43

Oliver S.J., Rowan-Robinson M., Saunders W., 1992, MNRAS, **256**, 15p

Oliver S. et al, 1995, in Wide-Field Spectroscopy and the Distant Universe, eds S.J.Maddox, A.Aragon-Salamanca, World Scientific(Singapore) p.274

Pearson C.P. & Rowan-Robinson M., 1996, MNRAS, *submitted*

Rowan-Robinson M. et al, 1991, Nat, **351**, 719

Rowan-Robinson M., 1992, MNRAS, **258**, 787-798

Rowan-Robinson M., Efstathiou A., Lawrence A., Oliver S. & Taylor A., 1993, MNRAS, **261**, 513-521

Rowan-Robinson M., Benn C.R., Lawrence A., McMahon R.G., Broadhurst T.J., 1993a, MNRAS, **263**, 123

Rowan-Robinson M., 1995, MNRAS, **272**, 737-748

Rowan-Robinson M., McMahon R.G., 1996, in preparation

Rowan-Robinson M., 1996, in Cold Gas at High Redshift, ed. P.van der Werf

Rush B., Malkan M., A. & Spinoglio L., 1993, ApJS, **89**, 1-33

Saunders W., Rowan-Robinson M., Lawrence A., Efstathiou G., Kaiser N., Ellis R.S. & Frenk C.S., 1990, MNRAS, **242**, 318

Thuan T.X., Condon J.J., 1987, ApJ, **322**, L9

Windhorst R.A., Dressler A., Koo D.C., 1987, in Observational Cosmology, eds A.Hewit, G.Burbidge, L.Z.Fang, Reidel (Dordrecht), p.573

Yoshii Y. & Takahara F., 1988, ApJ, **326**, 1-18

The LSA and Galaxy Surveys

A. W. Blain

Cavendish Laboratory, Madingley Rd., Cambridge, CB3 0HE, UK

Abstract. Observations of continuum and line radiation in the mm-band can be used to detect and study galaxies and QSOs at redshifts greater than 5. New instrumentation is making these observations easier, and soon the evolution and formation of galaxies, and the rates of metal enrichment and star-formation will become accessible to direct observation at redshifts beyond the reach of telescopes in other wavebands. Future large millimetre arrays will provide the sensitivity required to detect any normal star-forming galaxies at least to redshifts $z = 10$, combined with the resolving power required to investigate these sources in detail.

1 Introduction

Remarkable negative K-corrections are predicted for distant sources of dust continuum radiation in the mm-band (Franceschini et al. 1991, Blain and Longair 1993), and distant galaxies and QSOs at redshifts of order 5 have been detected in recent observations (e.g. Dunlop et al. 1994; Isaak et al. 1994). New instrumentation for single-dish telescopes will make such observations much easier, and the first blank-field surveys in the submm-band will soon be carried out (Blain and Longair 1996; Rowan-Robinson 1996). Sensitive arrays for mm-wave interferometry are also now available (e.g. Guilloteau et al. 1992). The proposed LSA will exploit these developments to provide such excellent sensitivities that a typical *IRAS* galaxy could be detected and imaged at redshifts greater than 10, to reveal detailed information about their internal dynamics and astrophysics. Here we discuss the prospects for detecting distant sources of both dust continuum emission, and of molecular and atomic fine-structure line emission (Loeb 1993), in order to test models of galaxy formation and evolution.

1.1 The LSA and a Deep Survey

Sensitivities for the LSA are given in the LSA study report. If we assume that the LSA incorporates 11 m antennae, which give a wider field-of-view for survey observations, and is equipped with χ-channel receivers, then the faintest sources that can be detected at a wavelength of 1.3 mm with a signal-to-noise ratio of 3 (3σ) in a survey covering an area A deg^2 in time t s have a continuum and peak line flux density of about $3.6(A/\chi t)^{1/2}$ mJy and $0.19(A/\chi t \Delta v)^{1/2}$ Jy respectively, where Δv is the velocity resolution of the spectrum in km s^{-1}. The area of the LSA primary beam at 1.3 mm is about 10^{-4} deg^2, so in a survey covering 1 deg^2 in 10^6 s the LSA should be able to detect a 10^{10} L$_\odot$ far-infrared continuum source with a dust temperature of 60 K at a redshift of about 11, or a source emitting $10^4 \, L_\odot$ in a $300 \, \mathrm{km \, s^{-1}}$ wide line at a redshift of 7.2, which corresponds to redshifting the strong $158 \, \mu\mathrm{m}$ [C II] line to $1.3 \, \mathrm{mm}$. Hence, the LSA could easily detect a typical star-forming galaxy during any reasonable epoch of galaxy formation. Models of source counts can now be combined with these sensitivities to assess the impact of a deep LSA survey on observational cosmology.

Because the LSA can observe galaxies and QSOs at very large redshifts, the predicted counts are rather uncertain, but *ISO* surveys (Rowan-Robinson 1996) will soon provide much more information. We have previously derived counts of continuum sources in simple models of galaxy formation (Blain and Longair 1993, 1996), and we use these models again to derive the counts shown in Fig. 1(a). To estimate the counts of line emitting sources we first assume that 1% of the bolometric far-infrared luminosity of a source is emitted in the [C II] line. Secondly, a luminosity-dependent line-width, which is close to $300 \, \mathrm{km \, s^{-1}}$ for an L^* galaxy, and a table of line intensities based on observations of CO, [C I] and [C II] lines (Scoville et al. 1995; Brown and Vanden Bout 1992; Stacey et al. 1991) are added: if the luminosity of the $\mathrm{CO}(J \rightarrow J - 1)$ transition is assumed to be J^2 then the luminosities of the $[\mathrm{C \, I}]_{492 \, \mathrm{GHz}}$, $[\mathrm{C \, I}]_{809 \, \mathrm{GHz}}$ and [C II] lines are 10, 100 and 7000 respectively. The resulting counts at $1.3 \, \mathrm{mm}$, which depend on the instantaneous bandwidth of the receivers, are shown in Fig. 1(b). A bandwidth of $8 \, \mathrm{GHz}$ allows sources to be observed within a range of redshifts $z \pm \Delta z$, where $\Delta z \simeq 1.5 \times 10^{-2}(1 + z)$. Note that above a certain value of J, the CO states will not be excited, and so the counts for the highest transitions shown in Fig. 1(b) will probably be much smaller. Estimates of the counts required in order to detect 100 sources at 3σ in a $10^6 \, \mathrm{s}$ integration with 7-channel receivers are also shown in Fig. 1. For single-channel receivers these estimates shift to flux densities which are larger by a factor of 2.6.

2 Detecting Distant Sources with the LSA

Figure 1 allows the number of sources expected in an LSA survey to be estimated, and it is clear that such a survey could detect 10^4 sources of either continuum or [C II] line radiation. Other fine-structure lines could also make a significant contribution, but molecular transitions are not sufficiently strong to contribute a significant number of sources at flux density limits below about $3 \, \mathrm{mJy}$. In order to maximise the signal-to-noise ratio in a deep survey, the densest possible antenna configuration must be used, giving a resolution of 2–$3''$ for the survey, but the survey map would still contain at least 10^6 pixels, and so source confusion would not be a problem. Follow-up observations would allow the detected sources to be mapped at $0.1''$ resolution; however, making high-resolution maps of 10^4 faint sources would be extremely time-consuming, and only a subsample could probably be observed in full detail.

The continuum counts predicted by our models are much larger at shorter wavelengths, and despite the smaller primary beam, a continuum survey should be carried out at the shortest wavelength for which there is good sensitivity; that

(a)

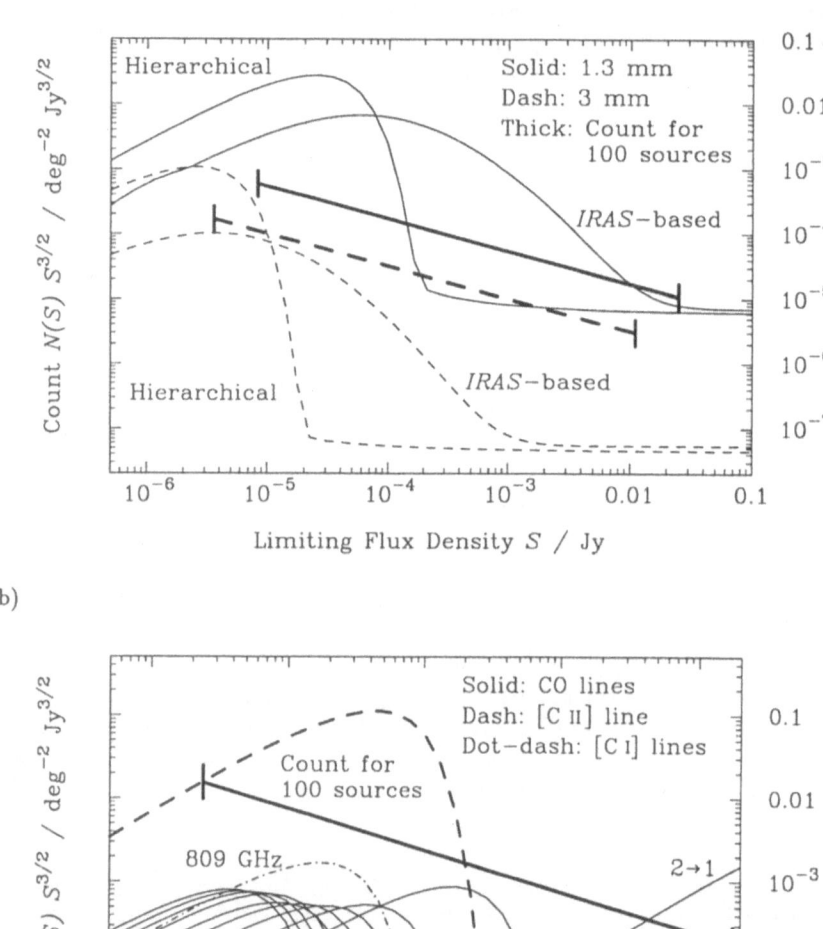

Fig. 1. (a) Continuum counts predicted by 2 different models of galaxy evolution. (b) Counts of various line sources at 1.3 mm in the hierarchical model, assuming an instantaneous bandwidth of 8 GHz. The 2 lowest CO transitions are labelled, and the highest transition shown is CO(11→10). The counts in the *IRAS*-based model are different in form, but the ratios of the counts in each line are similar. The heavy bars show the counts required in order to detect 100 sources above 3σ in a 10^6 s integration at the full range of plausible survey depths.

is, at 1.3 mm or at 0.85 mm if suitable receivers are available. A continuum survey would result in a flux-limited sample containing sources at very large redshifts, and it could be difficult to determine redshifts for many of them: either near infrared spectroscopy with 10 m telescopes, or perhaps an LSA search for [C II] emission could be used.

A deep [C II] line survey at 1.3 mm would detect sources at redshifts between 7.11 and 7.40, with the greatest number of detections expected at peak flux densities near 1 mJy. Further observations could be made at a range of adjacent central frequencies in order to map the distant universe in redshift-space, and so to investigate the evolution of star formation rate and metal abundance.

3 Conclusions

The LSA would be an extremely powerful tool for observational cosmology, extending the standard techniques of galaxy surveys in the optical and radio wavebands to much larger redshifts.

- The source count and redshift distribution of continuum sources, and the observed form of evolution of line sources could be used to investigate star formation activity and metal enrichment at early cosmic epochs.
- The two-point correlation function would give the power spectrum of density perturbations at large redshifts, and perhaps reveal its evolution.
- Any detections of distant clusters could be used to test theories of structure formation, and to provide targets for other instruments to detect the Sunyaev–Zel'dovich effect at large redshifts.
- The effects of the shear-field of weak lensing on the shapes of galaxies could perhaps be investigated to probe the dark matter density field (Blain 1996).

The sensitivity and resolving power of the LSA could answer many important questions in astrophysics and cosmology.

References

Blain, A. W. (1996): This volume
Blain, A. W., Longair, M. S. (1993): MNRAS **264**, 509–521
Blain, A. W., Longair, M. S. (1996): MNRAS, in press
Brown, R. L., Vanden Bout, P. A. (1992): ApJ **397**, L11–L14
Dunlop, J., Hughes, D., Rawlings, S., Eales, S., Ward, M. (1994): Nat **370**, 347–349
Franceschini, A., Toffolatti, L., Mazzei, P., Danese, L., De Zotti, G. (1991): A&AS **89**, 285–310
Guilloteau, S. et al. (1992): A&A **262**, 624–633
Isaak, K., McMahon, R., Hills, R., Withington, S. (1994): MNRAS **269**, L28–L 32.
Loeb, A. (1993): ApJ **404**, L37–L39
Rowan-Robinson, M. (1996): This volume
Scoville, N., Yun, M., Brown, R., Vanden Bout, P. (1995): ApJ **449**, L109–L112
Stacey, G. J., Geis, N., Genzel, R., Lugten, J. B., Poglitsch, A., Sternberg, A., Townes, C. H. (1991): ApJ **373**, 423–444

The LSA and Gravitational Lensing

A. W. Blain

Cavendish Laboratory, Madingley Rd., Cambridge, CB3 0HE, UK

Abstract. Gravitational lenses provide an opportunity to detect and observe galaxies at large redshifts, and can be used to both constrain cosmological parameters and to investigate the formation and evolution of large-scale structure. We predict that a significant fraction of distant mm-wave sources could be strongly-lensed, as in the case of the very luminous source *IRAS* F10214+4724, and we expect that in about 300 hr the LSA could detect between about 10 and 100 strongly-lensed sources, distributed out to very large redshifts. Because of its resolving power, the LSA should also offer excellent performance for imaging strongly-lensed sources and for detecting the lensed arcs of very distant sources in the backgrounds of clusters.

1 Introduction

Gravitational lenses have been found both serendipitously, as in the case of *IRAS* F10214+4724 (Rowan-Robinson et al. 1991), and in systematic surveys, as reported by Kochanek, Falco and Schild (1995) and Myers et al. (1995). A large carefully-selected sample would be a very useful tool for observational cosmology, because the distribution and evolution of distant galaxies and QSOs and the parameters of world models can both be constrained using the individual and statistical properties of lenses (e.g. Cen et al. 1994). The effects of lensing on both the detailed appearance of individual lensed objects (Blandford and Kochanek 1987), and on the observed populations of distant sources (e.g. Peacock 1982; Pei 1995) have been studied. In the mm-band we have found that not only can sources be observed at redshifts greater than 10 (Blain and Longair 1993, 1996), but that the fraction of lensed sources can be very much greater than that in other wavebands (Blain 1996b). Here we discuss the prospects for investigating a range of lensing phenomena with the LSA: strongly-lensed highly-magnified images, the shear field of weak lensing produced by the dark matter distribution, and lensed arcs from sources in the backgrounds of rich clusters.

2 Source Counts of Lensed Objects

The intrinsic luminosity function of galaxies Φ can be combined with a function $F(A, z)$, giving the distribution of lensing magnifications A expected for sources at redshift z, to yield an effective luminosity function Φ', which includes the effects of gravitational lensing, using the expression,

$$\Phi'(L, z) = \int_0^\infty \frac{F(A, z)}{A} \Phi\left(\frac{L}{A}, z\right) \, dA \ . \tag{1}$$

To ensure flux conservation the integrals of F and AF over A must both be equal to unity at all redshifts. F consists of a peak near $A = 1$, which has a redshift-dependent width, and a high-magnification tail of the approximate form $a(z)A^{-3}$, in which $a(z)$ increases with redshift (e.g. Peacock 1982; Pei 1993a); Pei (1993b) predicts that $a(3) \simeq 10^{-2}$. Only the tail of F is relevant for predicting the counts of strongly-lensed sources; therefore, including $F = a(z)A^{-3}$ for magnifications between the limits $A_t(> 1)$ and $A_{\max}(z)$ in (1) allows us to predict counts of highly-magnified sources, which depend on 3 factors: the intrinsic luminosity function, the mass distribution of lensing galaxies, and rather weakly on reasonable values of A_t and $A_{\max}$. Values of about 2 and 40 seem appropriate at $z = 2$, corresponding to a simple double image and to the results of recent observations of F10214 (e.g. Broadhurst and Lehar 1995) respectively. Peacock (1982) argued that the lens–source geometry imposes a limit $A_{\max} \propto D'/d$, in which D' is the angular diameter distance to the source, and d is the diameter of the lensing galaxy. Figure 1 shows the predicted counts of both lensed and unlensed sources of continuum and [C II] line radiation at 1.3 mm, in 2 different models of galaxy formation (Blain 1996a). It is interesting to note that at the continuum flux density of F10214, about 20 mJy, we would expect less than 10% of sources detected in the mm-band to be strong gravitational lenses.

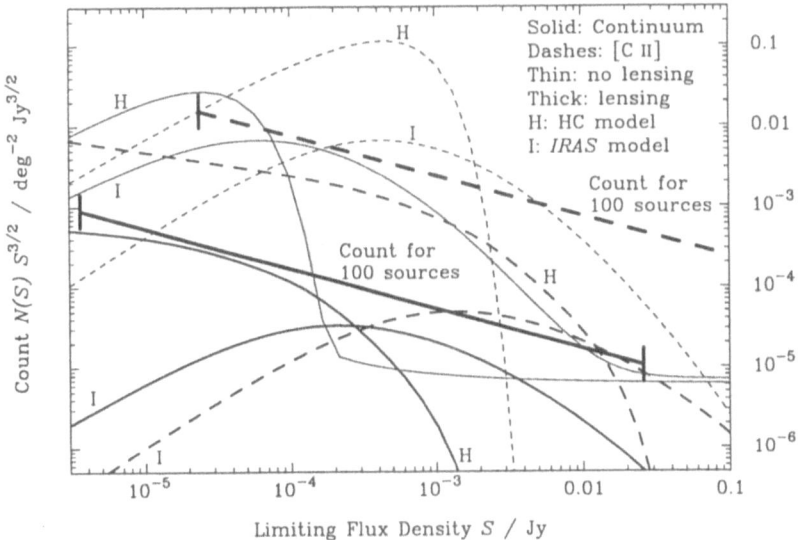

Fig. 1. Counts of strongly-lensed and unlensed sources predicted at 1.3 mm in 2 models of galaxy evolution: one based on an evolving population of *IRAS* galaxies, and another based on a model of mergers in hierarchical clustering. The heavy lines with terminals show the counts required in order to detect 100 sources at a signal-to-noise ratio greater than 3 in an integration time of 10^6 s with the LSA and 7 channel receivers.

3 Observations of Lenses with the LSA

Models of source counts can be combined with the specifications in the LSA study report to predict the number of gravitational lenses that should be detected in a deep LSA survey. The 0.1″ resolution of the LSA can be fully exploited to find features characteristic of strong lensing in the images of the detected sources, to search for faint lensed arcs from very distant background sources in rich clusters, and perhaps to detect the systematic shear-field of weak lensing imposed on field galaxies by the dark matter density field.

3.1 Observations in the Field

Figure 1 shows that the fraction of strong gravitational lenses present in the results of a blank-field survey in the mm-band is expected to peak at a flux density just above the value for which the counts are steepest. Both continuum and [C II] line observations at flux density limits of about 0.2–0.5 mJy could yield between 20 and 50 strongly-lensed sources at large redshifts in the hierarchical model, and a similar number of lenses would be expected from a continuum survey in the *IRAS*-based model. These sources would be accompanied by fewer than about 1000 unlensed galaxies in each case, and because of the fine angular resolution of the LSA there would be no problems with faint source confusion; however, lensed [C II] sources at redshift $z \simeq 7.2$ could be difficult to distinguish from faint sources of $CO(3{\rightarrow}2)$ and $CO(2{\rightarrow}1)$ emission at much lower redshifts, which we expect to be much more numerous at flux densities greater than about 10 mJy (Blain 1996a). The 0.1″ angular resolution of the LSA would provide a powerful tool to investigate the properties of the lens and source; however, careful analysis and interpretation could be required in order to distinguish lenses from irregular or merging galaxies, as the LSA will detect galaxies at large redshifts, and galaxies at redshifts of order unity have already evolved significantly (e.g. Cowie, Hu and Songaila 1995).

The resolving power of LSA could perhaps be exploited to investigate the dark matter density field, by detecting the weakly-lensed shear distortions that it imposes on the images of background galaxies on large angular scales (Kaiser 1992). LSA observations would complement investigations in the optical waveband because the galaxies involved would be selected by very different criteria. Kaiser calculated that to investigate the dark matter distribution on interesting scales of about 10 Mpc, 10^5 resolved galaxies would be required in an area of more than 1 deg^2 at a mean redshift $\bar{z} \simeq 1$, and expected a detection with a formal signal-to-noise ratio of about 100. Observations with the LSA could resolve galaxies at larger redshifts, but a deep survey over 1 deg^2 at high resolution would be very time-consuming. A reasonable LSA survey over 1 deg^2 in several 10^5 s could probably only detect about 5000 sources, but with a larger value of $\bar{z} \sim 3$. In this case, Kaiser's formulae predict that a signal-to-noise ratio of about 15 would be obtained. Hence, the LSA could probe the dark matter density field at redshifts greater than unity, and could probably use the sample of sources detected in a deep blank-field survey to do so (Blain 1996a).

3.2   Observations of Clusters

Faint lensed arcs, imaged from background galaxies, are detected in the cores of rich clusters by deep observations in the optical waveband (e.g. Pello et al. 1992). These regions cover areas smaller than the primary beam of the LSA, and so it could be possible to detect arcs from much more distant background galaxies in a very deep pointed observation with the LSA, which would offer an easy route to detecting the most distant galaxies. K-corrections boost the ratio of mm-to-optical flux densities for distant sources, and the lensed arcs of a typical star-forming galaxies would produce a greater flux density in the mm-band than in the B-band at any redshift greater than about 0.5. The LSA can reach flux density limits of about 40 and 250μJy in 3 hr for continuum and $60 \, \mathrm{km \, s^{-1}}$ resolution line observations respectively, at which the counts in Fig. 1 are almost saturated. Hence, it is practical to detect lensed arcs from all the star-forming galaxies in the background of a rich cluster using the LSA.

4   Conclusions

The LSA could be used to detect and classify a large sample of strongly-lensed sources at considerably greater redshifts than those accessible in surveys in the optical/infrared and radio wavebands, and it will be a very powerful tool for investigating strong lensing in the cores of clusters of galaxies. The weakly-lensed shear field caused by dark matter at redshifts from zero to well above unity could also perhaps be detected in a deep blank-field survey.

References

Blain, A. W. (1996a): This volume

Blain, A. W. (1996b): MNRAS, in prep

Blain, A. W., Longair, M. S. (1993): MNRAS **264**, 509–521

Blain, A. W., Longair, M. S. (1996): MNRAS, in press

Blandford, R. D., Kochanek, C. S. (1987): *Dark Matter in the Universe* (World Scientific, Singapore), 133–195

Broadhurst, T., Lehar, J. (1995): ApJ **450**, L41–L44

Cen, R. Y., Gott, J. R., Ostriker, J. P., Turner, E. L. (1994): ApJ **423**, 1–11

Cowie, L. L., Hu, E. M., Songaila, A. (1995): Nat **377**, 603–605

Kaiser, N. (1992): ApJ **388**, 272–286

Kochanek, C. S., Falco, E. E., Schild, R. (1994): ApJ **452**, 109–139

Myers, S. T. et al. (1995): ApJ **447** L5–L8

Peacock, J. A. (1982): MNRAS **199**, 987–1006

Pei, Y. C. (1993a): ApJ **403**, 7–19

Pei, Y. C. (1993b): ApJ **404**, 436–440

Pei, Y. C. (1995): ApJ **440**, 485–500

Pello, R., Leborgne, J. F., Sanahuja, B., Mathez, G., Fort, B. (1992): A&A **266**, 6–14

Rowan-Robinson, M. et al. (1991): Nat **351**, 719–721

Primordial Molecules

Pierre Encrenaz, and Roberto Maoli

Observatoire de Paris, DEMIRM, 61 avenue de l'Observatoire,
F-75014 Paris, France

Abstract. This topic covers two areas where L.S.A can make major contributions. The first one deals with observations of molecules in objects at high z, where emission can be detected between $0 \leq z \leq 5$. The continuum of quasars can be used to investigate the properties of intervening material seen in absorption; the observations of different transitions of the same molecules permit the direct measurement of the temperature of the Cosmic Microwave Background (CMB) at different redshifts.

The second area deals with the chemistry and observations of molecules at an epoch where only light elements were present in the Universe. Some of the key molecules could play a role in the cooling of the structures of the universe: LiH could for example attenuate CBR primary anisotropies through elastic resonant scattering and give birth to secondary anisotropies associated with protoclouds in motion relative to the Hubble flow.

1 Molecules in objects with $0 \leq z \leq 5$

1.1 Emission

Gravitational lensing may generate a huge amplification of the radiation emitted by far-away objects, and therefore permits the detection of highly redshifted rotational transitions of the most abondant molecules (carbon monoxyde CO, fig. 1; water vapor H_2O, fig. 2) in objects as far away as $z = 4.74$ (Omont 1996). The magnification factor may vary depending on the relative alignment of the objects (position relative to the caustic of the system), and a very high spatial resolution is necessary to assess the true properties of the molecular clouds and their thermodynamic properties.

1.2 Absorption

If the line of sight to a strong continuum source (quasar) crosses the spiral arms of a galaxy, absorption lines of the most common isotopes can be observed (Wiklind and Combes 1996). These observations give a unique opportunity to measure isotope ratios in high z objects, and to determine the temperature of the background radiation as a function of z. L.S.A. will permit very precise measurements of this brightness temperature, and test the CBR temperature as a function of z in the universe up to $z = 5$, perhaps $z = 10$ if strong continuum sources are detected with ISO.

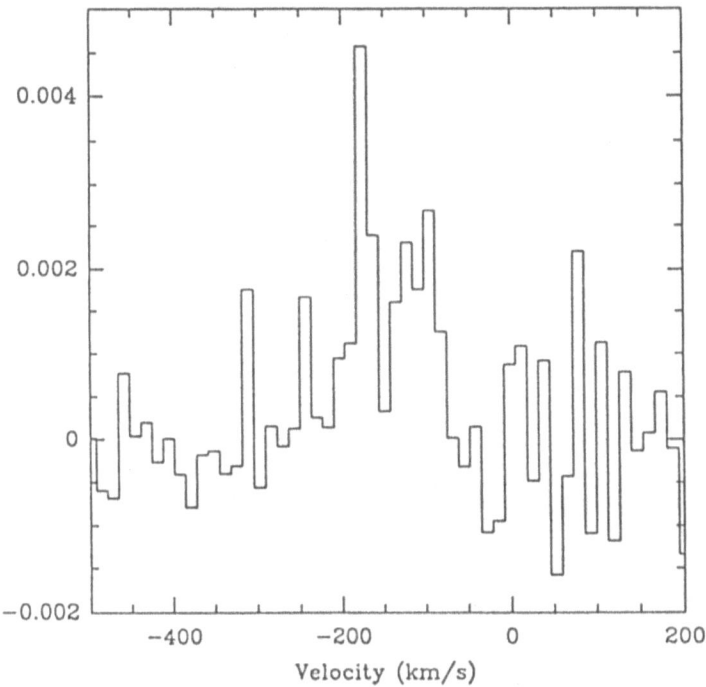

Fig. 1. Detection of CO in a lensed arc (Casoli et al. 1996)

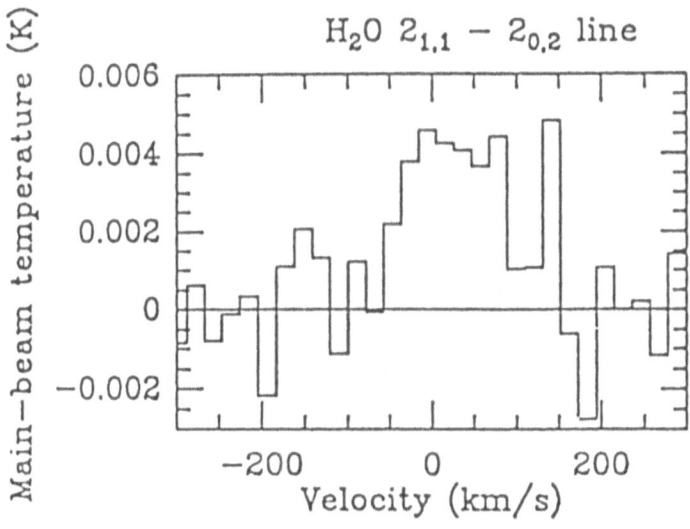

Fig. 2. Detection of water vapor at $z = 2.3$ (Casoli et al. 1994)

2 Molecules in objects with $5 \leq z \leq 400$

There is no direct observations of the universe in this domain. While COBE (Smoot et al. 1992) and balloon observations permit the measurement of the matter distribution at $z \approx 1000$, nothing is known for the epoch between $z = 5$ and $z = 1000$. With the hydrogen recombination the Universe became transparent to radiation and signals can only be produced by the evolving primordial perturbations via gravitational interaction (the Rees-Sciama effect). Nevertheless in the standard model ($\Omega = 1$), only perturbations in their non linear evolution and involving very high masses ($M \geq 10^{16} M_\odot$) can produce observable signals.

Observations of secondary anisotropies produced in the post-recombination Universe can help us to distinguish among different cosmological scenarios and to better understand the epoch of galaxy formation. Free electrons are not the only component that can guarantee the coupling between matter and radiation in the Universe. Starting at redshift $z = 200 \div 400$ the temperature was low enough to make the photo-dissociation processes ineffective allowing the formation of a molecular medium.

From the main constituents H, H^+, D, D^+, He, He^+ and ^7Li, the molecules formed are H_2, HD, H_2^+, HeH^+, LiH and LiH^+. The chemistry is quite different from the one of the interstellar medium due to the lack of grains and the lack of three body collisions.

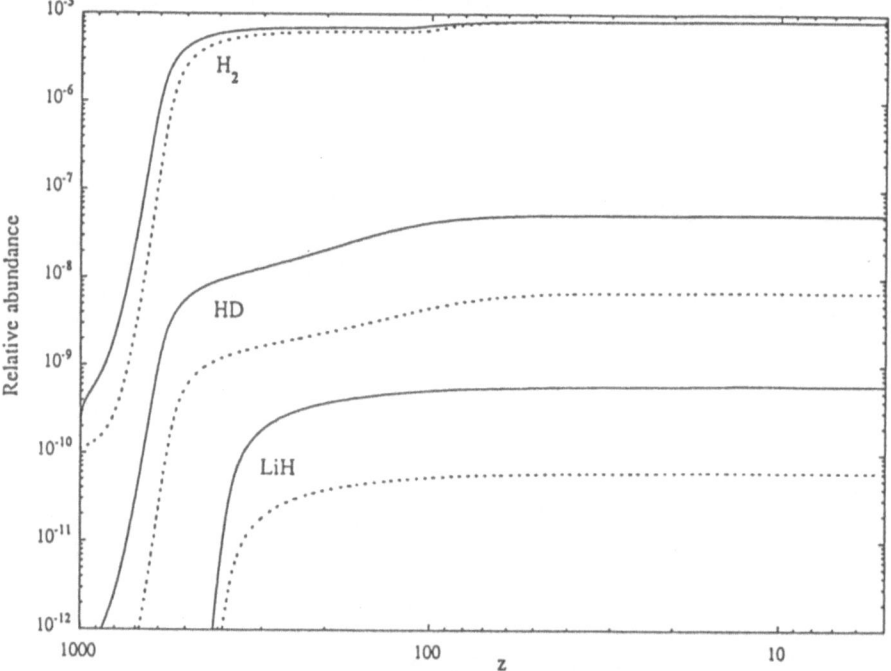

Fig. 3. Abundance of primordial molecules for the homogeneus (dotted line) and inhomogeneus (solid line) big bang nucleosynthesis models (Puy et al. 1993)

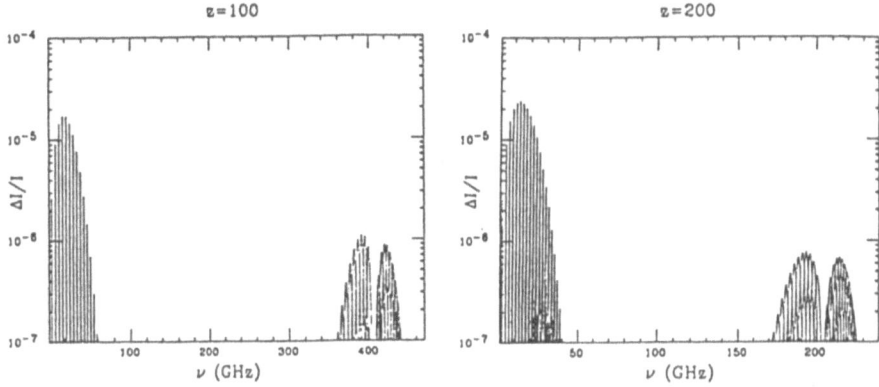

Fig. 4. Theoretical LiH spectrum for a primordial cloud at $z = 100$ and 200 (Maoli et al. 1994)

Maoli (1994) has shown that LiH is the most efficient molecule despite the relative low abundance of ^7Li due to its high dipole moment (5.9 Debye). It must be noted that a very strong line of Li has been observed during the collision of comet Shoemaker Levy with Jupiter, indicating a possible much higher abundance of Li than predicted by either standard big-bang or inhomogeneous big bang nucleosynthesis (fig. 3).

Primary anisotropies formed at the surface of last scattering may be smoothed out, and secondary anisotropics may be generated if these primordial molecular clouds have a peculiar velocity relative to the Hubble flow : these anisotropies appear as molecular lines (fig. 4).

The ro-vibrational lines of these molecules are shifted in the millimeter domain, and L.S.A. will be in a unique position to detect these transitions.

References

Casoli, F., Gerin, M., Encrenaz, P.J., Combes, F. (1994): A. & A. **287**, 716–718

Casoli, F., Encrenaz, P.J., Fort, B., Mellier, Y. (1996): A. & A. *accepted*

Maoli, R. (1994): *Utilisation des molécules primordiales pour l'étude des structures en formation dans l'Univers après recombinaison* Ph. D. Thesis, Université de Paris VI

Maoli, R., Melchiorri, F., Tosti, D. (1994): Ap. J. **425**, 372–381

Omont, A. (1996): *private communication*

Puy, D., Alecian, G., Le Bourlot, J., Leorat, J., Pinot de Foret, G. (1993): A. & A. **267**, 337–346

Smoot, G.F., et al. (1992): Ap. J. Lett. **396**, L1–6

Wiklind, T., Combes, F. (1996): Nature **379**, 139–141

1.3 mm Detection and Mapping of Radioquiet QSOs at Very High z

Alain Omont

Institut d'Astrophysique de Paris, C.N.R.S.

Abstract. We report the results of a double study of the 1.3 mm continuum emission of radio–quiet QSOs at z>~4 : i) Systematic bolometer observations of a sample of 25 sources with the IRAM 30m telescope with a sensitivity limit of 3–4 mJy; six sources at z>4 have been detected with fluxes ranging from 2.5 to 10 mJy. ii) Mapping the most prominent one, BR1202-0725 at z=4.7, with the IRAM Bure interferometer. The emission is extended and clearly double peaked, with the second peak 4″ NW of the QSO.

1 Introduction

IRAS has revealed a strong far infrared emission by bright radio–quiet quasars and hence the presence of large amounts of dust. Their far IR luminosity can compare to their huge UV luminosity. Since the far–IR–submillimeter emission spectrum of dust in the rest frame is very steep, it can considerably rise the detectability of high redshift sources in the submillimeter and millimeter ranges. The observed flux may even increase with redshift for constant luminosity (see e.g. Fig. 1 of McMahon et al. 1994). We have reported in McMahon et al. (1994) the success of our first bolometer search, with the detection of the bright QSO BR1202-0725, at z=4.7, with the IRAM 30m telescope. Its subsequent detection in the submm range at JCMT (Isaak et al. 1994) confirmed its large spectral index indicative of dust emission.

We report here first the results of the continuation of our program of systematic search for 1.25 mm detection of QSOs with z>4. The new detections show that the case of BR1202-0725 is not an exception, but that the rate of millimeter detections among such sources is relatively high. A more detailed report of this work is given in Omont et al. (1996a).

In addition, we show the preliminary results of the 1.35 mm observation of BR1202-0725 with the IRAM interferometer (Omont et al. 1996b). The map displays a second peak 4″ NW of the quasar whose origin is still unclear: second infrared luminous source or gravitational lensing.

2 Bolometer Results

The observations were performed with the IRAM 30m telescope equipped with MPIfR bolometer arrays (Kreysa 1993, see McMahon et al. 1994 and Omont et al. 1996a).

The results are detailed in Omont et al. (1996a). The confirmed detections are displayed in Table 1. Each of the detected sources was consistently detected at least at a 2 or 3 sigma level in several different days. Altogether the combination of these observations warrants a 5 sigma level for the six new detections.

Our main goal was a systematic study of the Cambridge APM sample of radio–

Table 1. 1.25 mm detections

Source	z	1.25mm Flux(mJy)
BRI0952-0115	4.43	2.78+-0.63
BR 1033-0327	4.50	3.45+-0.65
BR 1117-1329	4.00	4.09+-0.81
BR 1144-0723	4.14	5.85+-1.03
BR 1202-0725	4.69	12.6+-2.28
BRI1335-0417	4.45	10.3+-1.04
LBQS1230+1627	2.70	7.5 +-1.4

quiet QSOs with z>4 (Irwin et al. 1991, 1995). We observed 16 of them, i.e. about half the sample, with an rms sigma $\sim< 1.5$ mJy. In addition to the 5 new detections, there are thus 10 non detections among this sample with a 3 σ upper limit of 5 mJy (4 mJy for most of them). In addition, we observed with a comparable sensitivity, without any detection, 8 radio–quiet QSOs with z≥4 detected in the visible by various authors, mainly Schneider et al. (1991). Among these sources, are those reported as detected by Andreani et al. (1993). The non detection of PC2132+0126 (z=3.18) is quite puzzling since Andreani et al. reported 11.5 ± 1.7 mJy, while we measured 0.01 ± 1.20 mJy.

We have begun an exploratory programme to observe at 1.25mm luminous radio–quiet QSOs with z∼1–3.5. We have detected a strong source, Q1230+1627, at z=2.7, and we have three 3-σ tentative detections which should be reobserved with a better sensitivity.

Our new results rise the number of 1.25 mm detections of radio–quiet QSOs with z>4 from 1 to 6, and for those with z>1 from 3 to 9. For z>4, our study is relatively systematic since we observed with a good sensitivity about half of the objects known at the time of our observations. Accordingly, the general trends of the millimeter emission of the optically known z>4 QSOs can be inferred. Among color identified samples such as APM, the detection rate with an rms $\sim 1 - 1.5$ mJy, i.e. with a detection limit $\sim 3-5$ mJy, should be in between 20% and 30%. Sources with $S_{1.25} > 10$ mJy and even > 5 mJy are rare, with proportion $\sim 7\%$ and 10–15%, respectively.

The question of the frequency of a strong amplification by gravitational lensing of such objects remains a major issue. There is only one clear case known of strong lensing among the 6 mm detected objects with z>4, namely BR0952-0115. However, sensitive visible and near-IR searches are needed to definitely discard possible effects of lensing (including the case of BR1202-0725, see below).

Some trends begin to emerge from the relations of the strength of millimeter emission with other characteristics of the sources. All the millimeter detected QSOs are among those which have the largest blue-UV luminosities. Indeed, all the QSOs we have detected pertain to the color selected APM sample which privilegiates high luminosities. Not independent of a large luminosity is the fact that the visible spectra (Storie–Lombardi et al. 1996) of the millimeter detected QSOs have weak and broad emission lines. The presence of broad absorption lines (BAL) seems to increase the chance of a positive millimeter detection, but does not warrant it.

Waiting for submillimeter observations, we do not know the millimeter–submm spectral index of the new detected sources. However, as for BR1202-0725 (McMahon et al. 1994, Isaak et al. 1994), it is probable that it is large, 3–4, and characteristic of dust emission. The mass of dust M_D is probably at least $\sim 10^8 M_\odot$ for all the detected sources in the absence of lensing. There is no indication about the dust temperature, and hence the far infrared luminosity. However, it is likely that the latter is at least comparable to that of the strongest IRAS hyperluminous galaxies. Submillimeter and ISO observations will be essential to confirm that and to bring information about the dust temperature and heating.

3   Extended 1.35mm Emission of BR1202-0725

Almost no information on the spatial extent of the millimetre emission is derived from single disk observations. The $11''$ beam of the IRAM 30m telescope corresponds to ~ 30–150 kpc (depending on the cosmological parameters) at z=4.7. Conversely a compact 10 mJy source can be easily mapped with a beam of a few arcseconds with the IRAM Bure interferometer which has been recently upgraded to operation at 1.3 mm. We report here on the results of such a study showing that the emission of BR1202-0725 is extended (Omont et al. 1996b).

Continuum emission is well detected at 1.35 mm and extended with respect to the $2.2'' \times 1.7''$ beam (Fig. 1). The total integrated flux density is $\sim$ 15 mJy, which is reasonably consistent with the 30m value of Table 1, given the calibration uncertainties. There are clearly two main peaks, one coincident with the position of the quasar within the position uncertainty, and a second component about $4''$ to the North-West, with $\sim 35\%$ of the total flux. The latter is in the direction of the optical companion detected by Fontana et al. (1996), Hu et al. (1996) and Petitjean et al. (1996), but not coincident (Fig. 1).

Two interpretations are possible for this second source (Omont et al. 1996b): a massive hyperluminous infrared companion which could somehow be related to the optical companion emission and to an absorption system close to the QSO ; or a gravitational lens with rather constraining conditions to explain the absence of a second optical image.

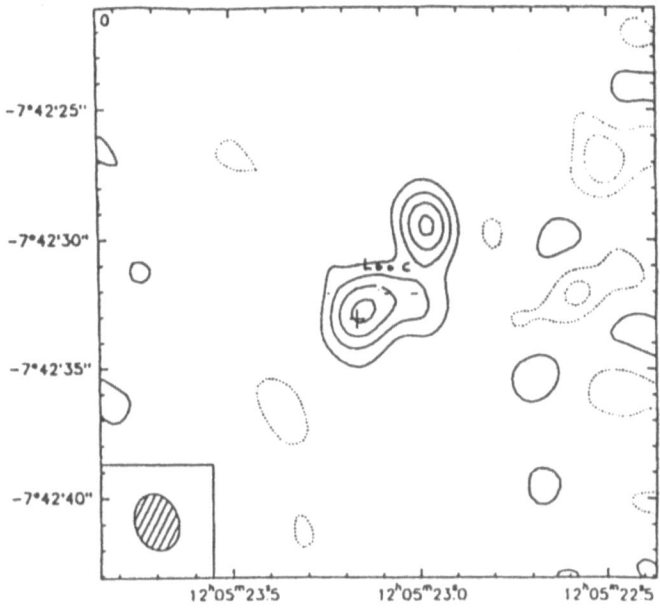

Fig. 1. 1.3mm continuum image. Superposed are the positions of the different visible features: Cross – QSO. Dots – Optical companion features (Hu et al. 1996): Continuum (C) and Lyα (L)

References

Andreani, P., La Franca, F. & Christiani, S. (1993): MNRAS, **261**, L35

Fontana, A., Cristiani, S., D'Odorico, S., Giallongo, E. & Savaglio, S. (1996): MNRAS, in press

Hu, E.M., McMahon, R.G. & Egami, E. (1996): ApJ Letters, in press

Irwin, M.J., McMahon, R.G. & Hazard, C., (1991): in *The Space Distribution of Quasars*, ASP Conference Series, Vol. 21, p. 117, (ed.) D. Crampton

Irwin, M.J., McMahon, R.G. & Hazard, C., (1995): in preparation

Isaak, K.G., McMahon, R.G., Hills, R.E. & Withington S. (1994): MNRAS, **269**, L28

Kreysa, E.,(1993): in *Proc. Int. Symp. on Photon Detectors for Space Instrumentation*, edited by T.D.Guyenne, ESA/ESTEC Noordwijk

McMahon, R.G., Omont, A., Bergeron, J., Kreysa, E. & Haslam, C.G.T. (1994): MN-RAS, **267**, L9

Omont, A., McMahon, R.G., Cox, P., Kreysa, E., Bergeron, J., Pajot, F. & Storrie-Lombardi, L.J. (1996a): A&A, in press

Omont, A., Petitjean, P., Guilloteau, S., McMahon, R.G., Solomon, P. & Pécontal, E. (1996b): Nat, in press

Petitjean, P., Pécontal, E., Valls-Gabaud, D. & Charlot, S. (1996): Nat, in press

Schneider, D.P., Schmidt, M. & Gunn, J.E. (1991): AJ, **101**, 2004

Storrie-Lombardi, L.J., McMahon, R.G., Irwin, M.J. & Hazard, C., (1996): ApJ, in press

Molecular Lines in Absorption (and Emission) From Distant Galaxies and Quasars

Tommy Wiklind[1] and Françoise Combes[2]

[1] Onsala Space Observatory, S–43992 Onsala, Sweden, tommy@oso.chalmers.se
[2] DEMIRM, Observatoire de Paris, 61 Av. de l'Observatoire, F–75014 Paris, France
bottaro@obspm.fr

Abstract. Observations of molecular gas in galaxies at high redshifts can give us information about the conditions for star formation at early epochs of the evolution of the universe. With present day instrumentation, emission lines can only be observed with the aid of gravitational magnification. Relatively detailed information can be obtained through molecular absorption lines, but only a few such absorption systems have been found. Here we discuss high–z observations of molecular gas, what we have learned and how a large millimeter array can improve our knowledge.

1 Introduction

Galaxy evolution is directly related to the star forming activity and the gas content of galaxies, more specifically the molecular gas component. Studying molecular gas at high redshifts will therefore give us an insight in the conditions for stellar formation when these may have been quite different from what is seen in nearby systems. However, observations have been hampered by lack of receivers sensitive enough to detect the weak emission lines of even the most abundant molecules. Nevertheless, a few high–z objects have been studied with the aid of gravitational magnification and, recently, molecular absorption lines have given us relatively detailed information on the ISM in a few galaxies at intermediate redshifts. Although these results are only first steps, they show that a ten–fold increase in sensitivity will allow us to probe the interstellar medium in galaxies existing at epochs when the universe was much younger and when the galaxies themselves were different from what we see in the nearby universe.

In this paper we present some of these early results, obtained with present day instrumentations, and extrapolate what could be obtained with a large millimeter wave array. The focus is on absorption line measurements, since emission lines will also be covered by others in these proceedings (cf. van der Werf).

2 Molecular Emission Lines

The quest for observing molecular gas at high redshifts started with the detection of CO(3–2) emission from the galaxy F10214+4724 at a redshift of z=2.29 (Brown & Vanden Bout 1991; Solomon et al. 1992). CO emission has subsequently been detected from the Cloverleaf quasar at a redshift z=2.56 (Barvainis et al. 1994). These two objects are gravitationally lensed and their CO luminosities, corrected for a magnification factor of 10–20, are comparable to nearby ultraluminous far–infrared (FIR) galaxies (cf. Sanders et al. 1991). CO emission from a third lensed galaxy has been detected in the bright arc in the cluster Abell 370 at a redshift z=0.72 (Casoli et al. 1996). With a magnification factor of ~50, this is intrinsically a relatively inconspicuous galaxy.

Despite several dedicated searches (cf. Wiklind & Combes 1994b; Evans et al. 1995), no CO emission from nonlensed objects has been observed at redshifts larger than z≈0.4. The main difficulty in reaching the very low noise levels necessary for detecting CO emission at higher redshifts is the intrinsic instability of heterodyne receiver systems, which cause curvature of the spectral baseline. This effect is to first order overcome by using interferometric systems, where only correlated signals are detected. This is demonstrated by the detection of very weak CO(1–0) emission from 3C48 at z=0.37 using the Owens Valley interferometer (Scoville et al. 1993).

3 Molecular Absorption Lines

Molecular absorption lines do not sample the same gas as molecular emission lines; whereas emission lines are biased towards warm and dense molecular gas, absorption is more likely to occur in cold diffuse gas. The observed quantity for a molecular absorption line is the optical depth integrated over the profile. Assuming that the population of rotational levels can be described by a single temperature T_{rot}, which is not necessarily equal to the kinetic temperature T_k, the integrated optical depth is:

$$\int \tau_\nu \, dV \propto \frac{N_{tot}}{T_{rot}} \, \mu^2 \left(1 - e^{-h\nu/kT_{rot}}\right) \; km \; s^{-1} \; , \tag{1}$$

where N_{tot} is the column density of the molecule in question, μ the electric dipole moment and where for simplicity we have assumed a linear molecule and a transition from the ground rotational level. From (1) it is clear that for equal column densities of warm ($T_{rot} \approx 20\,K$) and cold ($T_{rot} \approx 5\,K$) gas, the cold gas will have an opacity more than 10 times higher than the warm gas. Hence, an observed absorption line will be dominated by the cold gas along the line of sight. Equation (1) also shows that molecules with an abundance considerably smaller than CO can be observed if their dipole moments are large enough. Furthermore, the background continuum sources have angular extents which are less than a few milliarcseconds (mas), meaning that molecular absorption lines do not suffer from beam dilution in the way emission lines do, and can therefore be observed at any distance as long as the background source has a reasonably strong continuum flux at millimeter wavelengths.

A prerequisite for observing molecular absorption is a background continuum source, which can readily be identified from radio source catalogs. The absorption lines can originate in gas that resides either in the host galaxy to the radio source or in an intervening galaxy. Since molecular gas is usually found only at galactocentric distances less than 10–20 kpc, an intervening galaxy is likely to show molecular absorption only if the impact parameter[1] is less than this. The presence of molecular gas along the line of sight means that the background source will be heavily obscured, making it difficult to obtain redshift information from optical spectra. This is a severe limitation since millimeter wave spectroscopy is characterized by a small instantaneous bandwidth, although with a high spectral resolution. However, for a strong continuum source it is possible to 'scan' the 3– and 2–mm bands for absorption lines by sequential tunings of the receivers. Should there be any intervening molecular gas, the three lowest rotational transitions of CO and HCO[+], both which are likely to show strong absorption, cover the redshift range z=0–3.2. This method has successfully been used on one molecular absorption line system which previously had an unknown redshift (Wiklind & Combes 1996a). For the other three known absorption systems, optical imaging showed a galaxy at the position of the radio source, rather than a point source image of the AGN and the redshift of the galaxy was derived from optical spectroscopy. The necessity of a small impact parameter means that in the case of an intervening galaxy, the background source is likely to be gravitationally lensed. This is the case for two of the four known molecular absorption line systems.

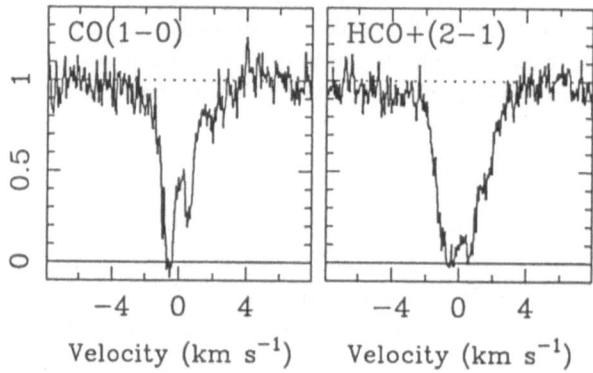

Fig. 1. The CO(1←0) and HCO[+](2←1) absorption lines from PKS1413+135 at z=0.247. The velocity resolution is 40 $m\ s^{-1}$. The continuum level has been normalized to unity.

4 Known Molecular Absorption Line Systems

4.1 Absorption in the host galaxy

In two absorption systems the molecular gas is likely to reside in the galaxy hosting the 'background' radio source. The main arguments for this is that the

---

[1] The projected distance between the center of the galaxy and the radio source.

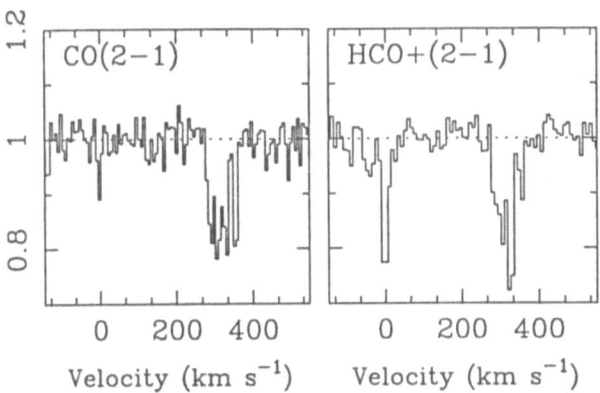

Fig. 2. The CO($2\leftarrow1$) and HCO$^+$($2\leftarrow1$) absorption lines from B3 1504+377 at z=0.673. The velocity resolution is $6\,km\,s^{-1}$.

center of the absorbing galaxy and the radio sources coincides to better than a few tenths of an arcsecond while there are no signs of gravitational lensing in interferometric images of the radio sources.

PKS1413+135. An optical image of this flat–spectrum radio source reveals a faint edge–on galaxy coincident with the radio source (McHardy et al. 1991), but lacking a point source which could be identified with the AGN. This suggests strong extinction, which is corroborated by the detection of 21 cm HI absorption (Carilli et al. 1992) with $N_{HI} = 1.3 \times 10^{21}\,(T_{sp}/100\,K)\,cm^{-2}$ (T_{sp} is the spin–temperature of the atomic gas) and a deficiency of low–energy X–ray photons (Stocke et al. 1992) implying an extinction of $A_V \approx 30\,$mag. A recent HST image (McHardy et al. 1994) shows a disk galaxy viewed almost perfectly edge–on, with a prominent dust lane. The CO($1\leftarrow0$) line was detected in absorption with the SEST telescope in 1993 (Wiklind & Combes 1994a). Several other molecular lines have now also been observed: the J= 2 $\leftarrow$1 and J= 3 $\leftarrow$2 transitions of HCN, HCO$^+$ and HNC. The CO and HCO$^+$ lines reached the zero level, suggesting that they are saturated, but since the isotopic lines, ^{13}CO and H^{13}CO$^+$ are not detected, the optical depth of the main lines are of order unity (Fig. 1). The detected lines are very narrow; the CO line is split into two components, with linewidths less than $1\,km\,s^{-1}$. The HCO$^+$ profile is wider than that of CO and the abundance of HCO$^+$ relative to CO is more than one order of magnitude larger than what is expected from models of diffuse gas chemistry (cf. Viala 1986; Herbst & Millar 1991). These effects have also been seen in Galactic molecular clouds when observed through molecular absorption lines (cf. Liszt & Lucas 1994). It has been suggested that high turbulence can modify the chemistry, causing both a larger abundance of HCO$^+$ and a higher velocity dispersion for this particular molecule (Hogerheijde et al. 1995; Falgarone et al. 1995). The small impact parameter between the optically invisible AGN and the center of the galaxy and the lack of gravitational lensing in VLBI images (Perlman et al. 1994), suggest that the BL Lac resides in the galaxy causing the molecular absorption lines.

B3 1504+377 is a flat–spectrum radio source similar to PKS1413+135. An R–band image shows an edge–on galaxy at the position of the AGN, but the image lacks a conspicuous point source and optical spectra reveals narrow galactic emission lines at a redshift z=0.674 (Stickel & Kühr 1994). VLBI images show the radio source to be of a typical core+jet morphology (Xu et al. 1995), with no indications of gravitational lensing. No 21 cm HI absorption has been reported for this object. We have detected several molecular transitions towards B3 1504+377, at the redshift of the galaxy; CO(2 ← 1), HCO$^+$(2 ← 1) and (3 ← 2), HCN(2 ←1) as well as HNC(2 ← 1) and (3 ← 2) (Wiklind & Combes 1996b). The molecular absorption lines differ from those towards PKS1413+135. First of all, they consists of two profiles, separated by about $330\,km\,s^{-1}$ (Fig. 2) and secondly, the individual components are quite broad: $15\,km\,s^{-1}$ and $70\,km\,s^{-1}$. The depth of the absorption lines is not equal to the continuum level. Since ^{13}CO(2 ← 1) is not detected it is likely that the opacity is intrinsically low, with $\tau \approx 0.1 - 0.3$ for CO and HCO$^+$. The implied column density of molecular hydrogen is $10^{19-20}\,cm^{-2}$. The radio jet extends ∼55 mas, corresponding to $300\,pc^2$. The jet has a steep spectral index and do not contribute to the continuum flux at millimeter wavelengths. Hence, the two absorption profiles are likely to be seen along a single line of sight towards the radio core. The velocity separation between the absorption profiles can be explained if one of them originates in the galactic disk, at a relatively large galactocentric distance, while the other comes from molecular gas situated on highly noncircular orbits, such as can be encountered in strongly barred galaxies.

4.2 Absorption in Gravitational Lenses

In the two molecular absorption line systems described below the absorption takes place in an intervening galaxy which acts as a gravitational lens for the background radio source. Since the absorption lines originate in the lensing galaxy they are not affected by gravitational magnification.

B0218+357 has been identified as a gravitational lens on basis of its radio structure (Patnaik et al. 1993). It consists of two compact flat–spectrum objects and a steep–spectrum ring. The two compact sources are separated by 0″.34, with one component situated inside the ring and the other outside. The compact components are interpreted as two images of a background compact core while the ring is the image of a jet structure. Optical spectroscopy shows narrow emission lines of at a redshift z=0.685 (Browne et al. 1993). Lens models put the center of the intervening galaxy at the center of the ring. An optical image shows the component outside the ring to be heavily obscured (Grundahl & Hjorth 1995). Carilli et al. (1993) have detected 21 cm HI absorption at the redshift of the lensing galaxy, with $N_{HI} = 4 \times 10^{20}\,(T_{sp}/100\,K)\,cm^{-2}$, assuming a covering factor of unity. The lensing galaxy is rich in molecular absorption lines. Several different transitions of CO, HCO$^+$, HCN, HNC, CS, CN, H$_2$CO and N$_2$H$^+$ has been

---

[2] $H_0 = 75\,km\,s^{-1}Mpc^{-1}$ and q_0 =0.5.

detected at a redshift z= 0.68466 ± 0.00001 (Wiklind & Combes 1995; Combes & Wiklind 1995). Most of the lines are heavily saturated, making it possible to observe the isotopic lines of $^{13}CO(2 \leftarrow 1)$ and $C^{18}O(2 \leftarrow 1)$. The $C^{17}O(2 \leftarrow 1)$ line is, however, not detected, indicating a $\tau \approx 1500$ for $^{12}CO(2 \leftarrow 1)$. This implies a column density $N_{H_2} \approx 10^{24} \, cm^{-2}$. Remarkably enough, the $H^{13}CO^+(2 \leftarrow 1)$ line is not detected, suggesting that HCO^+ is situated in a diffuse envelope with a small column density. Despite the high opacity, the absorption lines do not have a depth similar to the continuum level, consistent with only one of the compact cores being covered by molecular gas. The high opacity makes B0218+357 an excellent candidate for searching for molecular oxygen O_2. From chemical models, this molecule is expected to be of similar abundance as CO, but cannot be observed with groundbased telescopes due to atmospheric absorption. The redshift of B0218+357 shifts some of the O_2 lines into observable atmospheric windows. An upper limit to the O_2/CO ratio of 1.2×10^{-2} has been reached using the (N,J)=3,2-1,2 transition (Combes & Wiklind 1995), and recent results for the (N,J)=1,1-1,0 transition suggests a limit of 5×10^{-3}.

PKS1830−211 is a strong flat–spectrum radio source consisting of two compact radio components separated by 1" and connected by an elliptical ring (Subrahmanyan et al. 1990; Jauncey et al. 1991). It is interpreted as the gravitationally lensed images of a background quasar, with a core and a jet (Kochanek & Narayan 1992; Nair et al. 1993). The extended feature of the radio images makes PKS1830−211 ideally suited for modeling of the potential of the lensing galaxy. However, neither the foreground nor the background quasar has been optically identified and their redshifts have been unknown (Djorgovski et al. 1992). The lensing characteristics suggest that the line of sight to PKS1830−211 passes close to the center of an intervening galaxy and the nondetection of the quasar in the optical could therefore be due to heavy extinction in the intervening galaxy. Since PKS1830−211 has a strong continuum flux we could search for absorption lines by sequentially tuning the 3− and 2−mm receivers at the SEST telescope. We found one line, which after some trial tunings could be identified with HCN(1←2) at z= 0.88582 ± 0.00001. Further observations have shown that PKS1830−211 is a rich source of molecular absorption lines. The $J = 2 \leftarrow 1$ and $J = 3 \leftarrow 2$ of HCO^+, HCN, HNC and N_2H^+ as well as CS(3←2), (4←3) and $H_2CO(2_{02}-1_{11})$ has been observed (Wiklind & Combes 1996a). Some of the lines are heavily saturated, as shown by the detection of $H^{13}CO^+(2\leftarrow1)$ and $H^{13}CN(2\leftarrow1)$ (Fig. 3). PKS1830−211 represents the first case where a redshift has been derived from radio measurements.

5   Cosmography

Molecular absorption lines can be used to derive cosmographic parameters in two cases: (1) to put an upper limit to the temperature of the cosmic microwave background radiation, and (2) to measure the differential time delay in gravitationally lensed sources.

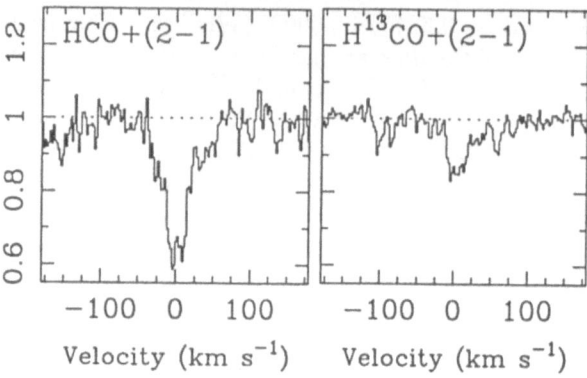

Fig. 3. The $HCO^+(2\leftarrow1)$ and $H^{13}CO^+(2\leftarrow1)$ absorption lines from PKS1830–211 at z=0.886. The velocity resolution is $3\,km\,s^{-1}$.

Cosmic microwave background temperature. According to the big bang theory, the temperature of the cosmic microwave background (CMB) radiation increases as $T_{CMB} = T_{bg}(1 + z)$, where T_{bg} is the measured background temperature at z=0. The energy levels of the lower rotational transitions in the most common linear molecules, such as CO, HCO^+ and HCN, correspond to 0 K for $J = 0$, 4–5 K for $J = 1$ and 12–16 K for $J = 2$. The $J = 1$ level of these molecules will therefore be significantly populated by radiative coupling to the CMB for $z\gtrsim0.5$. In the absence of significant collisional excitation, the absorption profiles of two or more transitions of the same molecule can therefore give an upper limit to T_{CMB}. Although the observed absorption lines are strongly biased towards cold gas, the amount of collisional excitation will remain unknown. The values of T_{CMB} will therefore always be upper limits.

The nonsaturated absorption lines towards PKS1830–211 appear to originate in molecular gas where the excitation is dominated by the CMB. The excitation temperatures of nonsaturated lines, such as N_2H^+, $H^{13}CO^+$ and CS, are $\sim 4\pm2\,K$, compared to the expected T_{CMB} of 5.2 K. Considering the uncertainties associated with the measurements, an upper limit to the CMB temperature at z=0.886 is $T_{CMB} < 6\,K$ (Wiklind & Combes 1996a).

Differential time delay. A value for the differential time delay in a gravitational lens, together with a model for the lens potential, allow a derivation of the angular size distance (cf. Narayan 1991), involving the Hubble constant H_0 and the deceleration parameter q_0. The molecular absorption lines in the lensing galaxies in B0218+357 and PKS1830–211 originate in gas which covers only one of the two lensed images. Although the lines are saturated, their depth do not equal the continuum flux. The flux contribution from each image can therefore be obtained from a single spectrum. Saturated molecular absorption lines can therefore be used to monitor the flux ratio and to derive the differential time delay. Optical measurements of the differential time delay are hampered by microlensing effects[3] and radio interferometry is usually done at long wavelengths, where the variable core component is blended with the steady state component of a jet (cf. Nair et al. 1993). This is not the case at millimeter wavelengths where the flux only comes from the compact flat–spectrum cores. Both B0218+357 and PKS1830–211 are known to change their millimeter continuum flux on time scales of months. By monitoring the continuum flux and the depth of the saturated molecular absorption profiles, it will be possible to derive the differential time delay to an accuracy of ~5%. Such monitoring programs are underway. The derivation of time delays in gravitational lenses may prove to be one of the more spectacular uses of high redshift molecular absorption lines.

6 The Need For LSA

Observations of molecular gas at high redshifts will greatly benefit from an increase in both sensitivity and angular resolution. From the discussion in Sect. 2 (see also van der Werf, these proceedings), it is clear that CO emission from galaxies similar to nearby merger systems such as Arp 220, will be observable to almost any redshift, given a sensitivity increase of a factor 10. Also, the detection of weak emission profiles will be facilitated with an interferometric instrument such as the LSA, since baseline curvature will to first order be canceled.

The detectability of molecular absorption lines is directly proportional to the observed strength of the continuum of the background source, which in turn is proportional to the effective area of the telescope. The proposed collecting area for the LSA of $10^4 \, \mathrm{m}^2$ corresponds to a sensitivity increase of a factor of ~55 compared to the SEST and a factor ~10 compared with the largest existing mm–wave telescopes. This means that the number of radio AGNs usable as continuum background sources will increase dramatically. The number of sources stronger than a given flux S_0 is $N(S > S_0) \propto S^{3/2}$. An increase in sensitivity by a factor of 10 means that the number of accessible sources will increase by a factor >30. If the flat–spectrum radio sources have an evolution with redshift similar to those of radio galaxies and quasars, the number of accessible background sources will be even larger, at least out to z≈3 (cf. Shaver 1995). Powerful optical telescopes, such as the VLT, will greatly increase the number of faint sources with known redshifts, further enhancing the efficiency of the LSA for high–z observations.

For observations of high redshift objects it is necessary to have a receiving system which is 'frequency agile', meaning that the accessible frequencies should be determined by the atmospheric windows rather than the receiving system. This includes the 2-mm band, which has been the most important frequency range for the hitherto observed molecular absorption lines. It is also desirable to have a fast tuning system for searching the mm–bands for absorption lines with unknown redshift, similar to what was done with the SEST for PKS1830–211.

---

[3] This can actually be the case for molecular absorption lines as well; if the molecular clouds have structures on very small scales and the background source is (1) small and (2) exhibits relativistic beaming. The latter effect may cause the line of sight to shift significantly on time scales comparable to the expected time delay.

References

Barvainis, R. et al. 1994, Nature 371, 586

Brown, R.L., Vanden Bout, P.A. 1991, AJ 102, 1956

Carilli, C.L., Perlman, E.S., Stocke, J.T. 1992, ApJ 400, L13

Carilli, C.L., Rupen, M.P., Yanny, B. 1993, ApJ, 412, L59

Casoli, F., Encrenaz, P., Fort, B., Boissé, Mellier, Y. 1996, A&A 306, L41

Combes, F., Wiklind, T. 1995, A&A 303, L61

Djorgovski, S., Meylan, G., Klemola, A., Thompson, D.J., Weir, W.N., Swarup, G., Rao, A.P., Subrahmanyan, R., Smette, A. 1992, MNRAS 257, 240

Evans, A.S., Sanders, D.B., Mazzarella, J.M., Solomon, P.M., Downes, D., Kramer, C., Radford, S.J.E. 1996, ApJ 457, 658

Falgarone, E., Pineau des Forets, G., Roueff, E. 1995, A&A 300, 870

Grundahl, F., Hjorth, J. 1995, MNRAS 275, L67

Herbst, E., Millar, T.J. 1993, in Molecular Clouds, eds. R.J. James, T.J. Millar, Cambridge University Press

Hogerheijde, M.R., de Geus, E.J., Spaans, M., van Langevelde, H.J., van Dishoeck, E.F. 1995, ApJ 441, L93

Jauncey, D.L. et al. 1991, Nature 352, 132

Kochanek, C.S., Narayan, R. 1992, ApJ 401, 461

Liszt, H.S., Lucas, R. 1994, ApJ 431, L131

McHardy, I.M., Abraham, R.G., Crawford, C.S., Ulrich, M.-H., Mock, P.C., Vanderspeck, R.K. 1991, MNRAS 249, 742

McHardy, I.M., Merrifield, M.R., Abraham, R., Crawford, C.S. 1994, MNRAS 268, 681

Nair, S., Narashima, D., Rao, A.P. 1993, ApJ 407, 46

Narayan, R. 1991, ApJ 378, L5

Patnaik, A.R., Browne, I.W.A., King, L.J., Muxlow, T.W.B., Walsh, D., Wilkinson, P.N. 1993, MNRAS 261, 435

Perlman, E.S., Stocke, J.T., Shaffer, D.B., Carilli, C.L., Ma, C. 1994, ApJ 424, L69

Sanders, D.B., Scoville, N.Z., Soifer, B.T. 1991, ApJ 370, 158

Shaver, P.A. 1995, in Proceedings of the Texas Meeting, Munich December 1994, in press

Scoville, N.Z., Radin, S., Sanders, D.B., Soifer, B.T., Yun, M.S. 1993, ApJ 415, L75

Solomon, P.M., Radford, S.J.E., Downes, D. 1992, Nature 356, 318

Stickel, M., Kühr, H. 1994, A&AS 105, 67

Stocke, J.T., Wurtz, R., Wang, Q., Elston, R., Jannuzi, B.T. 1992, ApJ 400, L17

Subrahmanyan, R., Narasimha, D., Rao, A.P., Swarup, G. 1990, MNRAS 246, 263

Viala, Y. 1986, A&A 64, 391

Wiklind, T., Combes, F. 1994a, A&A 286, L9

Wiklind, T., Combes, F. 1994b, A&A 288, L41

Wiklind, T., Combes, F. 1995, A&A 299, 382

Wiklind, T., Combes, F. 1996a, Nature 379, 139

Wiklind, T., Combes, F. 1996b, A&A in press

Xu, W., Readhead, A.C.S., Pearson, T.J., Polatidis, A.G., Wilkinson, P.N. 1995, ApJS 99, 297

Millimeter-VLBI with a Large Millimeter-Array: Future Possibilities

Thomas P. Krichbaum

Max-Planck-Institut für Radioastronomie, Auf dem Hügel 69, D-53121 Bonn, Germany

Abstract. We discuss possibilities and improvements which could be obtained, if a phased array with a large number (N= 50 − 100) of sub-millimeter antennas – like the planned large southern array (LSA) – is used for radio-interferometry with very long baselines (VLBI) at millimeter wavelengths. We find that the addition of such an instrument will push the detection limit and the imaging capabilities of a global mm-VLB-interferometer by 1-2 orders of magnitude.

1 Introduction

The Very Long Baseline Interferometry (VLBI-) technique (eg. Thompson, Moran & Swenson, 1986; Zensus, Diamond & Napier, 1995) is used to study compact radio sources. The classes of objects which can be studied at centimeter wavelengths range from the ultra-luminous nuclei of radio-loud quasars, blazars and BL Lac-objects and their emanating jets, to the moderately luminous radio-galaxies, Seyfert- and starburst galaxies. With milli-Jansky sensitivity at centimeter-wavelengths, also the mapping of peculiar binary stars (eg. Cyg X-3, SS 433), radio-stars and young supernova remnants (eg. SN 1993J) now is possible. Spectroscopic VLBI-observations allow to investigate compact maser emitting regions of various kinds (eg. OH–, CH_3OH, H_2O, SiO) in galaxies, star birth regions and stellar envelopes. Polarization-VLBI observations show the polarized structure of compact radio sources on milli-arcsecond scales. The phase-referencing technique allows to determine source positions with highest possible accuracy. In geodesy, VLBI is used to derive earth-rotation parameters and to measure the motion of tectonic plates.

At millimeter wavelengths ($\lambda \leq 0.7$ mm, $\nu \geq 43$ GHz) present VLBI-observations allow imaging with an angular resolution of up to 40 μas (1 μas $\cong 10^{-6}$ arcsec) at 3 mm wavelength (eg. Bååth et al., 1992). This gives the opportunity to study physical processes in regions of micro-arcsecond size, corresponding to a few up to a few hundred lightdays in a source at cosmological distance $z \geq 0.01$.

To date mm-VLBI is done mostly with telescopes designed for observations at longer wavelengths. This and the limitations set by not yet fully optimized receivers (T_{sys} in the range of 200 − 1000 K at 86 GHz), still limit the detection sensitivity to about 0.2 − 0.5 Jy in recent 3 mm-VLBI observations (eg. Standke et al., 1994, Schalinski et al., 1994). The addition of sub-millimeter telescopes (eg. the IRAM 30 m telescope on Pico Veleta, Spain) has helped considerably to push the detection thresholds below the 1 Jy level (eg. Krichbaum *et al.*, 1993 & 1994), but antenna apertures of typically 10^4 m^2 are needed to reach the milli-Jansky (mJy-) level, which now is accessible at the longer cm-wavelengths without major efforts.

2   Why millimeter-VLBI ?

Opacity: Most compact extragalactic radio sources are partially self-absorbed at cm-wavelengths. For a given brightness temperature T_B and flux density S of the radio source, the size θ of the emitting region only depends on the observing wavelength λ:

$$\theta \propto \lambda \cdot \sqrt{\frac{S}{T_B}} \tag{1}$$

Millimeter-VLBI observations therefore provide twofold advantage: high angular resolution and imaging of small scale regions, which are self-absorbed at longer wavelengths and which cannot be studied directly by other methods.

Recently the frequency band accessible for mm-VLBI was extended to $\lambda = 1.4$ mm ($\nu = 215$ GHz) (Greve *et al.*, 1995). This and earlier studies with US-american instruments at 223 GHz (Padin *et al.*, 1990) and at 86 GHz (Rogers *et al.*, 1984) indicate that the sources are sufficiently bright and compact and that the limitations set by the atmosphere (coherence) still allow VLBI observations up to frequencies of at least 300 GHz.

Assuming brightness temperatures below the inverse Compton-limit ($T_B \leq 10^{12}$ K) equation (1) yields a minimum size for a source of flux density S (in [Jy]) of $\theta > (1.22 \cdot S \cdot \nu^{-2})^{1/2} = 4 \cdot \sqrt{S}$ μas at $\nu = 300$ GHz. The rapid flux density variations of the 'intraday variable' (IDV-) compact radio sources suggest that the variable component in this class of objects is even more compact (eg. Wagner & Witzel, 1995). This has the important consequence that at millimeter wavelengths objects with the highest brightness temperatures will still appear unresolved on interferometer baseline lengths of $< 20\,000$ km. Less compact radio sources, however, with more typical brightness temperatures in the range of $T_B = 10^{9-11}$ K will be partially resolved by the interferometer beam. It is therefore possible to study their underlying structure with unprecedented accuracy.

Angular and spatial resolution: The angular resolution A of a radio interferometer is proportional to the observing wavelength λ and inversely proportional to the relative separation or baseline b between two antennas:

$$A \propto \frac{\lambda}{b} \text{ , or in convenient units : } A_{[mas]} = 3.1 \cdot 10^4 \cdot \frac{1}{\nu_{[GHz]} \cdot b_{[km]}} \tag{2}$$

Since b is limited by the earth's diameter, the only way to achieve higher angular resolution is to go to shorter wavelengths ($\lambda \rightarrow$ millimeter wavelengths) or to place one or more VLBI antennas in space (or on the moon). Efforts in both directions are underway:

λ [mm]	ν [GHz]	$A_{8000\,km}$ [μas]	$z=1$ [mpc]	$z=0.01$ [mpc]	$r=10\,kpc$ [μpc]
6.9	43	90	380	12.9	4.4
3.5	86	45	199	6.4	2.2
1.4	215	18	77	2.6	0.9
0.9	350	11	47	1.6	0.5

linear resolution in [cm]: $\quad 10^{17-18} \quad 10^{15-16} \quad 10^{12-13}$

in [R_S]: $\quad 10^{3-4}R_S^9 \quad 10^{2-3}R_S^8 \quad 10^{1-2}R_S^6$

Table 1. Angular and spatial resolution attainable with ground based mm-VLBI. The table gives the typical observing beam size A (equation (2)) for a 8000 km interferometer baseline (col. 3) depending on wavelength (col. 1) and frequency (col. 2). In columns 4–6 the corresponding spatial scales are given for some typical distances of the objects (using the redshift as distance indicator and $H_0 = 100$ km s^{-1} Mpc^{-1}, $q_0 = 0.5$). In the lower part of the table the spatial scale is shown in units of [cm] and Schwarzschild-radii [R_S]. With R_S^9 being the Schwarzschild-Radius for a $10^9 M_\odot$ mass black hole, a scale of 1 mpc ($= 3.1 \times 10^{15}$ cm) corresponds to $\simeq 10R_S^9$.

Recently fringes with significant signal-to-noise ratios of up to 10 have been detected for the quasars 3C273, 3C279 and 2145+067 at $\lambda = 1.4$ mm ($\nu = 215$ GHz) on the 1150 km baseline between the 30 m millimeter radio-telescope at Pico Veleta (near Granada, Spain) and a single antenna of the millimeter radio-interferometer at Plateau de Bure (near Grenoble, France) (Greve et al., 1995). Further VLBI test-experiments at frequencies $\nu \geq 150$ GHz are planned.

In table 1 the angular and spatial resolution, which now or in the near future could be achieved by ground based mm-VLBI is summarized. Depending on the distance to the source, spatial scales of order of $10^2 - 10^4$ Schwarzschild-radii of a $10^9 M_\odot$ mass black hole can be imaged in extragalactic sources. For galactic sources (eg. the Galactic Center Source Sgr A*) regions of 10-100 Schwarzschild-radii (assuming a $10^6 M_\odot$ mass black hole) can be mapped. This is close to the sizes of the expected central accretion disc !

Space VLBI at cm-wavelengths ($\lambda \geq 1.3$ cm) with an orbiting VLBI antenna is planned for the very near future (the Japanese satellite 'VSOP' will be launched in fall 1996). If successful, this will trigger further technical improvements and future missions. With an open mind for such future developments, it therefore is not unreasonable to assume that on timescales on which also a large sub-millimeter array like the LSA would be completed, space-VLBI observations might be quite common. One therefore should regard mutual observations between earth- and space-based telescopes, even at millimeter wavelengths. Millimeter space-VLBI would yield, for example, an angular resolution of ~ 4 μas, if interferometric observations on a 20000 km baseline are performed at $\lambda = 1$ mm ($\nu = 300$ GHz). Even if space-VLBI at the shortest millimeter wavelengths were not feasible by the time the LSA is operating, ground based mm-VLBI observa-

Station Location	Abbrev.	D [m]	T_{sys} [K]	η_A [%]	G_{eff} [K/Jy]
Large Southern Array, type 1	LSA1	50x16	150	0.6	2.16
Large Southern Array, type 2	LSA2	100x11	150	0.6	2.06
NRAO Millimeter Array	MMA	40x 8	150	0.6	0.43
Large Millimeter Telescope, Mexico	LMT	50	150	0.3	0.21
Plateau de Bure, France	Bure	5x15	150	0.3	0.086
Pico Veleta, Spain	Pico	30	150	0.3	0.077
Nobeyama, Japan	NRO	45	150	0.1	0.058
Nobeyama Millimeter Array, Japan	NMA	6x10	150	0.3	0.047
Owens Valley, California	Ovro	6x10.4	150	0.5	0.084
Hat Creek, California	Bima	10x6.1	150	0.5	0.050
Southern European Telesc., Chile	Sest	15	150	0.35	0.022
James Cark Maxwell Telesc., Hawai	JCMT	15	150	0.6	0.038
Kitt Peak, Arizona	KittP	12	150	0.4	0.016
Heinrich Hertz Telescope, Arizona	HHT	10	150	0.6	0.017
Caltech Sub-mm Observatory, Hawai	CSO	10.4	150	0.5	0.015

Table 2. A hypothetical future VLBI-array at $\sim 250 - 300$ GHz. Not all possible participants are listed. Being very optimistic we assumed for each antenna an effective system temperature $T_{sys} = 150$ K (including atmosphere). The effective antenna gain G_{eff} in [K/Jy] is shown in column 6, and follows from $G_{eff} \simeq 2.845 \cdot 10^{-4} \cdot \eta_A \cdot D^2$, with the telescope diameter D in [m]. Already existing mm-antennas which are expected to contribute with reduced gain are listed in the lower part of the table.

LSA	LSA/MMA/LMT	3–10 mJy
LSA	Ovro/Bure/Pico/Nobe/Bima/NMA	15–20 mJy
LSA	JCMT/Sest/HHT/KittP/CSO	20–35 mJy
MMA	LMT/Ovro/Bure/Pico/Nobe/Bima/NMA	20–40 mJy
LMT	Bure/Ovro/Pico/Bima/Nobe/NMA	40–60 mJy
	others	$\gtrsim 60$ mJy

Table 3. Typical 7σ VLBI detection thresholds. We assumed 2-bit sampling, 1 GHz bandwidth, 20 sec integration time, and a signal-to-noise ratio of the detection of at least SNR ≥ 7. The flux density limits were derived using table 2 and equation (3).

tions could complement space-VLBI observations at short cm-wavelengths with their *matching* angular resolution. Thus small scale regions could be imaged with nearly identical resolution at quite different frequencies (eg. 86 GHz space-VLBI and 215 GHz ground-VLBI observations yield similar observing beams).

3 Sensitivity

The single baseline 1σ-detection threshold (in [mJy]) of a VLBI-interferometer consisting of two antennas with diameters D_i and D_j (in [m]), aperture efficiencies η_i and η_j observing with receivers of system temperatures T_{sys}^i and T_{sys}^j (in

[K]) at a bandwidth $\Delta\nu$ (in [MHz]) is:

$$\sigma_{ij} = 2.485 \cdot 10^6 \cdot \frac{1}{C_l} \cdot \frac{1}{D_i \cdot D_j} \cdot \sqrt{\frac{T_{\text{sys}}^i}{\eta_i} \cdot \frac{T_{\text{sys}}^j}{\eta_j}} \cdot \frac{1}{\sqrt{\Delta\nu \cdot \Delta t}} \tag{3}$$

where C_l is a VLBI efficiency factor combining quantization and correlator losses (eg. $C_l = 0.88$ for 2-bit sampling), and Δt is the integration time (in [sec]). At mm-wavelengths coherence losses in the atmosphere limit the integration time Δt to a few up to a few tens of a second (note: incoherent averaging of the coherent segments in the initial fringe search allows detection of fringes somewhat beyond the atmospheric coherence time (Rogers *et al.*, 1995)). The limits for single antenna sizes ($D \leq 100$ m), aperture efficiencies ($\eta \lesssim 0.6$), and receiver performances ($T_{\text{RX}} \gtrsim 50 - 100$ K for $\nu \geq 50$ GHz) do not allow a substantial lowering of the detection threshold. Some improvement could be obtained from the extension of the observing bandwidth, but a major step could only be done with an increase of the effective collecting area.

The MK IV and VLBA data acquisition systems will provide data recording with extended bandwidths of up to $\Delta\nu = 512$ MHz in the foreseeable future. But even if observations with GHz-bandwidth were possible (and if the problems of phasing an array at such a large bandwidth are solved), the VLBI-detection limit can be pushed by not much more than a factor of ~ 3.

A large collecting area of order of 10^4 m^2 requires an aperture synthesis instrument. The addition of such a phased array to an existing network of smaller VLBI-telescopes would lower the VLBI-detection threshold at millimeter wavelengths drastically. As illustrating example, we list the station performances of a hypothetical future mm-VLBI array in table 2. We included both suggested configurations for the LSA (50 × 16 m antennas (LSA1) or 100 × 11 m antennas (LSA2)) and assumed also that the proposed NRAO Millimeter Array (MMA) and the UMASS/INAOE Large Millimeter Telescope (LMT) will be operating. From equation (2) and with some simplifying assumptions on antenna and receiver characteristics, we estimated the VLBI-detection limits, which we summarize in table 3. It is obvious that the combination of the LSA with other large sub-mm telescopes pushes the detection threshold towards the $5 - 10$ mJy level. More optimistic assumptions on observing bandwidth and receiver performances would even lower these sensitivity limits.

The addition of one very sensitive antenna to a VLB-array consisting mainly of smaller antennas produces a considerable improvement of the dynamic range of the images (see eg. C. Walker, 1989):

$$\frac{1}{\Delta S^2} = \sum_{i,j}^{\text{all baselines}} \frac{1}{\sigma_{ij}^2} \tag{4}$$

where ΔS (in [mJy]) is the noise in the map, σ_{ij} is taken from equation (3), and the sum is over all $N(N-1)/2$ baseline combinations of the interferometer array. Replacing for example one antenna in an array of 5 equal 30 m antennas by the phased LSA would cause a reduction in the map noise by a factor of 3 to about $\sigma = 1$ mJy. This and the fact that the single baseline 'a-priori' detection threshold between two smaller and less sensitive antennas can be reduced, if 'fringes' to a third more sensitive antenna are detected (the method of 'global fringe fitting' uses the closure relations in such antenna triangles, eg. Schwab & Cotton, 1983, Alef & Porcas, 1986), emphasizes the dramatical improvement of the data quality, which could be achieved, if the LSA is added to any preexisting VLB-array.

4 What can be observed ?

From equation (1) and (2) the lowest detectable brightness temperature can be written as:

$$T_B[K] \geq 1.27 \cdot \Delta S_{[mJy]} \cdot b^2_{[km]} \cdot r^{-2} \tag{5}$$

where the ratio r of source size to the interferometer beam ($r = \theta_{source}/A$) measures the source extent, b the maximum baseline length of the interferometer and ΔS the noise level in the map (equation (4)). With a detection limit of $\Delta S \simeq 1$ mJy and source sizes in the range $0 \leq r \leq 10$, the lowest detectable brightness temperature at $\nu = 300$ GHz is $T_B \geq 10^{2-4}$ K for a 100 km baseline ($A = 1$ mas), $T_B \geq 10^{4-6}$ K for a 1000 km baseline ($A = 0.1$ mas) and $T_B \geq 10^{6-8}$ K for a 10000 km baseline ($A = 0.01$ mas). In figure 1 this result is also displayed graphically.

Thus it is clear that most compact nonthermal radio sources and the brighter thermal objects ($T_B \gtrsim 10^{5-6}$ K) are accessible for mm-VLBI including the LSA. It can be expected that such observations substantially would improve our understanding of the various forms of nuclear activity in galaxies (eg. starbursts, mergers, central black holes) and quasars (accretion, production and propagation of jets, particle acceleration processes). High dynamic range mm-VLBI-imaging with micro-arcsecond resolution will reveal more insight in the central light-day size regions of these objects, self-absorbed at longer wavelengths. This should result in a much more detailed understanding of the still unsolved problem of energy production in active galactic nuclei (AGN).

VLBI imaging of high redshift quasars ($z \geq 3$) and gravitational lenses at mm-wavelengths can help in answering questions on the metric and structure of our universe. The extension of the angular size-distance relation towards smaller source sizes and larger distances, could help to more accurately determine the cosmological deceleration parameter q_0 (eg. Kellerman et al., 1993, Gurvits et al., 1994) or even the cosmological constant Λ (Krauss & Schramm, 1993).

With its inverted radio spectrum ($\alpha > 0$, $S_\nu \propto \nu^\alpha$), the flux density of thermally emitting objects increase from the cm- to the mm-regime. With a planned maximum antenna spacing of $\lesssim 10$ km and an angular resolution of $\gtrsim 10$ mas, the LSA will be particularly useful for detailed studies of low brightness temperature objects $T_B \lesssim 100$ K. At the mJy-level, radio stars with brightness temperatures in the range of $T_B = 10^{3-5}$ K have sizes in the milli-arcsecond range (see Altenhoff et al., 1994 for more details on radio stars). Thus studies of stars, stellar winds or even stellar surfaces would become possible, if the LSA and other

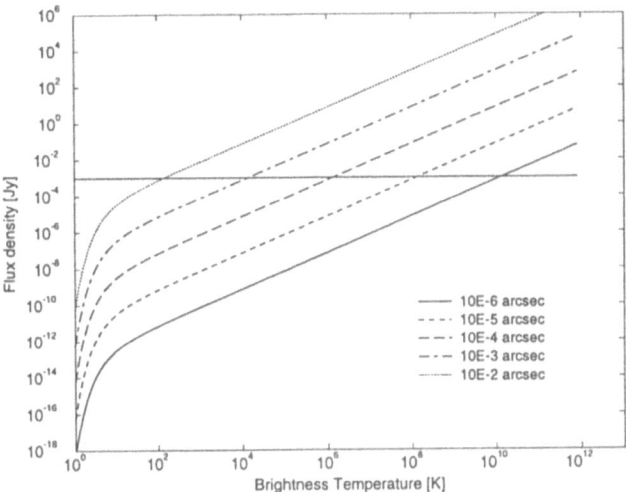

Fig. 1. Flux density of a Planck-black-body radiator plotted versus brightness temperature. The curves are for different source sizes ranging from 1μas (10^{-6} arcsec, lower solid line) to 10 mas (10^{-2} arcsec, upper dotted line). The horizontal line gives a hypothetical VLBI detection threshold of $S_{min} = 1$ mJy. For a given source size, the minimum detectable brightness temperature can be found at the position were horizontal line and radiation curve intersect.

sensitive antennas (eg. MMA, LMT) were used as a long baseline interferometer covering only intermediate baseline lengths in the range of typically $10 - 1000$ km. In the radio cm-bands the thermal part of the spectrum of a radio source (eg. a radio star) is too faint to be observable with VLBI. In the mm-bands this part may reach flux densities of order of up to a few mJy and sizes of less than a few or a few ten milli-arcseconds. Millimeter-VLBI observations with the LSA and possibly the MMA and other large antennas will for the first time allow direct imaging of such objects.

5 Summary

The planned large southern array will dramatically improve the quality of scientific research at millimeter wavelengths. With its large collecting area it could play an important role in future mm-VLBI observations, pushing the sensitivity to the milli-Jansky level. In VLBI, such high sensitivity at present is achieved only at the longer cm-wavelengths. Since thermal radiation becomes more dominant towards shorter wavelengths, mm-VLBI with milli-Jansky sensitivity will not only allow imaging of compact non-thermal radio sources, but also mapping of thermally emitting objects like stars or hot compact regions in extragalactic objects. With an angular resolution of micro-arcseconds, high sensitivity mm-VLBI or even space mm-VLBI (using an orbiting mm-antenna) will allow imaging of regions not directly accessible by any other method. It is therefore not unreasonable to assume that this observing method could yield spectacular scientific results, improving our present knowledge of many astronomical objects in the universe.

Acknowledgements: VLBI observations at millimeter wavelengths are a joint effort of numerous people, impossible to list here. We wish to express thanks to all of them. The following observatories were involved in recent 86 GHz VLBI campaigns: MPIfR (100 m Telescope), IRAM (Pico Veleta, Plateau de Bure), NRAO (Kitt Peak), Onsala, Sest, Haystack, Quabbin, the Owens Valley Caltech interferometer (OVRO), and the Berkeley Hat Creek interferometer (BIMA). The author appreciates financial support from the German BMFT-Verbundforschung.

References

Alef, W. & Porcas, R., 1986, *A&A*, **168**, 365.

Altenhoff, W.J., et al., 1994, *A&A*, **281**, 161.

Bååth, L.B., et al., 1992, *A&A*, **257**, 31.

Gurvits, L.I., et al., 1994, *ApJ*, **425**, 442.

Greve, A., et al., 1995, *A&A*, **299**, L33.

Kellermann, K.I., 1993, *Nat*, **361**, 123.

Krauss, L.M., & Schramm, D.N., 1993, *ApJ*, **405**, L43.

Krichbaum, T.P., et al., 1993, *A&A*, **275**, 375.

Krichbaum, T.P., et al., 1994, in: *Compact Extragalactic Radio Sources*, ed. J.A. Zensus and K.I. Kellermann (NRAO, Socorro), p. 39.

Padin, S., et al., 1990, *ApJ*, **360**, L11.

Rogers, A.E.E., et al., 1984, *Radio Science*, 19, 1552.

Rogers, A.E.E., et al., 1995, *AJ*, **109**, 1391.

Schalinski, C.J., et al., 1994, in: *Compact Extragalactic Radio Sources*, ed. J.A. Zensus and K.I. Kellermann (NRAO, Socorro), p. 45.

Schwab, F.R., & Cotton, W.D., 1983, AJ88,688.

Standke, K.S., et al., 1994, in: *VLBI Technology, Progress and Future Observational Possibilities*, ed. T. Sasao, S. Manabe, O. Kameya, and M. Inoue (Terra Scientific Publishing Company, Tokyo), p. 86.

Thompson, A.R., Moran, J.M., and Swenson, G.W. (eds.), *'Interferometry and Synthesis in Radio Astronomy'*, (Wiley & Sons: New York), 1986.

Wagner, S.J., & Witzel, A., 1995, *ARA&A*, **33**, 163.

Walker, C., 1989, in: *Very Long Baseline Interferometry, Techniques and Applications*, eds. M. Felli & R.E. Spencer, Nato ASI Series Vol. 283 (Kluwer: Dordrecht), p. 163.

Zensus, J.A, Diamond, P.J., and Napier, P.J., (eds.), *'Very Long Baseline Interferometry and the VLBA'*, ASP Conference Series, Vol. 82, 1995.

The variable microwave continuum of radio-loud AGN

Stefan J. Wagner

Landessternwarte, Königstuhl, 69117 Heidelberg, Germany,
swagner@lsw.uni-heidelberg.de

Abstract. It is shown that mm-variability measurements on short time-scales provide an efficient means to study structure in AGN on angular scales much smaller than those accessible even with mm-VLBI. A major drawback of present-day single dish antennae are measurement accuracies. Phased-array observations with the LSA would allow monitoring studies of *all* flat-spectrum radio sources in the universe.

1 Variability

Shortly after the identification of Quasars in the radio- and optical bands it was realized that these objects varied on time-scales as short as a few months. Causality hence constrains the diameters of the sources to be less than the light-travel time times the velocity of light, i.e. fractions of a parsec. This in turn implied very high luminosity densities. Less than a decade after the discovery the largest photon densities were found to exceed the brightness temperature limit of catastrophic Compton cooling, leading to the suggestion that the sources are beamed due to relativistic velocities of the radiating plasma. Almost a decade ago it was realized that in addition to variability introduced by scintillation, some flat-spectrum compact radio sources varied intrinsically on time scales as short as a day or less. The brightness temperatures derived in these 'Intra-day Variables' (IDV sources) exceed the Compton limit by many orders of magnitude (Wagner and Witzel, 1995). One of the suggestions to explain the variations without assuming extreme Doppler beaming factors appeals to lighthouse effects, similar to pulsar beams.

1.1 Lighthouse models

3 C 345 is one of the best-studied superluminous quasars in the sky. The well sampled light-curve was one of the first data sets to be interpreted in terms of a lighthouse model (Schramm et al., 1993). Following a steady decline in the radio, mm-, and optical frequencies throughout the late eighties, the source flared in the early nineties. At optical frequencies three distinct self-similar bursts, separated by about 220 days were observed. The mm light-curve as recorded by the Metsahovi-group is very similar, except for a severe blending of the discrete peaks which are well isolated in the optical data. At lower radio frequencies the distinction between the peaks weakens further and exhibits a delay increasing with wavelength. The overall pattern of the outburst and the shape of the three peaks are shown in figure 1.

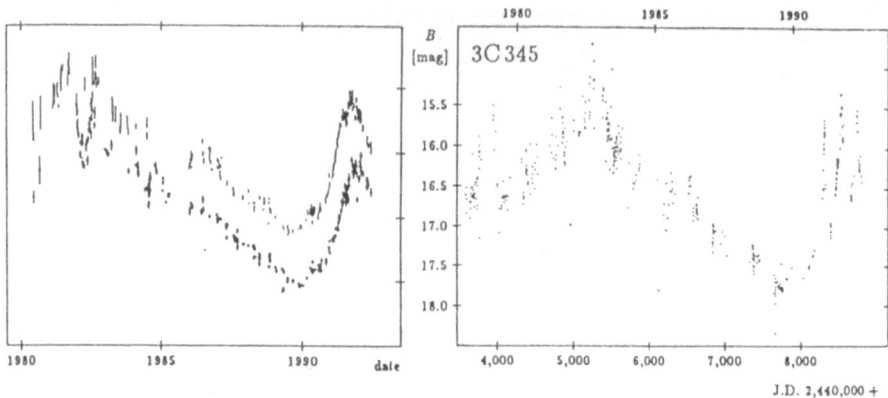

Fig. 1. Variability in 3 C 345 during the last decade. Trends in the mm and optical light-curves are very similar. In the early 1990s three distinct peaks are observed in the optical regime.

VLBI observations VLBI measurements had revealed very early on that 3 C 345 is ejecting knots at fairly regular rates which mostly tend to show super-luminal motion. The object was one of the first sources where curved trajectories of individual knots have been measured (Krichbaum et al., 1992). A full description of the properties were given e.g. by Zensus et al. (1995).

Extrapolating the trajectories of VLBI knots backwards to zero separation, Babadzhanyants and Belokon had realized as early as 1986 that the ejection of the knots coincides with the onset of a new major flare at optical frequencies. This nicely illustrates the correspondence between superluminal knots and variability.

Since the knots do not move on straight trajectories, the angle against the line-of-sight cannot be constant. Even if the source would have a constant flux in its rest-frame, a variable angle to the line-of-sight would introduce variations (Steffen et al., 1995). We hence explored the possibility that the changes of the local velocity vector are actually the main ingredient to the variations.

Jet Models It is often assumed that the jet plasma is injected from the accretion disk at very small jet radii. Since the matter at small radii still has a non-zero angular momentum, any plasma element injected into the jet preserves its angular momentum (Camenzind and Krockenberger, 1992). The trajectory of any fluid element will hence be a superposition of outflow and the rotation about the angular momentum vector. This superposition will lead to helical trajectory. Since all plasma elements travel on helical trajectories, perturbations such as the knots seen as VLBI knots will also follow this pattern. Travelling on helical trajectories, they will not only be seen to stream on curved paths as observed with VLBI observations, but are also bound to exhibit variability due to the changing angle to the line-of-sight towards a fixed observer. Due to the opening of the jet flow the helix will be more tightly wound in the central parts where variations are hence expected to be of higher amplitude, while less tightly wound helices at larger radii (when knots are seen in VLBI observations) would only correspond to moderate variations and slightly curved trajectories.

This model has been fit in detail to the case of 3 C 345 as described above (Schramm et al., 1993). The results of such a fit is shown in figure 2, showing (on the left-hand side) the helical path of a knot of 3 C 345 and (on the right-hand side) the resulting light-curve compared to the data (see figure 1).

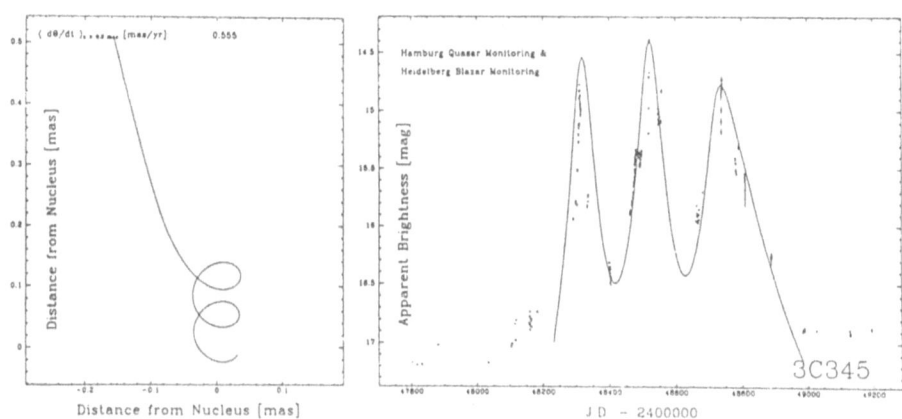

Fig. 2. a) the helical path of a knot within the jet of 3 C 345. The local angle to the line-of-sight of the velocity vector changes with time. The variable beaming leads to distinct outbursts, shown in panel b (superposed onto the optical data, taken from figure 1). For further details see Schramm et al. (1993).

1.2 Quasar structure and variability

As is obvious from the scale in figure 2a even the high angular resolution of LSA will not permit direct imaging studies of sources as 3 C 345. Nevertheless it is obvious from the match of the light-curves and the trajectories of the simple model shown in figure 2, that important constraints on the structure of the Quasars can be extracted from monitoring observations. This is especially interesting if shorter time-scales are investigated. IDV sources show quasi-periodic oscillations as seen in 3 C 345 (figure 1) on time-scales of several hours to a few days (Wagner and Witzel, 1995). Even if these intervals are relativistically compressed they still indicate that regimes are probed which are at most a few ten light-days across. Monitoring those events will be of high importance.

Why mm-observations? Flat spectrum radio sources are characterised by radio spectra $\nu \propto S^{\alpha}$ with $\alpha = 0 \pm 0.5$ throughout the cm- and mm regime. In the mm regime the sources become transparent and the spectra steepen into the IR and optical regimes. Obviously it is of crucial importance to remain in the optically thin regime to have a reasonably simple way of converting variability data into structural information.

Current Intraday Variability studies in the mm regime Intraday Variability in the cm- and optical regime has been discussed in detail by Wagner and Witzel (1995). Studies of rapid variability in the mm regime have always been handicapped from the low sensitivities and atmospheric instabilities at frequencies of about 230 GHz. Nevertheless it has been possible to detect IDV in a few flat-spectrum radio sources in a reliable way. Figure 3 illustrates the variations seen in PKS 0405-385 during a monitoring campaign at the SEST telescope. In excellent conditions other sources monitored at the same epoch showed variations of up to only 30 percent (Wagner, 1996). A close-by calibrator, PKS 0402-362 has been observed during the same epochs as 0405-385 and has shown variations of only 15 %. Since the source is very close to PKS 0405-385 it is likely to suffer from the same instrumental effects. Interpreting the scatter in the calibrator as a measure of the uncertainty, the factor of 3 change is clearly highly significant. While other sources have also shown significant variations during this and other runs with the bolometer at the SEST telescopes, the flare shown in figure 3 is one of the more spectacular outbursts.

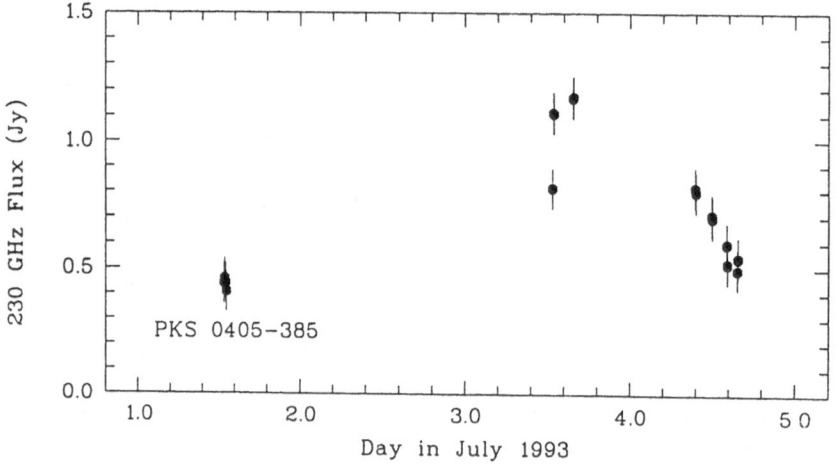

Fig. 3. 230 GHz light-curve of PKS 0405-385. Error bars represent the variations from a constant flux in the simultaneously observed nearby source PKS 0402-362.

Prospects for LSA Variability studies The number of flat-spectrum quasars beamed towards us is fairly limited. The fraction of such sources among the mJy sources is very small, suggesting that most of the sources will have flux-densities above 1 mJy at 3 mm. On the other hand, flat spectrum sources are a major constituent of all sources in the mm regime, i.e. in the range between strong thermal emitters (IRAS sources) and classical steep-spectrum sources (most of the 3C objects). Variability studies at acceptable levels of measurement accuracy with present-day mm-telescopes are possible only for the strongest sources. Even in these cases, however, noise and atmospheric instabilities limit the studies to a high degree.

Phased-array and simultaneous sub-array observations with a large array configuration such as LSA would permit both a much better control over atmospheric disturbances as well as an access towards fainter sources at the few mJy level and would bring essentially the whole population of flat-spectrum sources into reach. This would allow studies of the variability properties on a more regular basis. Repeated observations are required to confirm that a consistent picture of the intrinsic substructure can be deduced from variability properties. Extending the observations to well-defined samples will permit investigations on the statistical properties of the sub-pc structure and relations to other properties of the source. Finally simultaneous monitoring studies at other frequencies will be required to constrain radiation processes.

Acknowledgements: The work on Intraday Variability was partly supported by the DFG through Sonderforschungsbereich 328. Our mm IDV-monitoring programs benefited from support by the SEST staff (Drs. Lars-Ake Nyman and Lewis Knee). Discussions with Max Camenzind, Thomas Krichbaum, Arno Witzel, Anton Zensus and Esko Valtaoja are gratefully acknowledged. The author also thanks Peter Shaver for his patience.

References

Babadzhanyants, M. and Belokon, E.T., 1986, Astrophysics, 23, 639.
Camenzind, M. and Krockenberger, M., 1992, A&A, 255, 59.
Krichbaum, T. et al., 1992, in *Physics of AGN*, eds. Duschl, Wagner, Springer, p. 574.
Schramm, K.-J. et al., 1993, A&A, 278, 391.
Steffen, W., Zensus, J.A., Krichbaum, T.P., Witzel, A., Qian, S.J., 1995, A&A, 302, 335.
Wagner, S.J. and Witzel, A., 1995, Ann. Rev. Astron. Astrophys., 33, 163.
Zensus, J.A., Cohen, M.H., Unwin, S.C., 1995, ApJ, 443, 35.

Millimeter Radiation from Normal Galaxies and AGN

Rolf Chini , Endrik Krügel

Max–Planck–Institut für Radioastronomie, Auf dem Hügel 69, D-53121 Bonn, Germany

Abstract. During the last decade a variety of mm/submm continuum and line data has been collected for different types of galaxies and quasars. We illustrate the importance of this spectral region by the following examples (see Chini & Krügel (1996) for a more extensive discussion). *a)* The determination of the mass of the interstellar medium, either from dust emission or from the lower rotational lines of CO. Comparison of the two methods shows reasonable agreement. *b)* Once the mass has been determined, the ratio of luminosity over gas mass, L_{IR}/M_{gas}, yields the efficiency for converting gas into luminosity; it is thus an indicator for the star formation activity. *c)* For quasars the mm regime allows to discriminate between synchrotron radiation from the relativistic object and thermal emission from the host galaxy. *d)* Finally, mm/submm observations offer a promising tool to study the formation of galaxies in the early universe. We conclude that an increase in sensitivity and spatial resolution as provided by a large mm array would allow for a much deeper insight into the formation of stars, galaxies and quasars.

1 Determination of Dust and Gas Masses

Millimeter measurements offer the opportunity to trace in external galaxies the interstellar medium, either through continuum radiation of dust or through CO rotational lines. The dust emission originates from regions where hydrogen is in the form of HI as well as H_2. As it is optically thin, the received flux density S_{1300} at $1300\,\mu m$ is directly proportional to the total dust mass M_d. CO line emission, on the other hand, comes only from regions of molecular gas. Although the lines are optically thick, a correlation between CO luminosity L_{CO} and gas mass M_{gas} is well established for molecular clouds in the Galaxy. If we define the $1300\,\mu m$ luminosity as $L_{1300} = S_{1300} \cdot D^2$, where D denotes the distance, L_{CO} and L_{1300} are purely observational quantities with a completely different physical background, which both give the mass of the interstellar medium. The plot in Fig. 1 of L_{CO} vs. L_{1300} for a sample of normal spiral galaxies shows the compatibility of the two methods.

2 Thermal Emission from Quasars

Originally, quasars were discovered by their strong radio emission which could be identified as synchrotron radiation from a relativistic object. If optically selected, it was found that most QSOs ($\sim$ 95%) are radio–quiet (Condon *et al.* 1981).

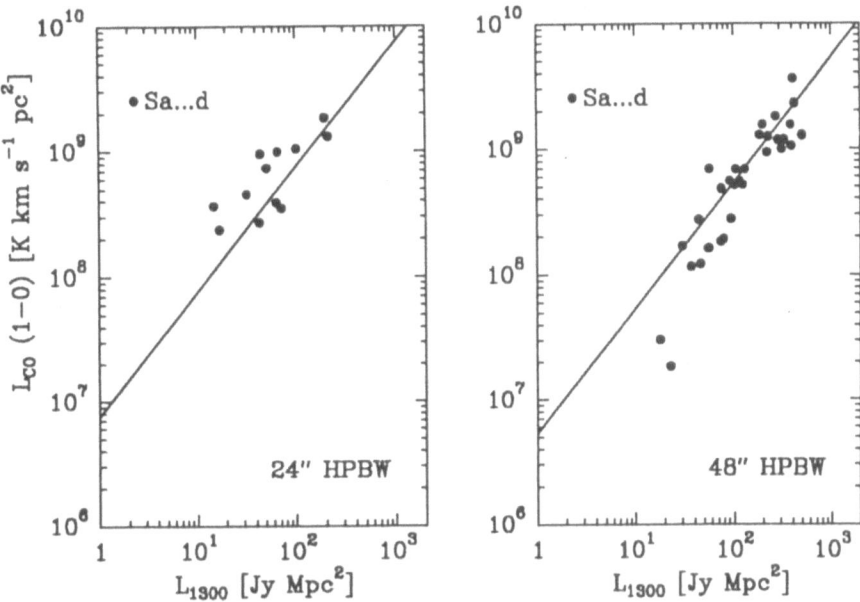

Fig. 1. L_{CO} vs. L_{1300} for normal spirals. In both boxes the dust and line observations refer to identical beams of 24 and 48″, respectively, directed toward the center of the galaxies. Data from Chini *et al.* (1996).

Furthermore, optical and IR imaging has demonstrated that QSOs reside in the centers of galaxies. Fig.2 shows the typical energy distribution (SED) of a radio–quiet QSO (IZw1) from optical to radio wavelengths. The emission between 0.3 and $10\mu m$ was interpreted as synchrotron radiation. Because it is neither polarized nor variable, and because *IRAS* data indicated a turnover around $100\mu m$ typical of galactic dust emission, doubts came up about the underlying physical process. Sensitive observations in the mm–regime yielded a steep spectral index $\alpha(100/1300)$, which exceeded the common theoretical limit of 2.5 for synchrotron self–absorption. It was concluded that the emission between NIR and mm wavelengths is thermal in origin (Chini *et al.* 1989a). There were subsequent attempts to rescue the synchrotron model for radio–quiet QSOs by introducing rather special electron energy distributions (de Kool & Begelman 1989, Schlickeiser et al. 1991) that could produce spectral indices $\alpha(100/1300) > 2.5$. Eventually, these models failed to interpret submm data at 450 and $800\mu m$ (Hughes *et al.* 1993) corroborating a thermal interpretation. The present view is that radio–quiet QSOs are encirlced by dust clouds, probably located in a disk. The dust grains are heated by energetic particles from the central engine and dominate the emission from 3 to $1300\mu m$ (Niemeyer & Biermann 1993). When converting the dust emission into a gas mass one obtains values similar to those found in active galaxies. This result is supported by CO observations (Aloin et al. 1992).

Radio–loud QSOs may be distiguished by their spectral appearance at cm wavelengths which either exhibits a "flat" or a "steep" slope. Fig.3 displays the

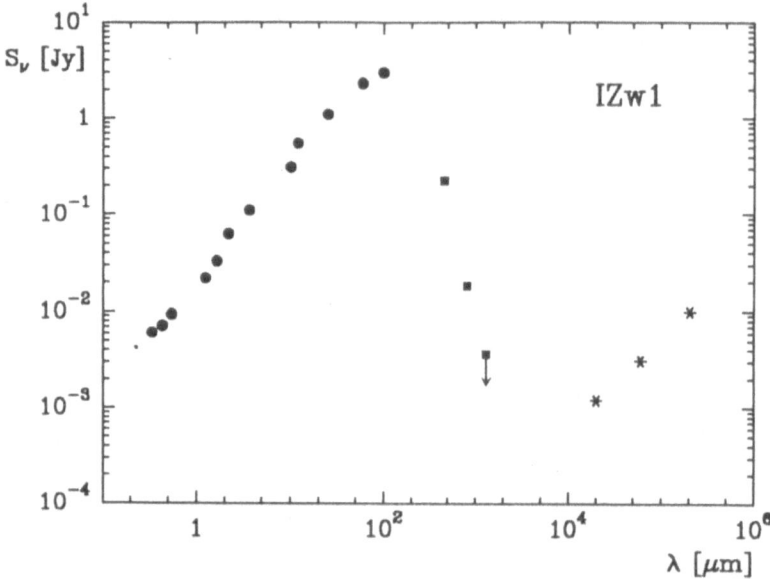

Fig. 2. Energy distribution of the radio–quiet quasar IZw1; data at 450 and 800 μm are from Hughes et al. (1993), the point at 1300 μm is from Chini et al. (1989a).

typical SED of a steep–spectrum source (0134+329). In the cm regime the synchrotron emission gives rise to a non–thermal component that extends into the mm region. A second component, very similar to the dust emission from galaxies and radio–quiet QSOs, is present between 0.3 and 100μm. It was suggested (Chini *et al.* 1989c) that also in the case of radio–loud QSOs the IR regime is dominated by thermal emission from the underlying galaxy.

3 Activity in Galaxies

In Fig. 4 we plot L_{IR} , defined as the luminosity from 12 to 1300 μm, versus M_{gas} for two samples of normal spirals and active Mkn galaxies. Although the scatter is large, both samples are clearly separated: the average value in solar units is L_{IR}/M_{gas} of 5 ± 2 for normal spirals and 92 ± 53 for Mkn galaxies. Obviously, L_{IR}/M_{gas}, which characterizes the efficiency of converting mass into luminosity is an indicator for activity (Chini *et al.* 1995). The radio–quiet QSOs are located around and above the strip where L_{IR}/M_{gas} is about 550; the source of luminosity here is, of course, of an entirely different origin.

Fig. 4 demonstrates that L_{IR} alone is not a unique signature for activity because normal spirals and active galaxies cover a similar luminosity range. Within the interval from 10^{11} to 10^{12} $L_\odot$ an extragalactic object may be a normal spiral an active galaxy or even a QSO. On the other hand, a galaxy of given gas mass M_{gas} may produce different amounts of luminosity. The range $100 \geq L_{IR}/M_{gas} \geq 3$ seems to be typical for star formation in the current samples. The

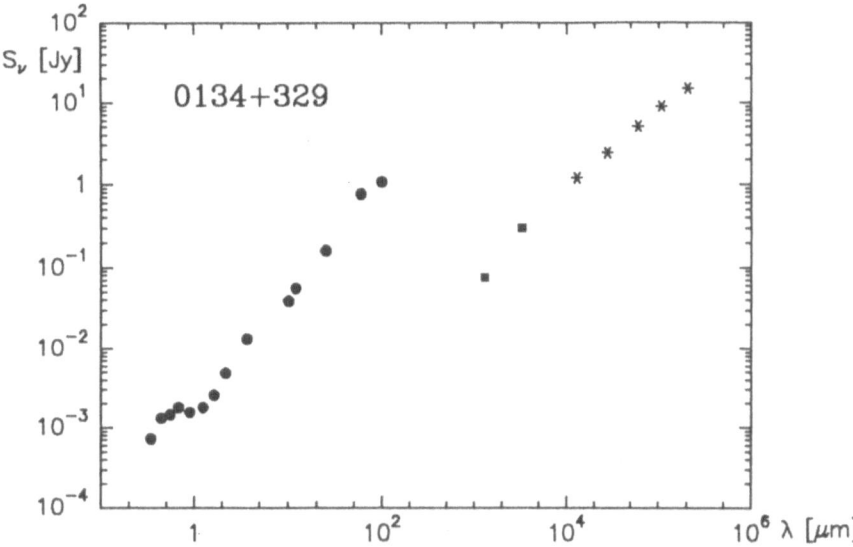

Fig. 3. Energy distribution of the steep spectrum quasar 0134+329; the data at 1.3 and 3 mm are from Chini et al. (1989c).

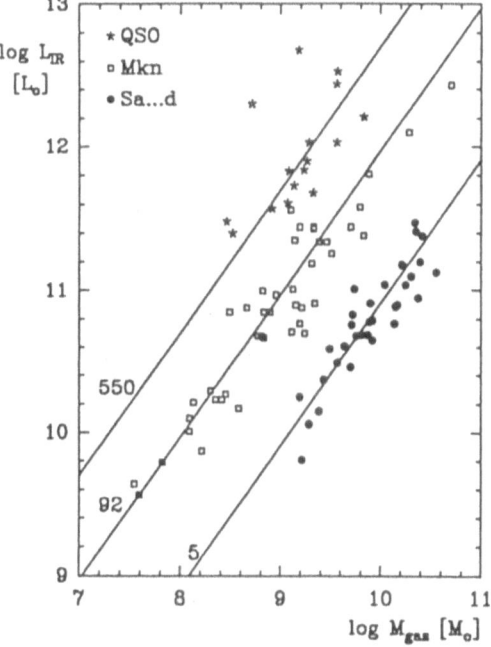

Fig. 4. L_{IR} vs. M_{gas} for two samples of galaxies: normal spirals (32 objects, symbol•) and active galaxies from the Markarian catalog (42 objewcts, symbol□), adapted from Chini *et al.* (1995). The solid lines through both samples are the loci where L_{IR}/M_{gas} in solar units is 5 (normal) and 92 (active), respectively. For comparison we also include some radio–quiet QSOs (⋆). Here data are from Chini *et al.* (1989a) and the solid line corresponds to $L_{IR}/M_{gas} = 550$.

different efficiencies reflect the violence of the process ranging from a quiescent steady state in galactic disks to starbursts in nuclei. The ratio L_{IR}/M_{gas} does not seem to change significantly with M_{gas}. Although the spread of the objects in Fig. 4 along the drawn lines is to a large degree a distance effect – both L_{IR} and M_{gas} contain a factor D^2 – a true proportionality is indicated between the two quantities. This suggests that both types of galaxies derive their luminosity from star formation, although for some active objects some contribution may come from a black hole.

4 Primeval Galaxies

Radio galaxies are currently known out to $z \approx 4$, corresponding to only 10% of the age of the universe. At this early stage one might expect objects that still contain a large reservoir of gas not yet converted into stars. The investigation of the interstellar medium at high z is therefore a step toward understanding the formation and evolution of galaxies. The performance of the MPIfR bolometer

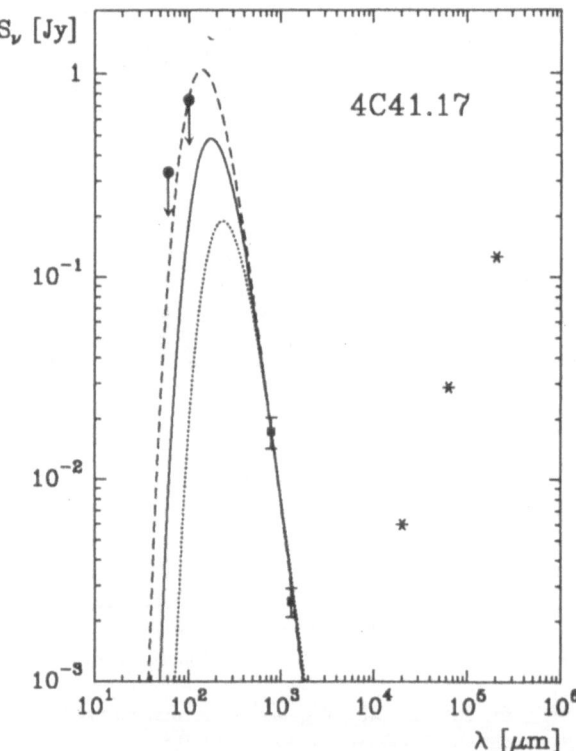

Fig. 5. The spectrum of the radio galaxy 4C41.17 from Chini & Krügel (1994). The solid curve corresponds to a dust component of 80 K at the rest–frame of the object.

arrays have reached a sensitivity that allows at the IRAM 30 m telescope the detection of $\sim 3\,$mJy signals at 1300 μm within 2 hours. As an example, 800 and 1300 μm emission from the radio galaxy 4C41.17 at $z = 3.8$ points towards hot dust at $80 \pm 20\,$K, a luminosity of $10^{14}\,L_\odot$ and a dust mass of $\sim 9 \cdot 10^7\,M_\odot$. The gas content has to be estimated by assuming some gas–to–dust ratio for objects at such an early epoch. Attempts to detect CO emission in this galaxy have so far failed. A big step forward in that area of research requires a large mm array to achieve the necessary spatial resolution and sensitivity.

References

Alloin, D., Barvainis, R., Gordon, M.A., Antonucci, R.R.J. (1992): CO emission from radio quiet quasars: new detections support a thermal origin for the FIR emission, A&A **265**, 429

Chini, R., Biermann, P.L., Kreysa, E., Gemünd, H.-P. (1989c): 870 and 1300 μm observations of radio quasars, A&A, **221**, L3–L5

Chini, R., Kreysa, E., Biermann, P.L. (1989a): On the nature of radio–quiet QSOs, A&A, **219**, 87–97

Chini, R., Krügel, E. (1994): Dust at high z, A&A, **288**, L33–L36

Chini, R., Krügel, E. (1996): Dust and gas in normal and active galaxies, Proc. *Cold Dust Morphology Conference*, Johannesburg, January 22–26, 1996, Kluwer, Dordrecht

Chini, R., Krügel, E., Lemke, R. (1996): Dust and CO emission in normal spirals, A&A Suppl., , in press

Chini, R., Krügel, E., Lemke, R., Ward–Thompson, D. (1995): Dust in spiral galaxies. II, A&A, **295**, 317–329

Condon, J.J., O'Dell, S.L., Puschell, J.J., Stein, W.A. (1981): Radio emission from bright, optically selected quasars, ApJ **246**, 624

de Kool, M., Begelman, M.C. (1989): Production of self–absorbed synchrotron spectra steeper than $\nu^{5/2}$, Nature **338**, 484

Hughes, D.H., Robson, E.I., Dunlop, J.S., Gear, W.K. (1993): Thermal dust emission from quasars – I. Submillimetre spectral indices of radio–quiet quasars, MNRAS **263**, 607

Niemeyer, M., Biermann, P.L. (1993): The emission spectra of radioweak quasars – I. The far–infrared emission, A&A **279**, 393

Schlickeiser, R., Biermann, P.L., Crusius–Wätzel, A. (1991): On a nonthermal origin of steep far–infrared turnovers in radio–quiet active galactic nuclei A&A **247**, 283

Submillimetre Observations of Low Ionization BAL Quasars

Dave Clements[1], Amanda Baker[2], and Paul Francis[3]

[1] European Southern Observatory, Karl-Schwarzschild Strasse 2, D-85748 Garching-bei-Munchen, Germany
[2] Institute of Astronomy, Madingley Road, Cambridge, UK
[3] Melbourne University, Parkville, Victoria 3052, Australia

Abstract. We present the first results of a programme of multiwavelength study of a sample of Low Ionization Broad Absorption Line Quasars (LIBALs) taken from the Large Bright Quasar Survey (LBQS). We make no clear detections, but coadding the data for all six observed objects yields a tantalizing 1.8σ result. We discuss the role of the Large Southern Array for future studies of the relationship between quasars and galaxies.

1 Introduction

Low ionization (MgII) broad absorption line quasars (LIBALs) represent about 1% of quasars in the LBQS and similar optically selected samples. They are distinguished by broad (> 30000 kms^{-1}) low ionization lines, especially MgII, thought to be coming from massive outflowing winds. At low redshift they seem to be bright in the far-IR, appearing as 3/7 IRAS selected quasars (with z high enough to see MgII) in the sample of Low et al. (1989). These low redshift objects have optical-IR ratios and IRAS colours that are different from normal quasars at 99% confidence.

LIBALs in general have very red optical-UV spectra (Sprayberry & Foltz (1992)) suggesting the presence of dust reddening. Allowing for this dust reddening, LIBALs might make up 15% of the parent quasar population. However, dereddening is still not sufficient to explain the strong far-IR emission seen by Low et al.. This emission must thus be associated with the host galaxy or central engine itself.

2 Theoretical Background

The most common theories suggest that all radio-quiet quasars have broad absorption lines visible from certain lines of sight. However, this does not explain the strong NV emission and excess radio flux seen in Broad Absorption Line (BAL) quasars (Francis et al. (1993)) or the strong far-IR emission seen in the Low et al. LIBALs.

An alternative explanation is that BAL and LIBAL quasars might be an evolutionary stage between ultraluminous IRAS galaxies (ULIRGs) and quasars.

Sanders et al. (1988) suggested that quasars start as ULIRGs, and only become revealed when the central engine has cleared away dense obscuring material. BALQSOs would then be an intermediary stage during the expulsion of this material. This picture is supported by the fact that one of the 10 original ULIRGs, Mrk 231, appears to be a LIBAL where we can see the host galaxy (Lipari et al. (1994)).

We aim to test this picture by observing LIBALs selected from the LBQS at submillimetre wavelengths in conjunction with ISO observations and ground-based optical and IR photometry. These data will be used to determine the strength and spectral energy distribution of their rest-frame UV to far-IR emission.

We here report the first results from this programme obtained at the SEST telescope.

3   Observations and Results

Observations were made using the SEST telescope and 1.3mm bolometer system. Data was taken from 9 to 14 July 1995 in adequate observing conditions (zenith optical depths ranging from 0.1 to 0.25). Each object was observed for a number of blocks consisting of 10 ON−OFF pairs, with an integration time of 10s each.

The results of initial data reduction are given in the following table.

Object	Flux
2350−0045	3.7±3.6 mJy
0335−339	2.2±2.2 mJy
0059−2739	2.8±2.5 mJy
2225−0534	1.9±3.1 mJy
1011+0910	−4.2±9.1 mJy
1232+1325	2.3±7.4 mJy

Table 1. Results of Observations

4   Discussion

At the redshifts of these quasars ($1.6<z<2.4$) an object like Mrk 231 would provide a 1.3mm flux of about 1 mJy, the exact value depending on the dust temperature and emissivity law. As can be seen, our observations are approaching the necessary sensitivity, but have not yet reached it.

However, combining all the fluxes to determine an 'average LIBAL flux' at 1.3 mm gives 2.4 ± 1.3 mJy, a tantalizing 1.8 σ result.

At this stage we can thus neither confirm nor reject the notion that LIBALs might be luminous in the rest-frame far-IR, and must await the results of ISO observations of two of these objects, which will be substantially more sensitive.

Further ground based observations in the submillimetre waveband will also be needed to determine the spectral energy distribution of any detected emission. The JCMT with SCUBA will be especially useful for this task, as would SEST with the addition of a bolometer array.

For larger samples examining the relationship between LIBALs and other quasars with ULIRGs, a large single-dish millimetre/submillimetre telescope, such as the 50m telescope proposed for Mexico (Wall & Carrasco, 1996), offers significant gains. The main advantage here is the combination of large collecting area with the large bandwidth offered by incoherent bolometric detectors, and access to the submillimetre waveband for the study of the spectral energy distribution near its peak ($\sim 100\mu m$ in the restframe). However, the Large Southern Array, with sensitivities of $10^{-4} - 10^{-6}$ Jy at 3mm, should be sufficient to detect the long wavelength continuum from even quite faint submm sources. It will also allow mapping of colder dust in the brighter sources at 0.1" resolution. If the dust responsible for this emission is extended through the quasar host galaxy, this might be very interesting.

The principle role for a Large Southern Array in these studies though will be to examine the molecular material, and to determine both its spatial and velocity structure. If quasars really do evolve from ULIRG-like objects, young quasars should have dense dust and molecular material in their nuclei which should make them ideal targets. The discovery of CO emission in the Cloverleaf quasar (Barvainis et al 1994), which is both a LIBAL and gravitationally lensed, would seem to confirm this suggestion. A large scale study of molecular material in a statistical sample of such objects, though, requires much higher sensitivity than is currently available, whilst a full understanding of the kinematics will need good spatial and spectral resolution. The Large Southern Array will provide excellent capabilities in both these areas, and will thus be the ideal instrument for these studies.

References

Barvainis et al (1994), Nature, 371, 586
Francis et al. (1993), AJ, **106**, 417
Low et al. (1989) Ap.J. **340**, L1
Lipari et al. (1994), Ap. J., **427**, 174
Sanders et al. (1988), Ap.J., **325**, 74
Sprayberry & Foltz (1992), Ap.J., **390**, 39
Wall & Carrasco (1992), this volume

Dynamical Studies of Spiral Galaxies

Francoise Combes

DEMIRM, Observatoire de Paris, 61 Av. de l'Observatoire, F–75014 Paris, France

Abstract. In nearby galaxies, the future large millimeter arrays (LMAs) will be able to resolve the main dynamical features, spiral arms, secondary bars and rings, with enough sensitivity to constrain theoretical scenarios of galaxy evolution. Particularly interesting is the fate of gas in the central regions of spiral galaxies, where hierarchical structures can be found and where the gas is mainly molecular. In the most nearby galaxies, the mass spectrum of molecular clouds can be determined, and large insight will be gained in the H_2/CO conversion ratio, through virial analysis, and also on large-scale star formation processes. At larger distances ($z \sim 1$), maps of comparable resolution as are obtained now with single dish in nearby galaxies as M51 could be obtained, which will provide direct information on galaxy evolution.

1 Introduction

Recent millimetric observations have revealed the huge molecular gas concentrations in central regions of galaxies. These regions cannot be investigated by the visible domain because of dust extinction, neither by atomic gas observations at 21cm, since HI is usually deficient in central regions: the gas is in the molecular phase. These central regions are yet of prime importance to understand the evolution of disks on which they have a profound influence, for instance they can destroy the primary bars, and are also at the origin of activity in nuclei. Future large mm arrays will map easily these central decoupled regions of 50pc sizes in nearby galaxies (less than 30Mpc away).

At larger scale in disks, more information could be gained in resolving the spiral structure, and comparing the various tracers of the density wave, HI, HII and H_2 regions. These studies presently are suffering from a lack of sensitivity, and progress will be gained with collecting areas an order of magnitude larger.

For the most nearby galaxies, fine studies could be done of the ensemble of molecular clouds, similar to what has been reached in the solar neighbourhood of the Milky Way disk. These are essential to investigate the complex fractal structure of the ISM, to determine variations of the H_2/CO conversion ratio, and tackle the triggering processes of star-formation.

Finally, we could extend to higher redshifts detailed studies of spiral galaxies, and determine the gross lines of evolution. Interacting galaxies have been observed with strongly enhanced CO emission, while bars gather molecular gas towards central regions, which makes them easy to detect at large distances. The LMA resolution and sensitivity will be sufficient to determine disk morphology until a redshift of $z = 1$ or 2.

2 Decoupled central regions

Gas can lose angular momentum in a few rotation times when non-axisymmetric features like bars exist in the galaxy disks: the gaseous central regions then dynamically decouple from the rest of the galaxy.

2.1 Observations

Nuclear disks are frequently observed in barred spiral galaxies, they are a strong mass concentration, mainly gaseous, in a fast rotating disk confined within the central 1kpc. When seen face-on, they correspond to a nuclear ring, often the site of star formation (e.g. IC342, Downes et al 1992, Rieu et al 1994 or NGC1068 Tacconi et al 1994 fig. 1). The rings correspond to the inner Lindblad resonance, where the spiral arms from corotation first wind up into a pseudo-ring: these correspond to the conspicuous leading dust lanes along the bar, and the shock at the intersection of the arms and ring produce the accumulation of CO emission seen as the "twin-peaks" (e.g. Kenney 1994). The apparent central peaks of CO emission are observed less frequently than the pseudo-rings, and they also could be central peaks only by lack of resolution (cf. Ishizuki 1994).

Since rings are often sites of enhanced star formation, it is not obvious that the CO line alone is a good tracer of the gas surface density. It is important to combine studies at different wavelength for several molecules (including high-density tracers such as HCN, CS..), to disentangle excitation or abundance effects. In particular, the CO(2-1) line that will provide the highest resolution with LMAs, is to be interpreted with caution, since it traces a different cloud layer than the CO(1-0) or ^{13}CO(1-0) lines (e.g. IC342, Turner et al 1994).

There appears in some cases an evolution of the ring in radius, the maximum surface density being at a smaller radius than the most star-forming ring (e.g. M82, Wild 1990, NGC 4314, Combes et al 1992), but this requires more spatial resolution. Theoretical scenarios of evolution and self-regulated star-formation could then be tackled: as soon as the cold gas has accumulated in the center and reaches the critical surface density, it is Jeans unstable to lumps formation, and is consumed through star-bursts. Does star formation evolves inside out or the reverse? (Telesco et al 1991, Kenney 1994). At which frequency are triggered nuclear starbursts?

At even smaller scales, one can some-times find a nuclear bar filling the ring, best seen in the near-infrared (e.g. Shaw et al 1993, Friedli 1995). When seen edge-on, these systems are conspicuous through their large central velocity gradients, delineating an almost detached independent system in the center. This is the case of the Milky Way, where the CO $l-v$ diagram reveals a striking width of 500km/s towards the central 200pc (Dame et al 1987), but also of NGC 891 (Garcia-Burillo et al 1992, fig. 2). HI absorption against the nuclear continuum source reveals also these nuclear disks (Koribalski et al 1993). A compilation of some of the nuclear disks discovered in spiral galaxies gives a typical diameter of 100- 200pc for the disk (Combes 1994a). But this is only an observational bias, due to the lack of resolution. Much smaller nuclear disks could exist that would be mistaken for a point mass concentration by the neutral gas tracers.

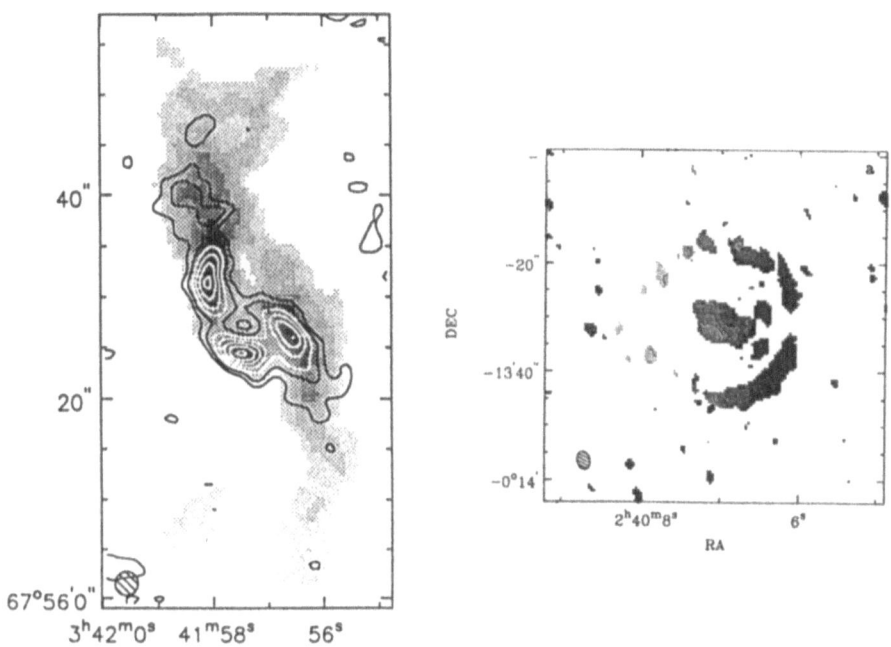

Fig. 1. Examples of nuclear rings seen almost face-on: IC342 (Rieu et al 1994) and NGC1068 (Tacconi et al 1994), observed with the IRAM interferometer in the HCO$^+$ and HCN line, respectively

2.2 Theory

On the theoretical side, nuclear mass concentrations are naturally expected in barred galaxies, since the non-axisymmetric potential produce torques, and transfer angular momentum outwards (e.g. Combes 1988). N-body simulations

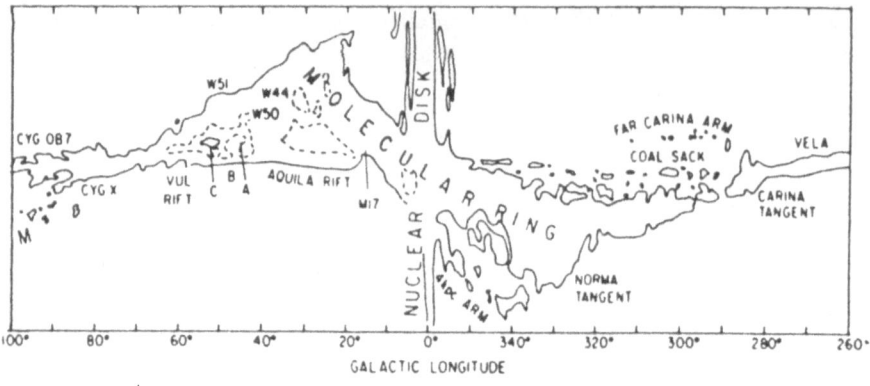

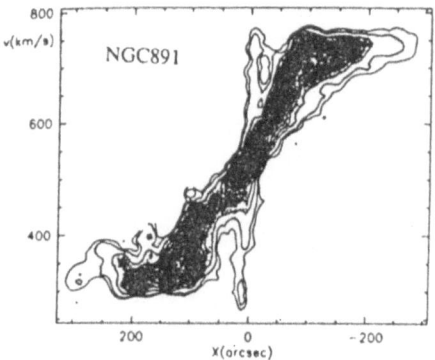

Fig. 2. Examples of nuclear disks or rings seen edge-on: the Milky Way (Dame et al 1987) and NGC891 (Garcia-Burillo et al 1992)

have shown that a nuclear disk is detached quite easily, and a double-bar pattern can develop (Friedli & Martinet 1993, Combes 1994b). Bars can help to concentrate the mass only until some threshold, however, since a too large mass concentration can destroy the bar (Hasan & Norman 1990, Hasan et al. 1993). This can be considered as a regulation process, in which the disk and the nucleus grow simultaneously (Combes 1996). The growth of a possible black hole can then be stopped, and vertical instabilities can lead to bulge formation (Pfenniger & Norman 1990).

Numerical simulations of a self-gravitating gaseous disk have shown that a large variety of phenomena can occur in the central regions: the large mass concentration accelerates the rotation, a dynamical time-scale being a few Myrs.

Due to the large value of the precessing rate for elliptical orbits $(\Omega - \kappa/2)$, there exist 2 Lindblad resonances for the main bar perturbation, and a family of perpendicular periodic orbits in between. The gas streams then with a large phase-shift with respect to the bar potential (Shaw et al 1993), and is driven quickly inwards. Its flow can decouple in two perpendicular rings, then in two corotating parallel bars, as is the case of NGC4321 (Knapen et al 1995). Decoupling of hierarchical structures, i.e. bars within bars (Shlosman et al 1989), rotating faster with a common resonance (the CR of the small bar being the ILR of the primary one for example, see Tagger et al 1987), is a way of slowing the flow of gas, since the latter is then more in phase with the potential. Recursive decoupling, like russian dolls, is possible and more spatial resolution is required to actually make the link with the gas flows fueling the active nuclei, at sub-pc scales. Already images from the HST have revealed such tiny dusty spiral or bar structures, of sub-pc scales (fig. 3). The decoupling is sometimes evident through the completely independent inclination of the nuclear disk (this concerns the smaller scales, where the disk has only the mass of one molecular cloud). A large range of scales remains to be explored, and the relation of the molecular gas to the nuclear activity (radio jets, radiation cones, etc..).

3 Spiral Structure and Cloud Ensemble

3.1 Spiral Structure

The spiral density wave is the main large-scale structure that can trigger starformation, propagate the tidal perturbation from nearby galaxies, and accelerate galaxy evolution. Yet, although a lot of controversial studies have tried to determine the chronology of events across a spiral arm (see e.g. Casoli 1991), we still do not know precisely the action of the density wave on the ISM. Comparison of several molecular lines with HI, Hα, NIR, etc.. would help to tackle the modifications to the cloud ensemble brought by the crossing of the potential well.

3.2 Cloud Ensemble

One major uncertainty in millimetric observations of the molecular component is the H_2/CO conversion ratio, and a constant value is adopted for sake of simplicity. However, we know that the hierarchical structure of the ISM is likely to change a lot across spiral arms, with radius, with metallicity, etc.. More insight would be gained with spatial resolution in external galaxies, when individual clouds could be identified, and their virial masses estimated. In M51, the spatial resolution of the future LMAs will be 1pc. In particular, we can hope to solve the present mystery of the two widely different radial distributions of atomic and molecular hydrogen in galaxies.

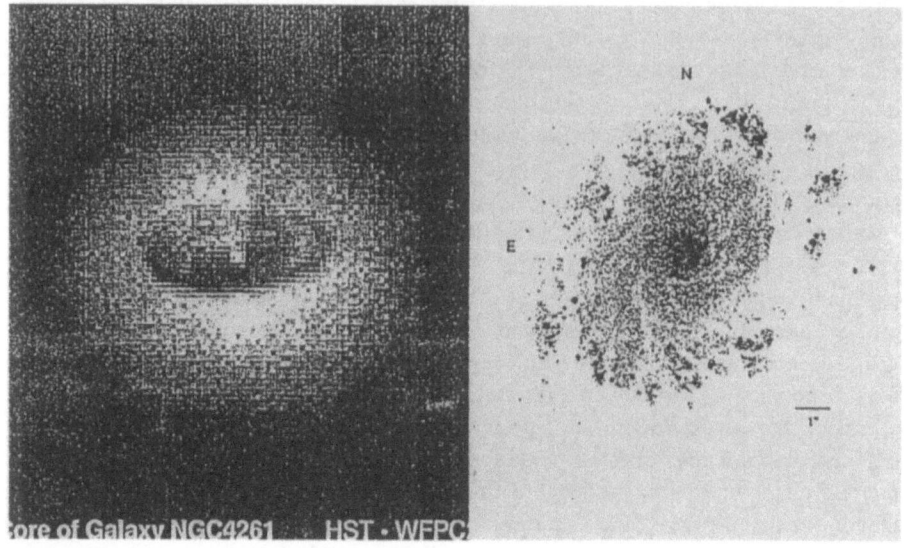

Fig. 3. Examples of nuclear disks of parsec scales, revealed by the HST: NGC4261 at the heart of an elliptical galaxy (Ford et al 1995) and NGC 6951 (Barth et al 1995)

4   Evolution at Moderate Redshift

The future LMAs will allow to determine the gaseous morphology of galaxies at moderate redshift, $z \approx 1$. This is made easier by the enhancement of CO emission in interacting galaxies (by a factor ≈ 4 e.g. Braine & Combes 1993), and the fact that interactions were more frequent in the past. At $z \approx 1$, the angular distance is $\approx 1200 Mpc$, and the luminosity distance $\approx 4700 Mpc$ (for $H_0=75$km/s/Mpc and $q_0=0.5$). The spatial resolution will then be of the order of 0.1" = 0.5kpc for a sensitivity of 1K at the 10σ level, with 12h integration. This corresponds to the resolution obtained today in the CO(2-1) line with the IRAM 30m on M51 (cf the map from Garcia-Burillo et al 1993), where signals of the order of 1K are detected. It will then be easy to recognize the essential morphological features of gaseous galaxies at that time, and through statistics determine the evolution with redshift.

For instance, one of the most important puzzle of observational cosmology is the large excess of faint blue galaxies at $z \sim 1$ (Tyson 1988). These galaxies have a space density that exceeds that of ordinary bright galaxies by a factor of 30 (Lilly, Cowie, & Gardner 1991; Babul & Rees 1992) and have no obvious bright counterparts in the local universe. Optical emission line observations indicate that these compact galaxies (diameters 2-4 kpc) have star formation rates comparable to present day spiral galaxies (1-20 $M_\odot$ yr^{-1}). Their blue luminosities are comparable to local field galaxies ($M_B \sim -21$), and are too large for them to be distant counterparts of the local blue compact galaxies ($M_B \sim -16$) (Koo et al. 1994). Since these systems are known to be forming stars, they must possess a rich gaseous content, although not specially metal-rich (Koo et al 1995). The Canada-France redshift survey (e.g. Lilly et al 1995) has also precised and detailed the strong evolution of the luminosity function of the blue galaxies between $z = 0.5$ and 1. Molecular spectroscopy will bring a lot of new information about the star-formation and gas content evolution of galaxies with redshift.

References

Babul, A. & Rees, M. J., 1992, MNRAS, 255, 346

Barth A.J., Ho L.C., Filippenko A.V., Sargent W.L.W. 1995, AJ 110, 1009

Braine J., Combes F. 1993, A&A 269, 7

Casoli F., 1991, in: Dynamics of Galaxies and their Molecular Cloud Distribution, ed. F. Combes & F. Casoli, IAU 146, p. 51

Combes F.:1988, in "Galactic and Extragalactic Star Formation" ed. R. Pudritz & M. Fich p. 475

Combes F., Gerin M., Nakai N., Kawabe R., Shaw M.A.: 1992, A&A 259, L27

Combes F.:1994a, in "Nuclei of Normal Galaxies", ed. R. Genzel & A. Harris, p. 65

Combes F., 1994b, in: Mass-Transfer Induced Activity in Galaxies, ed. I. Shlosman. Cambridge University Press, Cambridge, p. 170

Combes F.: 1996, in "Barred Galaxies", IAU157, ed. Buta, Elmegreen & Crocker, Tuscaloosa meeting May 1995, ASP Series, p. 286

Dame T.M., Ungerechts H., Cohen R.S. et al.: 1987, ApJ 322, 706

Downes D., Radford S.J.E., Guilloteau S., Guélin M., Greve A. & Morris D., 1992, A&A 262, 424

Ford H.C., Ferrarese L., Jaffe W.: 1995, HST Press Release

Friedli D.: 1995, in "Barred Galaxies", IAU157, ed. Buta, Elmegreen & Crocker, in press

Friedli D., Martinet L.: 1993, A&A 277, 27

Garcia-Burillo S., Guélin M., Cernicharo J., Dahlem M.: 1992, A&A 266, 21

Garcia-Burillo S., Guélin M., Cernicharo J.: 1993, A&A 274, 123

Hasan H., Norman C., 1990, ApJ 361, 69

Hasan H., Pfenniger D., Norman C., 1993, ApJ 409, 91

Ishizuki S. 1994, in: Astronomy with Millimeter and Submillimeter Wave Interferometry, ASP Conference Series, vol. 59, ed. M. Ishiguro and Wm. J. Welch, p. 293

Kenney J.D.P. 1994, in: Astronomy with Millimeter and Submillimeter Wave Interferometry, ASP Conference Series, vol. 59, ed. M. Ishiguro and Wm. J. Welch, p. 282

Knapen J.H., Beckman J.E., Heller C.H., Shlosman I., de Jong R.S., 1995, ApJ 454, 623

Koo, D. C., Bershady, M. A., Wirth, G. D., Stanford, S. A., & Majewski, S. R., 1994, ApJ, 427, L9

Koo, D. C., Guzman, R., Faber, S. M., Illingworth, G. D., Bershady, M. A., Kron, R. G., & Takamiya, M., 1995, ApJ, 440, L49

Koribalski B., Dickey J., Mebold U.: 1993 ApJ 402, L41

Lilly, S. J., Cowie, L. L., & Gardner, J. P., 1991, ApJ, 369, 79

Lilly S.J., Tresse L., Hammer F., Crampton D., Lefevre O. 1995, ApJ 455, 108

Pfenniger D., Norman C.A., 1990, ApJ 363, 391

Rieu N-Q., Viallefond F., Combes F. et al 1994, in: Astronomy with Millimeter and Submillimeter Wave Interferometry, ASP Conference Series, vol. 59, ed. M. Ishiguro and Wm. J. Welch, p. 336

Shaw M.A., Combes F., Axon D.J., Wright G.S.: 1993, A&A 273, 31

Shlosman I., Frank J., Begelman M.: 1989, Nat 338, 45

Tacconi L., Genzel R., Blietz M., Cameron M., Harris A.I., & Madden S., 1994. ApJ 426, L77

Tagger M., Sygnet J.F., Athanassoula E., Pellat R.: 1987, ApJ 318, L43

Telesco C.M., Campins H., Joy M., Dietz K, Decher R. 1991, ApJ 369, 135

Turner J.L., Hudson D., Hurt R.L. 1994, in: Astronomy with Millimeter and Submillimeter Wave Interferometry, ASP Conference Series, vol. 59, ed. M. Ishiguro and Wm. J. Welch, p. 300

Tyson, A. A., 1988, AJ, 96, 1

Wild W. 1990, PhD thesis, München Universität

Kinematics and Distribution of Molecular Gas in Galaxy Nuclei

Linda J. Tacconi and Reinhard Genzel

Max-Planck-Institut für extraterrestrische Physik, Postfach 1603, D-85740 Garching, Germany

1   Introduction

Among the outstanding problems in the study of activity in galaxy nuclei are (1) to determine the mechanism(s) which are bringing fuel to feed active nuclei (AGN) and central starbursts (*e.g.* Gunn 1979; Balick and Heckman 1982; Shlosman, Begelman and Frank 1990); (2) to understand the properties of the nuclear regions as related to the unified scenarios for AGN (*e.g.* Lawrence 1987; Antonucci 1993), and to view directly the pc-scale obscuring torus; (3) to determine how frequently AGN and circumnuclear starbursts are seen together, and to assess whether there is a dominant physical process responsible for both energetic phenomena. All of these problems can be addressed through high resolution, high sensitivity millimeter and submillimeter observations of molecular lines with the proposed LSA. The fact that progress is already being made towards answering some of these questions for the closest active systems attests to the crucial role that very sensitive millimeter interferometry will play in the study of galaxy nuclei. In this contribution, we present some of our recent results made with the IRAM interferometer, and point out specific cases where interferometric observations from the LSA will be crucial.

2   Distributions and Properties of Circumnuclear Molecular Gas

One of the best studied galaxy nuclei at any wavelength is that of NGC 1068, the archetypal Seyfert 2 galaxy. Located at a distance of 14.4 Mpc, a plethora of published molecular line observations at $2''$-$8''$ spatial resolution (a few examples are Helfer and Blitz 1995; Jackson *et al.* 1993; Planesas *et al.* 1991) clearly show a compact region of intense emission associated with the narrow line region (NLR), and a more extended spiral component associated with a 1 kpc radius starburst. In Figure 1 we show our ^{13}CO and HCN J=1$\rightarrow$0 images of NGC 1068 (Sternberg *et al.* 1996; Tacconi *et al.* 1994). At a spatial resolution of $\sim$4$''$ the ^{13}CO map clearly shows both the bright inner spiral arms and the central component, as well as some fainter emission associated with the NIR stellar bar at a position angle of 55° (Scoville *et al.* 1988). In contrast to the CO emission, the HCN map (at $\sim$2.5$''$ resolution) is dominated by very strong emission in the central few

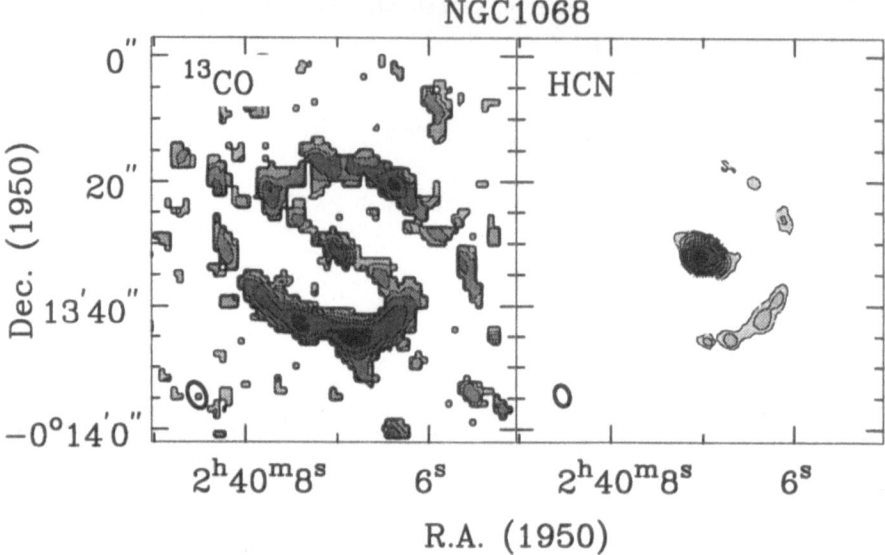

Fig. 1. Left panel: ^{13}CO J=1→0 image of NGC 1068 made with the IRAM interferometer at $4'' \times 3''$ resolution. The synthesized beam is shown in the bottom left corner. Right panel: The HCN J=1→0 image made with the IRAM interferometer from Tacconi *et al.* 1994. The synthesized beam is shown in the bottom left corner.

hundred pc, with much fainter emission associated with the starburst region. Very little structure is visible in the bright central source with ~3″ (200pc) resolution. The overall structure of the molecular gas associated with the NLR will be seen more clearly with the factor of 2–4 improvements in spatial resolution starting to be available with the existing interferometers.

However, to detect molecular structures close to the nucleus a powerful instrument like the LSA at its highest spatial resolution and sensitivity will be required. At the distance of NGC 1068 the LSA would be able to resolve a structure like the circumnuclear disk of the Galaxy seen in the HCN map of Güsten *et al.* (1987). Thus we would be able to **image directly the molecular torus in the closest AGN**. The existence of such a dense molecular torus is predicted by the unifying scenarios for AGN. In these schemes, most of the differences in properties between broad-line Seyfert 1 galaxies and narrow line Seyfert 2's can be largely explained by the angle at which we are viewing the torus. It will be critical to determine accurately the column densities and optical depths through the torus, and study the excitation conditions and abundances of molecular clouds which are exposed to extreme conditions not found anywhere in our Galaxy.

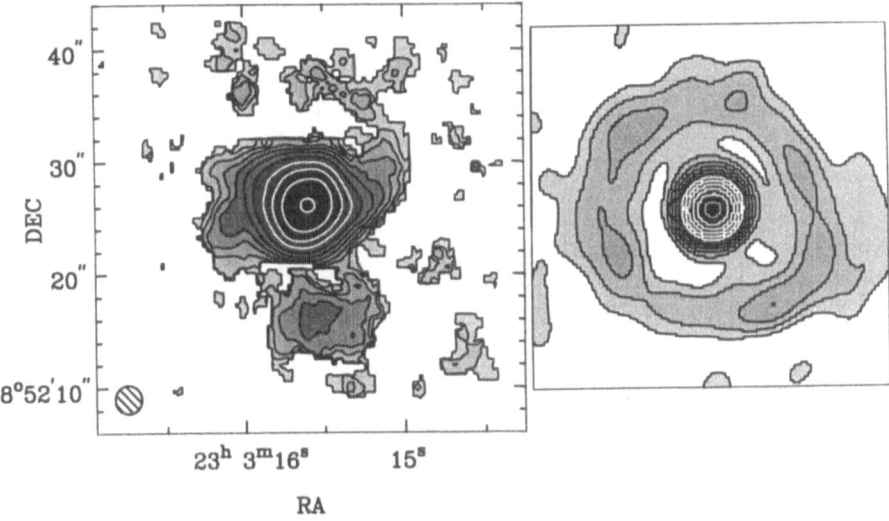

Fig. 2. Left panel: ^{12}CO J=1→0 image of NGC 7469 from Tacconi *et al.* (1996). The synthesized beam is shown in the bottom left corner. Right panel: The NIR K-band image of the central 5″ (edge-to-edge) of NGC 7469 from Genzel *et al.* (1995) made at 0.4″ resolution, and showing clearly separate central and circumnuclear components.

The study of molecular cloud physics and chemistry in the nuclei of galaxies of a wide range of activy will be another important area of study for the LSA. Again we use NGC 1068 as an example of what is currently achievable. The unusually high HCN/CO intensity ratio of ~1 in the central few hundred parsecs of NGC 1068 (Jackson *et al.* 1993; Tacconi *et al.* 1994; but see Helfer and Blitz 1995), has sparked much interest since it is the highest value found in any galaxy. There are currently several groups working on arriving at gas densities, optical produce extremely stron HCN emission relative to CO (*e.g.* Helfer and Blitz 1995; Sternberg, Genzel and Tacconi, 1994; plus work in progress). With the data currently available, these authors have concluded that on average the molecular gas is warm ($\geq$50 K) and dense ($\geq$10^5 cm^{-3}) throughout the central few hundred pc. There are still open questions regarding whether the intensities could be better explained by chemical effects, or what the detailed properties of the clouds in the NLR region are. In order to study the conditions of the molecular clouds in the NLRs of galaxies like NGC 1068 in detail we need to make another order of magnitude jump in sensitivity and an increase in resolution from what is currently available. Even for the brightest and closest case of NGC 1068 we are at the sensitivity limits of the current interferometers in mapping the emission from species other than ^{12}CO and HCN.

From the results of NGC 1068 it is clear that for studies of molecular gas associated with AGN and starburst regions in galaxies and for investigations of the AGN-starburst connection, it is necessary to separate spatially the emission from these components from eachother and from that of the larger scale galaxy disk. Another such galaxy harboring both an AGN and an intense circumnuclear starburst is the classical Seyfert 1 galaxy, NGC 7469. At a distance of 66 Mpc, it is roughly 4 times more distant than NGC 1068. We have studied the millimeter CO line (Tacconi *et al.* 1996) and NIR line and continuum emission (Genzel *et al.* 1995) in detail in this galaxy to get a benchmark for the stellar and gas properties of a typical Seyfert 1 galaxy. In Figure 2 we show the K-band continuum image (right panel) made at $0.4''$ spatial resolution with the MPE speckle camera SHARP at the ESO NTT. This image clearly shows both strong emission from a central component and the bright starburst ring. The brightest part of the CO image at $2.5''$ spatial resolution contains both the nuclear and the starburst ring regions, but it is not possible to separate spatially the molecular gas from these two components. Although the sub-arcsecond resolution necessary to observe the structural detail present in the NIR speckle data is now becoming feasible with the upgraded IRAM interferometer, the LSA is required to separate these regions in galaxies much more distant than NGC 7469. The proposed array would enable us to resolve spatially circumnuclear starbursts from AGN molecular components out to a redshift of ~ 0.2. This would include most of the known Seyferts, as well as many IRAS ultraluminous galaxies.

3 Kinematics

One of the most important aspects of millimeter interferometric line observations opportunity they provide for studying the detailed kinematic properties in the central regions of galaxies. Observations with very high signal-to-noise coupled with high spectral resolution of ~ 10 km s^{-1} make it possible to glean kinematic structures at scales which are significantly smaller than the spatial resolution. An example of this work is illustrated in the results of our CO study of NGC 7469. Spectra from the central $8''$ of the galaxy are shown in Figure 3.

We have constructed kinematic models which can explain the features observed in the spectra and in kinematic major and minor axis position-velocity maps. In these models we take an assumed axisymmetric density and kinematic distribution of the gas as a function of radius, and convolve these distributions spatially and spectrally with beams that are identical to those of the actual data. For the central regions of NGC 7469 the kinematic structures, the dynamic range of the observed intensities, and the spatial separation of the two line intensity peaks seen in the spectra are reasonably well fit by a $25°–30°$ inclined Gaussian ring density distribution with a radius of $1.5''$ and FWHM $= 2.5''$, combined with a compact central Gaussian source with FWHM $= 0.3''$. The central component has an intrinsic surface brightness which is 15–20 times greater than that of the ring. This intensity distribution is very similar to that seen in the $0.4''$ resolution

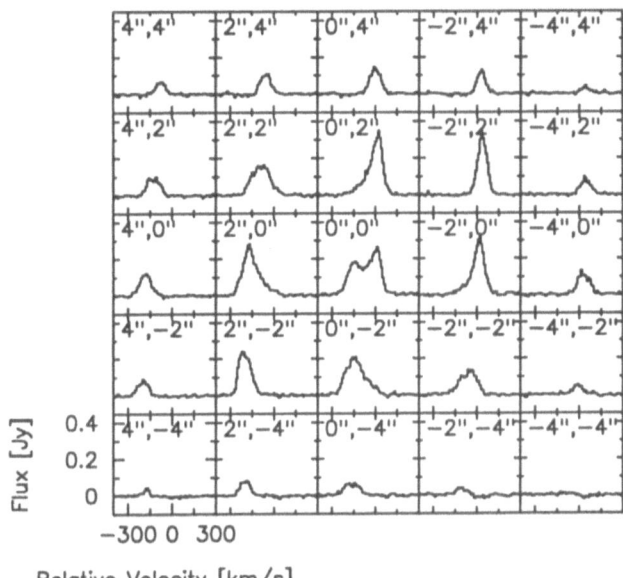

Fig. 3. Sample spectra from the central 8″ of NGC 7469 in the ^{12}CO J=1→ map at 2.5″ resolution. Spectral resolution is 10 km/s. The numbers in each box are offsets from the nucleus in arcsec.

near-infrared K-band map of Genzel *et al.* (1995) (Figure 2). The velocity field of this multi-component gas distribution has a rotation curve which is moderately constant at radii interior to ∼1″, then rises steeply in the starburst ring. From the inferred rotation curve and inclination angle we calculate a dynamical mass of 2.0×10^{10} M$_{\odot}$ in the central 5″, roughly the area containing the starburst ring. Since the stellar mass estimates of the region interior to and including the starburst ring are very well constrained by the analyses of Genzel *et al.* (1995) and are found to be ∼7×10^9 M$_{\odot}$, we are led to the very important conclusion that that the **gas mass is nearly twice the mass in stars over this region.**

Another important result is that we are able to *infer* the presence of a central CO component in NGC 7469, even though such a component would require a factor of 5 higher spatial resolution to detect it directly. The kinematical tools available to millimeter astronomers would become even more powerful at higher spatial resolution, however. To illustrate what we might see at spatial resolutions and sensitivities of the proposed LSA we have used our current kinematic model of NGC 7469 and have run several simulations. These are shown in Figure 4 as a series of position-velocity cuts taken along the kinematic major axis. The leftmost panel shows the p-v map resulting from our best model at the 2.5″ spatial resolution of the current data. Increasing the spatial resolution by a factor of 5 would enable us to observe the central component directly (central panel), seen as the steep contours at radii <1″, as well as get the rotation curve information

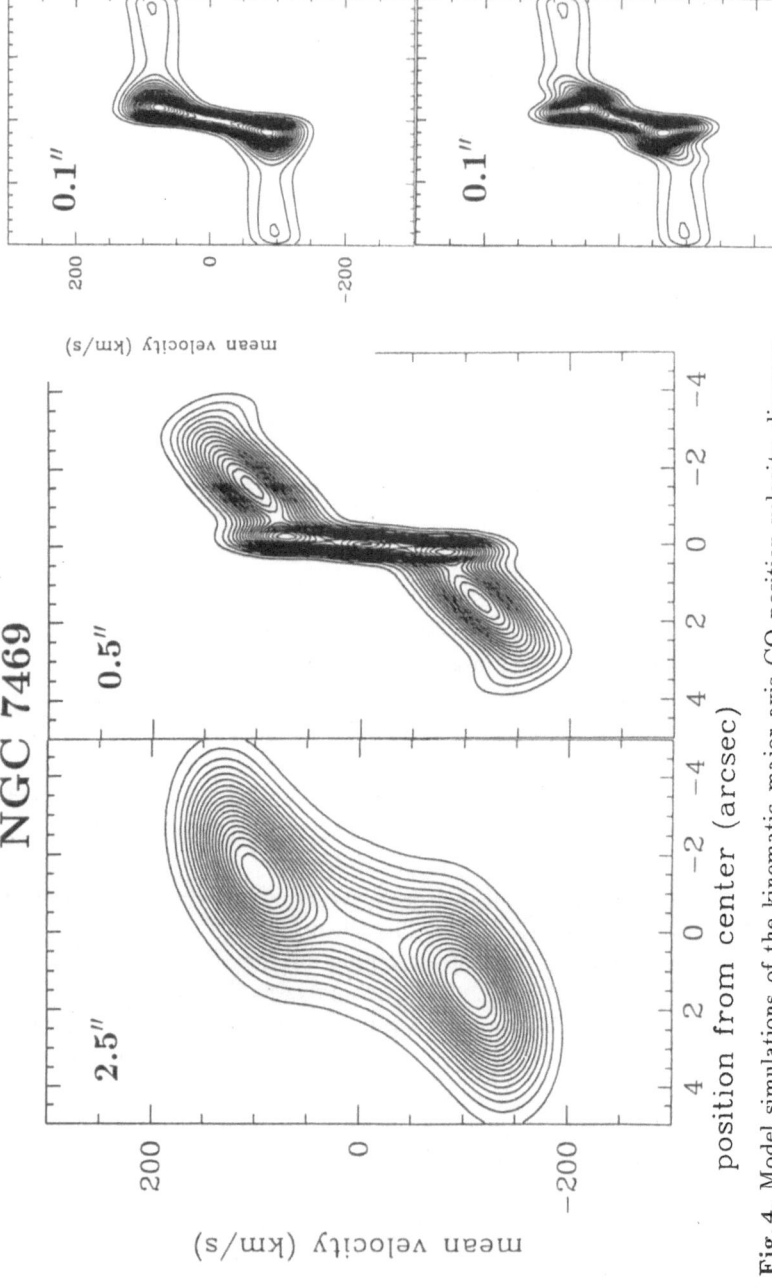

Fig. 4. Model simulations of the kinematic major axis CO position-velocity diagram for the Seyfert 1 galaxy, NGC 7469. The left panel shows the best fit model to our current observations at 2.5″ spatial resolution (see text). The center panel is the same model but with a resolution of 0.5″. To the right we show (top) the central 2″ of the same model with a spatial resolution of 0.1″, and (bottom), a model which includes a central massive object and observed at 0.1″.

in the ring. This resolution is achievable with upgraded present day interferometers, albeit only by compromising sensitivity. A factor of 5 increase again, or
the highest resolution achievable with the LSA, would be necessary to discern
fine details in this central component, however. The two left panels of Figure
4 illustrate this fact rather dramatically. The top right panel shows the central
component with the slowly rising rotation curve of the best NGC 7469 model.
The bottom panel shows the same model but with the addition of Keplarian
rotation due to the presence of a massive compact object. We would thus be
able to infer the presence of compact masses in the centers of galaxies out to
distances of ~70 Mpc through their CO rotation curves. One important thing to
point out is that in all of the above simulations we have assumed that the S/N
has remained constant with increasing resolution. That is structures detected on
scales of 0.1″ need to have the same S/N as the data at 2.5″ resolution (Figure 3). The only way to achieve this is to have a large collecting area on the
long baselines. **A large number of large antennas is thus critical for the
LSA to be able to tackle difficult kinematic problems in the centers
of galaxies.**

4 Summary

From the above examples it is clear that the most important consideration for a
large interferometer for the study of galaxy nuclei and circumnuclear starbursts
is that it must provide high sensitivity on the longest baselines, since merely
detecting structures at high resolution is insufficient to address the important
issues. For these studies it is therefore critical that the individual antennas have
15 meter diameters and that there be 40–50 of them. Particularly on the 10
km baselines, the large collecting area proposed by the LSA committee will be
crucial. Furthermore, the LSA should have receivers which would cover the range
in frequencies from 80–350 GHz to be able to study the physical properties of
the gas, although the emphasis should be on the 3 and 1 mm windows.

References

Antonucci, R.R.J. (1993), *ARA&A*, **31**, 473.

Balick, B. and Heckman, T. (1982) *ARA&A*, **20**, 431

Genzel, R., Weitzel, L., Tacconi-Garman, L.E., Blietz, M., Cameron, M., Krabbe,
A., Lutz, D., and Sternberg, A. (1995), *Ap.J.*, **444**, 129

Güsten, R., Genzel, R., Wright, M.C.H., Jaffe, D.T., Stutzki, J., and Harris, A.I.
(1987) *Ap.J.*, **318**, 124

Gunn, J. (1979):, *Active Galactic Nuclei*, ed. C. Hazard and S. Mitton, (Cambridge
University Press), 213

Helfer, T.T., and Blitz, L. (1995), *Ap.J.*, **450**, 90

Jackson, J.M., Paglione, T.A.D., Ishizuki, S., and Nguyen-Q-Rieu, (1993), *Ap.J.*, **418**,
L13

Lawrence, A. (1987), *PASP*, **99**, 309

Planesas, P., Scoville, N.Z., and Myers, S.T., (1991), *Ap.J.*, **369**, 364

Scoville, N.Z., Matthews, K., Carico, D.P., and Sanders, D.B. (1988), *Ap.J.*, **327**, L61

Shlosman, I., Begelman, M.C., and Frank, J. (1990), *Nature*, **345**, 679

Sternberg, A. *et al.* (1996), in preparation

Tacconi, L.J., Genzel, R., Blietz, M., Cameron, M., Harris, A.I., and Madden, S. (1994), *Ap.J.*, **426**, L77

Tacconi, L.J., Genzel, R., Tacconi-Garman, L.E., and Gallimore, J.F. (1996), *Ap.J.*, submitted.

Star Formation and Molecular-Gas Dynamics in Galaxies with the Nobeyama Array and LSA

Sumio Ishizuki

Nobeyama Radio Observatory, National Astronomical Observatory, Minami-maki, Minamisaku, Nagano, 384-13, Japan

For investigations of gas dynamics in bar or oval potentials of galaxies and its relation with star formation in the nuclear or circumnuclear regions, the CO ($J = 1 \rightarrow 0$) emission in nearby galaxies was mapped by using the Nobeyama Millimeter Array with angular resolutions 4–13 arcseconds. Refer to also ref. 1.

The observations revealed the variety of the CO distribution in barred galaxies; category [1]: the CO distribution is elongated and lies parallel to the large-scale stellar bar (NGC 7479 and NGC 2903), category [2]: there is an inner CO bar structure inscribed in the pair of the dust lanes which are along the large-scale stellar bar (NGC 5383, M83, NGC 1530 and NGC 2782), and category [3]: the CO emission concentrates on the nucleus (NGC 3504).

The category [1] is the case in which the x_2-to-x_1 coverage fraction, the fraction of the coverage by the orbits of the x_2-family (whose orbits' orientations are perpendicular to the large-scale stellar bar) over the coverage by the orbits of the x_1-family (whose orbits' orientations are parallel to the large-scale stellar bar) is very small[2]. This corresponds to the case where the radius of the inner Lindblad resonance (ILR) is much smaller than that of the corotation resonance or there is no ILRs. The category [2] corresponds to the case in which the x_2-to-x_1 coverage is larger than in the category [1]. In particular, the deviation angles of the inner bars of NGC 1530, M 83, and NGC 5383 with respect to the large-scale stellar bars are close to 45°. This may justify the damped-orbit-resonance theory by Wada[3] since it agrees with the theoretical anticipation.

There is variety of the CO distribution in the inner bars of the galaxies of category [2]. The inner gaseous bar of NGC 5383 (Figure 1) has local peaks of the CO integrated intensity at the apocenters and the Hα emission spreads over the inner gaseous bar. On the other hand, the inner bars of M 83[4,5] and NGC 1530 (Figure 2,3) have leading offsets of the CO integrated intensity and the star forming regions inscribed in the leading offsets. In addition, NGC 2782 (Figure 4) has a pair of CO peaks that are located far inside between the separation of the dust lanes along the large-scale stellar bar and the southern CO peak is elongated to the west toward the inner end of the straight dust lane. This variation is interpreted as that the galaxies represent different stages of dynamical evolution along which the molecular gas is transported toward the nucleus; NGC 5383 is in the earliest stage and NGC 2782 is in the latest stage. Note that the near-infrared image shows a weak elongation in the east-west direction[6] and the southern CO elongation of NGC 2782 is at the leading side of the near-infrared elongation. It is plausible that oval potentials inside the outer ILRs of the large-scale stellar bars exert torques on the gas in the leading offsets (like in NGC 1530 and M 83) to drive the gas toward the nucleus. If dynamical evolution of this kind is real, it is reasonable to think that NGC 3504 is in the stage after the molecular gas is transported on the nucleus. This means that high gas density in the nuclear region can be realized due to the dual bar system, the large-scale stellar bar and the inner oval potential (cf. ref. 7).

CO maps and radio continuum data at λ 20 cm have been compiled from literature (Figure 5). This revealed that galaxies are divided into two classes in terms of the gas depletion time due to star formation, τ_{SF}, as the reciprocal of the radio-to-CO intensity ratio[8]. Galaxies with higher radio-to-CO ratios have $\tau_{SF} \sim 10^8$ years while galaxies with lower ratios have $\tau_{SF} \sim 10^9$ years. The former τ_{SF} is similar to that for starburst galaxies M 82 and NGC 253. The latter τ_{SF} is similar to that for the whole of the normal galaxies[9]. Note that hot spot galaxies (NGC 5383, NGC 1530, NGC 3351, NGC 6951, etc.) have $\tau_{SF} \sim 10^9$ years, indicating that their apparent enhancement of star formation is the simply reflects the large amount of molecular gas. Starbursts are divided into two categories; the "ring starburst" which occurs in a ring with a diameter comparable to the width of the large-scale stellar bar (NGC 1068, NGC 1097, NGC 3310) and the "nuclear starburst" which occurs inside the diameter (NGC 2782, NGC 3504, NGC 7714, M82, NGC 253). The inner oval potential mentioned above is a plausible mechanism for fueling the nuclear starburst.

It is suggested[7] that after the nucleus is fueled with the ISM, the resulting mass concentration forms a ring. NGC 1097[10], NGC 3310[11], and NGC 4736[12] possess rings which resemble such a morphology. NGC 1068 has a similar ring with starburst[13]. The optical spectroscopy of the nuclei and the starburst rings of NGC 1097[14,15] and NGC 3310[16,17] suggests that the ages of the star formation are older in the nuclei than in the rings, supporting the scenario of dynamical evolution. This implies that a barred galaxy may experience starburst twice; nuclear starburst at first and ring starburst afterwards. It is also suggested[7] that the morphology of a galaxy changes to that of an early-type due to secular evolution. It is consistent with the statistics[18] that starburst galaxies are rich in early-type spirals. Note that NGC 1068 and NGC 1097 have active galactic nuclei. This is consistent with the other statistics that Seyfert galaxies tend to have outer and/or inner rings[19] and to be early-type spiral galaxies[20].

In the followings, I describe new probes that will become observable toward a large number of galaxies with LSA.

In my work, I used the 20 cm continuum as a measure of the local star formation rate. Although it does not trace the Lyman continuum either directly nor indirectly, it is the best probe today when an angular resolutions of about 5″ is needed. There are no other probes that satisfy the following three conditions: a high ($\lesssim 5''$) angular resolution, a large number of galaxies observed, and being free from interstellar extinction in the object galaxy. LSA will let *the thermal free-free continuum* replace the centimeter-wave continuum as a measure of local star formation rates; the merit of the free-free continuum is that the physical origin is clear and easily converted to the Lyman continuum flux. The estimates

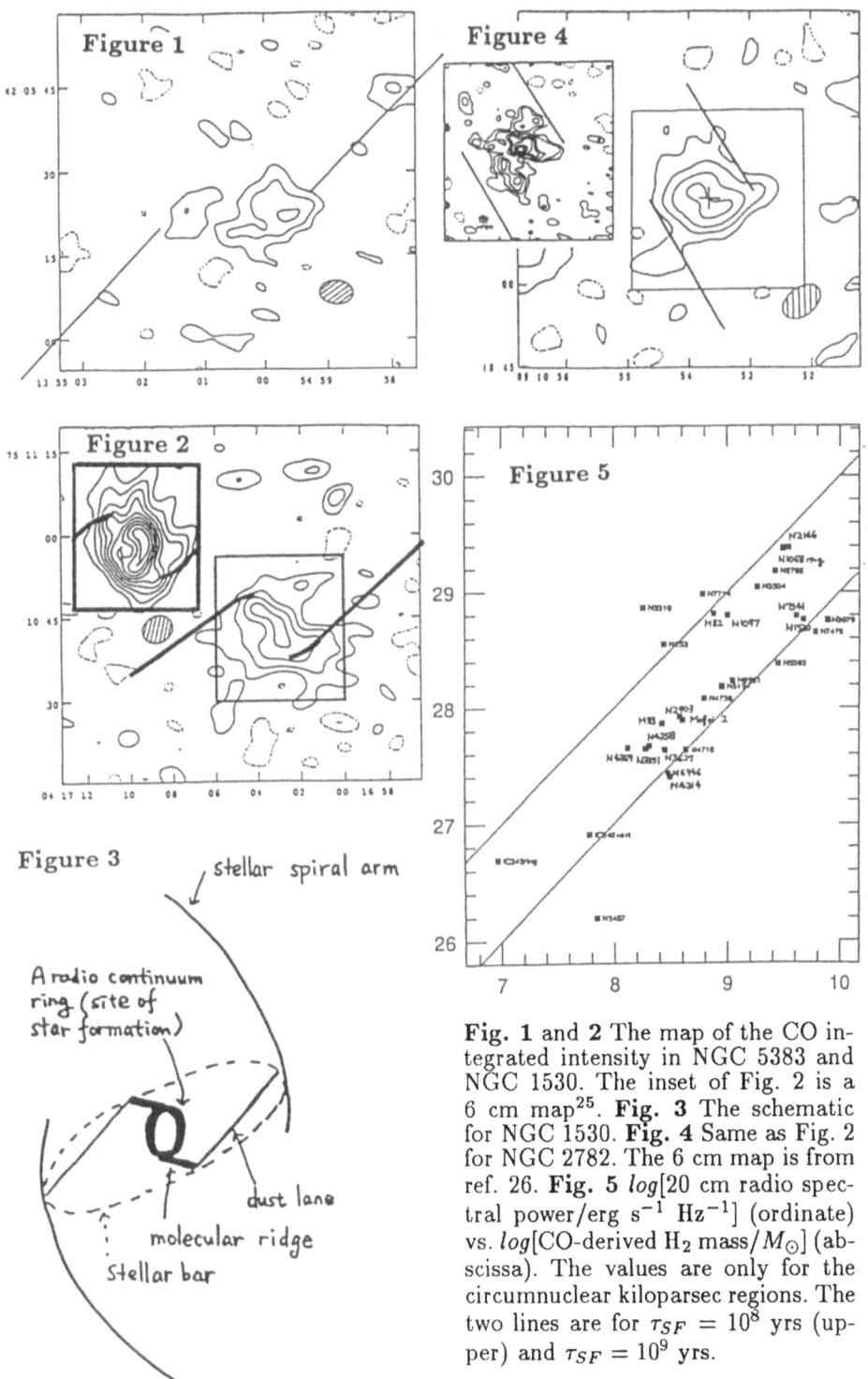

Fig. 1 and **2** The map of the CO integrated intensity in NGC 5383 and NGC 1530. The inset of Fig. 2 is a 6 cm map[25]. **Fig. 3** The schematic for NGC 1530. **Fig. 4** Same as Fig. 2 for NGC 2782. The 6 cm map is from ref. 26. **Fig. 5** $log[20$ cm radio spectral power/erg s^{-1} Hz$^{-1}]$ (ordinate) vs. $log[$CO-derived H$_2$ mass$/M_\odot]$ (abscissa). The values are only for the circumnuclear kiloparsec regions. The two lines are for $\tau_{SF} = 10^8$ yrs (upper) and $\tau_{SF} = 10^9$ yrs.

of the flux or brightness at 100 GHz are: 3 mJy per $2''.7$(25 pc)-beam and 32 mJy in $10''$(100 pc)-diameter in IC 342[21], 130 mJy per $6''.4$(100 pc)-beam and 590 mJy in $45''$(700 pc)-diameter in M 82[22], and $(0.1 - 1.7)$ mJy per $8''$(400 pc)-beam for H II regions in the disk of M 51 (calculated from ref. 23 correcting for 2-mag. extinction). Note that *millimeter recombination lines* are proposed as the tracer of compact H II regions in extragalaxies by Viallefond in this workshop[24].

The thermal dust continuum, which is optically thin, becomes a powerful measure of the interstellar medium at high angular resolutions and replace the 2.6 mm and 1.3 mm CO lines which have a problem of CO-to-H_2 conversion factor. In particular, *submillimeter wavelengths* are eagerly requested since the flux density becomes rapidly larger as one goes to the higher frequency. The estimated flux density of a giant molecular cloud of $10^6 M_\odot$ at the distance of M 51 (9.6 Mpc) are 16 mJy, 8 mJy, 1 mJy, 0.3 mJy, and 9 μJy at 800, 670, 350, 230, and 100 GHz, respectively, according to the Chicago model. In addition, the dust continuum will compensate for the Lyman continuum flux that are absorbed by the dust in H II regions when the star formation rate is measured with the thermal free-free continuum.

Typical GMCs with masses $10^{5-6} M_\odot$, sizes ~ 40 pc, and $\sigma_v \sim 4$ kms^{-1} in nearby galaxies like M 51 will be traced individually by basic molecular lines with LSA; ^{12}CO ($J = 2 \rightarrow 1$) will be used to measure the virial masses at $0''.1$-angular and 0.5 kms^{-1}-velocity resolutions: ^{13}CO and C^{18}O will be used to measure the column density since their optical depths are small: comparison with high-density tracers like HCN will shed light on the density evolution of GMCs in disks of extragalaxies. Also [C I] lines at 492 and 809 GHz will be important tracers of GMCs and ISM.

The *gas dynamics at radii 1 - 10 pc of the centers of galaxies* is the next target which succeeds my work. The LSA's angular resolution of $0''.1$ corresponds to 5 pc at a distance of 10 Mpc. Central regions of such small radii are expected that the self-gravity of the interstellar gas plays important roles in dynamics. LSA will test theoretical works and elucidate how AGNs are fueled with the ISM. The temperatures and densities are expected to high at the central regions, thus *submillimeter molecular lines* will be crucial probes.

REFERENCES

1. Ishizuki, IAU Symp 170, in press
2. Athanassoula 1993, AA, 259, 345
3. Wada 1994, PASJ, 46, 165
4. Handa et al. 1994, IAU Colloq, 140, 341
5. Cowan and Branch 1985, ApJ, 293, 400
6. Pompea and Rieke 1990, ApJ,356, 41
7. Friedli and Martinet 1993, AA, 277, 27
8. Wright et al. 1988, MN, 233, 1
9. Young et al. 1989, ApJS, 70, 699
10. Hummel et al. 1987, AA, 172, 32
11. Duric et al. 1986 ApJ, 304, 82
12. Pogge 1989 ApJS, 71, 433
13. Planeses et al. 1991, ApJ, 369, 364
14. Talent 1982, PASP, 94, 36
15. Phillips et al. 1984, MN, 210, 701
16. Terlevich et al. 1990, MN, 242, 48P
17. Kikumoto et al. 1993, AJ, 106, 466
18. Balzano 1983, ApJ, 268, 602
19. Simkin et al. 1980, ApJ, 237, 404
20. Terlevich et al. 1987, IAU Symp 121, 499
21. Downes et al. 1989, AA 262, 424
22. Carlstrom and Kronberg 1991, ApJ, 366, 422
23. van der Hulst 1988, AA, 195, 38
24. Viallefond, this workshop
25. Puxley (private communication)
26. Wilson and Willis 1980 ApJ, 240, 429

Radio Recombination Lines from External Galaxies

Francois Viallefond

Observatoire de Paris, 61 av. de l'Observatoire, F-75014 Paris, France

Abstract. With the improved sensitivity of the radiotelescopes both at centimeter and millimeter wavelengths, the study of radio recombination lines from external galaxies made an important breakthrough in the past several years. I review very briefly the recent results obtained with millimeter telescopes and show the complementarity of these measurements with lower frequency data obtained using the VLA or the ATCA to derive a full picture of the physical conditions encountered in starburst nuclei. These recent results provide some hints of what will be accessible observing radio recombination lines using new generation sensitive aperture synthesis arrays.

1 Introduction

It is almost two decades that Radio Recombination Lines (RRL) from external galaxies have been detected for the first time. These lines were detected using large single dishes at decimeter wavelengths from the starburst nuclei of M82 and NGC 253 (Shaver et al. 1977, Seaquist and Bell 1977). Because these nuclei are powerful radio continuum sources emitting synchrotron emission, stimulated emission was indeed expected from the low density ionized medium, component of the interstellar medium with a relatively large emission measure (EM $\sim 10^6$ pc cm^{-6}) and large beam filling factor (~ 0.1). Despite many efforts no RRLs from other galaxies have been detected during the 14 years following this pioneering observations. This lack of success is mostly due to the low sensitivity and the poor spectral dynamic range in the tentative experiments. This last point is critical because RRLs are broad (typical widths of 300 to 600 km s^{-1}) and, especially at low frequencies, weak relative to the continuum level (typically 1% or less). The next detection has been obtained at millimeter wavelength in the starburst nucleus of NGC 2146 (Puxley et al. 1991). A number of detections followed at centimeter wavelengths using the VLA and the ATCA and in the millimeter using both single dishes (the Nobeyama 45m antenna, the JCMT, the IRAM-30m and the SEST) and arrays (OVRO, IRAM-PdB).

2 Radio Recombination Lines observations

A compilation of the successful detections up to 1995 at centimeter and decimeter wavelengths using aperture synthesis arrays is given in Tab. 1. Table 2 reports the millimeter detections with single dishes and arrays.

M82 is the best studied galaxies with lines of quantum numbers ranging from 166 (in the L band) to 27 (in the 320 GHz band).These RRLs probe the physical

Table 1. List of galaxies detected in RRLs using aperture synthesis array at cm/dm wavelengths.

Source	RRL	Instrument	Reference
M82	H110α	VLA	Seaquist et al. 1985
M82	H166α	WSRT	Roelfsema 1987
M82	H92α,H110α	VLA	Roelfsema 1991
NGC 253	H166α,H110α,H92α	VLA	Anantharamaiah and Goss 1990
IC 694	H92α	VLA	Anantharamaiah et al. 1993
NGC 3628	H92α	VLA	Anantharamaiah et al. 1993
NGC 1365	H92α	VLA	Anantharamaiah et al. 1993
NGC 4945	H92α	ATCA	Anantharamaiah et al. 1994
Circinus	H92α	ATCA	Anantharamaiah et al. 1994
M 83	H92α	VLA	Zhao et al. 1996
Arp 220	H92α	VLA	Zhao et al. 1996
NGC 6240	H92α	VLA	Zhao et al. 1996
NGC 2146	H92α	VLA	Zhao et al. 1996

state of the ionized component of the interstellar medium under various conditions. The line strength and the line profile are affected by several effects. Slight differential departures of the equilibrium populations of the lower and upper levels may lead to population inversion and amplifications of the lines by stimulated emission. This non LTE effect is function of the principal quantum number and of the electron density. At low frequencies (large quantum numbers) the departure coefficient β_n is the largest for low densities while at high frequencies (low quantum number) it is the largest at higher densities (see e.g. Roelfsema and Goss, 1992). Selecting a frequency for the observations allow to filter a density regime in the plasma; a full picture of its physical state requires observations from meter to submilliter wavelengths. At millimeter and submillimeter wavelengths some difficulties in interpretating the observations are in principle avoided; at low densities pressure broadening can be ignored ($\Delta_{pb} \propto n^{7.4}$ where n is the quantum number) and the emission is spontaneous allowing to derive directly the the Lyc photon rate which is related to the star formation rate of the massive stars. However there is now direct evidence that a significant fraction of the millimeter emission originates from a compact HII regions leading to stimulated emission. In order to determine the contribution of stimulated emission in the total observed emission it is necessary to obtain high angular resolution images by using sensitive large aperture synthesis arrays; this allows to discrimate spatially these two components, these dense HII regions appearing in images as compact structures superposed on a diffuse large scale structure. Indeed, because the dust opacity is negligeable at millimeter wavelengths, the observation of millimeter

Table 2. List of galaxies detected in RRLs at millimeter wavelengths

Source	RRL	Instrument	Reference
M82	H40α,H53α	Nobeyama-45m	Puxley et al. 1989
NGC 2146	H53α	Nobeyama-45m	Puxley et al. 1991
M82	H30α	JCMT	Seaquist et al. 1994
NGC 4945	H36α,H42α	SEST	Anantharamiah et al. 1996
Circinus	H36α,H42α	SEST	Anantharamiah et al. 1996
NGC 253	H36α,H42α	SEST	Anantharamiah et al. 1996
NGC1365	H36α	SEST	Anantharamiah et al. 1996
M82	H26α,H27α,H29α,	JCMT	Seaquist et al. 1994
M82	H41α	OVRO	Seaquist et al. 1996
M82	H40α	IRAM-PdB	Viallefond et al. 1996

RRLs is a powerful method to probe the sites of massive star formation deeply embedded in the clouds of dusts. Since the lifetime of the compact HII regions is much shorter than the main sequence lifetime of the massive stars, the amount of stimulated emission relative to spontaneous give us some informations about the star formation process during the starburst phase.

2.1   A few illustrative observational results

With RRL measurements from decimeter up to submillimeter wavelengths, M82 is the best studied source. Figure 1 shows a position velocity image produced from IRAM-PdB observations of the H40α line which is in the 3mm band. Several features emerge above the noise in the data cube; three of them are identified in Fig. 1. The spatial coincidence (angular distance $\leq$ 1 arcsec) of these features with compact radio sources (marginaly or unresolved at 0.5 arcsec angular resolution) which have been identified on a VLA continuum image at 5 GHz. When comparing this 5 GHz image with an image at 1.4 GHz with a similar angular resolution (combination of VLA and MERLIN data, Muxlouw et al. 1994), it is found that these compact structures are sources with inverted spectra between these two frequencies (Sanders 1993). This result strongly suggests that the H40α emission detected with the IRAM-PdB array originates from compact HII regions which have densities of the order of 10^4 cm^{-3}. On the other hand with the IRAM-PdB only about 40% of the single dish line flux is observed. The missing emission must be extended as shown by single dish measurements or the OVRO data which includes shorter baselines. The 3 mm continuum image of the central part of M82 do not show much structures (see e.g. the image at about 2.5 arcsec obtained by Brouillet and Schilke 1993). To observe this population of compact HII regions in nearby ($<$ 10 Mpc) starburst nuclei in the millimeter a large sensitive array with sub-arcsecond resolution such as the LSA is required. Together with the turn-off of the continuum spectrum in the centimeter their physical properties could be well constrained.

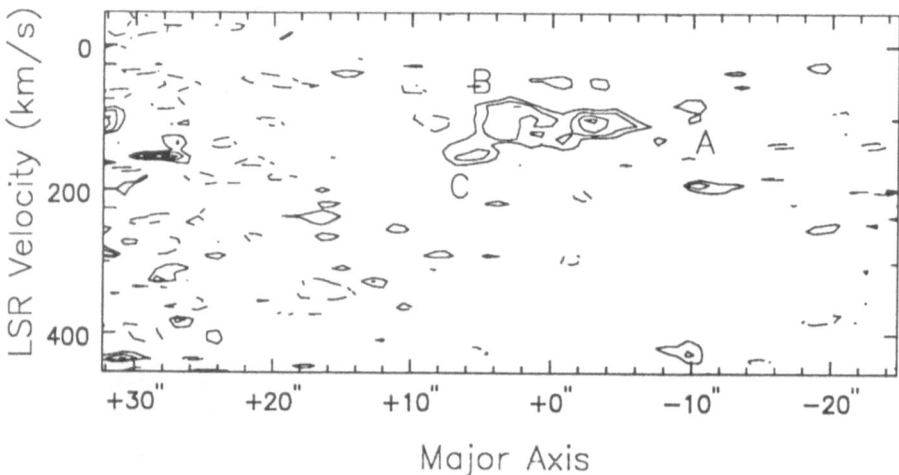

Fig. 1. H40α recombination line emission measured in M82 with the IRAM-PdB (Viallefond et al. 1996); the angular resolution is about 2.5 arcsec. A position-velocity unCLEANed image is presented along the major axis of M82. The noise level increases towards the western side due to the primary beam attenuation. The contours are -10 and 10 to 30 mJy by step of 5 mJy. The identified features A, B and C coincide with compact continuum sources and, in velocity as well as position, with dense cloudes traced by the HCN line (see text).

NGC 253 is the second best studied galaxy. To interpret the decimeter to millimeter observations, the models suggest a low density component responsible of the H166α line, mostly stimulated emission, and a high density region responsible of the H110α and H92α lines. Figure 2 shows the H36α line at 135 GHz observed using SEST (Anantharamiah et al. 1996); the line to continuum ratio is about 50 km/s corresponding to an electron temperature of 6000K if there is no stimulated emission at that frequency. The H40α to H36α line ratio is compatible with expectations if these millimeter lines originates entirely from spontaneous emission; this suggests that for NGC 253, in contrast to the case of M82, the contribution from stimulated emission in the observed millimeter line fluxes may not be important.

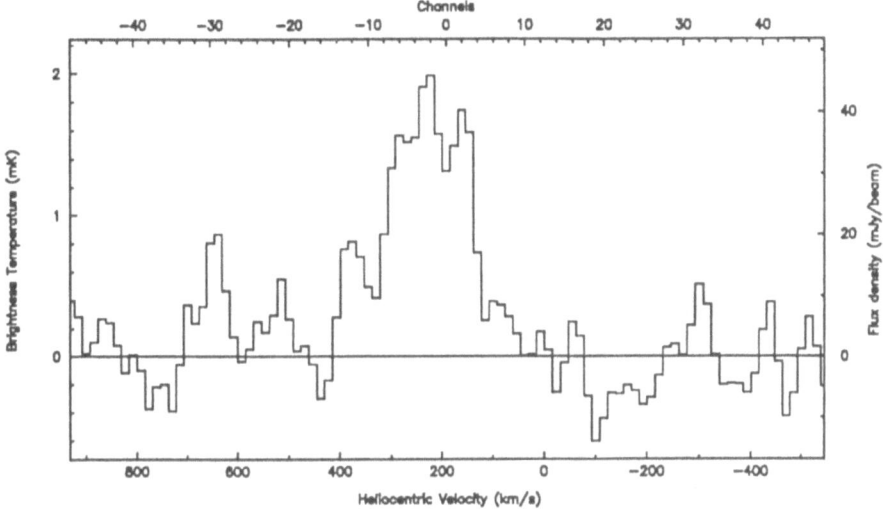

Fig. 2. H36α recombination line emission measured with the SEST (Anantharamiah et al. 1996). The temperature scale is in T_A^*. The continuum emission, at a level of 180 mJy, has been subtracted. The integration time to obtain this spectrum was 9.2 hours

3 Discussion

For the interpretation of the observed RRLs from starburst nuclei at centimeter wavelengths it has been postulated as a reasonable model the presence of a population of compact HII regions, much of the line emission coming from internal stimulated emission due to the continuum generated within these HII regions (Anantharamaiah et al. 1993). A similar model has been considered by Puxley et al. (1991) to explain the millimeter H53α line observed in the galaxy NGC 2146. For M82 we have now a direct evidence for the presence of these compact HII regions. The observation of both the dm/cm and mm RRLs are required to get a complete picture of the physical state of the ionized gas. The long wavelength lines originate from a low densy ($n_{e,rms}$ 1 to 30 cm^{-3}) extended medium and are stimulated by the extended synchrotron disk. In the millimeter high angular resolution (sub-arcsec) observations are required to discrimate between the stimulated line emission in the dense HII regions and the spontaneous emission from the lower density diffuse disk. Detailed studies for a number of nearby starburst nuclei are required to better understand how proceeds the star formation during these bursts. With a sensitive array such as the LSA statistical studies become feasible. Comparisons with images of the molecular component at this sub-arcsec resolution would allow to search for a link between the very young star forming regions and their parent neutral clouds out of which they form. For the RRLs projects a spectral resolution of ~ 20 km s^{-1} with a total bandwidth of the order of 600 km s^{-1} is necessary. In the millimeter, with a typical line to continuum ratio of 20% , a good spectral dynamical range is also required although this is less critical compared to the centimeter and decimeter where this ratio is much lower ($\leq 1\%$) and to the submillimeter because of the rapid rise of thermal continuum from dust. The RRLs from the compact M82 HII regions in the nuclear region of M82 could be detected in 1 hour with a S/N > 10 at 0.1 arcsec resolution in the 1 mm band with a LSA type array. These compact sources could be even imaged in the continuum at 0.03 arcsec with 10 km maximum baseline. For M82 or NGC 253 it would be necessary to mosaic and include total power measurements to the array data in order to image the extended emission; this is necessary for models which include both long and short wavelengths array data. Using a LSA type array with a single pointing and a full synthesis (12h) would be well adapted to study starburst nuclei up to distances of ~ 10 Mpc. A multi-wavelentgh capability would be also very useful to get the line ratios and the line to continuum ratio in the millimeter and sub-millimeter. Observations at 2 mm are the most suitable for the continuum since it is the spectral band where any contamination by synchrotron emission and by thermal dust emission are the lowest. In the submillimeter the emission could be stimulated by the background thermal emission from dust, e.g. in HII regions in front of dust clouds. The most distant starburst nucleus detected in RRLs so far is Arp 220. The strong infrared emission makes possible the detection of stimu-lated RRLs out to large distances to determine the physical conditions (electron density and temperature) in regions completely obscured in the visible.

References

Anantharamiah, Goss, W.M. (1990): *Radio Recombination Lines: 25 Years of Inves-tigations* eds M.A. Gordon and R.L. Sorochenko, (Kluwer, Dordrecht), p. 267

Anantharamiah, K.R., Zhao, Jun-Hui, Goss, W.M., Viallefond, F. (1993): ApJ 419, 585

Anantharamiah, K.R., Zhao, Jun-Hui (1994): in preparation

Anantharamiah, Goss, W.M., Viallefond, F., Zhao, Jun-Hui, (1996): in preparation

rouillet, ; N., Schilke, P. (1993) Brouillet, ; N., Schilke, P. (1993): A&A 277, 381

Muxlow, T.W.B., Pedlar, A., Sanders, E.M. (1994): *The formation of the Milky Way* (Joint workshop IAA(CSIC) - IAC - University of Pisa, Sep 1994, Granada, Spain)

Puxley, P.J., Brand, P.W.J.L., Moore, T.J.T., Mountain, C.M., Nakai, N., Ya-mashita, T. (1989): ApJ 345, 163

Puxley, P.J., Brand, P.W.J.L., Moore, T.J.T., Mountain, C.M., Nakai, N. (1991): MNRAS 248, 585

Roelfsema, P.R. (1987): Ph.D. Thesis, University of Groningen, The Netherlands

Roelfsema, P.R., Seaquist, E.R. (1991): unpublished

Roelfsema, P.R., Goss, W.M. (1992): A&ARev 4, 161

Sanders, E.M. (1993): *MSc thesis, University of Manchester*

Seaquist, E.R., Bell, M.B. (1977): A&A 60, L1

Seaquist, E.R., Bell, M.B., Bignell, R.C. (1985): ApJ 294, 546

Seaquist, E.R., Kerton, C.R., Bell, M.B. (1994): ApJ 429, 612

Seaquist, E.R., Carlstrom, J.E., Bryant, P.M., Bell, M.B. (1994): Submitted

Shaver, P.A., Churchwell, E., Rots, A.H.. (1977): A&A 55, 435

Viallefond, F., Anantharamiah, K., Pedlar, A. (1996): in preparation

Zhao, Jun-Hui, Anantharamiah, K.R., Goss, W.M., Viallefond, F. (1996): submitted
  to ApJ

Molecular gas in early-type galaxies

C. Henkel[1], T. Wiklind[2], F. Wyrowski[3] and F. Combes[4]

[1] Max-Planck-Institut für Radioastronomie, Auf dem Hügel 69, D-53121 Bonn, Germany
[2] Onsala Space Observatory, S-43992 Onsala, Sweden
[3] Universität zu Köln, I. Physikalisches Institut, Zülpicher Str. 77, D-50937 Köln, Germany
[4] DEMIRM, Observatoire de Paris, 61 avenue de l'Observatoire, F-75014 Paris, France

Abstract. CO observations have shown that many early-type galaxies contain a significant amount of cool dense gas. Here we briefly summarize recent results related to this interstellar gas component and outline scientific perspectives of a large interferometric facility operating at millimeter wavelengths.

1 Introduction

For more than half a century, it has been known that at least some early-type galaxies are not gas-free inert stellar systems but contain cold dust and warm ionized gas. This was inferred from the presence of dust lanes (Hubble 1936) and the detection of the [OII]$\lambda3227$ emission line (Mayall 1936, 1939). Interestingly, the detection of molecular hydroxyl (OH) and formaldehyde (H_2CO) absorption toward the elliptical galaxy Cen A (Gardner and Whiteoak 1976a,b, 1979) preceeds the first successful CO measurements by almost a decade. CO was observed by Verter (1985) and Stark et al. (1986) toward the early-type disk galaxies NGC 4438 and NGC 7371, by Wiklind and Rydbeck (1986) toward the dwarf elliptical system NGC 185, and by Phillips et al. (1987) toward Cen A. These CO detections provided the motivation for a large number of follow-up observations while the success of the Infrared Astronomical Satellite (IRAS) lead to a drastic improvement in our understanding of the interstellar dust component (e.g. Jura et al. 1987; Bally and Thronson 1989; Knapp et al. 1989). A detailed review summarizing the properties of the cool interstellar medium and its relation with other stellar and interstellar components is given by Henkel & Wiklind (1996).

There are several kinds of early-type galaxies. The 'latest' early-type systems are the lenticular or S0 disk galaxies which do not possess spiral arms. Among the nominally diskless systems we find giant, intermediate, and 'classical dwarf' or 'compact' ellipticals (summarized hereafter by 'E'), 'diffuse' or 'bright' dwarf ellipticals (hereafter 'dE') and 'extreme dwarf ellipticals' or 'dwarf spheroidal systems' (dSph). E and dE galaxies are morphologically distinct: E galaxies are more compact, show a higher optical surface brightness and populate a 'fundamental plane' in the three dimensional space defined by the central projected velocity dispersion, the 'core' or 'half-light' radius, and its associated surface brightness (cf. Wirth and Gallagher 1984; Binggeli et al. 1988; Kormendy and Djorgovski 1989; Bender et al. 1992). The dE and dSph galaxies appear to be more closely related to dwarf irregulars than to 'normal' ellipticals or to the bulges of disk galaxies (e.g. Kormendy 1987).

While the principal differences between early-type systems are well defined, it is often difficult to classify a given object. Even if we have a clearcut case, however, galaxy catalogs are not always consistent. Perhaps the most prominent case is the prototypical minor axis dust lane elliptical Cen A that is classified as a lenticular galaxy in the RC3 (de Vaucouleurs et al. 1991). There might also be a connection between dwarf irregular and dwarf elliptical galaxies, which can be studied via the molecular properties.

2   CO data from the last decade

CO J=1–0 or 2–1 line emission provides a unique tool to determine mass and kinematics of the low density molecular gas that is, in most sources, comprising the bulk of the molecular material. So far about 80 early-type galaxies have been detected in CO. IRAS selected samples were measured to maximize detection rates. The three most comprehensive CO line surveys for lenticular galaxies, those of Sage and Wrobel (1989), Thronson et al. (1989), and Wiklind and Henkel (1989) published almost simultaneously and agree in their main conclusions. Most of the molecular data from elliptical galaxies have been obtained from Lees et al. (1991) and Knapp & Rupen (1996) in the CO 2–1 and by Wiklind et al. (1995) in the CO 1–0 and 2–1 transitions. Some Blue Compact Dwarfs with elliptical outer isophotes, potentially dwarf elliptical systems, have been detected in CO by Sage et al. (1992). The following main results have to be mentioned:

- Molecular data from compact ellipticals (e.g. M 32) or dwarf spheroidals (e.g. the Sculptor system) have not yet been published, but far infrared data suggest that these classes of galaxies have little molecular gas. CO is reported from S0, E, and dE galaxies, with luminous ellipticals being seemingly more difficult to detect than dwarf ellipticals.
- The correlation between L_{FIR} and M_{H_2} (deduced from a standard N_{H_2}/I_{CO} conversion factor) is within $\pm$ an order of magnitude consistent with that in spirals. This implies, that most of the far infrared emission arises from dust.
- Early-type galaxies of a given blue luminosity exhibit weaker CO emission than spirals. There is a gradual decrease of CO luminosity relative to L_B when going from Sa to E galaxies, with CO intensities about a factor of ten weaker in lenticulars than in spirals. Furthermore, there seems to be a transition from a gaussian L_B/M_{H_2} distribution in lenticular to a power law distribution in elliptical galaxies. The lack of coupling of molecular (CO luminosity) and stellar (blue luminosity) components in elliptical galaxies suggests an external origin for the cool dense gas.
- The presence of large molecular clouds with normal star formation efficiencies in systems without spiral arms demonstrates that spiral structure is largely

irrelevant to the star formation efficiency and the molecular gas consumption time. Also, spiral arms are not needed to form large molecular clouds or complexes.

3 Individual sources

After having analysed some important global properties of early-type galaxies, we will provide brief sketches of the only two ellipticals which have so far been mapped in a molecular transition: NGC 759 and Cen A.

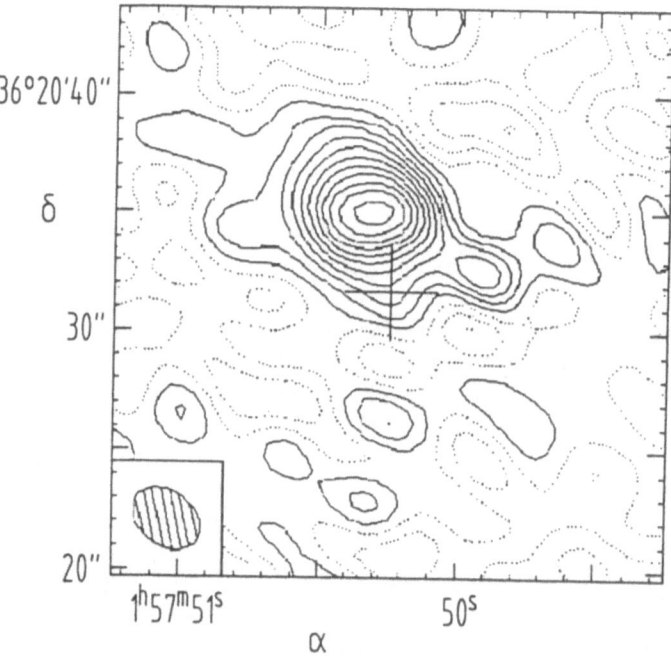

Fig. 1. Spatial distribution of the integrated CO emission from NGC 759, adding up all channels between –237 and +185 kms relative to $cz = 4665 \, \mathrm{km \, s^{-1}}$. The contour spacing is $0.8 \, \mathrm{Jy \, km \, s^{-1}}$ per beam.

3.1 A giant elliptical at a moderate distance: NGC 759

NGC 759, a moderately distant ($vz \sim 4665 \, \mathrm{km \, s^{-1}}$) giant elliptical, is a so far little known IRAS point source in the outer parts of the A 262 cluster. For an early-type galaxy, NGC 759 shows an unusual spectrum with strong Balmer absorption and emission lines (e.g. Vigroux et al. 1989). After CO emission, corresponding to an H_2 mass of $2 \times 10^9 \, M_\odot$, was detected by Wiklind et al. (1995), we have made high angular resolution measurements ($3\overset{''}{.}1 \times 2\overset{''}{.}3$, natural weighting;

this corresponds to 1000 × 700 pc at a distance of 65 Mpc) with the Plateau de Bure interferometer. Fig. 1 shows the spatial distribution of the integrated CO emission. The emission arises from an area which is slightly larger than the nominal beam. The extent at the 50% level, $5''.4 \times 4''.3$, has a major/minor axis ratio similar to that of the beam which may be a coincidence or an indication that the synthesized beam is larger than estimated. Channel maps indicate strongest CO emission at $\pm 185\,\mathrm{km\,s^{-1}}$ from the systemic velocity. The two peaks are spatially separated by a few arcsec. Using this kinematical information, it is possible to model the CO distribution in terms of a ring surrounding a central region devoid of CO. The gas surface density would then be a few $100\,\mathrm{M_\odot\,pc^{-2}}$. Such a scenario is, however, only indirectly inferred from the observational data. Measurements directly resolving the spatial distribution of the molecular gas are still needed.

NGC 5128

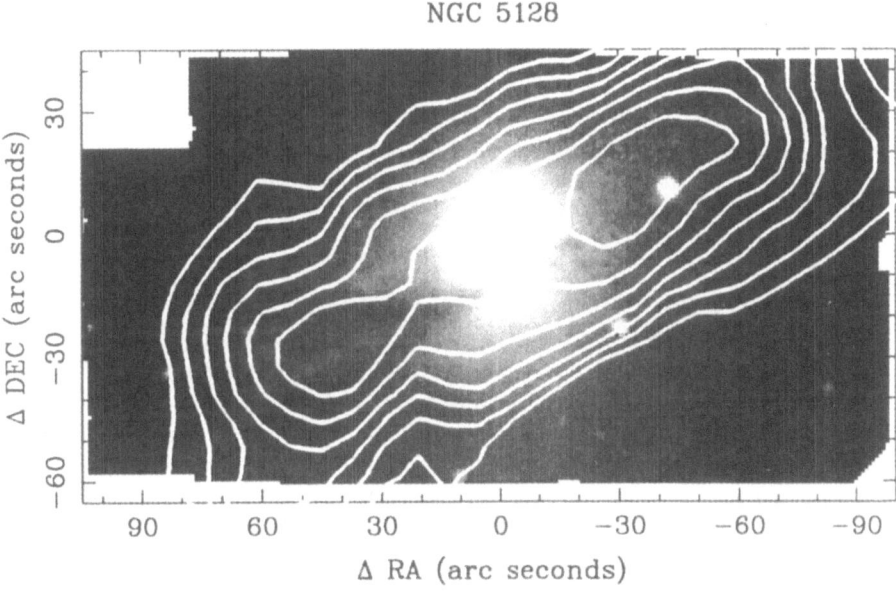

Fig. 2. Contour map of integrated ^{12}CO $J=1-0$ emission in Cen A (from Eckart et al. 1990a; lowest contour: $17.5\,\mathrm{K\,km\,s^{-1}}$, contour interval: $5\,\mathrm{K\,km\,s^{-1}}$) superposed on a K-band ($2.16\,\mu$m) mosaic showing the range 14.0–16.5 mag per arcsec^2 (from Quillen et al. 1993).

3.2 The nearby southern giant elliptical radio galaxy Cen A

The powerful elliptical radio galaxy Cen A, a member of the nearby Centaurus cluster, has already been studied extensively from radio to X-ray wavelengths.

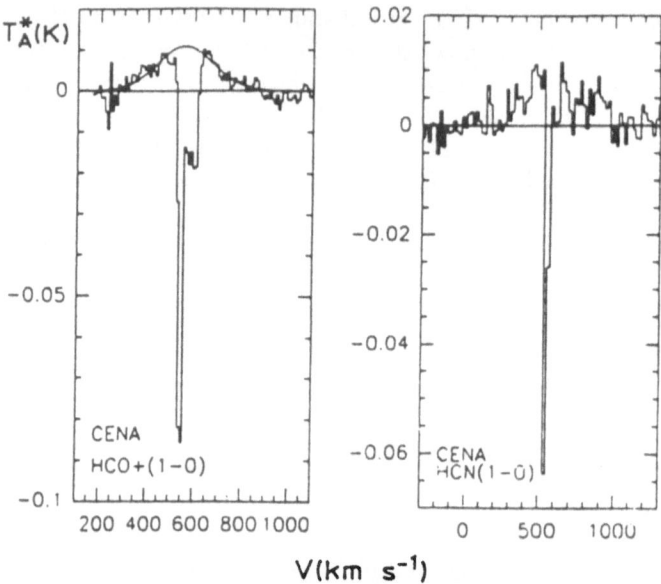

Fig. 3. Cen A HCN and HCO$^+$ spectra showing the broad nuclear emission and narrow foreground absorption components (from Israel 1992). Molecular excitation suggests that the latter spectral component is not associated with the nuclear region (Eckart et al. 1990b).

At a distance of a few Mpc only, it is the nearest and most suitable target for studies of the detailed dynamical, morphological, physical, and chemical state of the dense cool interstellar medium of a giant elliptical. While in the optical and near infrared, high resolution HST images have been obtained (Schreier et al. 1996), the center of the galaxy remains highly obscured. At least 14 molecular species have already been detected in up to a total of four distinct interstellar components:

- Narrow ($\Delta V < 10\,\mathrm{km\,s^{-1}}$) CO line emission follows the optically visible dust lane (see Fig. 2). The CO distribution, the CO line shapes, the optical appearance of the dust lane, and HI and near infrared data are well fitted by a warped disk ($r \sim 3\,\mathrm{kpc}$) in a prolate gravitational potential (see Quillen et al. 1992).
- Broad ($\Delta V > 100\,\mathrm{km\,s^{-1}}$) emission originates from the circumnuclear cloud(s) (see Fig. 3). This component is weak, because the size of the nuclear disk is much smaller than the primary beam of a 15-m single-dish telescope. Rydbeck et al. (1993) find a diameter of 15″ for the nuclear disk (see Fig. 4), but the deduction is somewhat indirect involving the use of Statistical Image Analysis to improve the angular resolution of the data.
- If the frequency resolution is high, three systemic velocity lines at $V_{\mathrm{LSR}} = 539$, 544, and $550\,\mathrm{km\,s^{-1}}$ can be distinguished that are usually seen in absorption. While being located along the line of sight toward the central

continuum source, the low degree of molecular excitation indicates a gas component that is not associated with the circumnuclear region (cf. Eckart et al. 1990b; Seaquist and Bell 1990).

– Red-shifted absorption lines ($570\,km\,s < V < 620\,km\,s^{-1}$) are observed in a variety of molecular species toward the nuclear continuum source. This gas *might* be falling toward the nucleus. Abundance ratios are consistent with both Galactic absorption lines (Liszt & Lucas 1994) and absorption lines seen at redshifts z=0.25–0.89 (cf. Wiklind & Combes, these proceedings).

The extended narrow emission line component has been completely mapped and its physical properties are roughly known. Our knowledge of the other components is however much poorer. In particular the nuclear disk of Cen A (see Fig. 4) deserves follow-up studies. Thus both NGC 759 and Cen A will provide interesting targets for high resolution studies with a large millimeter-wave interferometer.

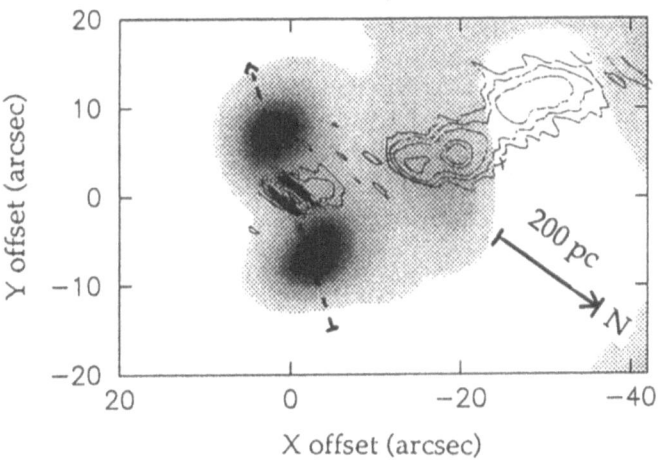

Fig. 4. Deconvolved CO 2–1 emission line intensities toward the Cen A nuclear continuum source. The angular resolution of the map was improved by the application of a Statistical Image Analysis technique. Radio contours are given as solid lines (from Rydbeck et al. 1993).

4 Future perspectives

With a sensitive array of mm-wave telescopes, many open problems related to the cool gas and dust components of early-type galaxies could be attacked. Among the various types of large early-type galaxies, no molecular detection of compact ellipticals and only a few molecular detections of giant elliptical galaxies have yet been obtained. Note that most of the prominent giant ellipticals, most notable

Cen A, are located in the southern hemisphere. The following problems may deserve some attention:

- What is the detailed distribution of the dust and gas in early-type galaxies? So far, only two elliptical galaxies has been investigated in some detail. Are there many more ellipticals with a compact nuclear source of cool dense gas and dust similar to those seen in Cen A and NGC 4261 (cf. Israel 1992; Möllenhoff & Bender 1987; Jaffe et al. 1993, 1996; Jaffe & McNamara 1994)?
- N_{H_2}/I_{CO} conversion factors have not yet been determined for any lenticular or elliptical galaxy. As for spiral galaxies, not only ^{12}CO but also ^{13}CO, $C^{18}O$, and 1 mm continuum data are needed in order to estimate molecular column densities by several independent methods and to compare the results. Are the N_{H_2}/I_{CO} conversion factors in large early-type galaxies consistent with those in spirals? And are the N_{H_2}/I_{CO} factors in dust lanes at large galactocentric radii different from those closer to the nuclear region? Such a difference appears to exist in the Galaxy and in some nearby spirals (Dahmen 1995; Mauersberger et al. 1996a,b).
- The study of decoupled molecular gas in early-type galaxies is still in its infancy. What is its detailed kinematical state, its excitation, chemistry, and star forming rate? Interesting insights into astrophyical phenomena are provided by the interface between the decoupled gas and the co-rotating interstellar component(s). In addition, we need more observational results which allow a detailed kinematical modelling of the data (see Quillen et al. 1992, 1993 for Cen A).
- The study of the high density gas in compact molecular cores or near the nuclear region is presently confined to only a few elliptical or lenticular galaxies. A thorough study of sources like NGC 759 with its compact molecular core or detailed mapping of the molecular disk or torus near the center of Cen A would be important to elucidate the nuclear properties of early-type galaxies and to compare the results with those obtained for spirals. Detailed observations of molecular gas in the central regions of giant elliptical galaxies may be the best way to determine the mass distribution and, possibly, to study the infall mechanism of matter into a central super massive black hole. This might be related to megamaser sources, which are believed to be associated with active galactic nuclei (Braatz et al. 1994, 1996; Koekemoer et al. 1995). Searches for other molecular maser transitions or for absorption lines (quasithermal emission lines will be too weak to be detectable in such spatially confined regions) would greatly enhance our knowledge on the compact, optically highly obscured nuclear regions of AGNs.

CO line and (sub)mm-wave continuum surveys of dwarf elliptical galaxies are also urgently needed to enlarge the so far small sample of detected sources and to start projects which are not only astronomically but also astrophysically worthwhile. If dwarf ellipticals evolve in a similar way as other dwarf galaxies, it would be desirable to measure:

- the distribution of the H_2 component of such galaxies in the Blue Compact Dwarf (BCD) stage of evolution where they undergo a brief 'burst' of star formation. Detailed comparisons of HI, CO, and $H\alpha$ emission provide information on the extent of the star forming region(s) and would delineate the boundary between the neutral atomic and the neutral molecular gas. In addition, conversion factors between CO intensity and H_2 column density could be derived.
- the distribution of the dust and H_2 component in Red Compact Dwarfs (RCDs). The molecular gas is believed to be built up during long quiescent periods between the rare bursts of star formation (the BCD phase). Thus searches for CO and dust continuum in RCDs are needed to verify and follow evolutionary models (see e.g. Gerola et al. 1980).
- a variety of molecular transitions in some selected dwarfs of low metallicity. To date, such studies are possible only for the Magellanic Clouds (Johansson et al. 1994; Chin et al. 1996a,b) but it remains to be seen whether these nearby galaxies are typical.
- intergalactic molecular clouds or complexes in groups of interacting galaxies. Such clouds might form dwarf galaxies in a later stage of evolution. To date it is not known how common such theoretically predicted objects (e.g. Barnes & Hernquist 1991) actually are. Furthermore, no detailed molecular study could yet be obtained for the single known source belonging to this class of objects (see Brouillet et al. 1992; Henkel et al. 1993). Because of observational time constraints, systematic searches for more such souces will only be feasible with a sensitive array and a large primary beam.

References

Bally, J., Thronson, H.A. (1989): AJ 97, 69

Barnes, J.E., Hernquist, L. (1992): Nat. 360, 715

Bender, R., Burstein, D., Faber, S.M. (1992): ApJ 399, 462

Binggeli, B., Sandage A., Tammann, G.A. (1988):, Ann. Rev. Astr. Ap. 26, 509

Braatz, J.A., Wilson, A.S., Henkel, C. (1994): ApJ 437, L99

Braatz, J.A., Wilson, A.S., Henkel, C. (1996): ApJ Suppl., in press

Brouillet, N., Henkel, C., Baudry, A. (1992): A&A 262, L5

Chin, Y.-N., Henkel, C., Millar, T., Whiteoak, J.B. (1996a): A&A, submitted

Chin, Y.-N., Henkel, C., Whiteoak, J.B., Millar, T.J., Hunt, M., Lemme, C. (1996b): A&A, in press

Dahmen, G. (1995): The large Scale Distribution of Molecular Gas in the Galactic Center Region from $C^{18}O$ and the Nitrogen Isotope Abundance in the Galaxy, Ph.D. Thesis, Bonn

de Vaucouleurs, G., de Vaucouleurs, A., Corwin, H.G., Buta, R.J., Paturel, G., Fouqué, P. (1991): Third Reference Catalogue of Bright Galaxies (RC3), Springer-Verlag, New York

Eckart, A., Cameraon, M., Rothermel, H., Wild, W., et al. (1990a): ApJ 363, 451

Eckart, A., Cameron, M., Genzel, R., Jackson, J.M., Rothermel, H., Stutzki, J. (1990b): ApJ 365, 522

Gardner, F.F., Whiteoak. J.B. (1976a): MNRAS 175, 9p

Gardner, F.F., Whiteoak, J.B. (1976b): Proc. Astr. Soc. Austr. 3, 63

Gardner, F.F., Whiteoak, J.B. (1979): MNRAS 189, 51p

Gerola, H., Seiden, P. E., Schulman, L.S. (1980): ApJ 242, 517

Henkel, C., Stickel, M., Salzer, J.J., Hopp, U., Brouillet, N., Baudry, A. (1993): A&A 273, L15

Henkel, C., Wiklind, T. (1996): Sp. Sci. Rev., submitted

Hubble, E. (1936): The Realm of the Nebula, Dover, New York

Israel, F.P. (1992): A&A 265, 487

Jaffe, W., Ford, H.C., Ferrarese, L., van den Bosch, F., O'Connell, R.W. (1993): Nat. 364, 213

Jaffe, W., McNamara, B.R. (1994): ApJ 434, 110

Jaffe, W., Ford, H.C., Ferrarese, L., van den Bosch, F., O'Connell, R.W. (1996): ApJ 460, 214

Johansson, L.E.B., Olofsson, H., Hjalmarson, Å, Gredel, R., Black, J.H. (1994): A&A 291, 89

Jura, M., Kim, D.-W., Knapp, G.R., Guhathakurta, P. (1987): ApJ 312, L11

Knapp, G.R., Guhathakurta, P., Kim, D.-W., Jura, M. (1989): ApJ Suppl. 70, 329

Knapp, G.R., Rupen M.P. (1996): ApJ 460, 271

Koekemoer, A., Henkel, C., Greenhill, L.J., Dey, A., van Breugel, W., Codella, C., Antonucci, R. (1995): Nat 378, 697

Kormendy, J. (1987): Nearly Normal Galaxies: From the Planck Time to the Present, ed. S.M. Faber, Springer-Verlag, New York, p163

Kormendy, J., Djorgovski, S. (1989): Ann. Rev. Astr. Ap. 27, 235

Lees, J.F., Knapp, G.R., Rupen, M.P., Phillips, T.G. (1991): ApJ 379, 177

Liszt, H.S., Lucas, R. (1994): ApJ 431, L131

Mauersberger, R., Henkel, C., Whiteoak, J.B., Chin, Y.-N., Tieftrunk, A.R. (1996a): A&A, in press

Mauersberger, R., Henkel, C., Wielebinski, R., Wiklind, T., Reuter, H.-P (1996): A&A 305, 421

Mayall, N.U. (1936): Pub. Astr. Soc. Pac. 48, 14

Mayall, N.U. (1939): Pub. Astr. Soc. Pac. 51, 282

Möllenhoff, C., Bender, R. (1987): A&A 174, 63

Phillips, T.G., Ellison, B.N., Keene, J.B., Leighton, R.B., Howard, R.J., Masson, C.R., Sanders, D.B., Veidt, B., Young, K. (1987): ApJ 322, L73

Quillen, A.C., de Zeeuw, P.T., Phinney, E.S., Phillips, T.G. (1992): ApJ 391, 121

Quillen, A.C., Graham, J.R., Frogel, J.A. (1993): ApJ 412, 550

Rydbeck, G., Wiklind, T., Cameron, M., Wild, W., Eckart, A., Genzel, R., Rothermel, H. (1993): A&A 270, L13

Sage, L.J., Salzer, J.J., Loose, H.-H., Henkel, C. (1992): A&A 265, 19

Sage, L.J., Wrobel, J.M. (1989): ApJ 344, 204

Schreier, E.J., Capetti, A., Macchetto, F., Sparks, W.B., Ford, H.J. (1996): ApJ, in press

Seaquist, E.R., Bell, M.B. (1990): ApJ 364, 94

Stark, A.A., Knapp, G.R., Bally, J., Wilson, R.W., Penzias, A.A., Rowe, H.E. (1986) ApJ 310, 660

Thronson, H.A., Tacconi, L., Kenney, J., Greenhouse, M.A., Margulis. M., Tacconi-Garman, L., Young, J.S. (1989): ApJ 344, 747

Verter, F. (1985): ApJ Suppl. 57, 261

Vigroux, L., Boulade, O., Rose, J. (1989): AJ 98, 2044

Wiklind, T., Combes, F., Henkel, C. (1995): A&A 297, 643

Wiklind, T., Henkel, C. (1989): A&A 225, 1
Wiklind, T., Rydbeck, G. (1986): A&A 164, L22
Wirth, A., Gallagher, J.S. (1984): ApJ 282, 85

Magellanic Cloud Studies with Large mm Arrays

Frank P. Israel

Sterrewacht Leiden, Postbus 9513, 2300 RA Leiden, the Netherlands

Abstract. Prospects for millimetre observations of the Magellanic Clouds with arrays of large collecting area (typically 10 000 m^2) are discussed. Subparsec resolutions can already be obtained with compact array configurations. The most important application of such an array, a comparative study of dense (molecular) clouds in the low-metallicity, high radiation-density environment of the Cloud and Galactic clouds observed at identical linear resolutions, is reviewed in some detail. In addition, the use of array observations to study circumstellar shells around late-type stars, (SiO) masers, and mass loss of luminous early-type stars is briefly discussed.

1 Importance of the Magellanic Clouds

The Magellanic Clouds are the nearest extragalactic systems at distances of 53 kpc (LMC) and 63 kpc (SMC) respectively, an order of magnitude closer than other members of the Local Group of galaxies such as M 31, M 33 and NGC 6822 and more than fourty times closer than other major galaxies.

Because of their extragalactic nature, we command a full view of the Magellanic Clouds allowing the construction of complete samples of objects, all at essentially the same distance. Despite their extragalactic nature, their vicinity adds to this the advantage of linear resolutions unparalleled in the field of extragalactic astronomy. For example, 1″ corresponds to a linear dimension of 0.25 pc in the LMC and 0.30 pc in the SMC respectively. The Large Magellanic Cloud in particular extends a large angle of 8° (7.5 kpc) on the sky; the Small Magellanic Cloud extends over 3.5° (3.8 kpc). Statistical studies therefore generally require either a large field of view or a large number of individual pointings. For instance, a million SEST 115 GHz J=1–0 CO beams (40″) are required to fully cover the LMC. If we would integrate for only 5 minutes on each pointing for 6 hours per day, we would need 35 years to complete our survey. These numbers equally apply to an array consisting of SEST-like (15 m diameter) dishes. Thus, efficient and succesful use of mm arrays in Magellanic Cloud studies requires lower-resolution 'pathfinder' surveys, such as the 'Mini' and SEST CO surveys (Cohen et al. 1988; Israel et al. 1993; Rubio et al. 1993; 1996; Kutner et al. 1996) and the IRAS surveys (Schwering & Israel 1990).

In some respects, studies of Magellanic Cloud objects offer possibilities superior to those of similar objects in the Milky Way galaxy. Although the Milky Way will always offer linear resolutions one to two orders of magnitude better, our location inside it hampers a good overview and frequently introduces distance ambiguities absent in Magellanic Cloud observations. More importantly, Magellanic Cloud studies are complementary to those of Milky Way objects. Ambient radiation fields in the Magellanic Clouds are one to ten times stronger than those in the Solar Neighbourhood (cf. Schwering 1988), while at the same time LMC and SMC metallicities are only about 30 per cent and 10 per cent respectively of the local Galactic value (Dufour 1984). Thus, comparative studies of objects in the very different environments of the Milky Way, the LMC and the SMC enable us to determine the important effects of environment on the properties of these objects. In this respect, the Magellanic Clouds are *unique and essential*: neither the Milky Way by itself, nor the more distant other galaxies present such a large range of environmental conditions observable with sufficiently high linear resolution. This is especially important for studies of the interstellar medium which by its nature is extremely susceptible to the effects of e.g. radiation fields and metallicities. Stellar astronomers have long recognized the Magellanic Clouds for the unique laboratory they are; now it is the turn of the interstellar astronomers.

2 Millimetre Array Requirements and Capabilities

Array observations of the Magellanic Clouds have the invaluable advantage that, within limits, the linear resolution in the Clouds can be matched to that of Galactic counterparts observed with present-day instruments by choosing the appropriate array configuration. In the same way, observations at different frequencies can be performed with identical resolutions.

In order to perform comparative studies, we wish to be able to observe Magellanic Cloud sources with *linear* resolutions comparable to those of Galactic objects observed with present-day millimetre telescopes. Single dish instruments now provide *angular* resolutions of typically $20''$, while mm interferometers are getting down to $1''$. Nearby Galactic objects (distance 300 pc) thus require Magellanic Cloud counterparts observed with angular resolutions reaching $0.1''$ - $0.005''$ translating into maximum array baselines B_{max} of 3 to 60 km at 230 GHz, while Galactic objects in the ten times more distant Perseus Arm relax the angular resolution requirement to $1''$ - $0.05''$ or maximum baselines of 0.3 to 6 km, again at 230 GHz. However, increasing the resolution by going to longer baselines at constant collecting area results in increasingly poor (u,v) plane filling. While resolution increases as B_{max}, flux-per-beam noise ΔS remains constant but temperature noise ΔT_b increases as B_{max}^2. For a more detailed discussion of millimeter interferometer techniques we refer to Downes 1989.

As a consequence, increasing resolution rapidly becomes a losing proposition for resolved sources. In the following discussion, we will explore the possibilities offered by a large mm array by basing ourselves on the LSA concept subject of this meeting. For the Magellanic Clouds (adopted mean distance 60 kpc), the linear resolutions and array sensitivities given as a function of maximum baseline B_{max} on page 9 of the LSA report (Downes et al. 1995) are modified to the values given in Table 1. Sensitivities assume dual-polarization receivers operating at a system temperature $T_{sys} = 100$ K (including the sky) at a frequency of 230 GHz ($\lambda = 1.3$ mm) and an effective collecting area $A = 10^4$ m^2. The r.m.s. noise in line measurements refers to a channel width of 1 km s^{-1} ($\Delta\nu = 0.77$ MHz at 230

GHz) and an 8-hour integration; it marks the surface brightness sensitivity to an extended source. Continuum noise refers to a total bandwidth of 2 GHz and a shorter integration time of one hour. It marks the sensivity to a point source.

Table 1. Millimetre Array Capabilities (Adapted from LSA Report Oct 95)

Baseline	Beam Size	Linear Resolution at 60 kpc	Sensitivity	
			Line $1\sigma/8$ hrs 1 km s^{-1} Band	Continuum $1\sigma/1$ hr 2 GHz Band
(km)	('')	(pc)	(K)	(mJy)
0.3	1	0.3	0.005	0.02
1	0.3	0.09	0.05	0.02
3	0.1	0.03	0.5	0.02
10	0.03	0.009	5	0.02

Table 1 shows that even a compact array configuration already provides subparsec resolutions in the Magellanic Clouds, simultaneous with very good line sensitivities. For extended sources, the signal-to-noise ratio drops rapidly with increasing resolution.

In principle, observations may sample a variety of objects in the Magellanic Clouds: molecular clouds, shells around late-type (AGB) stars, circumstellar (protoplanetary) disks and circumstellar masers. An impression of LSA capabilities for molecular line observations of such objects is given in Table 2. The third column contains typical surface brightnesses that will be encountered for various molecular species of interest, the fourth lists the baseline at 230 GHz B_{res} that yields a resolution corresponding to the characteristic size of the object, the fifth column gives the signal-to-noise ratio in an 8-hour integration at all baselines $\leq B_{res}$, i.e. as long as the source is unresolved, while the last column gives the signal-to-noise ratio as a function of B_{max} (in km) $> B_{res}$ resolving the source.

3   Interstellar Molecular Clouds

The structure of molecular clouds observed in the Galaxy reflects the as yet poorly understood interplay of gravity, rotation, magnetic fields, pressure and radiation. Newly formed luminous stars inject significant amounts of energy into nearby molecular clouds both by radiation and mass loss. As these clouds are hierarchical in structure and inhomogeneous on all observed scales, Galactic clouds must be observed up to very high linear resolutions in order to disentangle the effects of chemistry, radiation, dynamics and geometry. A variety of sometimes elaborate models exist to interpret the observations (e.g. van Dishoeck & Black 1988; Lequeux et al. 1994), but the Galactic environment provides only a limited range of conditions hampering attempts to fully test and refine them. Although

Table 2. LSA Line Capabilities Magellanic Clouds

Type of Object	Typical Size (pc)	Surface Brightness (K)	Resolving Baseline (km)	S/N Ratio in 8 hours Unresolved	Resolved
Orion BN/KL Region	0.05	100	1.8	620	$2000/B^2$
Molecular Core	0.5	15	0.2	8300	$330/B^2$
		1-4		1000	$40/B^2$
AGB Shell	0.005-0.5	1-10	>0.2	<1750	$<70/B^2$
		0.1-3		<300	$<12/B^2$
Circumstellar Disk	0.0005-0.005	10	>18	<1	----
SiO Maser	0.0005	---------- See Text ----------			

the Magellanic Clouds can never be observed on the smallest scales accessible with the same telescope in the much closer Galactic objects, images with sub-parsec resolution (sampling masses down to a few hundredths of a solar mass) will already provide *most of the necessary information* on the response of molecular clouds to low metallicities and a variety of radiation fields that cannot be obtained with Galactic observations.

Low metallicities and dust abundances have the major effect of diminishing or even removing the shielding and selfshielding of molecular material. The hierarchical and filamentary structure of molecular clouds implies a relatively high surface-to-volume ratio. Consequently, physics and chemistry of molecular clouds in the LMC and especially the SMC are much more susceptible to the influence of ambient radiation fields than those in the Milky Way. For instance, average gas kinetic temperatures will be higher in Magellanic objects than in otherwise comparable Galactic objects. The temperature dependence of chemical processes may thus lead to LMC and SMC chemistries rather different from the Galactic chemistry. As the large-scale radiation fields impinging on Magellanic objects vary by more than an order of magnitude (Schwering 1988) comparative, multispecies studies of molecular clouds in the Milky Way, the LMC and the SMC will thus provide a grid of cloud properties as a function of both metallicity and radiation fields that can serve as the basic database to test and improve physical and chemical models of the interstellar medium. Detailed understanding of the behaviour of the dense interstellar medium under such varying conditions is *essential* to interpret molecular line observations of disks and centers of galaxies at much greater distances.

Studies of the molecular medium in the Magellanic Clouds on scales of 10 pc are already in progress. As an example we mention the ESO-Swedish Key Programme on CO in the Magellanic Clouds. The CO clouds in the LMC and especially in the SMC are much less luminous than those in the Galaxy, because they are smaller and because they appear to have a smaller volume filling factor. The data suggest that, in contrast, the associated amounts of molecular hydrogen, hence molecular complex masses, may not differ much from their Milky Way counterparts, resulting in a much higher H_2 column density to CO luminosity conversion factor than applicable to the Galaxy. The important influence of low metallicity and varying but strong radiation fields is illustrated by a comparison of the SEST J=1–0 ^{12}CO map at 10 pc resolution of the starforming complex N11 in the LMC, shown in the LSA report on page 21 (Downes, 1995), with e.g. the CO map at 1.3 pc resolution of the Orion starforming region published by Maddalena et al. (1986). The overall aspect of both maps as regards structure and clumpiness is very similar despite the difference in scale, but N11 rather conspicuously lacks the large amount of interclump gas that characterizes the Orion clouds. This is a real effect, and not an artefact of limited sensitivity. Evidently, the Galactic environment provides the shielding enabling the relatively tenuous interclump CO gas to survive the radiative onslaught of the Orion OB association, whereas the LMC environment does not provide similar protection against the N11 OB stars so that only the denser CO concentrations remain. The more extreme SMC environment exhibits the same behaviour in a more pronounced way. Even the limited SEST observations of the lower two transitions of ^{12}CO and ^{13}CO in six SMC clouds already provide valuable first insights in the effect of metallicity and radiation on the properties of molecular clouds of varying density (Lequeux et al. 1994). Likewise, the far-infrared C$^+$ observations of a sample of SMC and LMC cloud complexes illustrate the dominant effect of strong radiation fields in the low-metallicity Magellanic molecular cloud complexes (Israel & Maloney, 1993; Israel et al. 1996).

These molecular line observations (as well as the far-infrared observations) just resolve the observed cloud complex components which have sizes of 10 - 40 pc. Thus, only *global* averages of cloud parameters can be determined, whereas these parameters almost certainly vary a great deal on scales not resolved. LSA mosaicing observations – obviously an essential technique for molecular clouds – at subparsec resolutions will bring these variations to light and finally allow us to match the observations to more than just crude models. The first two lines in Table 2 illustrate the LSA capability for 'normal' cloud cores and for objects such as the very bright and very compact Orion A core. Normal molecular cloud cores are resolved, with high signal-to-noise ratios in strong molecular lines such as ^{12}CO even in relatively short integrations by compact array configurations starting at 200 m maximum baseline. They can be mapped to resolutions of about 0.02 pc (5 km baseline). Bright cores similar to the Orion BN/KL region can be mapped in strong lines with resolutions down to 0.006 pc (1300 AU). This resolution corresponds to that provided by *current* mm interferometers on Orion *itself*, and is sufficient to just resolve structures similar to the 'hot core' south of IRc 2 (cf. Figures 6 and 7 of Genzel & Stutzki 1989).

In Table 3, we illustrate LSA capabilities by listing (sub)parsec resolution molecular line maps of a few well-known Galactic molecular cloud complexes and indicating the baseline required to obtain a 230 GHz (e.g. J=2–1 CO) map of a Magellanic Cloud complex at *the same linear resolution*.

Table 3. Galactic Counterparts for Magellanic Clouds Molecular Complexes

Object	Distance (pc)	Transition	Linear Resolution (pc)	Reference	Required Baseline (m)
Taurus Clouds	140	$^{12}CO(1-0)$	1.22	Ungerechts & Thaddeus (1987)	72
Orion OMC-1	500	$^{13}CO(1-0)$	0.24	Bally et al. (1987)	360
		HDO	0.03	Plambeck & Wright (1988)	10000
M17SW	2200	$C^{18}O(2-1)$	0.14	Stutzki & Güsten (1990)	630
		$C^{17}O(3-2)$	0.16	Hobson et al. (1994)	545
W3	2300	$C^{18}O(2-1)$	0.25	Oldham et al. (1994)	360

References: 1. Ungerechts & Thaddeus (1987); 2. Bally et al. (1987); 3. Plambeck & Wright (1988); 4. Stutzki & Güsten (1990); 5. Hobson et al. (1994); 6. Oldham et al. (1994).

Continuum measurements of the (molecular) cloud complexes in the Magellanic Clouds are an important complement to molecular line observations, because they provide information not otherwise accessible on both the column density of dust and the temperature structure of the clouds, as do to a lesser extent high-resolution (ATCA) radio continuum observations. Such information provides important constraints on the input parameters of the models to be matched to molecular line observations. IRAS far-infrared fluxes of moderately bright dust associated with molecular clouds in the the LMC and SMC (Schwering & Israel 1990) can be extrapolated to yield expected *mean* surface brightnesses at 1.3 mm continuum of the order of 0.03 mJy per arcsec^2, which corresponds to a mean 5σ signal in 7 hours for a compact array configuration ($B_{max} = 300$ m). Actual 1.3 mm surface brightnesses may be higher, because the extrapolation assumes a uniform dust temperature of 30 K (with an emissivity coefficient n $= 1.5$) throughout the cloud. If colder dust is present, it will increase the 1.3 mm signal with respect to the observed $100\mu m$ intensity. In addition, the inhomogeneous structure of the clouds will yield a significantly better image than suggested by the mean S/N ratio, in particular showing hot compact cores in rather shorter integration times. However, the extent of the clouds will require mosaicing partly offsetting these time advantages. The LMC with its larger number of bright dust clouds offers better perspectives than the SMC. Although subparsec resolution imaging with compact array configurations is quite feasible, it is unlikely that longer, km-length baselines will yield useful images in reasonable integration times for any but the most luminous and compact clouds such as N159 and N160 in the LMC. However, the situation significantly improves if one observes at higher frequencies because of the ν^3 dependence of the dust emissivity.

4 Circumstellar Matter and Stars

4.1 Shells around Late-Type Stars

Stars on the Asymptotic Giant Branch (AGB) suffer significant mass loss. Expanding envelopes are accelerated by resonant absorption of stellar photons by dust and elements such as carbon, nitrogen, oxygen. The details of these mechanisms are not yet fully understood. It appears that the lower-metallicity stars in the Magellanic Clouds have mass-loss rates similar to their Milky Way counterparts but appreciably lower terminal velocities of the expanding envelopes, so that generally higher shell densities are to be expected. The low metallicities of the Magellanic Clouds are precisely what is needed for further progress. Comparative studies of such stars in the Milky Way and in the Clouds, and especially mapping the envelope velocity fields and abundances, promise to bring closer an understanding how AGB stars lose mass and how they enrich the interstellar medium, thereby fulfilling a crucial function in the chemical evolution of galaxies. In addition, the relative geometrical simplicity of the spherical shells compared to e.g. molecular cloud cores makes them very attractive objects to test the models referred to in the preceeding section, before applying them to the more complex molecular cloud geometries. For a very extensive up-to-date review, the reader is referred to Habing (1996).

As Table 2 shows, line emission from the brightest shells around Magellanic Cloud AGB stars are easily detectable, but only the largest shells can be mapped with resolutions down to about 0.04 pc (i.e. 10 resolution elements across the shell). In the continuum, very bright shells around late-type supergiants have been found in the IRAS LMC database (Elias et al. 1986). Scaling the $100\mu m$ flux-density of 10 Jy of IRAS 05436-6949 by a factor of 1000, we expect a 1.3 mm flux-density of 10 mJy, easily detectable with very short integration times. Its brightness suggests large shell dimensions, so that it is probably resolved by baselines less than a kilometer in length. A good, high S/N image may be obtained with an integration time of a few hours, allowing analysis of the shell structure. However, this is the brightest object of its kind in the LMC, and none has so far been identified in the SMC. Good mm-wavelength flux measurements of objects one to two orders of magnitude fainter require integration times no longer than a few hours, and fairly detailed images can still be obtained with integration times of several hours of objects up to an order of magnitude weaker than IRAS 05436-6949. Once again, we note that the strong frequency dependence of dust continuum emission considerably improves detection *and* imaging prospects at higher frequencies, providing a powerful rationale for equipping a prospective array with such a capability: even a modest increase in frequency coverage to 345 GHz already produces a signal more than three times stronger.

4.2   Circumstellar Disks

Circumstellar disks of gas and dust offer the exciting prospect of learning more about the formation of stars and the origin of planetary systems such as our own. Galactic observations show that resolving such disks in the Magellanic Clouds requires baselines well in excess of 10 km; moreover scaling the line flux of 10 Jy beam^{-1} km s^{-1} measured for the brightest such objects in Taurus an Chamaeleon (cf. Koerner & Sargent 1995) indicates that a 1σ detection would require of the order of 2000 hours with the LSA. Scaling the observed 1.3 mm continuum flux-densities of the same objects (cf. Henning & Thamm

1994) to Magellanic Cloud distances leads us to expect at most 3μJy, which again would lead to unrealistically long integration times to merely obtain a detection. Circumstellar disks in the Magellanic Clouds require a collecting area one (continuum) to two (line) orders of magnitude larger than provided by the LSA concept. However, ignoring possible confusion, we wish to point out that circumstellar disks may in fact become just detectable in the continuum if the LSA would be able to operate at higher frequencies, and could be equipped with a larger bandwidth.

4.3 Masers

Maser sources, in contrast, may be easily detectable. For instance, the 86 GHz $v=1$, $J=2$-1 SiO maser in the Orion A star formation region would have a flux-density of 104 mJy observed at 1 km s^{-1} resolution. This resolution is sufficient to resolve the maser line profile. A high signal-to-noise detection of such a maser in the Magellanic Clouds could be obtained in minutes. The compact nature of the maser would, however, require VLBI baselines of more than 200 km to resolve it. SiO masers around late-type stars (Mira stars, OH/IR stars) should likewise be detectable, but not spatially resolvable. In Table 4 we have listed Galactic SiO masers (taken from Nyman & Olofsson 1986 and Nyman, Hall & Le Bertre 1993). The fourth column gives the peak flux these sources would have at 60 kpc distance, and the last column gives the LSA signal-to-noise ratio that such a source would yield in one hour observed at 1 km s^{-1} resolution. The brighter masers around OH/IR stars are easily detectable in an hour, but masers around other late-type stars would require more effort. For instance, if it were located in the Magellanic Clouds, a 5σ detection of Mira (o Ceti) itself would require over 2000 hours, while R Cas would require 115 hours. However, brighter members of this class, such as NML Tau and R Leo are detectable at reasonable signal-to-noise ratios in a few hours. Obviously, somewhat better results can be obtained by sacrificing spectral resolution for sensitivity.

4.4 Early-type Stars

Continuum emission from mass-losing early type stars is of interest, for instance because of the energy these objects deposit in the interstellar medium. Knowledge of their mm spectrum allows good determination of their mass loss rates. Typical flux-densities of bright Galactic OB supergiants (R. Waters, private communication) lead us to expect 1.3 mm flux-densities of the order of 0.08 mJy for their Magellanic Cloud counterparts. Reasonably accurate flux densities can thus be determined in a few hours per object; prospects once again brighten if observations can be made at shorter wavelengths and with larger bandwidths. Equally interesting Be stars are, however, weaker by a factor of about 150. These are essentially undetectable by an array such as the LSA.

Table 4. Observability of Galactic 86 GHz SiO masers at Magellanic Distances

Object	Luminosity SD^2 (Jy kpc^2)	Velocity Width (km s^{-1})	Peak Flux at 60 kpc (mJy)	S/N Ratio in 1 hr 1 km s^{-1} Band
Orion-A	375	5	104	63
Mira	0.75	13	0.2	0.1
NML Tau	13	17	4	2
R Leo	28	15	8	5
R Cas	3	16	0.8	0.5
OH285.1+0.1	164	7	45	27
OH286.5+0.1	866	3	240	145
OH341.1-0.0	18	6	5	3

5 Conclusion

1. The most important application of a Large Southern Millimetre Array on the Magellanic Clouds will be mapping of molecular clouds in a variety of species in order to determine the influence of metallicity and radiation environment on structure, chemistry and evolution of molecular complexes. Such studies are essential to fully test and develop models that are not only of interest in themselves, but also needed to properly interpret observations of more distant galaxies. Molecular cores and warm dust clouds can routinely be mapped with such an array at subparsec resolutions in both line and continuum. Mosaicing techniques are essential.

2. Circumstellar shells of molecular gas and dust in the Magellanic Clouds are likewise of importance to determine in particular the effect of metallicity on mass loss phenomena and to test models on sources with relatively simple geometries. Such shells can be detected with the LSA throughout the Magellanic Clouds in both line and continuum. The larger and brighter shells can be imaged.

3. The presence or absence of Orion-type 86 GHz SiO masers can succesfully be determined by LSA observations throughout the Clouds. The brighter SiO masers around late-type stars are likewise detectable.

4. OB supergiants suffering significant mass loss can be detected in the continuum throughout the Magellanic Clouds. Their mm spectrum, important to determine mass loss parameters can be established. Other stellar emission is, however, not detectable. Circumstellar disks in the Magellanic Clouds are not detectable in either continuum or line.

5. The dramatic improvement in continuum signal-to-noise ratios at higher frequencies (e.g. 345 GHz) and larger bandwidths (e.g. 8 GHz), important especially for studies of circumstellar shells and disks, fully warrants an effort to equip a Large Southern Array with such a capability.

References

Bally, J., Langer, W.D., Stark, A.A., Wilson, R.D., 1987 ApJ **312**, L45

Cohen, R.S., Dame, T., Garay, G., Montani, J., Rubio, M., Thaddeus, P., 1988 ApJ **331**, L95

Downes, D., 1989 in: *Evolution of Galaxies and Astronomical Observation* Eds. I. Appenzeller, H.J. Habing & P. Léna (Springer, Berlin), p. 351–383

Downes, D. et al, 1995: *LSA: Large Southern Array: a 21st Century Millimeter Array with 10000 m² Collecting Area* (IRAM, Grenoble).

Dufour, R.J., 1984 in: *Structure and Evolution of the Magellanic Clouds*, Eds. S. van den Bergh & K.S. de Boer, IAU Symposium 108 (Reidel, Dordrecht), p. 353

Elias, J.H., Frogel, J.A., Schwering, P.B.W., 1986 ApJ **302**, 675

Genzel, R., Stutzki, J., 1989 ARA&A **27**, 41

Henning, T, Thamm, E., 1994 Ap&SS **212**, 215

Hobson, M.P., Jeness, T., Padman, R., Scott, P.F., 1994 MNRAS **266**, 972

Israel, F.P., et al. 1993 A&A **276**, 25

Israel, F.P., Maloney, P.R., 1993 in: *New Aspects of Magellanic Cloud Reserarch*, Eds. B. Baschek, G. Klare & J. Lequeux (Springer, Berlin), p. 44

Israel, F.P., et al. 1996 ApJ, in press

Koerner, D.W., Sargent, A.I., 1995 AJ **109** 2138

Kutner, M.L, et al. 1996 A&A, in preparation

Lequeux, J., et al. 1994 A&A **292**, 371

Maddalena, R.J., Morris, M., Moscowitz, J., Thaddeus, P., 1986 ApJ **303**, 375

Nyman, L.-A., Olofsson, M., 1986 A&A **158**, 67

Nyman, L.-A., Hall, P.J., Le Bertre, T., 1993 A&A, **280**, 551

Oldham, P.G., Griffin, M.J., Richardson, K.J., Sandell, G., 1994 A&A **284**, 559

Plambeck, R.L., Wright, M.C.H., 1987 ApJ **330**, L61

Rubio, et al. 1993 A&A **271**, 1

Rubio, et al. 1996 A&A, in press

Schwering, P.B.W, 1988, Ph.D. Thesis Leiden University, Chapter V

Schwering, P.B.W., Israel, F.P.: 1990 *Atlas and Catalogue of Infrared Sources in the Magellanic Clouds* (Kluwer, Dordrecht)

Stutzki, J., Güsten, R., 1990 ApJ **356**, 513

Ungerechts, H., Thaddeus, P., 1987 ApJS **63**, 645

van Dishoeck, E.F., Black, J.H., 1988 ApJ **334**, 771

Will the LSA detect continuum or line emission from AGB stars in the LMC ?

Martin Groenewegen

Max-Planck Institut für Astrophysik, Karl-Schwarzschild Straße 1,
D-85740 Garching, Germany

Abstract. I predict the (sub-)mm continuum and line emission fluxes expected from the reddest and brightest AGB stars in the Large Magellanic Cloud.

Flux-densities at 1300 μm for AGB stars are expected to be less than 0.05 mJy in a 5″ beam. This will make the detection of continuum emission in AGB stars virtually impossible. There are however a handful of red supergiants which are expected to be detectable.

Main-beam peak temperatures as high as 0.6 K are expected for the CO 2-1 transition of the reddest and most luminous AGB stars in the LMC. This implies that CO 2-1 could be detected in a few dozen AGB stars in a reasonable integration time.

1   Introduction

The LSA (Large Southern Array) will have unprecedented sensitivity and angular resolution. In the field of late-type (AGB) stars it will e.g. be possible to resolve the dust and molecular shell around nearby AGB stars and study the dust formation process and the acceleration of the dust and gas close to the central stars in detail. In the continuum it will be possible to map the dust shells out to large distances (with lower angular resolution) and address the question of mass loss rate variations with time.

The question I address here is whether AGB stars in the LMC can be detected in the continuum and/or in CO with the LSA. Continuum observations in the mm-range would give a handle on the dust opacity in that wavelength regime. Line emission data would allow first off all an estimate of the total mass loss rate. Since modeling of the spectral energy distribution (SED) gives the dust mass loss rate and the distance to the LMC is known this would allow a direct measurement of the dust-to-gas (DTG) ratio. It would therefore be possible to measure the DTG ratio as a function of position in the LMC. In addition, line emission data provides the expansion velocity of the envelope, a quantity which is thought to be lower than for Galactic AGB stars because of the lower metallicity.

2   CO line emission

To get an idea of the expected strength of the CO J = 1-0, 2-1 and 3-2 lines for AGB stars in the LMC I take the well known carbon Mira IRC +10 216 (= CW Leo = AFGL 1381) as an representative example.

Previously I have modeled the CO lines of this star (Groenewegen 1995). For an assumed luminosity of 15000 $L_\odot$ (based on a period-luminosity relation) the distance to CW Leo is 135 pc. For a mass loss rate of 2.2 10^{-5} $M_\odot$ yr^{-1} and an expansion velocity of 14.5 km s^{-1} the radius where the CO abundance has dropped to 1% of its value close to the star is $\sim$ 7.4 10^{17} cm (about twice the photodissociation radius). At a distance of 50 kpc this corresponds to 1.0''. This is very well matched to the spatial resolution of the LSA at 230 GHz using the smallest baseline. In Table 1 the expected main-beam peak temperatures and integrated line intensities are listed if CW Leo is placed at a distance of 50 kpc. The beam at all three wavelengths is assumed to be a Gaussian with a FWHM of 1''. The expected CO J = 2-1 line strength is 0.56 K. This calculation assumes a DTG ratio of 0.005. Table 1 also includes calculations with a DTG ratio of 0.0025 and FWHM beams of 2''. Both effects would decrease the expected flux.

Table 1. CO line emission: IRC +10 216 placed in the LMC

Transition	T_{peak} (K)	$\int T$ dv (K km s^{-1})	T_{peak} (K)	$\int T$ dv (K km s^{-1})	T_{peak} (K)	$\int T$ dv (K km s^{-1})
	FWHM = 1, Ψ = 0.005		FWHM = 1, Ψ = 0.0025		FWHM = 2, Ψ = 0.005	
1-0	1.03	23.3	0.47	10.4	0.33	7.3
2-1	0.56	11.9	0.16	3.4	0.13	2.6
3-2	0.25	5.2	0.068	1.1	0.096	0.79

3   Continuum emission

To estimate the continuum dust emission of AGB stars in the LMC I consider two approaches. First I will use IRC +10 216 as a template again and then I will extrapolate model fits to actual AGB stars in the LMC.

Regarding IRC +10 216, the size of the dust envelope, calculated as the radius where the dust temperature has dropped to 20 K, is 2.4 10^{18} cm which corresponds to 3.2'' at the distance of the LMC. Again, the synthesised beam size of the LSA at the smallest baseline is reasonably well matched to the expected physical size of circumstellar dust envelopes in the LMC. The calculations are summarized in Table 2. They are monochromatic flux densities in a Gaussian beam with a FWHM of 1''. The expected flux level at 1300 μm (230 GHz) is 0.018 mJy. The fluxes scale approximately linearly with the DTG ratio, for which a value of 0.005 is assumed. In a beam of 5'' FWHM the flux is about a factor of 2 higher (see Table 2).

Luminous and red AGB stars have been identified in the LMC, mainly using IRAS data (see e.g. Zijlstra et al. 1996, Wood et al. 1992 and references therein).

Groenewegen et al. (1995) obtained 8-13 μm spectra and identified the first silicate features in AGB stars outside our Galaxy. We also made model fits to the SEDs from 1.25 - 25 μm. We are now in the process of extending these calculations to a larger sample of stars (Groenewegen et al., in preparation). We obtain good quality fits and therefore it is reasonable to extrapolate these results to longer wavelengths. This basically assumes that the dust opacity which we know to work well for Galactic O-rich AGB stars also holds for AGB stars in the LMC. Based on the results so far I calculated the (sub-)mm fluxes for all stars in our sample. The one with the highest flux is LMC 1164 (Schwering & Israel 1989), which has an estimated luminosity of 46100 $L_\odot$ and a mass loss rate of 4.4 10^{-5} $M_\odot$ yr^{-1}. Its predicted (sub-)mm fluxes are listed in Table 2.

Table 2. Expected continuum emission

Wavelength (μm)	CW Leo scaled to LMC		LMC 1164		LMC 181
	FWHM = 1″	FWHM = 5″	FWHM = 1″	FWHM = 5″	FWHM = 1″
811	0.074 mJy	0.13	0.13	0.26	0.61
1300	0.018	0.034	0.023	0.048	0.12

4 Discussion

The continuum emission of AGB stars in the LMC at 1300 μm in a 5″ beam is expected to be less than 0.05 mJy. The quoted 1σ sensitivity for a 1 hour integration with the LSA is 0.02 mJy. This implies integration times of more than 16 hours to get a S/N of better than 10. This is prohibitively long.

Although the prospect to detect continuum emission of AGB stars in the LMC appears poor, the situation is better for red mass-losing supergiants. The most extreme case is LMC 181 (= IRAS 04553–6825; see Elias et al. 1986). I made a fit to its SED and 8-13 μm spectrum. It has a M_{bol} of -9.2 and mass loss rate of about 10^{-4} $M_\odot$ yr^{-1}. The predicted (sub-)mm fluxes are listed in Table 2. This object should be detectable with a S/N of 10 in 3 hours of integration.

The CO J = 2-1 main-beam peak temperature expected for IRC +10 216 if it were in the LMC is about 0.6 K. The 1σ sensitivity of the LSA at 230 GHz is 0.01 K in 1 hr of integration with a velocity resolution of 2 km s^{-1}. This implies that a S/N of 10 can be obtained in 2 minutes.

In summary, in its presently proposed configuration the LSA will be able to detect CO line emission in possibly up to 30 AGB stars (based on the number of luminous and red candidate objects, see Zijlstra et al. 1996) and red mass-losing supergiants in the LMC. This will allow an accurate determination of the DTG ratio and the mass loss rate. Regarding continuum emission the number of potential targets is much smaller.

References

Elias J.H., Frogel J.A., Schwering P.W.B., 1986, ApJ 302, 675

Groenewegen M.A.T., 1995, in: "Circumstellar Matter", eds. G.D. Watt & P.M Williams, p. 471

Groenewegen M.A.T., Smith C.H., Wood P.R., Omont A., Fujiyosji T., 1995, ApJ 449, L119

Schwering P.B.W., Israel F.P., 1989, A&AS 79, 105

Wood P.R., Whiteoak J.B., Hughes S.M.G., Bessell M.S., Gardener F.F., Hyland A.R., 1992, ApJ 397, 552

Zijlstra A.A., Loup C., Waters L.B.F.M., Whitelock P.A., van Loon J.Th., Guglielmo F., 1996, MNRAS 279, 32

High-resolution Submm and MIR imaging of the central parsec of the Milky Way

Peter G. Mezger

MPI für Radioastronomie, Auf dem Hügel 69, D-53121 Bonn, FRG

Abstract. High resolution imaging of the Galactic Center at radio, mm/submm and MIR wavelengths allowed to determine the radio through IR spectrum of the compact synchrotron source Sgr A*, which is the best candidate for an underfed Black Hole close to or at the dynamical center of our Galaxy. The ensuing spectrum of Sgr A* could be typical for the non-thermal part of AGN spectra. But it requires - especially at submm and IR wavelengths and in the case of weak activity - a sufficiently high angular resolution to separate this spectrum from secondary contributions, such as optical/UV emission from a central cluster of hot stars and of starlight reprocessed by dust.

The compact synchrotron source Sgr A*, located at or close to the dynamical center of our Galaxy, is the best candidate for a starving Black Hole (BH) of a few $\sim 10^6 \, M_\odot$. Its presence has been predicted on theoretical grounds by Lynden-Bell and Rees (1971); it was definitely detected three years later by Balick and Brown (1974) using the Green Bank interferometer.

Since then Sgr A* became a favorite object of VLBI observations. At $\lambda 1$ cm its apparent size decreases $\propto \lambda^2$, indicating source broadening due to electron scattering. Recent observations at mm wavelengths may actually have detected the true apparent size of Sgr A* of a few milliarcseconds, corresponding to a linear size of $\sim 2.5 - 4 \, 10^{13}$ cm (Krichbaum et al., 1993).

Figure 1 shows the most recent radio/IR spectrum of Sgr A* (Beckert et al., 1996). The flux densities at $\leq \lambda 3$ mm are variable. For wavelengths $\leq \lambda 350 \, \mu$m there are only upper limits available. The spectrum has a high frequency cut-off at $\sim 2 - 4 \, 10^3$ GHz and a low-frequency turnover at ~ 0.8GHz. In between the spectrum increases $S_\nu \propto \nu^{1/3}$. Beckert et al. explain this spectrum as being optically thin synchrotron emission from quasi-monoenergetic electrons with the low-frequency turnover caused either by free-free absorption or by synchrotron self absorption.

The lack of data in the submm/FIR/MIR regime is due to an effect intimately connected to the main theme of this symposium, i.e. high angular resolution imaging at mm/submm wavelength.

Figure 2 (from Zylka et al.,1995) shows three images of the central 3x3 pc² together with cuts through Sgr A*. One clearly recognizes that the features of the background emission representing cold dust increase dramatically with decreasing λ relative to the flux density of the compact synchrotron Sgr A*. Therefore, using the CSO 10-m-telescope at 350μm with an angular resolution of $\sim 10''$, only an insignificant upper limit of ~ 10 Jy for the flux density of Sgr A* could be obtained. The reason is that the surface brightness of optically

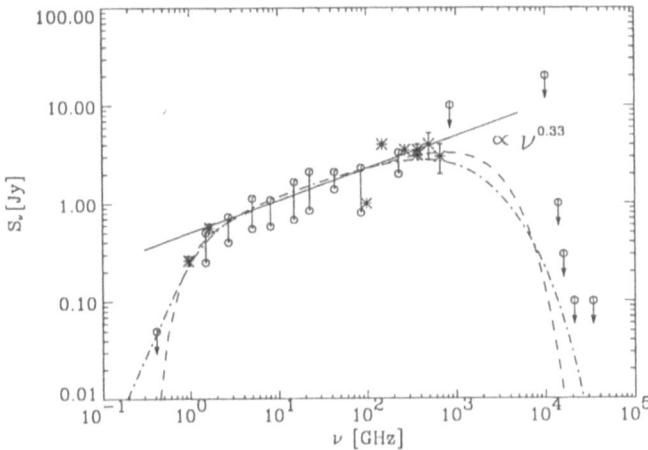

Fig. 1. A comparison of the observed radio spectrum of Sgr A* with best-fit model spectra (symbols as in Zylka et al., 1995; note that bars with symbols at both ends denote the variability range and not error bars of individual observations; additional rhombs at low frequencies are from Davies, Walsh and Booth 1976). The low-frequency turnover can be either due to free-free absorption or to synchrotron self-absorption.

thin submm dust emission $S_\nu/\Theta_A = I_\nu$ increases with frequency $\propto \nu^4$ while the flux density of the point source Sgr A* increases with $S_\nu \propto \nu^{1/3}$ only. In the IR beyond $\lambda 60\mu$m, the dust surface brightness actually begins to decrease and the angular resolution of ground-based large telescopes in the MIR is $\leq 1''$ so that background emission from hot dust should be efficiently suppressed. Nevertheless, todate only upper limits of MIR flux densities of Sgr A* have been obtained which, however, significantly restrain the IR part of the spectrum, which appears to decrease exponentially. The integrated luminosity of this part of the Sgr A* spectrum amounts to a few $10^2 L\odot$.

Interstellar gas and dust, amounting to $A_v \sim 31$mag, makes the Galactic Center inaccessible to direct observations between $\lambda 1\mu$m and the soft X-ray regime $E \geq 1$KeV. At $\lambda 2.2\mu$m, Eckart et al. (1995) find a cluster of 6 star-like objects one of which - with $S_{2.2\mu m} \sim 0.01$ Jy - could coincide with Sgr A* (Radio). In the soft X-ray regime 1.2-2.5KeV Predehl and Truemper detected a point source with a - for $A_v = 31$ mag dereddened - flux density of $3\,10^{-7}$ Jy.

If Sgr A* were a massive but underfed Black Hole surrounded by an accretion disk we would expect its spectrum to be composed of a synchrotron spectrum (possibly associated with a jet-like object) and a thermal spectrum originating mainly close to the inner edge of the accretion disk. While the synchrotron spectrum is well observed (see Fig.1) the thermal disk emission - if it exists at all - is only vaguely defined by the NIR and X-ray flux densities mentioned above. Fig.3 (from Mezger, Duschl and Zylka, 1996) shows a possible disk model

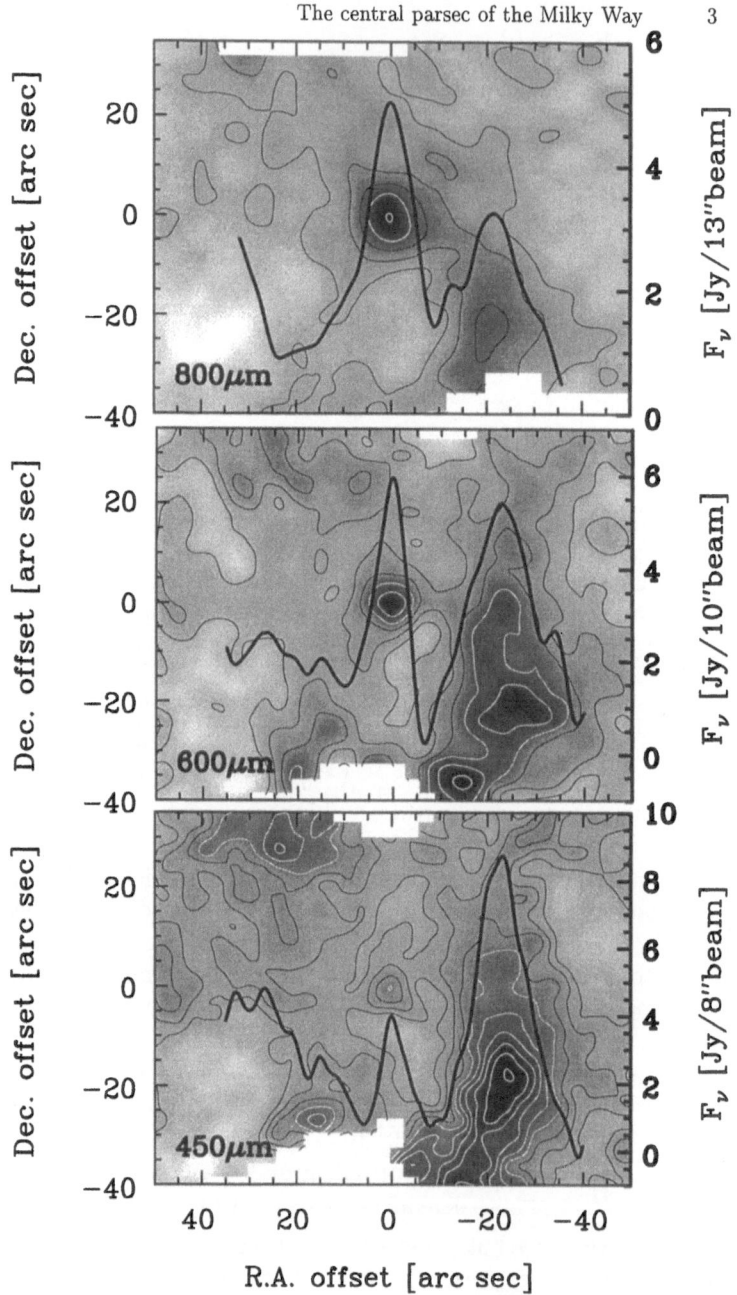

Fig. 2. Averaged JCMT images of the central $\sim 80''$ obtained with the UKT14 bolometer at 800, 600 and 450μm. Contours are drawn in equidistant steps of 1Jy/beam. Cuts through the images at a position angle of 45°, i.e. from SW to NE, are shown as black solid lines. The intensity scales for these cuts are indicated on the right image ordinate. $\Delta\alpha$, $\Delta\sigma$ are coordinates off-set relative to the position of Sgr A*, $\alpha=17^h42^m29\overset{s}{.}314$, σ=-28°59'18''3 (epoch 1950, see, e.g. Rogers et al., 1994).

spectrum fitted to the NIR point which is found to be the much more stringent limit. The corresponding integrated luminosity is $\leq 5\,10^6 L\odot$.

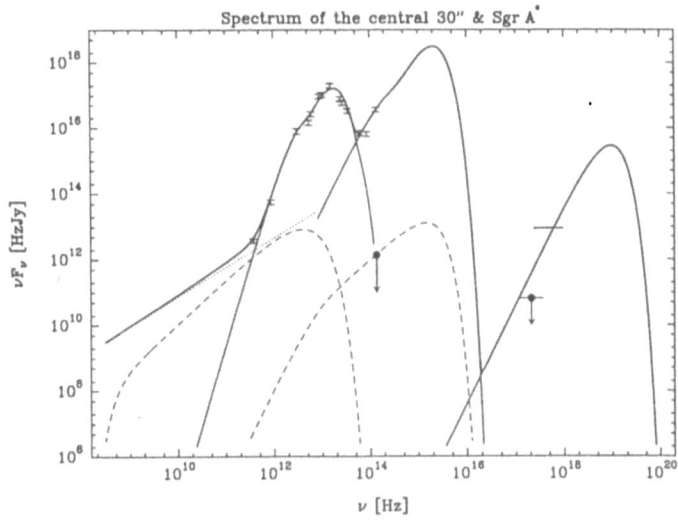

Fig. 3. The radio through UV spectrum of the central $30''(=1.2\mathrm{pc})$ as seen from the Galactic Poles (heavy curve). Free-free emission dominates the spectrum for $\nu < 2\,10^{11}\mathrm{Hz}$, dust emission for $2\,10^{11} < \nu/\mathrm{Hz} < 3\,10^{13}$, stellar radiation for $\nu > 3\,10^{13}\mathrm{Hz}$ and a diffuse X-ray emission of unknown origin for $\nu > 10^{16}\mathrm{Hz}$. Note that the stellar flux densities relate to the central parsec only. The observed radio/IR spectrum of Sgr A* together with a disk spectrum fitted to the upper limit determined by the observed $\lambda 2.2\mu\mathrm{m}$ flux density are shown as dashed curves.

Figure 3 compares the spectrum of Sgr A* with the spectrum integrated over the central parsec. Details how this spectrum was constructed can be found in Mezger, Duschl and Zylka (1996). Note that here we use a νS_ν representation of both spectra. The integrated spectrum with $S_{15\mathrm{GHz}} \sim 7.9\mathrm{Jy}$ is dominated by free-free emission for $\nu < 10^{11}$ Hz, by optically thin dust emission for $\nu < 10^{14}$ Hz, by black-body emission (mainly from stars with $T_{\mathrm{eff}} \sim 4000\mathrm{K}$ and $\sim 25\,000\mathrm{K}$) for $\nu < 10^{16}$ Hz and by diffuse X-ray emission of $3\,10^{-5}$ Jy in the energy range 0.8-3KeV ($\nu > 10^{16}\mathrm{Hz}$) whose origin is not yet clear. The integrated luminosity of the central parsec amounts to $\sim 10^8 L\odot$ and is dominated by a central cluster of hot stars with a core radius of $\sim 0.17\mathrm{pc}$, rather than by the central source Sgr A* which probably represents an underfed $10^6 M\odot$ Black Hole. About $\sim 20\%$ of the stellar emission is absorbed by dust and reradiated in the MIR/FIR.

Observed from M31 (D~ 700 Kpc) with an angular resolution of $\sim 1'' \cong 3.4\mathrm{pc}$ the spectrum of the nucleus of our Galaxy would look similar to the integrated spectrum in Fig.3, and thus would mimick a thermal (or radio quiet) spectrum similar to that of a weak Seyfert 1 AGN, rather than the obviously non-thermal spectrum associated with the compact synchrotron source Sgr A*.

I want to thank my colleagues W.J. Duschl and R. Zylka for helpful discussions and remarks and for preparing the Figures.

References

Balick B., Brown R.L., 1974 ApJ 194, 265

Beckert T., Duschl W.J., Mezger P.G., Zylka R., 1996, A&A (in press) Davies R.D., Walsh D., Booth R.S., 1976 MNRAS 177, 319

Eckart A., Genzel R., Hofmann R., Sams B.J., Tacconi-Garman L.E., 1995 ApJ 445, L23

Krichbaum T.P., Zensus J.A., Witzel A., Mezger P.G., Standke K.J., Schalinski C.J., Alberdi A., Marcaide J.M., Zylka R., Rogers A.E.E., Booth R.A., Rönnäng B.O., Colomer F, Bartel N., Shapiro I.I., 1993 A&A 274, L37

Lynden-Bell D., Rees M.J., 1971, MNRAS 152, 461

Mezger P.G., Duschl W.J., Zylka R., 1996 AAR (in prep.)

Predehl P., Truemper J., 1994 A&A 290, L29

Rogers A.E.E., Doeleman S., Wright M.C.H. et al., 1994 ApJ 434, L59

Zylka R., Mezger P.G., Ward-Thompson D., Duschl W.J., Lesch H., 1995 A&A 297, 93

New evidence for interaction of a molecular cloud/HII region with the G359.54+0.18 nonthermal filaments

J. Staguhn[1], J. Stutzki[1], F. Yusef-Zadeh[2] and K.I. Uchida[3]

[1] Universität zu Köln, 1. Physikalisches Institut Zülpicher Str. 77, 50937 Köln, Germany

[2] Dearborne Observatory, Northwestern University, 2131 Sheridan Road, Evanston, IL 60208, USA

[3] Max Planck Institut für Radioastronomie, Auf dem Hügel 69, 53121 Bonn, Germany

Abstract. We present a study of the Galactic Center nonthermal filament system G359.54+0.18 located to the north of the Sgr C region. We find evidence in support of the assertion by Serabyn & Morris that the nonthermal filaments are the manifestations of large-scale vertical magnetic field lines illuminated by collisions with molecular clouds. Included in the study are observations of, (1) the 3 mm emission lines of CS, HCO^+ and other molecular species with the SEST, (2) several lines of ^{12}CO and ^{13}CO with the 3-m KOSMA antenna, (3) 5 GHz radio continuum emission with the VLA, and (4) H79α recombination line emission with the 100-m antenna at Effelsberg.

The G359.54+0.18 nonthermal filaments and its (potentially) associated molecular cloud is a prime candidate for this study: It is the system furthest off the Galactic plane and thus suffers least from source confusion. Our high resolution 5 GHz continuum image of the G359.54+0.18 system (Fig. 1, contours) reveals a diffuse and somewhat clumpy emission structure near the easternmost tip of the filaments. Additionally, the molecular maps of the region, made with the SEST [1], show a localized cloud (identified as "Cloud A", seen east of the radio feature in Fig. 1 and in Fig. 2) east of, and directly adjacent to, the clumpy continuum component. H79α recombination line emission, at velocities ($V_{lsr} = 106$ km s^{-1}) consistent with the velocity gradient of the nearby molecular gas, is detected at the position of the diffuse continuum component ($\alpha,\delta = 17{:}41{:}10.0$, -29:14:00), indicating that it is thermal in nature and linking it kinematically to the adjacent molecular Cloud A. A strong velocity gradient in Cloud A, directed towards the interface region, implies interaction between the two (Fig. 2). The cloud emitts stronger in the high density tracing transition of HCO^+(1-0) than in CS(2-1). At the central position of Cloud A we have detected emission from the HNCO 5(0,5)-4(0,4) transition, a transition possibly pumped by the IR emission from warm dust or tracing high density clumps within the cloud with $n_{crit} \sim 10^5$ cm^{-3} (Armstrong & Barret, 1985).

On larger scales, both the position and velocity of Cloud A suggests that it is part of the northern edge of the "negative velocity feature" (Bally et al., 1988).

---

[1] The SEST telescope is operated by the Swedish National Facility for Radio Astronomy, Onsala Space Observatory and by ESO.

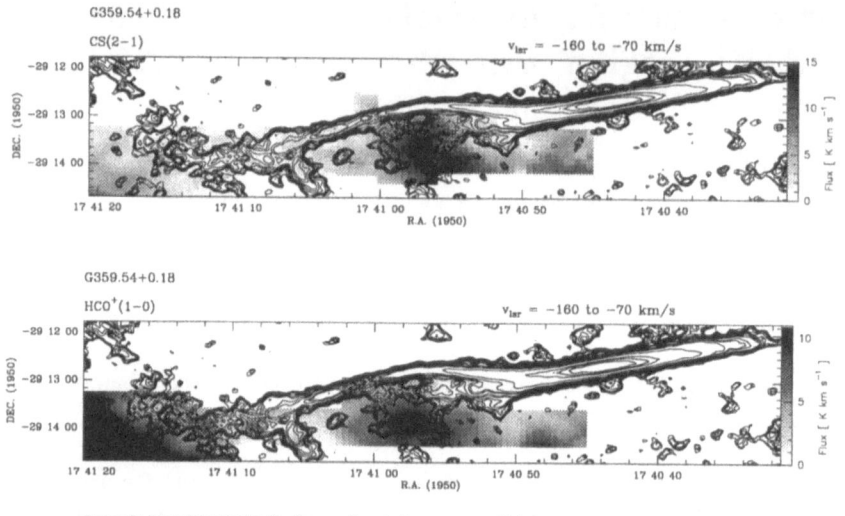

Grey scale: Integrated molecular line flux over the velocity range as specified above
Contours: Continuum flux at 5GHz: Levels [mJy/beam]: .15, .2, .25, .3, .4, .5, .6, .8, 1.0, 1.2, 1.4, 1.6, 2.0, 3.0, 5.0, 6.0, 7.0

Fig. 1. 5 GHz continuum flux ranging from 0.3 to 7 mJy/beam (contour lines) super-imposed on CS(2-1) flux (top) and HCO$^+$(1-0) flux (bottom) integrated between -160 and -80 km s^{-1} (grey scale). Cloud A with the associated clumpy radio continuum emission can be seen at the eastern tip of the radio structure, cloud B is situated at a position where the nonthermal filaments appear to bend.

The negative velocity feature has highly "forbidden" velocities and displays one of the steepest velocity gradients in the Galactic Center region.

A second molecular cloud (identified as "Cloud B", lower center of Fig. 1 and Fig. 3) is observed along the curved portion of nonthermal filament. This cloud also exhibits a velocity gradient towards the nonthermal filaments. The CS(2-1) emission from Cloud B is relatively strong. In fact, the brightness ratios of CS to HCO$^+$ are very different in the two clouds described here. Keeping the slightly lower critical density of HCO$^+$(1-0) in comparison to CS(2-1) in mind, this finding could indicate a different chemistry in both clouds. D. Jansen (1995) e.g. argues that HCO$^+$ emission in comparison to most other neutral molecules, appears to be more enhanced in diffuse molecular clouds which can be penetrated by ambient UV fields. Sternberg (1995) points out that HCO$^+$ is efficiently produced in the hot HI/H$_2$ regions of PDRs whereas the formation of CS is more efficient in the more inner cold SI layers of PDRs in which the molecules are more shielded from the ambient UV field. Cloud A, with the smaller of the CS to HCO$^+$ intensities, is situated closely adjacent to the diffuse HII region at the filaments tip, whereas Cloud B is located near the nonthermal filaments itself.

Cloud B also appears to be physically associated with the filaments. It has

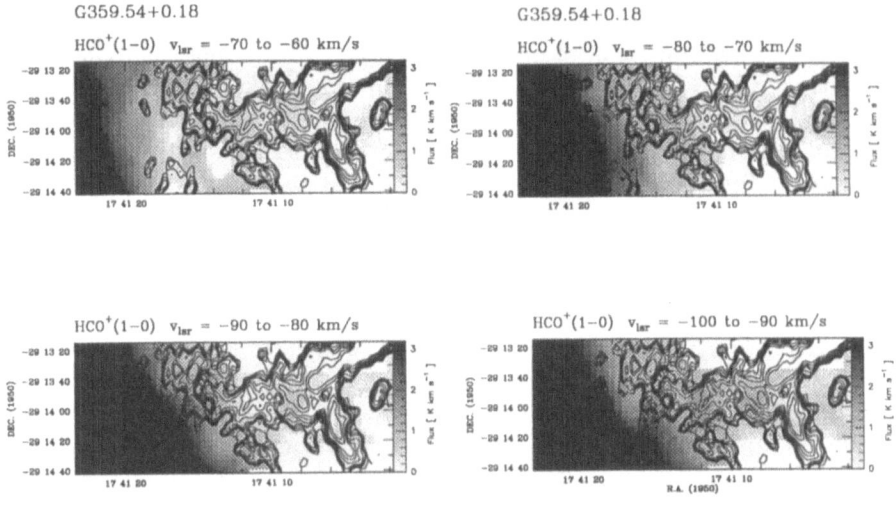

Grey scale: Integrated molecular line flux over the velocity range as specified above
Contours: Continuum flux at 5GHz: Levels [mJy/beam]: .15, .2, .25, .3, .4, .5, .6, .8, 1.0

Fig. 2. 5 GHz continuum flux (contour lines) superimposed on $HCO^+(1-0)$ channel maps of cloud A, integrated in the intervals as indicated. Note the velocity gradient in the molecular line emission towards the interface region of the 5GHz continuum source. The H79α recombination line velocity of -106 km s^{-1} fits well to the the observed molecular velocity gradient.

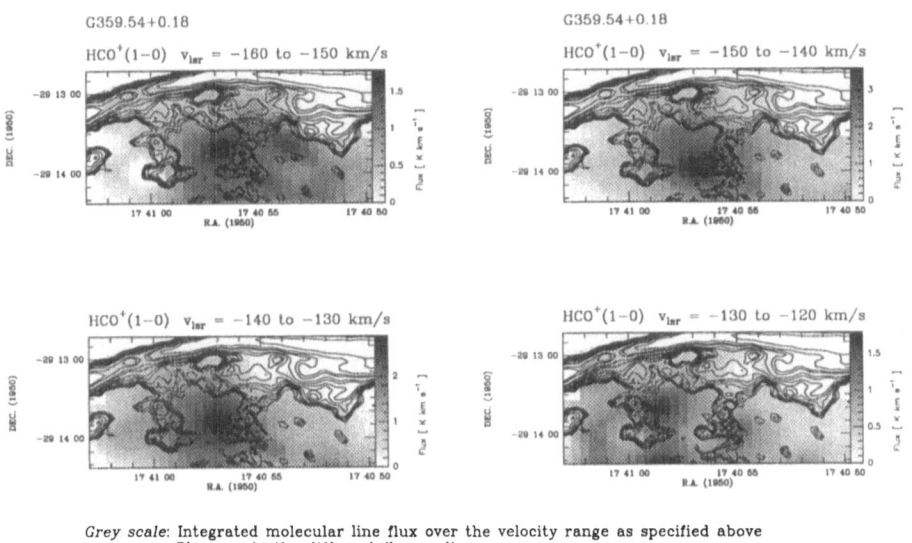

Grey scale: Integrated molecular line flux over the velocity range as specified above
Please note the different flux scaling
Contours: Continuum flux at 5GHz: Levels [mJy/beam]: .15, .2, .25, .3, .4, .5, .6, .8, 1.0, 1.2, 1.4, 1.6, 2.0, 3.0, 5.0, 6.0, 7.0

Fig. 3. 5 GHz continuum flux (contour lines) superimposed on $HCO^+(1-0)$ channel maps of cloud B, integrated in the intervals as indicated. The molecular cloud appears to be situated at a position where the nonthermal filaments appear to bend.

been previously suggested by Bally & Yusef-Zadeh (1989) that the bend in the nonthermal filaments is in response to a collision with a molecular cloud. The location of Cloud B within the bend of the filaments makes it a prime candidate for the colliding partner. As with the case of Cloud A, the molecular line data towards this region shows possible kinematic evidence of a collision — a steep velocity gradient, directed toward the nonthermal filaments, is observed within Cloud B (Fig. 4). The role that Cloud B possibly plays in the illumination of the nonthermal filaments, however, is yet to be explored.

Our observations of the G359.54+0.18 filament system support the scenario, proposed by Serabyn & Morris (1994), whereby the nonthermal filaments are illuminated by synchrotron emitting electrons accelerated by the process of magnetic field line reconnection instigated by colliding molecular clouds. HII regions are also involved in this process — as is observed toward at least three other potential cloud/filaments pairs. Some degree of preionization is possibly needed in order that the cloud is able to effectively interact with the magnetic field component and/or, perhaps, a nearby HII region is required to provide enough free electrons for subsequent acceleration along the filaments. We find evidence of all three components requisite for the cloud-collision/magnetic-reconnection scenario: (1) a set of nonthermal filaments supposedly delineating large scale GC magnetic field lines, (2) two molecular clouds adjacent to the filaments, and (3) an HII region at the tip of the filaments, evidenced by structure in the radio continuum images and by the detection of H79α recombination line emission. Moreover, a direct association between molecular Cloud A and the HII region is suggested by their similar H79α recombination line and molecular emission line velocities.

What is missing is a detailed insight to the presumed interaction region with sub-arcsecond resolution. At a resolution of 0.1", which would be possible to achieve with the *LSA*, one could resolve molecular structures smaller than 1,000 AU in the Galactic Center. One would be able to observe e.g. streamers of gas, evaporating from single molecular clump condensations, a mechanism which most likely happens at the interface region seen in Fig. 2. A comparison with sub-arcsecond resolution NIR observations and high resolution Radio observations would doubtless be a powerful tool to reveal the physical processes responsible for the existence of such phenomena as the nonthermal filaments in the Galactic Center.

References

J.T. Armstrong & A.H. Barret, 1985, APJS, 57, 535;

J. Bally, A.A. Stark, R.W. Wilson, & C. Henkel, 1987, APJS, 65, 13;

J. Bally & F. Yusef-Zadeh, 1989, APJ, 336, 173;

D. Jansen D. Jansen, Thesis, Leiden 1995;

A. Sternberg, 1995, in "The Physics and Chemistry of Interstellar Molecular Clouds', Proceedings of the 2nd Zermatt Conference, ed. by G. Winnewisser and G.C. Pelz, Springer Verlag, Heidelberg 1995;

E. Serabyn & M. Morris, 1994, APJ, 424, L91;

Galactic Molecular Clouds

T.L. Wilson[1,2], R.A. Gaume[3], K.J. Johnston[3], J. Schmid-Burgk[1]

[1] Max-Planck-Institut für Radioastronomie, Auf dem Hügel 69, D-53121 Bonn, Germany

[2] G.A. Miller Professor, Astronomy Dept., University of Illinois, 1002 W. Green St., Urbana, Ill., 61801, USA

[3] U.S. Naval Observatory, 3450 Massachusetts Av. NW, Washington, D.C. 20392-5420 USA

Abstract. Molecular cloud studies allow determinations kinetic temperatures, H_2 densities, **B** fields and abundances. These results allow a determination of the chemical and isotopic content of the the interstellar medium (ISM). Data for clouds in our galaxy, especially those near the galactic center are needed to interpret results obtained for other galaxies. Molecular clouds are the birthplaces of stars. For investigations of star formation and the interaction of young stars with molecular clouds, measurements of abundant polar molecules, such as carbon monoxide, on scales of $< 10^{15}$ cm are needed.

1 Introduction

Molecular clouds are composed of $\sim 70\%$ H_2 by mass. The H_2 is formed on the surface of dust grains from atomic hydrogen, and destroyed by ultraviolet spectral line radiation from the interstellar radiation field. The H_2 in the inner part of a molecular cloud are shielded by both dust and H_2 in the outer parts of the molecular cloud. This is a complex balance of production and destruction processes, typical of interstellar chemistry: *one cannot separate molecular clouds and chemistry.* Usually, H_2 does not emit spectral lines, so the properties of molecular clouds must be determined by measurements of gas phase polar species or dust continuum. Because interstellar chemistry can drastically affect the abundances of rarer species, studies of the most abundant polar molecule, CO, are needed to provide a reliable mass tracer (see, e.g., the articles in Wilson & Johnston 1994).

2 Classifications and Cloud Physics

Wilson & Walmsley (1989) have summarized molecular cloud properties. There are two large groups of clouds in the disk of our galaxy: Giant Molecular Clouds (GMC's) and Dark Dust Clouds (DDC's). GMC's are actually complexes; these are the birthplaces of higher mass stars, and have kinetic temperatures, T_k of ~ 20K, caused by heating by embedded high mass stars. The observed linewidths, $\Delta V_{1/2}$, measured using the CO molecule are a few $\mathrm{km\,s^{-1}}$, far larger than the width caused by thermal broadening. The DDC's are the birthplaces of lower mass stars. T_k in DDC's are ~ 10K; it is believed that these clouds are heated by cosmic rays. Both the GMC's and DDC's are thought to be close to virial equilibrium, although the lower density envelopes may not be. There are two additional categories: (1) Cirrus–like emission seen at high galactic latitudes; these clouds are not self gravitating, but may be partly stabilized by external pressure (see, e.g., Blitz 1990), and (2) Molecular clouds within a few hundred parsecs of the center of our galaxy.

In regard to heating, one example is the densest part of the Orion KL cloud, also known as the 'Hot Core' (see, e.g., Genzel & Stutzki 1989). In Orion, dust is heated by embedded sources, and the gas is heated by collisions with hot dust. At the H_2 densities in the Orion Hot Core, $\sim 10^7 cm^{-3}$, the dust and gas temperatures should be equal. For clouds without embedded high mass stars, the heating is caused by cosmic rays. If cosmic rays originate in the inner region of our galaxy, the cosmic ray rate will be lower in the outer parts of our galaxy. Thus, clouds there may be cooler. Lequeux et al. (1993) report two examples of clouds for which $T_k \sim 3K$, but Wilson & Mauersberger (1994) have found at least one case which contradicts this picture. Completely different is the heating of galactic center clouds. These are characterized by strikingly larger linewidths, $\Delta V_{1/2}$, usually of order 10 to 20 $km\,s^{-1}$. In many cases, galactic center clouds show radial velocities which are forbidden by the usual rotational velocities of our galaxy. These clouds are much hotter than normal with significant amounts of material having $T_k = 100K$ over very large regions. There are a number of possible heating processes: (1) Embedded stars, (2) A vastly enhanced cosmic ray rate, (3) Ambipolar diffusion (ion–slip heating), or (4) Cloud–cloud collisions. The first heating process requires a large number of high mass stars, and thus radio continuum emission. The level of such continuum emission is too low at the positions of these clouds. In addition, embedded stars should also heat the dust. This is the case for the core of SgrB2, where >57 HII regions are found, but *not* for other galactic center clouds. For these, the dust is *cold*, while the gas is *hot*. Processes (2) to (4) are direct, that is, would heat the gas but not the dust. However, in the case of Process (2), the cosmic ray rate must be > 100 times the local value; with this cosmic ray flux, most of the H_2 would be destroyed. Heating process (3) can easily function if the clouds are weakly ionized, with a fractional electron abundance of 10^{-7}. However, the measured magnetic field strengths appear to be too small in galactic center clouds. Heating process (4) is given strong observational support from the detection of widespread emission from the J=2–1 rotational transition of the silicon monoxide molecule, SiO, in galactic center clouds (Hüttemeister et al. 1996). The connection between cloud collisional heating and the presence of gas phase SiO is that collisions destroy dust grains which are composed of graphite and silicon. The gas phase Si can then then combine to form SiO. The most simple picture involves the collision of clouds with speeds of $\sim 10 km\,s^{-1}$, which is also consistent with the observed linewidths. The cloud–cloud collisions are enhanced in the galactic center region because of the deep potential well caused by the high stellar mass concentrated near the center. This is not the case for the disk of our galaxy, so the effect of heating through cloud–cloud collisions is negligible.

3   The Need for the Large Southern Array

So far, molecular clouds have been investigated mostly with single radio telescopes, so that the *best* angular resolution is 8″, with a more typical value of 12″. Compared to single telescopes, interferometers have higher angular resolutions and better pointing accuracies. With a single telescope, the angular resolution and collecting area are directly related. For unfilled apertures the sensitivity in temperature is usually low. If all other factors are kept the same, sensitivity improves at shorter wavelengths. However, on the best sites and with the best receiver systems, the sensitivity can be improved only by enlarging the collecting area. In addition, one also needs high dynamic range which requires many antennas. The LSA design is similar to that used successfully with the Very Large Array (VLA) of the U.S. National Radio Astronomy Observatory, so dynamic range will be excellent. The collecting area of the LSA is somewhat less than that of the VLA, so that the flux density sensitivity is less than that of the VLA. However, because of the Rayleigh-Jeans relation, the *temperature* sensitivity of the LSA will be better than that of the VLA. In the following, a few clouds containing high mass stars are discussed. These are a distances of ∼500 pc, so $0.1″ = 2.4\ 10^{-4}$ pc$=7\ 10^{14}$ cm. For lower mass star forming regions, the distances are 160 pc. Thus to reach the same linear scale, the angular resolution could be ∼0.3″. However, the observed $\Delta V_{1/2}$ in cores of DDC's are ∼0.1 $km\,s^{-1}$, so the velocity resolution at $\lambda = 3$mm must be 25 kHz. For LSA images of extended clouds, mosaicing will be needed; also, additional measurements are needed to sample structures larger than >30″. In addition, the possibility of polarization measurements are needed for B field determinations based on the Zeeman effect in the line radiation of certain molecular species such as SO, or from polarized dust continuum emission.

4   A Few Selected Results

4.1   Molecular Clouds and 'Protostars' in the NGC2024 Region

There are two groups of molecular clouds associated with the HII region NGC 2024. The clouds with higher densities are located behind the HII region. These were mapped with a 12″ resolution in the 1.3 mm dust continuum (Mezger et al. 1988). Six maxima (cataloged as FIR 1 to 6), were interpreted as ultra dense molecular regions, thought to be in the process of collapse. Follow up measurements in spectral lines and continuum were made by Mauersberger et al. (1992) and Mezger et al. (1992). To measure the column density of H_2, Mauersberger et al. (1992) used measurements of the rare isotopomer $C^{18}O$ in order to avoid saturation effects. With the IRAM 30m telescope, measurements of $J = 2-1$ line give a 12″ resolution; these measurements provide an excellent estimate of the total column density of $C^{18}O$ over a wide range of T_k and n(H_2). The results showed that $X(CO) = \frac{CO}{H_2}$ in the densest regions is lower than the usually accepted ratio $X(CO) = 10^{-4}$. However, the positions of line and continuum maxima were not

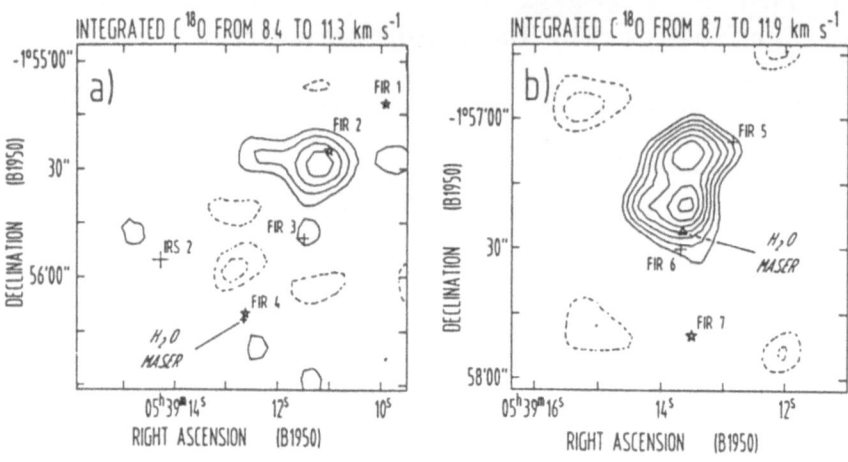

Fig. 1. The solid and dashed lines are contours of the integrated intensity of the $J = 1 - 0$ line of $C^{18}O$ in the NGC 2024 region. This image was produced with the BIMA array (angular resolution $\sim 10''$ at 2.7 mm; Wilson et al. 1995). The stars indicate the positions of 1.3 mm dust continuum emission sources (labelled FIR) which were found by Mezger et al. (1988), but not detected in the 2.7 mm continuum with the BIMA array. Those FIR sources detected by Wilson et al. (1995) at 2.7 mm are indicated by crosses; in all cases the spectral indices of the FIR sources, α are $\geq \nu^4$. The triangles are H_2O maser positions, newly determined using the VLA by Gaume et al. (1996) with a $0.1''$ resolution.

completely certain. Simultaneous mapping of $C^{18}O$ and dust continuum, carried out with the BIMA array (Wilson et al. 1995) showed that the dust and $C^{18}O$ maxima are indeed spatially separated. There are two possible interpretations of these results. Mezger et al. (1988) argued that the densities obtained from dust measurements are $\sim 10^9 \text{cm}^{-3}$, later revised to $n(H_2) \sim 10^8 \text{cm}^{-3}$. Even at the lower density, molecules will not be in the gas phase, but frozen onto dust grains, since the time scale at a density of $\sim 10^8 \text{cm}^{-3}$ for CO–grain collisions for $T_k = 20K$ is 230 years whereas the free fall time is $6\ 10^3$ years (see Chapter 14 of Rohlfs & Wilson 1996). However, some of the molecules on grain surfaces *must* be returned to the gas phase since cosmic rays will occasionally strike grains, heat these and free these molecules. Then the relative abundance of CO will be 10% of the usual value, but still easily detectable. In the BIMA array images, there are compact molecular clouds near the dust maxima, but there is *no* CO toward the dust maxima. In these molecular clouds, the H_2 density is 10^7cm^{-3}. In the extended molecular gas, with $n(H_2) = 10^5 \text{cm}^{-3}$, $X(CO) = 10^{-5}$. The same X is found in gas with $n(H_2) = 10^7 \text{cm}^{-3}$. However, in the FIR sources, with a density of 10^8cm^{-3}, the limit to X is a factor of 10 lower. It seems *highly unlikely* that the factor of 10 increase in H_2 density can cause a sudden decrease in the relative abundance of CO, while there is no such decrease from H_2 densities ranging from 10^5cm^{-3} to 10^7cm^{-3}. Thus 'freezing out' of CO in the FIR regions is unlikely. Support for this conclusion is based on an analysis of the spectral indices, α, of the FIR dust emission. All α values are $\sim \nu^4$, which is usually taken to be a signature of dust grains with normal sizes. This is the result one would expect for regions of low, but not very high, H_2 density. Taken together, these results allow the following interpretation: The dust maxima are the locations of embedded low mass stars. The molecules in these regions have been destroyed by shock waves from the stars; in the neighborhood of these stars there are dense molecular clouds in which stars may form or are in the process of forming, but these star forming regions are weak infrared emitters, which is what one would expect if the dust is cool, as expected for clouds *without* embedded stars. Richer et al. (1992) have determined that FIR 5 and 6 seem to be outflow sources; this strengthens the notion that these FIR regions contain stars.

New results (Gaume et al 1996) provide extremely accurate positions for the H_2O masers in NGC 2024. The northernmost maser is nearly coincident with the dust continuum source FIR 4. This region is though to be associated with a reflection nebula. The southernmost H_2O maser is definitely north of the location of FIR6. These masers may mark the locations of highly obscured, newly formed stars. However, these are more likely the locations where outflows, such as from FIR6, interact with quiescent clouds. Chernin (1996) has mapped the CO emission in NGC 2024 with the BIMA array, and finds line wings which are centered at locations offset from the FIR sources. The CO linewings near FIR6 are close to our position for the H_2O maser. Chernin (1996) has interpreted the CO data in terms of highly obscured embedded young stars or protostars. Sensitive, high dynamic range, $<1''$ resolution images of the CO high velocity line wings should lead to a resolution of this question.

4.2   The Orion South Region

In the 400 μm continuum map of the OMC 1 cloud by Keene et al. (1982) there are two prominent maxima. The more intense source is the Orion KL nebula. The second peak is located $\sim 100''$ to the south. Since there is no radio continuum maximum at cm wavelengths near this position (Garay et al. 1986), the mm wavelength continuum cannot be caused by free-free emission, but rather by hot dust (Mundy et al. 1986). Interferometer studies provide an accurate position and show that this source has a FWHP size of $11''$ by $7''$ (Mundy et al. 1986 and McMullin et al. 1993). Ziurys et al. (1990) found bipolar emission in lines of SiO oriented approximately N-S from this region, with a velocity full width to zero power (FWZP) of $25 \mathrm{km\,s}^{-1}$. In CO there is a highly collimated outflow lobe extending $2'$ to the SW from this maximum (Schimd-Burgk et al. 1990). The CO outflow has at most only a relatively small extension in the NE direction, which may indicate that the NE flow is hindered by the presence of the Trapezium. If so, the outflow source is close to the H II/H_2 interface. Observations of transitions of SiO, SO_2, CH_3OH and HC_3N at $\lambda = 3$ mm with a resolution of $10''$ have been presented by McMullin et al. (1993). From the presence of the CO outflow, and the small sizes of the infrared sources, there must be a compact energy source. From 100-m observations there are a number of H_2O masers in this source. These were followed up with VLA measurements, in order to determine the

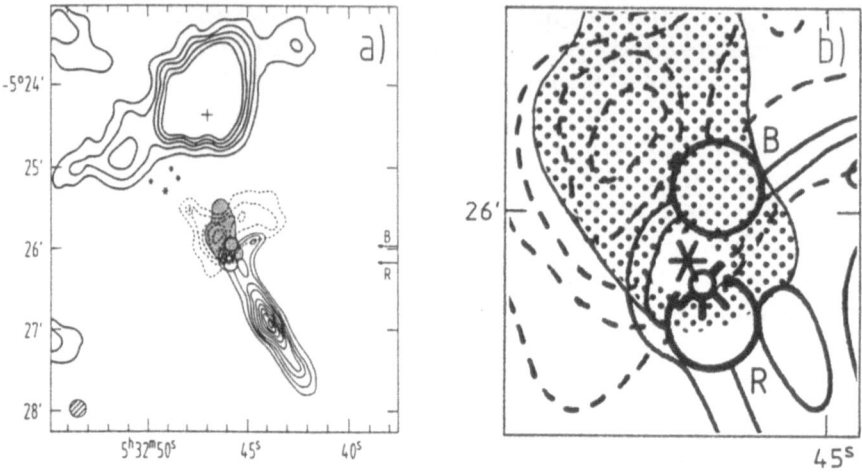

Fig. 2. a) South of the Trapezium, denoted by 4 stars, are the CO J=2–1 integrated intensities for OMC S (red shifted V_{lsr}=13 to 16 km s^{-1}, thin solid, and blue shifted, 0 to 3 km s^{-1}, dashed contours). The units are the same for both (Schmid-Burgk et al. 1990). The 1.3 mm dust emission (Mezger et al. 1990) is shown shaded. **b)** An expanded version of **a)**; we use a star to mark the position of the H$_2$O maser center (Gaume et al. 1996). The two open circles 'B' and 'R' show the location of the the SiO quasi–thermal emission (Ziurys et al. 1990). The open circle with rays marks the position of the dust continuum maximum of Mezger et al. (1990), which is (-1.5″, -4″) from the interferometric position of McMullin et al. (1993).

center of the outflow with high accuracy, and to provide new information about the exciting source and the nearby molecular regions. These results show that the largest concentration of masers, with the largest spread in V_{lsr}, is centered on $\alpha = 5^h\ 32^m\ 45.96^s$, $\delta = -05°\ 26'\ 04.94''$ (1950.0). We identify this H$_2$O maser center with the source of the CO outflow and the heating source in OMC S. This position is $(-2.5'', -3.2'')$ from the most accurate dust continuum position (McMullin et al. 1993). Although this H$_2$O maser is not centered on the dust continuum peak, the dust emission is extended over about 10″. McMullin et al. (1993) find that the most intense quasi-thermal emission from SiO is, as are the H$_2$O masers, SW of the dust continuum maximum. OMC S does *not* show the richness of molecular emission found in the Orion Hot Core: McMullin et al. (1993) interpreted the differences in abundance between OMC S and Orion KL as being due to age. In their view the chemical evolution of OMC S is less advanced than in Orion KL. Since kinematic evidence indicates that the outflow in Orion KL is less than 10^4 years old, the OMC S source must be even younger. On this basis, OMC S would be one of the youngest stars in the galaxy. The images of quasi-thermal molecular emission have angular resolutions of $\sim 10''$. To match the H$_2$O maser results, and to adequately image the star formation region in the OMC S cloud, a resolution of 0.1″ is needed.

4.3 The Orion KL Region

The Orion KL region is the nearest rich source of complex molecules in the ISM. Downes et al. (1981) had argued that IRc2 is the sole heating source of the Hot Core molecular cloud and is the cause of the CO outflow in the Orion KL nebula. Averaged over the 24″ beam of the 100 meter telescope, Wilson et al. (1994) found a T_k gradient in NH_3 emission, with T_k values from 165K to 400K. In Fig. 3, we show the distribution of the high excitation NH_3 from VLA measurements of the $(J,K)=(10,9)$ inversion line, with positions of infrared/radio continuum sources. According to Menten & Reid (1995), the position of their source 'I' coincides with the SiO maser peak, but *not* with the mid–IR source IRc2; they suggested that source 'n' may contribute to the heating of the Hot Core. From 1.3 mm dust continuum data, Mezger et al. (1990) have also argued that the heat source is located 2″ south of IRc2, near the center of the Hot Core. In contrast, our VLA data show that molecular material requiring higher excitation is closer to IRc2, or source 'I', but not 'n'. Thus, it would appear that either IRc2 or 'I' (or both) provide the excitation for the hottest molecular gas.

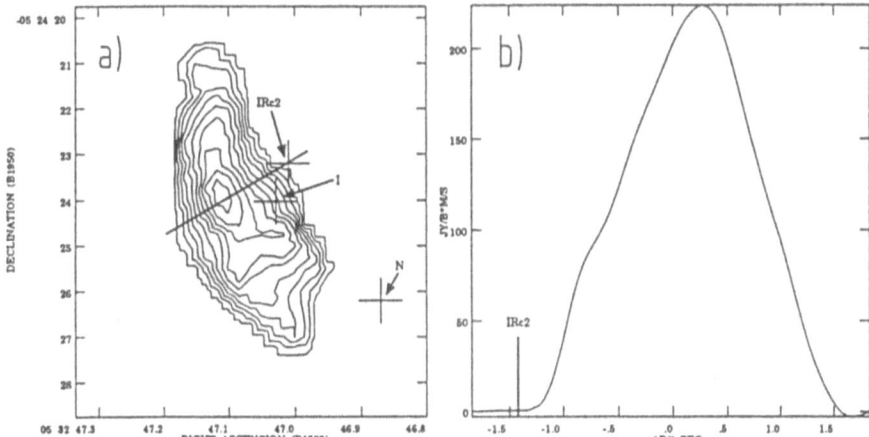

Fig. 3. a) A superposition of our VLA map of the integrated intensity of the $(J,K)=(10,9)$ inversion line of NH_3 on the locations of the near IR sources (Menten & Reid 1995) 'I', which has SiO maser emission, the peculiar radio double source 'n' and the mid–IR source IRc2. b) The fall off of NH_3 intensity toward IRc2 may be caused by a decline in the abundance of this fragile species, due to destruction. The steep decline of emission away from IRc2 is caused by a complex excitation effect, not simply a decrease in T_k. The (10,9) emission is north of the center of NH_3 emission from lower excitation lines (cf. with data of Migenes et al. 1989).

On the basis of VLA measurements of the $(J,K)=(4,3)$ line of NH_3, and assumptions about geometry and clumping, Hermsen et al. (1988a) showed that the southern part of the Hot Core is too warm to be heated by IRc2 alone. Thus, additional heat sources must be present. A more complete investigation of Hot Core heating involves a multi-line NH_3 study, since the excitation of the (10,9) line requires both high T_k and large $n(H_2)$ or intense IR fields. The (10,9) VLA data shown in Fig. 3 arise from a region of FWHP size $\sim 5''$ by $\sim 1.5''$, which is considerably smaller than the size of the Hot Core as observed in lower excitation metastable inversion lines of NH_3. The centroid of the (10,9) region is shifted to the NE of the center of the Hot Core region as seen in the (3,2) inversion line of NH_3 (Migenes et al. 1989). The peak temperature of the (10,9) inversion line is $\sim 90K$. If the excitation temperature is 165K (the cooler component in the Hot Core, from Hermsen et al. 1988b) the line optical depth is ~ 0.8. This is an *upper* limit to $\tau(10,9)$. In addition the the T_k map provided by the NH_3 images would be a $<1''$ image in the J=2–1 line of $C^{18}O$.

References

Blitz, L. 1990 in IAU Symp. 144, *The Interstellar Disk-Halo Connection in Galaxies* ed. H. Bloemen, Kluwer, Dordrecht, p. 41.

Chernin, L. 1996 Ap. J. (in press)

Garay, G., Moran, J.M., Reid, M. 1987 Ap. J. , 314, 535

Gaume, R.A., Wilson, T.L., Johnston, K.J. 1996 (in prep)

Genzel, R. & Stutzki, J 1989, Ann Rev A&A , 27, 41

Hermsen, W., Wilson, T.L., Bieging, J.H. 1988a A&A 201, 276

Hermsen, W., Wilson, T.L., Walmsley, C.M., Henkel, C. 1988b A&A 201, 285

Hüttemeister, S., Wilson, T.L. et al. 1996 (in prep)

Keene, J., Hildebrand, R.H., & Whitcomb, S.E. 1982, Ap. J. , 252, L11

Lequeux, J., Allen, R.J., Guilloteau, S. 1993 A&A 280, L23

Mauersberger, R., Wilson, T.L., Mezger, P.G., Gaume, R.A., Johnston, K.J. 1992 A&A 256, 640

McMullin, J.P., Mundy, L.G., Blake, G.A. 1993, Ap. J. , 405, 599

Menten, K. M., Reid, M.J. 1995 Ap. J. 445, L157

Mezger, P.G., Chini, R., Kreysa, E., Wink, J.E., Salter, C. 1988 A&A 191, 44

Mezger, P.G., Wink, J.E., Zylka, R. 1990, A&A , 228, 95

Mezger, P.G., Sievers, A. et al. 1992 A&A 256, 631

Migenes, V., Johnston, K.J., Pauls, T.A., Wilson, T.L. 1989 Ap. J. 347, 294

Mundy, L.G., Scoville, N.Z., Baath, L.B., Masson, C.R., Woody, D.P. 1986, Ap. J. , 304, L51

Richer, J.S., Hills, R.E., Padman, R. 1992 MNRAS 254, 525

Rohlfs, K., Wilson, T.L. 1996 *Tools of Radio Astronomy*, Springer-Verlag, Heidelberg, in press.

Schmid-Burgk, J., Guesten, R., Mauersberger, R., Schulz, A., Wilson, T.L. 1990, Ap. J. , 362, L25

Wilson, T.L., Walmsley, C.M. 1989 A & A Rev. 1, 141

Wilson, T.L., Henkel, C. et al. 1993 A&A 276, L29

Wilson, T.L., Mauersberger, R. 1994 A&A 282, L41

Wilson, T.L., Johnston, K.J. 1994 (eds.) *The Structure and Content of Molecular Clouds*, Springer Lecture Notes in Physics, 439, Springer-Verlag, Heidelberg

Wilson, T.L., Mehringer, D.M., Dickel, H.R. 1995 A&A 303, 840

Ziurys, L.M., Wilson, T.L. & Mauersberger, R. 1990, Ap. J. , 356, L25

Millimeter-wave absorption spectroscopy of molecular clouds

Robert Lucas

Institut de RadioAstronomie Millimétrique, 300, rue de la Piscine, Saint-Martin d'Hères, France

Abstract. We review recent observations of millimeter wave absorption line measurements in front of compact extragalactic radio sources. The availability of a new instrument with a sensivity gain of an order of magnitude would considerably imcrease the number of available sources, with the potentiality of improving our knowledge of the molecular phase of the interstellar medium, and to boost the study of chemistry in diffuse/translucent and possibly dense molecular clouds regions. Our survey of HCO$^+$ absorption shows that the population of detected clouds has statistical properties resembling that of HI clouds. Isotopic abundances in the CO molecules show strong fractionation, giving and indication of rather low kinetic temperatures. A close relationship between OH and HCO$^+$ abundances is found, which is not fully understood.

1 Introduction

Observations of interstellar line absorption in front of distant radio sources provide direct measurements of the line optical depth, averaged over the solid angle of the source. Many such measurements have been made at centimeter wavelengths, and have provided valuable information on the abundances and excitation conditions of interstellar molecules observable in this domain. Absorption measurements are even more important at millimeter waves, where many more molecular transitions occur. Direct measurements of CO optical depths are of considerable interest, since the process of CO line formation, 25 years after its discovery, is far from being well understood. Ideally statistics on the optical depths in front of point sources should give information on cloud structure, through both area and velocity filling factors, and thus give clues to line formation mechanisms.

Absorption lines also provide a direct measurement of molecular abundances, in low density regions where rotational transitions are in radiative equilibrium with the cosmic background radiation. In these conditions *emission measurements are impossible*, since the number of emitted photons is compensated by an equal number of absorbed background photons. This is the case for most molecules, which have dipole moments higher than ~ 1 debye. If some collisional excitation occurs, as for CO and its isotopomers, then several transitions have to be observed to sample different levels in the rotational ladder. But in most cases collisions may be neglected, and column densities are accurately derived by assuming an excitation temperature of 2.73 K. The interest for interstellar chemistry is obvious: one may explore the low density molecular regions, and compare the derived abundances with those measured in the optical window by observing interstellar absorption lines in front of stars. Furthermore the abundances have a better accuracy than those derived from emission measurements, since the excitation is expected to be accurately known.

Several studies of molecular absorption in front of extragalactic radio sources have already been published by Marscher et al. (1991), Lucas and Liszt (1993, 1994), Lequeux et al. (1993), Hogerheijde et al. (1995), Kobulnicky et al. (1995). They show some common characteristics: although optical extinction in those lines of sight are moderate (typically 1 mag. or lower), many molecules are present. Due to its high dipole mement, HCO^+ is easy to detect, with optical depths of order of unity. HCN, CN, HNC, C_2H are also found. ^{13}CO is also detected in some lines of sight. The HCO^+ absorption line spectra are generally richer, broader, more complex than the corresponding ^{12}CO emission spectra. The derived molecular abundance ratios are similar in magnitude to those in dark clouds (Lucas and Liszt 1993, 1994), while the absolute column densities and, when available, interstellar extinction, imply that the ultraviolet radiation field should be the dominant factor for chemistry.

2 Instrumental requirements

In this section we summarize observational constraints on millimeter wave absorption studies. Most extragalactic sources have very small extent at millimeter wavelengths, since the extended decimeter wave halos generally have steep synchrotron spectra. Although extended galactic sources can in principle be used, they are quite often associated with local emission from dense gas, and separating absorption from emission seems almost an unfeasible task (Liszt and Lucas 1995a). Even in the case of point-like sources (VLBI objects), interferometers with long baselines are preferable to resolve out any emission; single dishes such as the 30-m cannot be used when strong emission is present, such as in the CO transitions. This is illustrated in Fig. 1.

The flux of the background source should be strong enough to measure the absorption spectrum in a reasonable time: in Table 1, b is the bandwidth in MHz, t the integration time in hours. The second column refers to the current Plateau de Bure array, with its 4 antennas. We assume here a line width of 1 km s^{-1}, typical of the observed absorption lines. In 3 hours, with Plateau de Bure, absorption spectra may be observed in front of sources of flux larger than 2 Jy. Table 1 gives the number of those sources, the solid angle in which one such source should be present on average, as well as the corresponding numbers for the proposed large array.

Apart from sensitivity requirements, these observations are quite simple: the background source is observed continuously, and the phase in a continuum bandwidth (1 GHz) is measured in less than 1 minute if the source is strong enough to be studied. Thus self calibration may be applied, and an atmosphere coherence time of typically one minute is only needed. With atmospheric radiometric phase

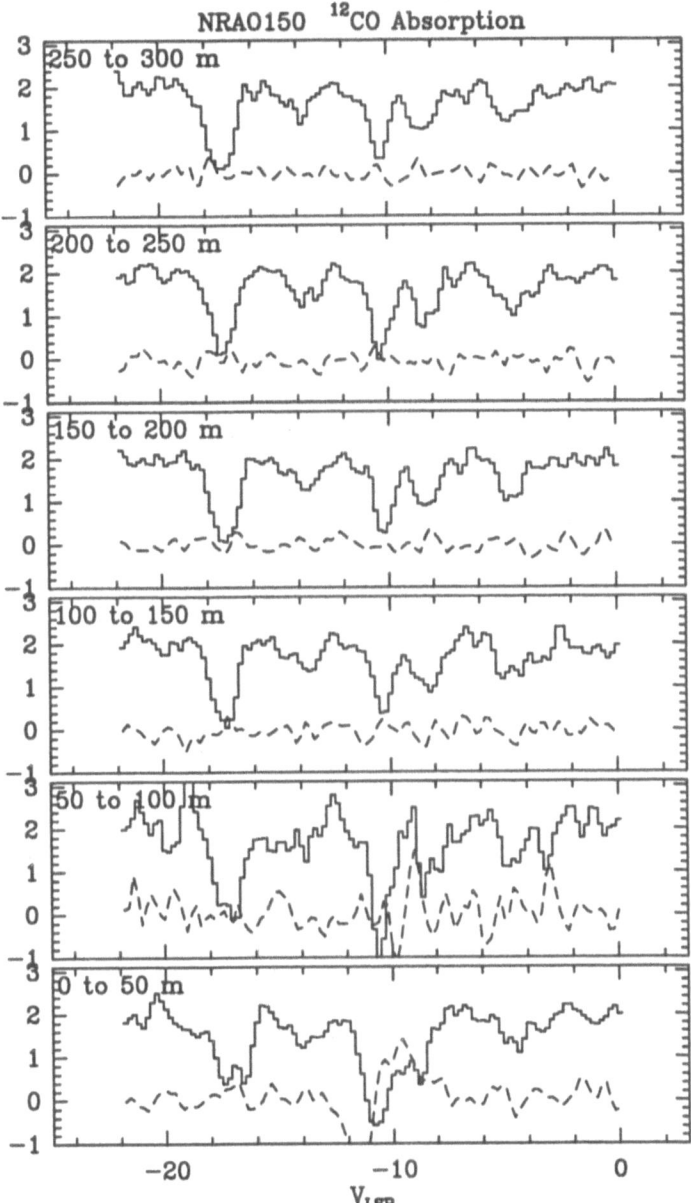

Fig. 1. Absorption spectra of CO ($J = 1-0$) in front of B0355+508 (NRAO 150), split in various baseline length intervals. The continuous line is the real part of the visibility, the dashed line its imaginary part, which should be zero in the absence of emission. One sees that foreground emission is not totally resolved out for baselines smaller that about 150m. B0355+508 is at a galactic latitude $b = -1.6°$.

Table 1. Observational parameters

	Plateau de Bure (4)	LSA
ΔS [mJy]	$\frac{20}{\sqrt{b}\,t}$	$\frac{0.9}{\sqrt{b}\,t}$
Limiting flux [mJy]	~ 2000	~ 100
Number of sources	~ 40	~ 4000
Solid angle	$\sim (30°)^2$	$\sim (3°)^2$

correction (Bremer et al., this volume), the weather requirements are even more relaxed: a coherence time larger than about 1 second is enough.

3 A Plateau de Bure HCO⁺ absorption survey

Lucas and Liszt (1996) have systematically searched, with the Plateau de Bure interferometer, for HCO$^+$ absorption in front of a sample of 30 sources, including all known sources with 3mm flux larger than 2.5 Jy at $b > 15°$, and larger than 2.0 Jy for $b < 15°$. The sample also included sources in the direction of which CO emission had been found by Liszt and Wilson (1993). A bandwidth of 100 MHz (336 km s^{-1}) was searched with a resolution of 140 kHz (0.47 km s^{-1}). All sources but one below $b = 15°$ showed HCO$^+$ absorption. A parallel search for HCO$^+$ emission only occasionally resulted in some weak lines (~ 0.05 K), even where optically thick absorption lines were present, implying that excitation temperatures must be very close to 2.73 K. Accurate column densities could thus be deduced for about 50 absorption features.

Fig. 2 shows some statistics on the HCO$^+$ absorption features, compared with similar data for 21cm HI absorption. In summary, absorption occurs in HCO$^+$ about $\sim 30\%$ as often as for 21cm HI, and optical depth distributions for HCO$^+$ and 21cm HI have very similar slopes.

4 Carbon monoxide

To further characterize the clouds observed in absorption it is important to obtain information on the CO content of those clouds. CO emission data have been obtained for most of these clouds, and CO absorption data for some of them (work in preparation). When both emission and absorption data are available in a given line of sight, one may derive excitation temperatures. The values found are in the range 3–8 K. These values are indicative of excitation conditions where $n(\mathrm{H}_2)T_{\mathrm{K}} \sim 3000 - 10000$ cm^{-3}K. Further constraints from HCO$^+$ emission, which is excited by electrons, imply $T_{\mathrm{K}} > 20 - 40$ K for most sources. At these excitation temperatures, about half of the CO molecules are in the lowest rotational state; the CO column densities can be accurately deduced from the

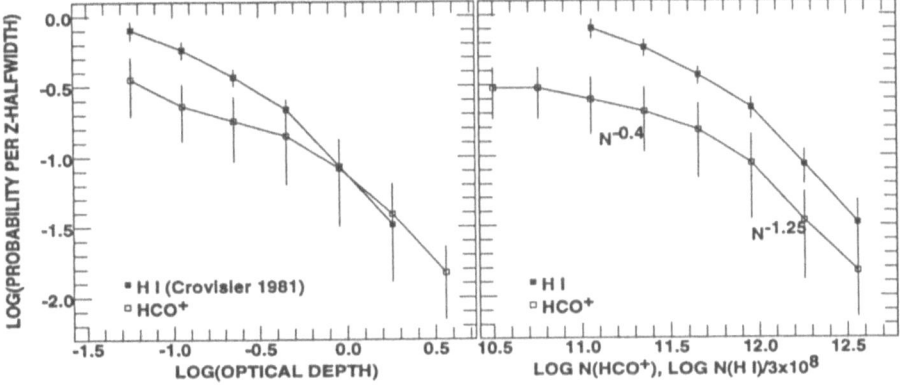

Fig. 2. *Left:* Statistical probability of encountering a feature of the specified optical depth over a path equivalent traversing half the galactic layer. The (very similar) distribution for H I clouds, taken from Crovisier (1981) is shown for comparison. *Right:* Same for column density. The distribution for H I, assuming constant linewidth and spin temperature, is shown for comparison.

$J = 1 - 0$ optical depth after correcting for the excitation. Fig. 3 shows a plot of ^{12}CO versus HCO$^+$ column densities.

Clearly, there is an underabundance of ^{12}CO for N_{HCO^+} lower than $\sim 10^{12}$ cm^{-2}; in the same range the strength of the ^{12}CO emission increases abruptly (Lucas and Liszt 1996). The same phenomenon is observed for CO column densities deduced from UV absorption studies at $N_{H_2} \sim 4\,10^{20}$ cm^{-2}. We believe we are witnessing here the onset of dust- and self-shielding at extinctions below 1 mag, in diffuse clouds and outer regions of dark clouds.

Further information on carbon monoxide may be deduced from isotopic data. In emission, the ^{13}CO to C^{18}O is generally much larger than 5.5 (the terrestrial ratio); in front of one source (B0415+379 = 3C 111), this ratio may be measured in absorption and is actually larger than 25. Direct measurement of the ^{12}CO to ^{13}CO ratio in absorption is possible in a few sources, and averages to about 20. It thus appears that ^{13}CO is quite overabundant in those clouds. We may invoke here chemical fractionation, resulting from the equilibrium reaction between ^{13}C$^+$ and ^{12}CO: this process increases the ^{13}CO abundance for T_K lower than 35 K. Selective photodissociation is expected in these regions of low extinction: ^{12}CO is self-shielded, resulting in a C^{18}O to ^{12}CO abundance ratio lowered by $\sim 5 - 8$; but the same effect should also apply to ^{13}CO. Very possibly both processes are active, with chemical fractionation overcoming the selective photodissociation; the implication is that the kinetic temperature should be rather low.

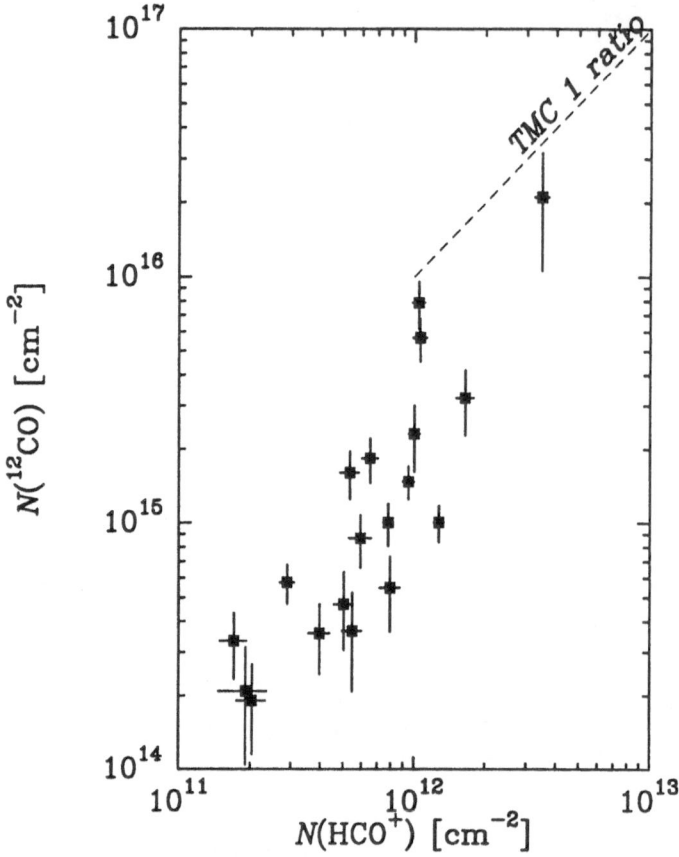

Fig. 3. CO column densities determined from absorption measurements, as a function of HCO$^+$ column densities.

5 Other molecules

We have measured 18 cm OH absorption with the VLA on some of the sources from the HCO$^+$ survey. To derive column densities from optical depth measurements at decimeter wavelengths one needs to know the excitation temperature. Absorption/emission observations by Dickey et al. (1981) in front of similar sources have led to surprisingly low excitation temperatures, exceeding the cosmic background temperature by typically 1 K. We have checked this in several directions with new emission measurements. It is obvious from Fig. 4 that OH and HCO$^+$ are closely related. The HCO$^+$ to OH abundance ratio varies from 0.03 to 0.05; this ratio is about the same as in dark clouds.

One may understand the close relation between these two species by considering the probable HCO$^+$ formation route:

$$C^+ + OH \rightarrow CO^+ + H \quad (k_1 = 8 \times 10^{-10} cm^3 s^{-1})$$
$$CO^+ + H_2 \rightarrow HCO^+ + H$$

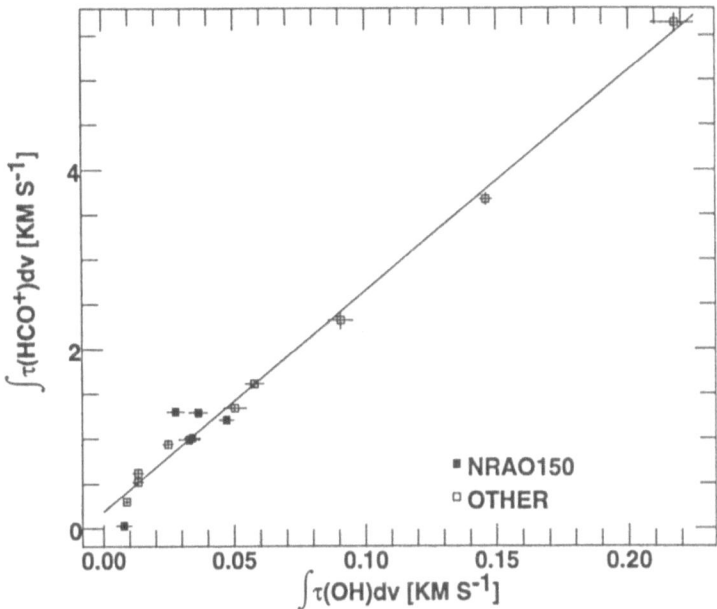

Fig. 4. Line profile integrals for OH and HCO$^+$. Column densities may be derived using: $N(\text{HCO}^+) = 1.03 \times 10^{12}\text{cm}^{-2} \int \tau(\text{HCO}^+)dv$, $N(\text{OH}) = 8.3 \times 10^{14}$ cm^{-2} $\int \tau(\text{OH})dv$ for $T_{\text{EX}} - T_{\text{CMB}} = 1$ K, a typical, but poorly-understood value.

Then HCO$^+$ recombines ($k_2 = 1.1 \times 10^{-7}(T/300)^{-1}$). Assuming $n(e) = n(C^+)$, one obtains $X(\text{HCO}^+)/X(\text{OH}) = k_1/k_2 = 0.0073(T/300)$. This is much lower than observed. New chemical models of diffuse clouds involving supersonic turbulence, and its dissipation, succeed to produce CH$^+$, and HCO$^+$ with endothermic reactions (Falgarone et al. 1995, Spaans et al. 1996). It not clear however whether the high observed HCO$^+$ to OH ratio is compatible with these models.

Liszt and Lucas (1995b) have shown, using 2mm and 6cm H$_2$CO absorption, that H$_2$CO turns on at $N_{\text{HCO}^+} \sim 10^{12}$ cm^{-2}. Similar studies of absorption lines of HCN, CN, HNC, C$_2$H, C$_3$H$_2$, CS, SO, H$_2$S, and SiO are in progress with the Plateau de Bure interferometer and the 30m telescope.

6 Variability

Marscher et al. (1993) and Moore and Marscher (1995) have found significant secular variations in 6 cm H$_2$CO absorption profiles in front of two sources, B0415+379 (3C 111) and B0355+508 (NRAO 150). This is interpreted as structure in the dense gas on scales of order 10 AU. If the variations reported in H$_2$CO are indeed density variations, rather than abundance variations, they should be visible in other species. We have started a search for such variability in the absorption lines of C$_2$H and HNC in front of the same sources, with negative results so far, over a 10-month time interval.

7 Conclusions and prospects

Recent experience on absorption spectroscopy of molecular clouds has shown absorption profiles with numerous line components, some of which have no counterpart in emission. An HCO^+ absorption survey gives evidence that these components sample a population of clouds with a distribution typical of those of HI clouds. ^{12}CO absorption reveals a transition in the ^{12}CO to HCO^+ abundance ratio at $N_{HCO^+} \sim 10^{12}$ cm^{-2}, which we identify as the photodissociation of ^{12}CO at $A_V \sim 1$. The ^{13}CO and $C^{18}O$ isotopomers have relative abundances strongly affected by selective photodissociation and carbon ion exchange. Comparison with OH absorption shows a very strong correlation of both species, indicating a very close chemical relationship, although not yet quantitatively understood. Absorption studies do reveal that the densest among diffuse clouds and/or the outer parts of dark clouds have a rich chemistry, although the UV field is strong.

These studies require a large collecting area: a gain in sensitivity by a factor of ~ 10 should enable statistical studies, and extend the sampled regions to denser clouds, since the number of available sources would be multiplied by about 100, and thus to more complex molecules. Molecules in diffuse clouds, however, have very low excitation temperatures, close to that of the cosmic background, and thus only levels with low rotational numbers are populated. Thus for species species containing more than one heavy atom, the millimeter window will be most useful.

References

Crovisier, J. (1981): A&A, 94, 162

Dickey, J. E., Crovisier, J., Kazès, I., (1981): A&A, 98, 271

Falgarone, E., Pineau des Forêts, G., Roueff, E. (1995): A&A 300, 870

Hogerheijde, M. R., de Geus, E. J., Spaans, M., van Langevelde, H. J., van Dishoeck, E. F. (1995): ApJ, 441, L93

Kobulnicky, H. A., Dickey, J. M., Akeson, R. L. (1995): ApJ, 443, L45

Lequeux, J., Allen, R. J., Guilloteau, S. (1993): A&A, 280, L23

Liszt, H. S., Lucas R. (1995a): A&A, 295, 811

Liszt, H. S., Lucas R. (1995b): A&A, 299, 847

Liszt, H. S., Wilson, R. W. (1993): ApJ, 403, 663

Lucas, R., Liszt, H. S. (1993): A&A, 276, L33

Lucas, R., Liszt, H. S. (1994): A&A, 282, L5

Lucas, R., Liszt, H. S. (1996): A&A, in press

Marscher, A. P., Bania, T. M., Wang, Z. (1991): ApJ 371, L77

Moore, E.M., Marscher, A.P. (1995): ApJ 452, 671

Marscher, A.P., Moore, E.M., Bania, T.M. (1993): ApJ 419, L01

Spaans, M., Black, J. H., van Dishoeck, E. F. (1996): ApJ, in press

Astrochemistry

Peter Schilke[1], Malcolm Walmsley[1,2]

[1] I. Physikalisches Institut, Universität zu Köln, Zülpicherstrasse 77, D-50937 Köln, Germany
[2] Osservatorio di Arcetri, Largo E.Fermi 5, I-50125 Firenze, Italy

Abstract. We discuss the impact upon Astrochemistry of recent single dish spectral scans towards prominent regions of star formation and the progress expected by the use of interferometers.

1 Introduction

Understanding molecular abundances is a challenge in itself but it also is of importance for many other purposes. One obvious example is that of determining the cooling rate in the various regimes of importance in star–forming regions. Another is that of estimating the degree of ionization in molecular cloud gas which in turn is critical in determining the time scale for ambipolar diffusion. One clearly needs to develop reliable methods of determining molecular abundances in a wide variety of astrophysical circumstances.

The most straightforward approach towards achieving this aim has been traditionally to carry out spectral scans towards sources such as Orion, SgrB2, and IRC+10216, which are judged of special interest. Early studies of this type were carried out at Onsala (Johansson et al. 1984), at Bell Labs (Cummins, Linke, and Thaddeus 1986) and at Caltech (Sutton et al. 1985, Blake et al. 1986). A useful analysis of the results towards OMC-1 was given by Blake et al. (1987). More recently, both the frequency range and the variety of sources examined have been considerably extended (Jewell et al. 1989, Turner 1989, 1991, Sutton et al. 1991, Avery et al. 1992, Ziurys and McGonagle 1993, Groesbeck et al. 1994, Kawaguchi et al. 1995). Rightly or wrongly however, most attention has been given to analysis of the Orion-KL spectrum whose complexity and rich spectrum have been considered a challenge. We will in this review focus upon studies of high temperature star–forming regions such as Orion but note that cool dust clouds (TMC-1) and circumstellar shells (IRC+10216) are likely to be a more promising area of research for those interested in complex organic species.

Our main aim in this brief review is both to demonstrate that high angular resolution is required even for nearby objects and also to summarize our present understanding of abundances in regions of high mass star formation. We also demonstrate that there are important advantages in covering a large frequency range. In section 2, we consider the angular resolutions and sensitivities required when observing hot dense regions such as those implicated in interstellar shocks. In section 3, we summarize recent work with the CSO in the 325-360 GHz and 607-725 GHz ranges. Finally, in section 4, we consider future prospects.

2 Size scales of interest for astrochemistry

Chemical reactions proceed in general (there are exceptions) most rapidly at high temperatures. On the other hand, cooling times for the interstellar gas in general and molecular clouds in particular decrease rapidly at high temperature. A consequence of this is that the regions in which "interesting chemistry" takes place tend to be those regions or interfaces where energy is released giving rise to local heating. Such energy can drive the chemistry leaving a chemical "footprint" which persists long after the original energy source is extinguished. Examples of energy release which can cause such effects are interstellar shocks and photon dominated regions (PDRs).

An example of a set of chemical reactions which are sensitive to temperature are the neutral–neutral reactions which in many situations are responsible for the production of interstellar water. These are :

$$O + H_2 \rightarrow OH + H - 2980\,K \tag{1}$$

and

$$OH + H_2 \rightarrow H_2O + H - 1490\,K \tag{2}$$

The activation energies have the consequence that one requires temperatures of above $200\,K$ to produce water efficiently in this way. Once one has such temperatures however, one rather rapidly transforms free oxygen atoms into water which explains why essentially all current shock models predict large water and OH abundances. The water once produced acts as a rapid coolant thus stifling its own production. However, breakdown of the water into other oxygen containing species in the post-shock (cold) gas is a slow process and so the water and other stable species produced in the shock can often remain for (relatively) long periods of time subsequent to the energy input.

The consequence of all this for observational chemistry is that one almost certainly observes in "normal" molecular cloud material an abundance mix which is representative of the past history of the cloud or clump being examined. As with all fossils, this tends to be an ambiguous situation. It is much easier to study history as it is happening. This is one reason why there is considerable current interest in studying the abundance mix in gas which is either at high temperature or which we believe had been recently heated. In the next section, we briefly review recent observational efforts to derive the abundance distribution in the hot gas surrounding Orion-KL. Here we discuss what size scales are likely to be of interest in the hot regions where we believe that chemical changes (such as water production) are taking place. It is clear in general that such regions will be small simply because of the energy required to keep gas at temperatures above, say, $100\,K$ (note incidentally that the situation becomes even worse if the dust temperature is coupled to the gas temperature).

One example for energy being released rapidly in a small region is what is currently described as a PDR or Photon Dominated Region (see Sternberg and Dalgarno 1995, Genzel 1992 for recent discussions). Here, radiation in the wavelength range 900-1500 Å (typically from hot stars) heats the surrounding dense neutral gas up to temperatures of between 100 and 5000 K. This occurs due either to photo-electric emission from small dust grains or collisional de-excitation of vibrationally excited H_2. The size scale over which this transformation occurs is basically defined by the absorption by dust of the incident radiation and thus by a visual extinction (depending somewhat on the radiation field) of a few visual magnitudes (roughly a hydrogen column density of 5×10^{21} cm^{-2}). It follows that (in 1 dimension), the angular resolution that one requires to resolve PDR phenomena is $15/(n_4 D_k)$ arc seconds where n_4 is the H_2 density (in units of 10^4 cm^{-3}) and D_k is the distance in kiloparsec. For typical galactic HII regions therefore at distances of 5-10 kpc, one requires roughly arc second resolution. In the above, we have implicitly assumed an edge-on geometry. However, as various recent studies of the Orion Bar have shown (e.g. Tielens et al. 1993, Hogerheijde et al. 1995), edge–on geometries are observationally favored for purely geometrical reasons (larger column density). So we conclude that galactic studies of the future will require arc second resolution or better. For extragalactic studies, one loses a factor typically of 10^3 in the distance but this may often be partially compensated by the fact that lower densities are involved. However, clearly, one requires still better resolution than in the galactic case.

Molecular gas heated by interstellar shocks also cools over short time scales and correspondingly short length scales (see e.g. Draine and McKee 1993 and references therein). One can get an idea about what size scales are interesting by considering current attempts to understand the observations of molecular hydrogen and other shock tracers towards nearby outflows (see e.g. Smith 1991, Smith 1994, Gredel 1996). One fairly general finding from these studies is that small dense "bow shock" structures are responsible for much of the H_2 emission. At the head of the bow, the shock velocities are high and molecules are dissociated. In the wings, the shocks become more oblique and molecules can survive. The H_2 emission line intensities from such a structure depend on the shape of the bow and on the pre-shock abundances. The observations of regions such as the Orion flow show fairly clearly that the bows are not being resolved with current apertures and that the size–scales involved are less than 0.002 pc $(1'')$.

High angular resolution implies high sensitivity. If one is investigating abundances, it is necessary to be able to observe weak optically thin lines. In the hot dense regions discussed above, it is likely that transition excitation temperatures will be of the order of 10-100 K and thus line intensities of the order of 1 K are likely to be of low optical depth. A 1 K line observed with arc second resolution towards (say) a region where one is investigating the shock chemistry corresponds to roughly 70 mJy at a wavelength of 1 mm or 7 mJy at 3 mm. The current RMS sensitivity for an hours integration with the Plateau de Bure Interferometer (5 Antennae) is roughly 10 mJy at 3 mm and 60 mJy at 1.3 mm $(3 \, \text{km s}^{-1}$ bandwidth assumed). Thus current sensitivities are on the borderline of allowing reliable abundance estimates. The proposed LSA would have 10 times the collecting area, and hence 10 times the flux sensitivity, i.e. 1 mJy at 3 mm and 6 mJy at 1.3 mm. For unresolved sources, the 10 times higher resolution provides an additional factor of 100 in sensitivity to line temperature. Hence, the LSA should allow vastly improved chemical abundance analyses on the arc second scale or better.

What is likely to limit the LSA in fact is line confusion or blending. In particular in the hot core regions, discussed in the next section, this is already being reached with current single dish telescopes. Interferometry will help here (the hot core and compact ridge regions in Orion are 5″ apart) but it is unclear how much. Model calculations are needed to estimate the importance of this effect. Interferometry would also favor observations of highly excited lines, since they are emitted by very small, hot regions, while low excitation lines, emitted by a larger region, may be resolved out, thus diminishing the line blending problem.

A first interferometric line survey of Orion-KL has been carried out by Blake et al. (1996) at 1.3 mm with the OVRO array. The survey has only a limited frequency coverage (4 GHz in total), but a beam size of $1.0″ \times 1.5″$. First investigations concentrated on the spatial relationship between dust emission (which is determined using line-free channels, thus avoiding any line contamination problems broad-band maps have) and high excitation lines from various species. Preliminary results indicate an anti-correlation between dust and line emission, suggesting that the clumps are externally heated. Again, the high resolution and high sensitivity are necessary to draw these conclusions. These first results illustrate that much can be learned by high spatial resolutions.

3 Recent high frequency spectral scans of Orion KL and SgrB2

Most chemical abundance studies carried out until now have used single dish telescopes with angular resolutions of order 10-60″. However, there is no doubt that a real understanding of the chemical abundance distribution in regions such as Orion-KL and SgrB2 requires angular resolutions of the order of one arc second or better. One can illustrate this by considering the tangled problem of understanding the physical conditions and chemical make-up of Orion.

It has been traditional in past work on Orion to assume that the observed spectra (as seen at single dish resolution) consist of the superposition of different velocity components emanating from spatially distinct regions in the Orion Kleinmann-Low nebula. To add some folk lore, these components have received names: the Plateau, Ridge, Hot Core and even the "Warm quiescent core". The boundaries (spatial and in velocity) between these regions has never been entirely clear as has been their relationship and physical origin (see however Genzel and Stutzki 1989 for a general discussion). It is clear however that high spatial resolution is needed to advance much further and that instruments of the type being discussed at this meeting could play a decisive role. We illustrate this point by figure 1 which shows the superposition of two recent maps made with the extended BIMA array (Wright et al. 1995, Plambeck et al. 1995). The 3.4 mm continuum map here is thought to represent the dust emission from the high density high temperature clump known as the "hot core". The SiO ($v = 0, J = 2 - 1$) emission on the other hand should be representative for the outflow gas associated with the "plateau". Evidently one needs angular resolutions on the order of an arc second in order effectively to be able to distinguish the two spatially. Future work and in particular future spectral scans will have to bear this in mind.

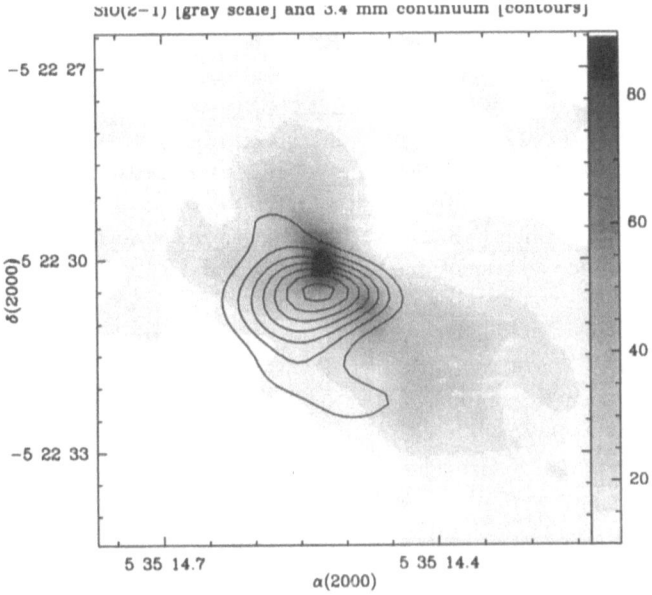

SiO(2−1) [gray scale] and 3.4 mm continuum [contours]

Fig. 1. Superposition of integrated SiO v=0 J=2-1 emission (gray scale) and 3.4 mm continuum emission contours towards the Orion Kleinmann-Low region. The data have been taken from the BIMA maps discussed by Wright et al. (1995) and Plambeck et al. (1995). The contour interval of the continuum map (resolution 1.2″) is 0.77 K. The SiO map has 0.5″ resolution.

On the other hand, we have already learned a considerable amount about Orion chemistry from the (in comparison) low angular resolution work done to date. This is in part due to the recent capability to carry out sensitive submm observations. Recent advances in SIS technology (for recent reviews see Phillips 1994, Carlstrom and Zmuidzinas 1996) have made submm extensions of the spectral scans earlier carried out at millimeter wavelengths into a realistic possibility. As one goes to higher frequencies, one gains information about species such as hydrides which have no observable long wavelength transitions. One also obtains a much more complete measurement of the population distribution for many species thus facilitating column density determinations and understanding of the excitation. These facts have led to a series of studies of the Orion-KL region and SgrB2 regions at increasingly shorter wavelengths.

In particular, there have been a series of studies in the 350 GHz window. Jewell et al. (1989) used a Schottky receiver (1800 K SSB) in conjunction with the 12-m Kitt Peak telescope to detect 160 spectral features towards Orion-KL in the range 330–360 GHz down to a limit of 0.5–1 K. Sutton et al. (1991) on the other hand used the JCMT to observe SgrB2-M and detected lines down to 0.3 K. Most recently, Schilke et al. (1996a) have used the CSO (SIS with $\approx$ 1000 K SSB) to observe the frequency range 325–360 GHz. They detected more than 900 lines down to a sensitivity of $\approx$ 0.2 K. One interesting point which emerges from these studies is the relatively small fraction of unidentified features. Schilke et al. (1996a) are able to identify 786 lines out of a total of 923 or 85%. Moreover, as new laboratory data on transitions from rare isotopomers and vibrationally excited states of known species become available, the number of unidentified lines is rapidly shrinking. Schilke et al. (1996a) for example made use of new laboratory work by Klisch and Winnewisser (priv. comm.) to assign 5 U-lines to $^{33}SO_2$ which had not previously been detected in interstellar space. They also identified 36 transitions from the v_t=1 and 2 torsionally excited states of methanol.

An overall view of the Orion spectrum has been provided by Serabyn and Weisstein (1995) who used a Fourier Transform Spectrometer on the CSO to cover the atmospheric windows between 200 and 900 GHz with a resolution of 0.2 GHz. This corresponds to a spectral resolution of 309 kms^{-1} at 200 GHz and 69 kms^{-1} at 900 GHz. They detected 182 lines from 17 species of which approximately half were above 400 GHz. An interesting feature of the Serabyn and Weisstein results is the dominating role of the sulfur oxides SO_2 and SO which turn out to be important "coolants".

In order to spectrally resolve lines in Orion however, one requires heterodyne techniques. Harris et al. (1995) have made the first attempt at a really high frequency line survey using the JCMT to cover the range 685-692 GHz towards Orion-KL. They detected 13 lines in this band down to a limit of 5 K. They also mapped the 8-7 transition of $H^{13}CN$ and conclude that the emission mainly emanates from the "hot core" with a diameter of 5″ (2300 AU) and a temperature of 200 K.

That this is just the tip of the iceberg however is demonstrated by the CSO survey of Schilke et al. (1996b) which covers the frequency range 607-725 GHz. In figure 2, we show the "line forest" in the 650 GHz window as compared with that detected by the earlier work at lower frequencies. Schilke et al. (1996b) detected $\approx$ 1300 features down to a limit of 0.5 K. This corresponds to 11 lines per GHz as compared to 26 lines per GHz in the 350 GHz window based on the Schilke et al. (1996a) results.

Do such studies teach us anything about the characteristics of the star form-ing regions in Orion and SgrB2? The conclusion of Serabyn and Weisstein men-tioned above that the sulfur oxides can be important coolants provides one par-tial answer. Spectral scans give us a relatively unbiased summary of the energy emitted in a variety of cooling transitions (however, they inevitably miss impor-tant transitions of H_2O and other species blocked by our atmosphere). Another important point is that one is enabled to detect or place limits upon the abun-dances of hydrides and other species which do not have strong transitions at millimeter and centimeter wavelengths. Schilke et al. (1996b) for example have

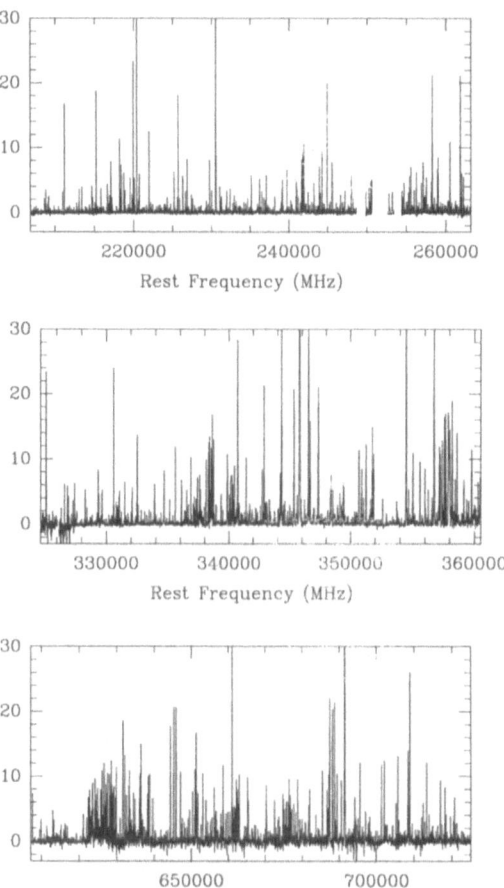

Fig. 2. The Orion-KL spectrum between 607 and 725 GHz (bottom panel, Schilke et al. 1996b) is shown in comparison to the results from the CSO survey between 325 and 360 GHz (Schilke et al. 1996a, center panel) and the Owens Valley surveys between 215 and 260 GHz (Sutton et al. 1985 , Blake et al. 1986, top panel)

tentatively identified SiH at a level which suggests that the hot core abundance of this species may be comparable to that of SiO. Theory suggests (Herbst et al. 1989) that in a steady state situation, SiO should be the most abundant silicon–bearing molecule in the gas phase. However for time scales of order 10^5 years, it is possible to maintain silicon hydrides with abundances higher than that of SiO (MacKay 1995). The detection of SiH, if real, suggests that this may be happening in practice.

4 Future Prospects

In this brief review, we have taken one area of "Astrochemistry" where there has been recent progress (high frequency spectral scans) and made a brief examination of what sort of results are being obtained. One may reasonably ask what is the most suitable strategy for the future. It is quite clear that obtaining an improved understanding of the Orion-KL region and *a fortiori* of other high mass star formation regions requires one to be able to distinguish regions which are separated by distances of the order of thousands of AU's.

Carrying the search further out, the question of extragalactic chemistry arises. It can be said that a factor 10 improvement in resolution would allow studies in the Magellanic Clouds with linear resolutions comparable to interferometric studies in the inner Galaxy now. Even for galaxies outside the local group one gets down to size scales interesting for chemistry: observations of M82 with $0.1''$ resolution at 3 mm would correspond to observations of our Galactic center region with a 20 m telescope and have a linear resolution on a parsec scale. While this is certainly insufficient to resolve hot cores, it permits going from scales of GMC complexes to GMC cloud cores. This is a size scale where chemical differences do show up in our Galaxy and it would be rewarding to investigate this phenomenon in other types of galaxies, e.g. starburst galaxies. Results with present day interferometers suggest that for M82 such abundance variations do indeed exist (Schilke & Brouillet 1996), but it is obvious that higher resolution is needed to quantify such results.

References

Avery, L.W. et al. 1992 , *Astrophys. J. Suppl.* ,**83**, 363.

Blake, G.A., Sutton, E.C., Masson, C.R., Phillips, T.G. (1986): *Astrophys. J. Suppl.*, **60**, 357.

Blake, G.A., Sutton, E.C., Masson, C.R., Phillips, T.G. (1987): *Astrophys. J.*, **315**, 621.

Blake, G.A., Mundy, L.G., Carlstrom, J.E., & Woody, D. 1996, in preparation

Carlstrom, J.E., & Zmuidzinas, J. 1996, in: *Reviews of Radio Science 1993-1995*, ed. W.R. Stone, Oxford, The Oxford University Press

Draine, B.T., McKee, C.F. 1993 *Ann. Rev. Astron. Astrophys.*, **31**, 373.

Genzel, R., Stutzki, J. 1989 *Ann. Rev. Astron. Astrophys.*, **27**, 41.

Genzel, R. 1992 p275 in: *The galactic interstellar medium*, (Burton, W.B., Elmegreen, B., Genzel, R.), publ. Springer.

Gredel, R., 1996, *Astron. Astrophys.*, **305**, 582.

Groesbeck, T.D., Phillips, T.G, Blake, G.A. 1995, *Astrophys. J. Suppl.*, **94**, 147

Harris, A.I., Avery, L.W., Schuster, K.-F, Tacconi, L.J., Genzel, R. 1995, *Astrophys. J.*, **446**, L85.

Herbst, E., Millar, T.J., Wlodek, S., Bohme, D.K. 1989, *Astron. Astrophys.*, **222**, 205.

Hogerheijde, M.R., Jansen, D.J., van Dishoeck, E.F. 1995, *Astron. Astrophys.*, **294**, 792.

Jewell, P.R., Hollis, J.M., Lovas, F.J., Snyder, L.E. 1989, *Astrophys. J. Suppl.*, **70**, 833.

Johannson, L.E.B et al. *Astron. Astrophys.*, **130**, 227.

Kawaguchi K., Kasai Y., Ishikawa S., Kaifu N. 1995, *Pub. Astron. Soc. Japan*, **47**, 853.

MacKay, D.D.S. 1995 *Monthly Notices Roy. Astron. Soc.*, **274**, 694.

Phillips, T.G. 1994, in: IAU Colloquium 140, *Astronomy with Millimeter and Submillimeter Wave Interferometry*, eds. M. Ishiguro and W.J. Welsh, ASP Conference Series 59

Plambeck, R.L., Wright, M.C.H., Mundy, L.G., Looney, L.W. 1995 *Astrophys. J.* , **455**, L189.

Schilke, P., Groesbeck, T.D., Blake, G.A., & Phillips, T.G. 1996a, *Astrophys. J. Suppl.* (in press)

Schilke, P. et al. 1996b (in preparation)

Schilke, P. & Brouillet, N. 1996, in preparation

Serabyn, E., Weisstein, E.W. 1995, *Astrophys. J.*, **451**, 238.

Smith, M.D. 1991 *Monthly Notices Roy. Astron. Soc.* **253**, 175.

Smith, M.D. 1994 *Monthly Notices Roy. Astron. Soc.*, **266**, 238.

Sternberg, A. & Dalgarno, A. 1995 *Astrophys. J. Suppl.*, **99**, 565.

Sutton, E.C., Blake, G.A., Masson, C.R., Phillips, T.G. 1985 *Astrophys. J. Suppl.*, **58**, 341.

Sutton, E.C., Jaminet, P.A., Danchi, W.C., Blake, G.A. 1991, *Astrophys. J. Suppl.*, **77**, 255.

Tielens, A.A.G.M. et al. 1993 *Science*, **262**, 86.

Turner, B.E. 1989, *Astrophys. J. Suppl.*, **70**, 539.

Turner, B.E. 1991 *Astrophys. J. Suppl.*, **76**, 617.

Wright, M.C.H., Plambeck, R.L., Mundy, L.G., Looney, L.W. 1995 *Astrophys. J.*, **455**, L185

Ziurys L.M., McGonagle D. 1993, *Astrophys. J. Suppl.*, **89**, 155.

Abundance and Origin of Galactic Water

R. Mauersberger[1], P. Gensheimer[2], and T.L. Wilson[2]

[1] Steward Observatory, The University of Arizona, Tucson, AZ 85721, U.S.A.
[2] MPI für Radioastronomie, Auf dem Hügel 69, D 53121 Bonn, Germany

Abstract. The abundance water in Galactic hot cores is discussed, addressing the observational challenges. The HDO/H_2O abundance ratio in various molecular cores has only a small scatter, and its value is similar to that found in comet Halley, suggesting an interstellar origin of cometary matter. A cometary origin of ocean water is consistent with the high its deuterium fraction. If this can be confirmed it may be speculated that also the first prebiotic molecules on Earth are of cometary, and, hence, interstellar origin.

1 Interstellar water: an observational challenge

Water plays a special role among the more than hundred molecules which have been detected in interstellar space. It is not only nessecary for the evolution and existence of life, it also shaped the surface of the Earth, determines its climate and helped to concentrate many ores and minerals which are necessary for our civilization into deposits. Since the line emission of water vapor is thought to be an important coolant of the interstellar gas (e.g. Brown et al. 1988), water may have even aided in the collapse forming our Sun.

Because it *is* so abundant in the Earth's atmosphere, its line emission is also notoriously difficult to observe (with the exception of highly excited maser lines, which, unfortunately, bear little information about the topic of this paper). The lowest lying transitions of water can only be observed from balloons or from satellites, such as ISO, ODIN or SWAS. The disadvantage is not only the low angular resolution and the short duration of such experiments but also the high costs of orbiting observatories, which can easily exceed the cost of the proposed Large Southern Array (LSA). At the frequency of the $3_{1,3} - 2_{2,0}$ line of H_2O, 183 GHz, the Earth atmosphere can, under exceptional circumstances, become partly transparent for ground based observations. For the excellent sites which have been proposed for the LSA (Bååth et al. 1996) such conditions may become even the rule. Although this transition has a high excitation energy of ~ 200 K, its emission is extended toward the Orion star forming region (Cernicharo et al. 1994). As for the mid-infrared transitions, this line may be highly optically thick, and, hence, difficult to interpret in terms of H_2O column densities, especially if this line is a strong maser.

2 The abundance and deuteration of interstellar water

One solution to circumvent the need for extraordinary weather conditions and to ensure optically thinness is to look for rare isotopes of water. The $H_2^{18}O$

counterpart of the 183 GHz line lies at 203.4 GHz, outside the atmospheric water absorption and is therefore readily observable. ^{18}O is 500 times (250 times in the Galactic center) rarer than main isotopic oxygen. There is probably no chemical fractionation of isotopic O. This ensures that H_2O lines are in many cases optically thin and are a measure of the H_2O column density. This may be the *only* water line that can be observed from the ground and that is optically thin and not a strong maser (in fact there is another more difficult to observe, and yet to detect transition of $H_2^{17}O$). Jacq et al. (1988, 1990) have detected the 204.3 GHz line toward a small number of hot molecular cores and Gensheimer et al. (1996) used the narrow beam of the IRAM 30-m telescope to search for sources of water in the Milky Way. They detected $H_2^{18}O$ in 13 out of 20 hot cores. In order to estimate reliable column densities one has to get an idea about the excitation conditions, which of course is difficult with just one line. Therefore they observed a number of transitions of deuterated water. A multilevel study by Helmich et al. (1996) suggests that the rotational temperature is determined by the infrared field and is around 200 K.

Gensheimer et al. (1996) concluded that the relative abundance of water in the observed hot cores scatters with an order of magnitude around a value of 10^{-5}, which makes water less important as a coolant of molecular gas than previously thought. The high uncertainty is mainly due to the uncertainty in the column density of H_2 and may not reflect actual variations in the abundance of water. A ratio of $HDO/H_2O \sim 3\,10^{-4}$, i.e. 30 times higher than the "cosmic" D/H abundance ratio, has been measured using lines of similar excitation. The scatter of the HDO/H_2O ratio is only about a factor of two. Both the high HDO/H_2O ratio and its low scatter are surprising. Enrichment of deuterated gas phase compounds is expected only for relatively cool gas. The high isotopic ratio can be best explained if a large part of the gas phase molecules has recently evaporated from dust grains, maybe under the influence of a newly formed star (Walmsley et al. 1987). This does not explain why the HDO/H_2O ratio is so constant, since in view of a timescale for a deuterium equilibrium chemistry of only a few thousand years (Walmsley et al. 1987) a much larger scatter is expected. Of course the sample is biased toward warm, high density clouds, but it is possible that the column density of HDO is a better indicater of the H_2O column density than previously thought, at least for hot cores.

The origin of comets, the ocean and life

If the polar molecules we see in hot molecular cloud cores have recently evaporated from grains, the exciting consequence is that hot cores may be representative of the composition of comets, which are thought to be the product of "cold" agglomeration of protosolar dust grains. This implication seems to be confirmed by the striking similarity of the HDO/H_2O abundance ratio measured in situ toward comet Halley and Hyakutake ($3\,10^{-4}$, Eberhardt et al. 1995, Lis et al. 1996) and also of organic matter in meteorites ($5.4\,10^{-4}$). This idea about the common origin of comets and hot core molecules can be tested by comparing their chemical composition. There is, however, a considerable scatter in the chemical abundance of molecules for a sample of different comets (Eberhardt et al. 1994). A better indication could be the deuterium enrichment in different molecules. A number of deuterated molecules have been observed in the ISM with various degrees of deuterium fractionation (see e.g. the lists in Howe et al. 1994, Jacq et al. 1990, 1993, Mauersberger et al. 1988). If the D/H abundance ratio in the ice mantles does not change, one expects a similar fractionation in cometary molecules as in hot cores.

There is another reservoir of highly deuterated water, namely the Earth's oceans (D/H=$1.5\,10^{-4}$, Hageman et al. 1970). Traditionally this has been explained by selective evaporation of H (Donahue et al. 1982). An alternative explanation is that a comet rain onto the young Earth brought a large amount of highly deuterated water which now is forming the oceans (Anders & Owen 1977). The similarity between the HDO ratio in comets and the oceans seems to support this hypothesis.

The first traces of life are found already in the oldest sedimentary rocks, which have an age of $3.5 - 3.8\,10^9$ years (see the discussion in Horneck 1994). If one assumes that no life could have formed on Earth during its first few hundred million years because of the intense bombardement with planetesimals (Maher & Stevenson 1988) evolution had only a couple of $100\,10^6$ years to form. One can speculate that comets brought with the water also the first prebiotic molecules to the Earth and thus triggered the rapid evolution of life.

3   The role of the LSA

At the proposed LSA sites the atmospere will be partly transparent for the 183 GHz line of H_2O. With the LSA, this important line as well as the corresponding line of $H_2^{18}O$ will be observable not only toward many sources in our Milky Way but they can be detected in external galaxies and comets and mapped in planetary atmospheres. Therefore the receivers of the LSA should be designed to cover the wavelength range down to 183 GHz. Perhaps at the upper frequency range of the LSA one might be able to measure another important line, namely the $1_{01} - 0_{00}$ ground transition of HDO.

While the D/H abundances are known for a number of interstellar molecule, we have little information from comets. Future spaceprobes will make in situ measurements, but if we want systematical data for a larger number of comets we have to investigate them from Earth. Due to the compact size of the molecular emission and the weak emission of deuterated molecules, the collecting area and resolving power of the LSA will be needed.

Acknowledgement. R.M. was supported by a Heisenberg fellowship by the DFG.

References

Anders, E., Owen, T., 1977, Sci 198, 453

205

Bååth, L., 1996, these proceedings

Brown, P.D., Charnley, S.B., Millar, T., 1988, MNRAS 231, 409

Cernicharo, J., González-Alfonso, E., Alcolea, J., Bachiller, J., John, D., 1994, ApJ 432, L59

Donahue, T.M., Hoffman, J.H., Hodges, R.R., Jr., Watson, A.J., 1982, Sci 216, 630

Eberhardt, P., Meier, R., Krankowsky, D., Hodges, R.R., 1995, A&A 302, 301

Gensheimer, P.D., Mauersberger, R., Wilson, T.L., 1996, A&A, in press

Hageman, R., Nief, G., Roth, E., 1970, Tellus, 22, 712

Helmich F.P., Van Dieshoeck, E.F., Jansen, D.J., 1996, A&A, in press

Horneck, G., 1994, Planet. Space Sci. 43, 189

Howe, D.A., Millar, T.J., Schilke, P., Walmsley, C.M., 1994, MNRAS 267, 59

Jacq, T., Jewell, P.R., Henkel, C., Walmsley, C.M., Baudry, A., 1988, A&A 199, L5

Jacq, T., Walmsley, C.M., Henkel, C., Baudry, A., Mauersberger, R., Jewell, P.R., 1990, A&A 228, 447

Jacq, T., Walmsley, C.M., Mauersberger, R., Anderson, T., Herbst, E., de Lucia, F.C., 1993, A&A 271, 276

Lis, D., Keene, J., Young, K. et al. 1996, IAU Circular 6362

Maher, K.A., Stevensen, D.J., 1988, Nat 331, 612

Mauersberger, R., Henkel, C., Jacq, T., Walmsley, C.M., 1988, A&A 194, L1

Walmsley, C.M., Hermsen, W., Henkel, C., Mauersberger, R., Wilson, T.L., 1987, A&A 172, 311

Interferometric Observations of Sagittarius B2: Evidence for Grain Chemistry

Yi-Jehng Kuan[1], Lewis E. Snyder[2], Yanti Miao[2], and David M. Mehringer[2]

[1] Institute of Astronomy and Astrophysics, Academia Sinica, Taipei, Taiwan, R.O.C.
[2] Astronomy Department, University of Illinois, Urbana, IL 61801, U.S.A.

Abstract. Various complex molecules were only detected toward the Sgr B2(N) dense core. Together with other observational evidence for the existence of warm core-mantle grains in Sgr B2(N), we believe this is due to abundance variations as a result of enhanced grain-surface chemistry.

1 Introduction

1.1 Grain Chemistry

Gas-phase chemistry cannot produce the high abundance observed (a column density $\sim 10^{15}$ cm^{-2}, and a fractional abundance $\sim 10^{-9}$) of various complex molecules and some simple molecular species such as HNCO (isocyanic acid). Recent gas-grain chemistry suggests that a dramatic increase of the abundance of large molecules in gas phase may be driven by grain mantles evaporation during the collapse phase of molecular clouds, as big molecules could be mostly formed in mantles (cf. Caselli, Hasegawa, & Herbst 1993; Charnley 1994).

1.2 Sgr B2: The Best Candidate to Test Grain Chemistry

The giant molecular cloud Sgr B2 is known for its rich chemistry and active star-forming activities. Sgr B2 consists of nearly 60 ultracompact HII regions (Gaume et al. 1995) and has more detected molecular species than any other galactic source. Of the 109 interstellar/circumstellar molecular species known to date, more than 80 were found in the cloud and more than one half were first detections. In addition, strong far-infrared and millimeter emission indicates the presence of dust. Sgr B2 is thus *the* ideal laboratory for astrochemistry study.

2 Complex Molecules in Sgr B2(N)

High-resolution interferometric observations using the BIMA (Berkeley-Illinois-Maryland Association) millimeter array showed that complex molecules NH$_2$CHO (formamide, Fig. 1), HCOOCH$_3$ (methyl formate) (Kuan & Snyder 1996), CH$_2$CHCN (vinyl cyanide), CH$_3$CH$_2$CN (ethyl cyanide) (Miao et al. 1995), CH$_3$COOH (acetic acid, tentative), and (CH$_3$)$_2$CO (acetone, Fig. 2) were only detected toward the northern compact cloud core Sgr B2(N)[1] at $\alpha(1950) = 17^h\ 44^m\ 10^s.2$ and $\delta(1950) = -28^\circ\ 21'\ 13''$. However, none of these big molecules was detected in the more evolved Main cloud core Sgr B2(M) at $\alpha(1950) = 17^h\ 44^m\ 10^s.6$ and $\delta(1950) = -28^\circ\ 22'\ 05''$.

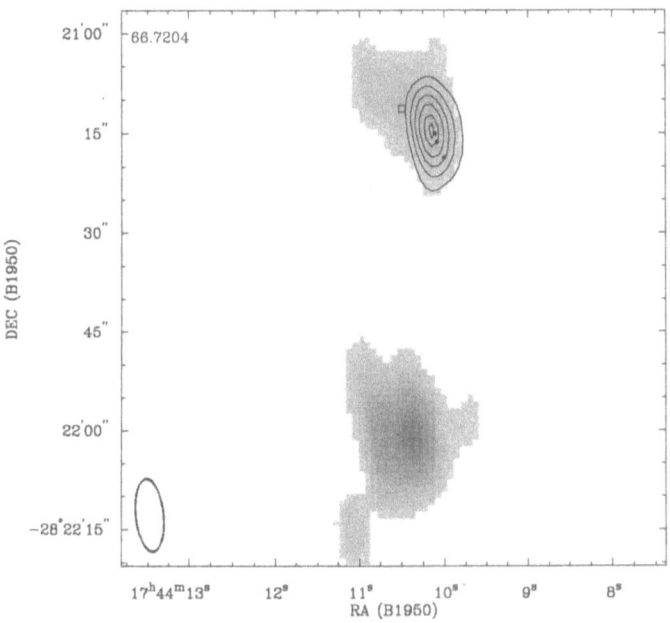

Fig. 1. The map of NH$_2$CHO averaged over the LSR velocity range from 60.6 to 72.8 km s^{-1}. The emission contains two partially blended lines $J_{K_{-1}K_1} = 4_{32}\text{-}3_{31}$ and $4_{31}\text{-}3_{30}$ at 84.89 GHz. The contour levels are 0.3, 0.5, 0.7, 0.9, 1.1 and 1.2 Jy beam^{-1} with $\sigma = 0.09$ Jy beam^{-1}; no negative contours are seen stronger than -0.3 Jy beam^{-1}. The beam size is $10''.8 \times 4''.6$ (HPBW) as shown on the lower left-corner. The underlying grey scale map is the 85 GHz continuum; the LSR velocity is in the upper left-hand corner. Filled circles mark the positions of the ultracompact HII regions K1, K2, and K3 from south to north in Sgr B2(N); and the square denotes the HII region K5. No emission was seen from Sgr B2(M).

3 Evidence for Grain-Surface Chemistry

3.1 Existence of Abundant Dust Grains in Sgr B2(N)

The 1.3-mm and far-infrared continuum excess in Sgr B2 is known to be due to warm dust grains heated by embedded (proto)stars. Among the dense cloud

---

[1] A concentration of CH$_3$OH (methanol), CH$_3$CH$_2$OH (ethanol), and (CH$_3$)$_2$O (dimethyl ether) toward Sgr B2(N) was also observed (M. Ohishi, 1995, private communication)

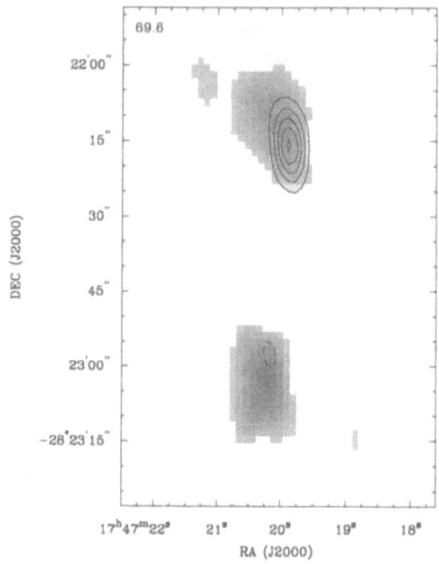

Fig. 2. The $(CH_3)_2CO$ velocity map averaged over LSR velocity range from 58.3 to 80.9 km s^{-1}. The line is a blend of two transitions, $J_{K_{-1}K_1} = 8_{08}\text{-}7_{17}$ EE and $8_{18}\text{-}7_{07}$ EE, at 82.917 GHz. Contour levels start at 0.2 Jy beam^{-1} with an increment of 0.2 Jy beam^{-1}; the dashed line denote a negative contour at -0.2 Jy beam^{-1}. The HPBW is 12$\farcs$8 × 5$\farcs$5. The LSR velocity is in the upper left-hand corner. The underlying grey scale map is the 83 GHz continuum. Note the epoch of the coordinates is J2000. No emission was seen from Sgr B2(M).

cores in Sgr B2, Sgr B2(N) appears to be most dusty as it shows the strongest 1.3 mm emission in the cloud (cf. Gordon et al. 1993).

In addition, the $J_{K_{-1}K_1} = 4_{23}\text{-}3_{22}$ and $4_{22}\text{-}3_{21}$ transitions of HNCO were *only* detected toward Sgr B2(N) (Kuan & Snyder 1996). The $K_{-1}= 2$ energy levels are 181.1 K above the ground state and are likely radiatively excited ($\sim$ 330 μm for $K_{-1}= 0\rightarrow1$ and $\sim$ 110 μm for $K_{-1}= 1\rightarrow2$; in contrast, the critical densities $\sim10^{11}$ cm^{-3} for $3_{12}\text{-}4_{23}$ and $3_{13}\text{-}4_{22}$ transitions). The presence of a strong IR field in Sgr B2(N) clearly suggests the existence of warm dust grains in abundance in the dense core.

Furthermore, continuum studies indicate that Sgr B2(N) is dominated by thermal dust emission with a steep spectral index ($S_\nu \propto \nu^\alpha$) $\alpha = 4.6\pm0.5$ near 3 mm. The estimated dust mass column density is thus $\lesssim$ 2.5 g cm^{-2}. In contrast, a moderate spectral index $\alpha = 1.0\pm0.3$ was measured in Sgr B2(M) which yields a dust column density only $\sim$0.07 g cm^{-2} (Kuan, Mehringer, & Snyder 1996).

3.2 Ice Coated Core-Mantle Grains in Sgr B2(N)

The high spectral index, hence the grain emissivity exponent $\beta = 3.7\pm1.3$, implies that Sgr B2(N) contains ice coated core-mantle grains (Kuan et al. 1996). The short life time of the core-mantle phase ($\sim10^4$ yrs) indicates that Sgr B2(N) is in a very early stage of star formation[2].

3.3 Other Supporting Evidence

Grain chemistry predicts the simple molecule HNCO is formed mainly via surface reactions. Observations of the $K_{-1} = 0$ transitions (4_{04}-3_{03}, 5_{05}-4_{04}) indicate HNCO is more abundant toward Sgr B2(N) (Kuan & Snyder 1996).

4 Conclusion

The sole detection of big molecules toward Sgr B2(N), which is closely related to core-mantle grains, provides a strong observational evidence in support of grain surface chemistry.

5 A Few Thoughts

The old wisdom says a detection work is always the job for a single dish. This statement may be true in the past but it can no longer be applied to interferometers with very high sensitivity and spatial resolution, such as LSA. As we have shown, the complex molecule *heimat* in Sgr B2(N) is extremely compact. An interferometric array with its small beam is thus an excellent tool to study gas-grain chemistry and to search for prebiotically important heavy molecules. Observations of compact sources with high angular resolution will not only yield better determined column densities but also improved S/N ratios. And above all, interferometers can easily filter out extended, diffuse components along the line-of-sight. As a consequence, the inevitable *confusion limit* would be lowered which is especially crucial in detecting weak molecular lines.

References

Caselli, P., Hasegawa, T.I., & Herbst, E. 1993, ApJ, **408**, 548

Charnley, S.B. 1994, in *Molecules and Grains in Space*, ed. Irene Nenner, 155

Gaume, R.A., Claussen, M.J., De Pree, C.G., Goss, W.M., & Mehringer, D.M. 1995, ApJ, **449**, 663

Gordon, M.A., et al. 1993, A&A, **280**, 208

Kuan, Y.-J., Mehringer, D.M., & Snyder, L.E. 1996, ApJ, **459**, 619

Kuan, Y.-J., & Snyder, L.E. 1996, ApJ, submitted

Miao, Y., Mehringer, D.M., Kuan, Y.-J., & Snyder, L.E. 1995, ApJL, **445**, 59

---

[2] This is consistent with the estimated outflow age of Sgr B2(N) ($\sim 6 \times 10^3$ yrs), which is a factor of ~3 younger than the outflow age of Sgr B2(M) ($\sim 2 \times 10^4$ yrs) (Kuan & Snyder 1996).

A hot ring in the Sgr B2 molecular cloud

Jesús Martín-Pintado[1], Pablo de Vicente[1], Tom L. Wilson[2], and Ralph Gaume[3]

[1] Observatorio Astronómico Nacional (IGN), Campus Universitario, Apartado 1143, E-28800 Alcalá de Henares, Spain
[2] Max-Planck Institut für Radioastronomie, Auf dem Hügel 69, D53121 Bonn, Germany
[3] U.S. Naval Observatory, 3450 Massachussets Ave., NW, Washington D.C. 20392-5420, USA

Abstract. Recent single dish and interferometric observations of the Sgr B2 molecular cloud are presented. The results of these observations illustrate how the combination of high angular resolution and high sensitivity can be used to derive the kinetic temperature and density structure of this prototypical giant molecular cloud in the Galactic center. The data reveal the presence of hot (70-120 K) shell-like structures at different scales (from 0.5 to 5 pc). The heating of the warm molecular clouds in the galactic center is probably related to the energetic events which also create the hot molecular shells. Observations with the new generation of millimeter arrays will be fundamental to determine the physical properties of the small scale structure of the molecular clouds in the Galactic center and to dilucidate the heating mechanism.

1 Introduction

The Sagittarius B2 molecular cloud is one of the most active regions of star formation in the Galaxy. Massive star formation is occurring in two regions, Sgr B2M and Sgr B2N, which are very strong IR emitters (Thronson & Harper 1986) and contain all signposts of this activity like ultracompact HII regions (Martin & Downes 1972; Gaume & Claussen 1990), hot cores (Vogel, Genzel & Palmer 1987) and maser emission in OH, H_2O, H_2CO and SiO (Gaume & Claussen 1990; Kobayashi et al. 1989; Gaume & Mutel 1987). The star-forming regions, which contain 10^5 $M_\odot$, are embedded in a giant molecular cloud with a total mass of $7 \cdot 10^6$ $M_\odot$ (Lis & Goldsmith 1989).

In spite of the large mass of the envelope, little is known about its physical properties. Large scale studies of this remarkable molecular clouds have only been made in ^{12}CO and $C^{18}O$ (Scoville et al. 1975; Lis & Goldsmith 1989). Absorption line studies for the line of sight toward the continuum sources (Wilson et al. 1982, Henkel et al. 1983) indicate the presence of a warm ($T_k \geq 100$ K) (Wilson et al. 1982; Hüttemeister et al. 1993) and moderate density envelope ($10^3 - 10^4$ cm^{-3}). Unlike the molecular clouds in the disk, the gas kinetic temperatures in the envelope are above the dust temperatures (~ 20K), and cloud-cloud collisions have been proposed to heat the gas in the molecular clouds in the galactic center (Wilson et al. 1982).

2 The kinetic temperature distribution. The hot ring

Using the unmatched sensitivity and angular resolution of the IRAM 30-m telescope we have determined the kinetic temperature and the H_2 density structure
of the molecular gas in the Sgr B2 cloud at scales of several parsecs. We have
mapped with an angular resolution of 12-24″ the J=5-4, J=8-7 and J=12-11
lines of the symmetric top rotor CH_3CN. The kinetic temperature and density
distributions have been derived from a LVG analysis of the CH_3CN data. (The
detail of the model calculations are described in de Vicente et al. 1996.) The kinetic temperature distribution is shown in Fig. 1. The temperature varies from
300 K in the dense hot cores associated to Sgr B2M and Sgr B2N to 50 K for
the warm envelope up to distances of 8 pc. The density in the envelope observed
in CH_3CN is $\sim 10^5$ cm^{-3} and contains $3 \cdot 10^6$ $M_\odot$. The most remarkable feature
in the warm envelope is the presence of a hot ring with a kinetic temperature of
$\sim$120 K which surrounds the star forming cores. The ring has a radius of $\sim$4 pc
and a thickness of $\sim$1.4 pc.

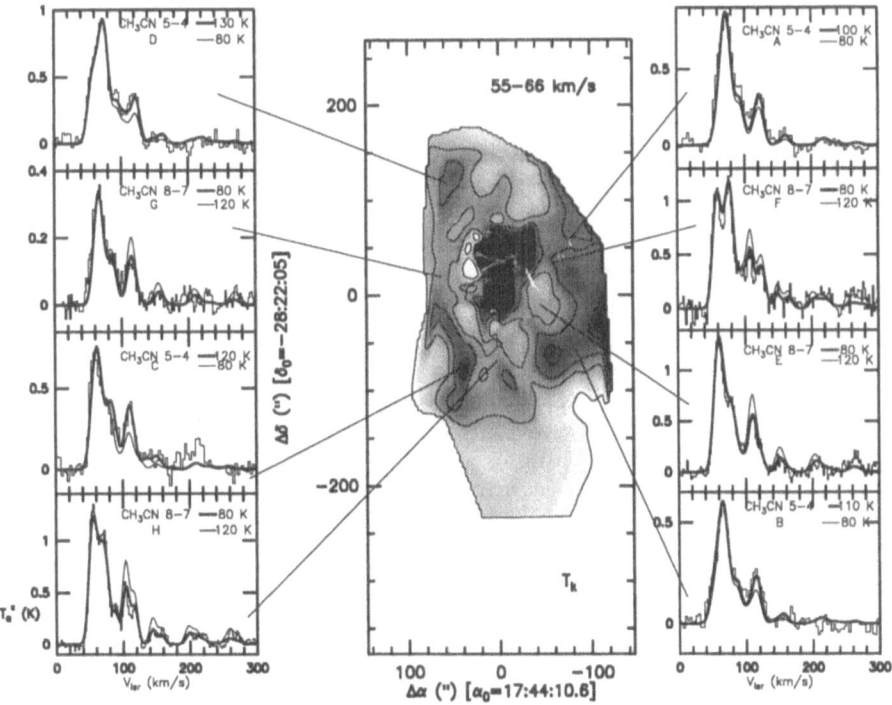

Fig. 1. Kinetic temperature structure in Sgr B2. The contour levels are 400, 200, 120
80 an 40 K km s^{-1}. At both sides CH_3CN spectra with the predicted profiles for 80
and 120 K towards selected positions in the warm envelope

The analysis of the thermal balance of the molecular cloud in Sgr B2 (de Vicente et al 1996) shows that the temperature of the hot cores and the kinetic temperature distribution for distances to the IR sources smaller than 1 pc can be accounted for by gas-dust collisions. However, the dust temperatures in the warm envelope are too low (10–20 K) to heat the molecular gas by this mechanism. Heating by dissipation of turbulent motions in the envelope of Sgr B2 can explain the gas kinetic temperature derived from the CH_3CN data. The presence of the hot ring requires the existence of another heating mechanism. It has been proposed (de Vicente et al. 1996) that the hot ring might be associated to the interface between the warm envelope and the ionized bubble created by the OB stars recently formed in the Sgr B2 core. In this interface, heating by UV photons and/or shock fronts produced by the expansion of the ionized gas could explain the hot ring.

3   Hot molecular shells within the hot ring

The understanding of the origin of the high temperature in the warm envelope and the presence of a hot ring requires the knowledge of the morphology and the physical properties of the molecular gas at smaller scales. VLA[1] interferometric observations (angular resolution of $2''$) of the southern part of the hot ring in the (3,3) and (4,4) lines of NH_3 reveal the presence of several shell-like structures at different radial velocities (see Fig. 2). The typical radii of the shells in the hot ring are 0.5-1 pc and the thickness 0.1-0.2 pc. The rotational temperatures derived form the (3,3) and the (4,4) lines of NH_3 in the shells are 70-100 K. From the analysis of the NH_3 column densities and the thickness of the shells, we conclude that, either the wall of these structures is composed of high density gas ($\sim 10^7$ cm^{-3}) or the NH_3 abundances are enhanced by three orders of magnitude. Though the origin of the hot shell is unknown, both high density and enhancement of the NH_3 abundance are expected in regions undergoing shocks (Flower et al. 1995). The origin of the hot shells in the massive envelope of Sgr B2 is probably related to energetic events like stellar winds from WR stars. From this limited high resolution data it is unclear whether there is an appreciable amount of hot shells whose origin could also be considered as a possible heating mechanism for the warm molecular envelope

4   Prospectives for large millimeter arrays

The understanding of the properties and the chemical composition of the giant molecular clouds (GMCs) in the galactic center, require high angular resolution large scale mapping of density and kinetic temperature. The large millimeter array will provide angular resolutions better than $1''$ which are needed to measure the kinematic and the physical conditions of the hot shells observed in NH_3. In addition to that, the large collecting area will allow to detect the weak

---

[1] The NRAO is operated by Associated Universities, Inc., under cooperative agreement with the National Science Foundation.

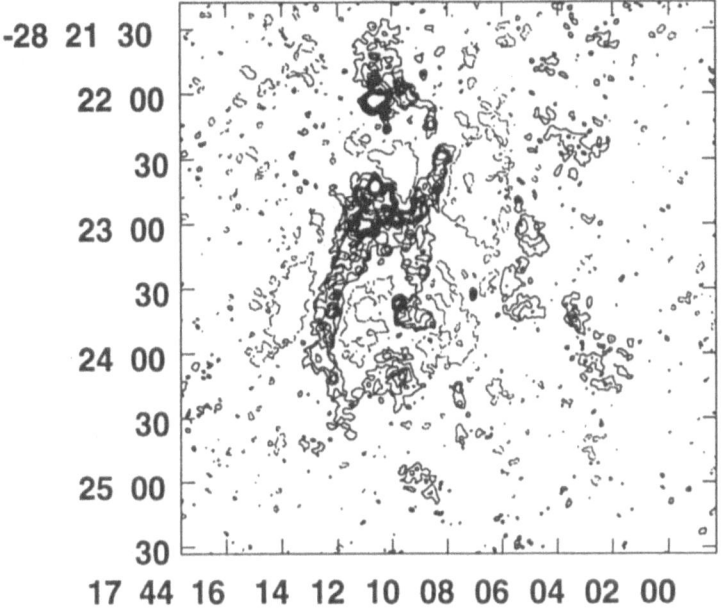

Fig. 2. VLA image of the (3,3) line in Sgr B2. The contour levels are:-6, 6, 12, 18, 24, 30, 36, 42, 48, 54 and 60 mJy/beam

lines arising in the envelopes of the GMCs which contain most of the mass in these objects. For instance, high sensitivity is required to measure the weak K components of CH_3CN which are fundamental to derive the kinetic temperature distribution. Furthermore, the millimeter array should have the capability of mapping large areas (mosaicing) in several transitions simultaneously.

References

de Vicente, P., Martín-Pintado, J., Wilson, T.L. 1996, A&A, in press

Flower D.R., Pineau des Fôrets G. Walmsley C.M. 1995, A&A 294, 815.

Gaume R.A., Mutel R.L., 1987, ApJS 65, 193.

Gaume R.A., Claussen M.J., 1990, ApJ 351, 538

Huettemeister S., Wilson T.L., Henkel C., Mauersberger R., 1993, A&A 276, 445

Kobayashi H. et al., IAU 136, *The Center of the Galaxy*, 181-187, Ed.: Morris M., Kluwer Academic Publishers

Lis D.C., Goldsmith P.F., 1989, ApjJ 337, 704

Martin A.H.M., Downes D., 1972, Astrophys. Letters 11, 219

Scoville N.Z., Solomon P.M., Penzias A.A., 1975, ApJ 201, 352

Thronson H.A., Harper D.A., 1986, ApJ 300, 396

Vogel S.N., Genzel R., Palmer P., 1987, ApJ 316, 243

Wilson T.L., Ruf K., Walmsley C.M., Martin R.N., Pauls T.A., Batrla W., 1982, A& A 115, 185

High-density filaments in the photodissociation region (PDR) associated with NGC 7023

Jesús Martín-Pintado ,Asunción Fuente

Observatorio Astronómico Nacional (IGN), Campus Universitario, Apartado 1143, E-28800 Alcalá de Henares, Spain

Abstract. We present a study of the structure and the chemical composition of the photodissociation region (PDR) associated with NGC 7023. The favorable geometry of this PDR has allowed to measure chemical changes across the PDR. Interferometric observations of the strong emission of HCO^+ shows that the dense molecular gas is highly filamentary. The morphology of the molecular filaments is very similar to that of the filaments detected at 2.1 μm (IR filaments), suggesting that the IR filaments are indeed high density structures in the PDR. The new generation of millimeter arrays will provide the angular resolution and the sensitivity to measure the chemistry and the small scale structure across PDRs.

1 Introduction

The physical and chemical structure of photodissociation regions (PDRs) is still poorly known. There is strong indirect evidences that most PDRs must be clumpy. The observations of the 158 μm C^+ and the high-J CO lines that cannot be accounted by uniform cloud models, and they are better explained by assuming a clumpy medium which allows a deeper penetration of UV radiation into the molecular cloud (Stutzki et al. 1988, Howe et al. 1991; Jaffe et al. 1990). Optical and infrared images of reflection nebulae present a highly filamentary structure (Witt & Malin 1989, Sellgren et al. 1992, Field et al. 1994). The nature of the IR filaments is, however, controversial. These features have been interpreted as high density filaments (Sellgren et al. 1992) which probes the fragmented structure of PDRs, but also as the consequence of a geometrical effect in a uniform cloud (rugosities on the walls of the nebula which produce regions with large column densities of hot dust and gas along the line of sight) (see e.g. Field et al. 1994).

Because of the UV radiation, PDRs are expected to show different chemistry than the shielded cores in molecular clouds (Fuente et al. 1990; Sternberg & Dalgarno 1995). High angular resolution molecular spectroscopy of selected molecules would be an essential tool to determined the chemical composition and to measure the small scale structure in dense PDRs. NGC 7023 shows the strong emission in the fine structure lines of O and C^+ typical of a dense PDR (Chokshi et al. 1988). Due to its proximity ($\sim$ 440-600 pc) and the edge-on geometry, the PDR associated with NGC 7023 is well suited for this kind of studies.

2   The overall structure of a prototypical reflection nebula

Large scale maps of NGC7023 in the J=1→0 and J=2→1 transitions of ^{12}CO
and ^{13}CO reveal that the molecular cloud has a ring like structure with a cavity
around the exciting star HD200775. The walls of the molecular cavity perfectly
delineate the borders of the reflection nebulosity as traced in the optical image
(see Fuente et al. 1992). The HI emission fills the molecular cavity delineated
by ^{13}CO (Rogers et. al 1995). Moreover, the HI column density does not peak
toward the star, but along elongated rim adjacent to the walls of the molecular
cavity (see Fig. 1). The morphology of the HI and molecular emission clearly
show the atomic-molecular interface expected in a PDR.

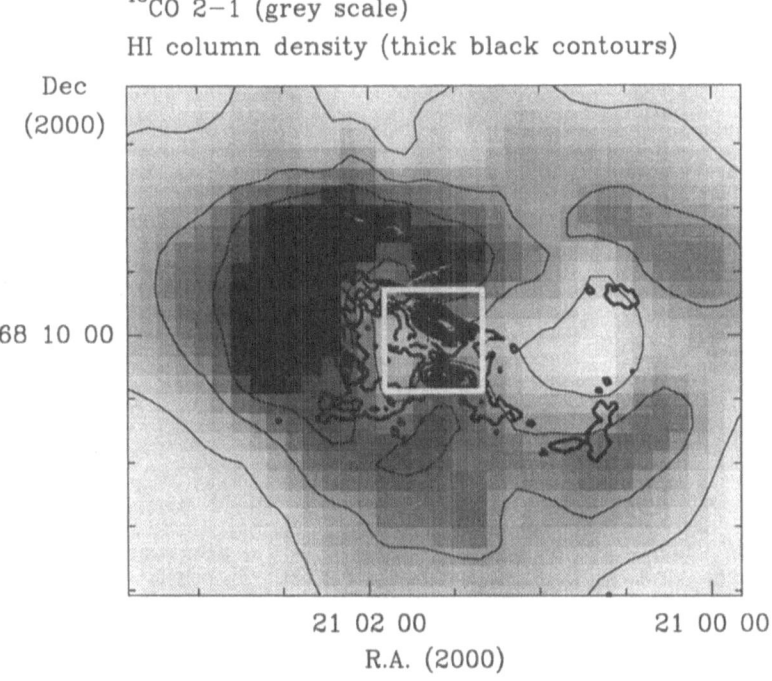

Fig. 1. Contours of the total HI column density (thick line) derived after combining
DRAO and VLA data superposed on the integrated intensity map of the ^{13}CO J=2→1
line (grey scale) reported by Fuente et al. (1992). Contours of ^{13}CO are 2 K kms^{-1} to
10 K kms^{-1} by 2 K kms^{-1}. Contours of HI column density begin with 3.2 10^{20} cm^{-2}
and increase to 1.6 10^{21} cm^{-2} by steps of 1.6 10^{20} cm^{-2}. The white rectangle shows
the region shown in Fig. 2.

3 A chemical study of the PDR associated with NGC 7023

Single dish mapping (angular resolution 10-40") of the atomic-molecular interface in several molecular species (^{12}CO, ^{13}CO, C^{18}O, HCO$^+$, HCN, N$_2$H$^+$, NH$_3$,CS, CN and C$_2$H) shows that while some molecules like CN, HCN and HCO$^+$ present strong emission coincident with that of HI, others, like HNC, N$_2$H$^+$ and NH$_3$, are intense only beyond the HI emission. In fact, the CN/HCN and the HCO$^+$/N$_2$H$^+$ abundance ratios increase by a factor of $\sim$ 10, and the (CN+HCN+HNC)/NH$_3$ ratio increases by a factor of $\sim$ 30 toward the star. The results suggest that CN and HCO$^+$ are extremely useful molecules to study the structure and the physical properties of the molecular gas in the atomic-molecular interface. Furthermore, the CN/HCN ratio can be used as a tracer of regions with enhanced UV fields.

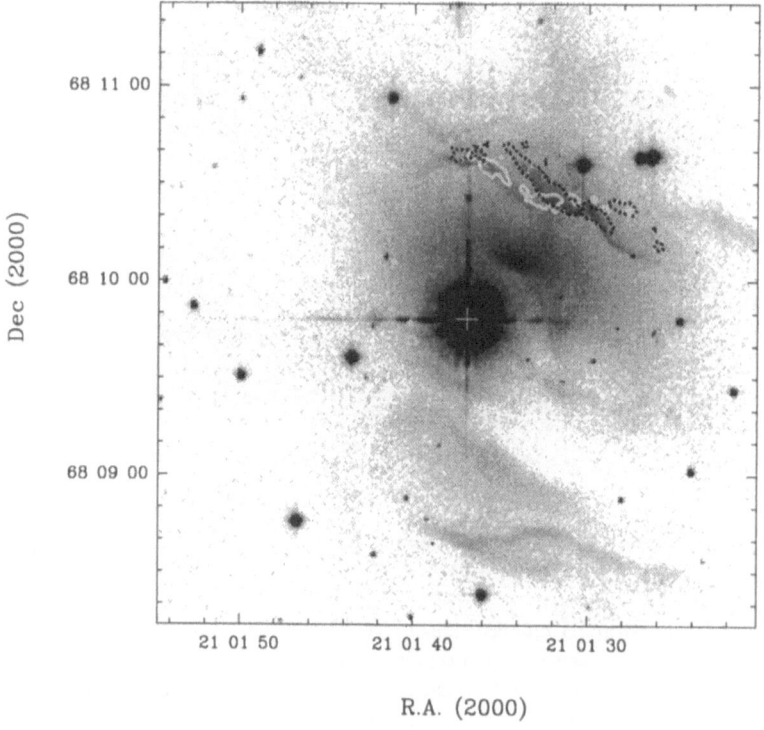

Fig. 2. Molecular filaments superposed on the 2.1 μm image (Sellgren et al. 1992).

4 Small scale structures in PDRs.
High density molecular filaments

To investigate the small scale structure of the PDR associated with NGC 7023 we have mapped with an angular resolution of ~3" the interface between the molecular and atomic phases in the HCO^+ J=1→0 line using the IRAM interferometer. The data show the existence of several molecular filaments with a thickness of ~6" which are located at the sharp edge that delimits the HI and the molecular emission (Fig. 2). Detailed comparison of the single-dish and the interferometric data reveals that the molecular filaments are actually regions of enhanced hydrogen density ($n \geq 10^5$ cm^{-3}) located in the interface between the atomic and molecular phases. While one of the filaments is very likely embedded in an atomic medium, the others seem to be located in a mostly molecular medium (Fuente et al. 1996). The molecular filaments found in HCO^+ match perfectly the filaments seen in the 2.1 μm image of NGC 7023 (see Fig. 2). This strongly suggests that both emissions trace the same regions and that the 2.1μm filaments frequently found in PDRs are in fact high density regions.

5 Prospectives for large millimeter arrays

The study of the structure and of the physical and chemical properties of the molecular material in PDRs requires both high angular resolution and sensitivity. The combination of high angular resolution and sensitivity will allow to determine the degree of clumpiness and to measure the molecular abundances across the molecular filaments found in the atomic-molecular interface in PDRs.

References

Chokshi A., Tielens A.G.G.M., Werner M.W., Castelza M.W., 1988 ApJ 334, 803

Howe J.E., Jaffe D.T., Genzel R., Stacey G.J., 1991, ApJ 373, 158

Jaffe D.T., Genzel R., Harris A.I., Howe J.E., Stacey G.J., Stutzki J., 1990, ApJ 353, 153

Field D., Gerin M., Leach S., Lemaire J.L., Pineau des Forets G., Rostas F., Rouan D., Simons D., 1994, A&A 286, 909

Fuente A., Martín-Pintado J., Cernicharo J., Bachiller R., 1990, A&A 237,471

Fuente A., Martín-Pintado J., Brouillet N., Duvert G., 1992, A&A 260, 341

Fuente A., Martín-Pintado J., Cernicharo J., Bachiller R., 1993, A&A 276, 473

Fuente A., Martín-Pintado J., Neri R., Rogers C., Moriarty-Schieven G.,1996, A&A, in press

Rogers C., Heyer M.H., Dewdney P.E., 1995, ApJ 442, 694

Sellgren K., Werner M.W., Dinerstein H.L., 1992, ApJ 400, 238

Sternberg A. & Dalgarno A., 1995, ApJS 99, 565

Stutzki J., Stacey G.J., Genzel R., Harris A.I., Jaffe D.T., Lugten J.B., 1988, ApJ 322, 379

Witt A.N. & Malin D.F., 1989, ApJ 347, L25.

Searching for Star Formation Regions

Marcello Felli

Osservatorio di Arcetri, Largo E. Fermi 5, I-50125 Firenze, Italy

Abstract. The objective of this contribution is to present examples of Star Formation Regions (SFRs) in order to show their complexity and to stress the importance of large mm arrays to separate different evolutionary phases of massive stars in the same SFR.

1 The importance of mm arrays for SFRs studies

A reference working model for the cradle of newborn luminous stars has been around for many years. Since the star itself is not visible, all the attention is focused on the cradle itself, i.e.: UC HII regions, molecular clouds, hot molecular cores, cool dust envelpes responsible for the FIR emission, hot dust cocoons bright in the near IR, bipolar outflows, H_2 jets, masers, etc. However, these features occur on widely different scales and/or have been observed with different resolutions, and a direct comparison is often very difficult or even misleading.

Different and complementary approaches have been used to search for the earliest phases of massive stars. We have given preference to H_2O masers as the guiding indicator for selecting our target fields. In particular, we have studied the association of H_2O masers with UC HII regions, near IR sources with strong excess, hot molecular cores and bipolar outflows.

Why H_2O masers? Even though originally H_2O masers were discovered in diffuse HII regions, it has now become clear from surveys of IRAS selected sources (Palla et al. (1993)) that H_2O masers *not* associated with diffuse HII regions may be the majority, up to 80%, suggesting that the maser emission occurs in the earliest phases, much before the onset of an (UC) HII region (Codella and Felli (1995), Codella et al. (1994)). Also, on simple energetic arguments, H_2O masers must be very close to the stellar source of energy (from 10^{14} to 10^{16} cm), so that with the ~ 0.1" accuracy available with the VLA at 22 GHz the field to be searched can be considerably narrowed. In a large number of cases, H_2O masers occur near the center of bipolar outflows (Felli et al. (1992)) and of high density hot molecular cores (Cesaroni et al. (1994)).

With this in mind, we explored the cores of 22 molecular outflows with the VLA in the 8.4 GHz continuum and in the 22 GHz H_2O line (Tofani et al. (1995)). The results show that in all the cases a cluster of H_2O masers is found at the center of the bipolar outflow, but only in 40% small diameter continuum sources (size ≤ 1") with flux density ≥ 0.1 mJy are found. The new scenario that emerges is that in the very early stages bounded ionized winds or UC HII regions maybe present around the star and close to the H_2O maser, but so optically thich that they are undetectable in the radio continuum. These early stages maybe observable in the near IR thanks to the emission of hot dust near the star. As the ionized region expands it becomes detectable and the H_2O maser disappears. More than one episode of massive star formation and associated bipolar outflow may be present in the same SFR. The more extended ones may be related to the oldest star formation events, while smaller ones to the youngests. However, the small outflows may be masked by the present insufficient resolution of mm arrays if larger outflows are also present. We shall now show 6 examples that illustrate the obove points.

1.1 W75N

As shown by Hunter et al. (1994), the continuum emission from the center of the outflow is resolved into three small diameter (≤ 0.6 ") components, either UC HII regions or ionized winds. The H_2O maser emission is composed of five spots, all clustered around the two brightest radio continuum components. Each maser presents a spectrum with several velocity components. No K band source is present at the same position of the cluster, but a diffuse polarized emission is found few arcsec south of it (Moore et al. (1991)). The polarization pattern indicates that the source of radiation must coincide with the cluster, which must suffer a large extinction ($A_v \geq 90$). A hot molecular core is located at the position of the cluster. While the entire complex clearly represents a recent episode of star formation, expecially when compared with more diffuse HII regions in the same area, it is still not clear if the large outflow is a remnant of older episodes of star formation, or if there is a smaller outflow related to one (and which?) of the three continumm components. Similarly, it would be of extreme interest to see the structure of the hot molecular core on scale sizes comparable to those of the UC components, as well as to obtain deeper and higher resolution K band images of the cluster.

1.2 AFGL2591

A similar situation, but perhaps susceptible of a clearer interpretation, can be found in AFGL2591 (Tofani et al. (1995)). Also in this case at the center of the outflow there is a cluster of UC HII regions. N. 1 is the more extended one, it is not associated with a maser and it is emitting very weakly at K band. N. 2 is smaller and weaker and is associated with a weak maser. N. 3 is the weakest of all, but is associated with two powerful masers and a strong source at K band. The velocity pattern of the two masers associated with n. 3 and placed on opposite sides of it have blue-shifted and red-shifted spectra, suggesting a bipolarity on a scale ≤ 1". The polarization pattern at K band points to n. 3 as the center of radiation for the scattered emission. An H_2 jet points towards n. 3. Finally, to match radio and $Br\gamma$ emission, an ionized wind at the position of n. 3 is the most plausible explanation. In conclusion, the time sequence seems to be much clearer here, proceeding from n. 3 to n. 1 in order of increasing age. However, with the present resolution it cannot be stated which is the source of the molecular outflow, even though all the above indications seem to suggest that n. 3 is the most probable candidate.

1.3   AFGL5142

Single dish observations of the CO bipolar outflow in AFGL5142 suggest that it is composed of two separate flows: a more extended ($\sim$2') and a smaller one ($\leq$0.5'), both originating from approximately the same center (Hunter et al (1995)). Four H_2O masers and a very weak continuum source (most probably an ionized wind) are found at the centre of the two outflows. As for AFGL2591, two maser components (C3 and C4) are located on opposite side of the continuum source and have blue-shifted and red-shifted spectra. The orientation of this bipolarity (on a scale of $\leq$1") is in the same direction of the smaller CO outflow. Near IR observations reveal the presence of a stellar cluster at the same position. The source with larger H - K colour (but not the brigthest) coincides with the continuum source, while the brightest K source lies $\sim$20" to the west of it, near the position of an IRAS point source. H_2 jets pointing towards C3 and C4 and in the same direction of the smaller outflow were also found. In summary, the morphology of the SFR viewed at arcsec resolution points out to more than one episode of star birth in the same area. To separate the different episodes, arcsec resolution at mm wavelength is required.

1.4   G035.20-1.74

We shall now consider SFRs with no radio continuum emission near the position of the H_2O maser (down to $\leq$ 1 mJy). Examples of this type were found during a survey of H_2O masers with accurate VLA position made in the near IR. Out of 17 masers observed in J, H and K, almost all had a close-by IR source with high IR excess (H - K $\geq$2), but only in three cases there were radio continuum sources within few arcseconds (Testi et al. (1994)). A typical result from this survey is G035.20-1.74. Coincident with the IRAS position there is a rich stellar cluster and some diffuse emission, the brightest part of which coincides with a radio UC HII region. However, the H_2O maser is about 20" away from the HII region. It is associated with a K source with high IR excess. Most probably this represents a second generation of star birth. No accurate molecular observations of this source are available.

1.5   IRAS20126+4104

IRAS20126+4104 is an example of the CO outflow/H_2O maser observed by Tofani et al. (1995) with no radio continuum. The suggestion that this is an earlier evolutionary phase is confirmed by new molecular and near IR observations (Felli et al (1995a)). In fact, using high density tracers such as CH_3OH(3-2) and CH_3OH(5-4), the densest part of the molecular core coincides with the H_2O masers. With ^{13}CO(2-1) a bipolar outflow with total extension of few arcseconds is found at the same position. Narrow band near IR observations in the H_2 line show that there is a jet of H_2 pointing towards the maser. The broad band near IR observations show a cluster of sources around the maser. The closest to the maser is highly reddened and may harbour a massive newborn star.

1.6   G9.62+0.19

The last example, G9.62+019, summarizes most of the aspects that were shown previously and also stresses the power of mm arrays, since it is the only case for which there are such observations (Cesaroni et al. (1994), Hofner et al. (1995)). Radio continuum observations at 1.3 cm show a cluster of UC HII regions with different sizes (components A,B,C,D and E), which suggests the co-spatial existence of different evolutionary phases in the same SFR. However, interferometric 2.7 mm continuum observations show that there is an other component (F) a few arcseconds from D, but clearly disctinct from it. Component F (basically dust emission) also coincides with a peak in NH_3 and CH_3OH emission. H_2O masers are also present in the same area and the brigthest ones are associated with component F. Finally, a $C^{18}O$ broad profile was observed at the position of component F, suggesting a bipolar outflow on a scale size of the order of 1". Our near IR observations (Felli et al (1995b)) show a K band source with strong IR excess at the position of component F, but no emission from component D. This cannot be due to an absorption effect, since A_v is much greater in component F than in component D. We are currently trying to interpret these results in terms of an UC HII region of varying diameter. Very crudely, when the HII radius is small, of the order of the dust survival distance ($\sim 4\ 10^{14}$ cm), the free free emission will be strongly self-absorbed, and consequently no UC HII region will be observable, while there will be a strong near IR emission because of the high dust temperature. With the expansion of the HII region, the radio continuum emission increases because of the reduced optical depth, while near IR emission decreases because of the lower dust temperature.

References

Cesaroni, R., Churchwell, E., Hofner, P., Walmsley, C.M., Kurtz, S. (1994): A&A, **288**, 903

Codella, C., Felli, M. (1995): A&A, **302**, 521

Codella, C., Felli, M., Natale, V., Palagi, F., Palla, F. (1994): A&A, **291**, 261

Felli, M., Palagi, F., Tofani, G., (1992): A&A, **255**, 293.

Felli, M., Cesaroni, R., Testi, L., Walmsley, C.M. (1995a): in preparation

Felli, M., Persi, P., Testi, L. (1995b): in preparation

Hofner, P., Kurtz, S., Churchwell, E., Walmsley, C.M., Cesaroni, R. (1995): ApJ, in press

Hunter, T.R., Taylor, G.B., Felli, M., Tofani, G., (1994): A&A, **284**, 215

Hunter, T.R., Testi, L., Taylor, G.B., Tofani, G., Felli, M., Phillips, T.G. (1995): A&A, **302**, 249

Moore, T.J.T, Mountain, C.M., Yamashita, T., McLean, I.S. (1991): MNRAS, **248**,377

Palla, F., Cesaroni, R., Brand, J., Caselli, P., Comoretto, G. Felli, M. (1993): A&A, **280**, 509

Testi, L., Felli, M., Persi, P., Roth, M. (1994): A&A, **288**, 634

Tofani, G., Felli, M., Taylor, G.B., Hunter, T.R. (1995): A&AS, **112**, 299

Infrared and mm Data for class I candidates

Martin Osterloh , Thomas Henning and Ralf Launhardt

Max-Planck Research Unit "Dust in Star Forming Regions",
Schillergäßchen 2-3, D-07745 Jena, Germany

Abstract. We obtained near-infrared, CO (2-1), CS (2-1) and 1.3 millimeter continuum data for thirty-one southern objects known to have extremely red IRAS colors.
The data set is similar to northern surveys performed by Wilking *et al.* (1989) and Mc-
Cutcheon *et al.* (1991). K-band near-infrared counterparts to the IRAS point sources
were detected in 22 out of 25 good K' images. The three non-detections are too far
away for our limiting magnitude. Nineteen are multiple systems, *e.g.* embedded clusters, and five are single sources. 19 out of 21 CS observations were detections; the two
non-detections are, again, explained by large distances. CO wings suggesting outflows
were detected in almost all sources of the total sample. Most systems are luminous (10^3-
10^5 $L_\odot$) and presumably harbour at least one massive young star. For several clusters,
differing degrees of embeddedness are deduceable, and the dust must be concentrated
around individual stars. High-resolution millimeter mapping would help to deduce the
evolutionary status of individual members for a large number of such embedded, young
clusters.

1 Observations and Data Reduction

1.3 millimeter continuum and CO (2-1) line emission were measured with the
15m Swedish-ESO Submillimetre Telescope (SEST) during March 1994. Detectors were the SEST bolometer and an SIS receiver, respectively. The lines were
calibrated using the Penzias & Burrus (1973) chopper wheel method. For continuum photometry, Neptune and Uranus served as calibrators.
The infrared images were taken during November 1994 and March 1995 with
the ESO infrared camera IRAC2, equipped with a NICMOS3 HgCdTe array.
The optics were set to $0''.51$/pixel. Integration times were one minute per frame.
Coordinates were assigned by identification with field stars of the *Digital Sky
Survey*, of which were four or more in each frame; the positional uncertainty in
the field should be about $1''$.

2 Results

The central beam mass M is computed from the mm continuum datum via the
usual formula

$$M = \frac{F_\nu D^2}{\kappa_\nu B_\nu(T)}.$$ (1)

Here, $B_\nu(T)$ is the Planck function at frequency ν=230 GHz, D is the distance, and F_ν denotes the flux density. The absorption coefficient κ_ν was set to

Table 1. Parameters of the objects

IRAS Name (1)	D (2)	L_{bol} (3)	T (4)	M_{Dust} (5)	M_{CS} (6)	$\tau_{\nu_0}(CS)$ (7)	N_{H_2} (8)	A_V (9)
05155+0707	0.47	23	57	0.65	10	0.72	3.3	18
05338-0624	0.47	110	40	2.8	18	2.01	14	74
05491+0247	0.47	20	67	0.48	6.7	1.00	2.5	13
06084+0611	0.89	5500	53	12	110	2.56	17	90
07299-1651	1.00	2100	64	4.3	120	0.54	4.9	26
07427-0930	7.6	61000	57	300	16000	0.58	5.7	30
09591-5724	17.1	24000	40*	<140	*	*	*	< 1.0
10286-5838	6.5	38000	63	16	5500	0.09	0.42	2.3
10337-5710	0.40	53	41	0.40	11	0.33	2.7	14
10366-5745	18	1900	*	*	*	*	*	*
12091-6129	3.5	13000	55	54	1700	1.01	5.1	27
12326-6245	1.0	12000	56	55	*	*	63	330
12383-6128	4.4	35000	44	125	*	*	7.3	39
12405-6238	3.4	25000	72	65	*	*	6.5	35
15360-5554	5.5	45000	41	210	*	*	8	42
15520-5234	3.1	190000	50	560	*	*	64	340
16019-4903	1.9	2500	62	10	345	0.50	3.1	16.5
16060-5146	6.5	850000	45	860	21000	1.28	23	122
16228-5014	5.8	3600	40*	<18	2200	*	*	< 0.2
16445-4459	1.1	2400	41	1.5	*	*	1.5	8.0
16579-4245	19	290000	44	870	*	*	2.6	14
16594-4137	0.70	610	47	3.1	*	*	7.0	37
17016-4124	1.3	13000	44	34	*	*	23	120
17233-3606	2.2	95000	52	140	*	*	33	175
17419-2907	3.6	7700	14	16	3100	0.04	1.4	7.4
17431-2846	0.88	1000	61	<0.21	160	*	*	<0.26
18079-1756	2.5	19000	47	82	1500	1.80	14	77
18145-1707	3.0	5200	49	<3.15	840	*	*	<0.89
19035+0641	2.2	5400	50	21	490	0.58	5.1	27
19088+0902	4.6	17000	44	61	2000	0.51	3.2	17
19095+0930	9.2	280000	48	646	24000	0.59	8.6	46

Notes to the table columns: (2) - in kpc, (3) - in $L_\odot$ from the integrated infrared to 1.3mm luminosity,(4) - in K from the IRAS 100μm/60μm flux density ratio (5),(6) - in $M_\odot$, (8)in 10^{26} m^{-2}, (7),(9) - from the gas column densities calculated fromM_{Dust}
* - 40 K (marked with an asterisk) have been assumed as the dust temperature for two sources since only an IRAS 100μm upper limit is available.

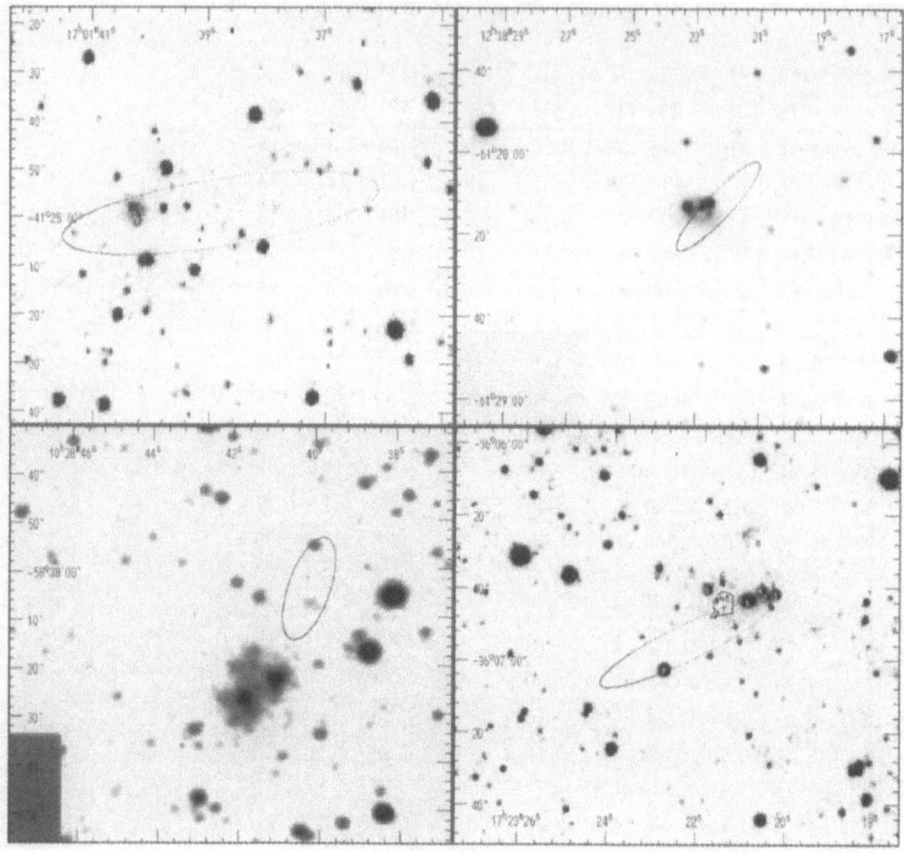

Fig. 1. Example K' images of four of our survey's sources. Embedded stars are found single (IRAS 17016-4124, top left), multiple (≤4, IRAS 12383-6128, top right), or in clusters (IRAS 10286-5838, IRAS 17233-3606, bottom left and bottom right). In IRAS 17233-3606, sources 1,4-6 are much deeper embedded than 1,2, tracing an increasing degree of embeddedness which culminates in the VLA-detected ionizing stars (crosses). Note that S of source 3, two additional VLA sources were found by Hughes and MacLeod (1993). Source # 7 is foreground, and # 8's nature is not clear. To trace the dust concentrations around single members of young clusters (which are most frequently found in our survey), millimeter mapping with sub-arcsecond resolution is required.

$0.01 cm^2/g$ at 1.3 mm (Ossenkopf & Henning 1994). The distances D to were taken from the literature or, if not found, estimated from the central CS or, if absent, the CO Gaussian center velocity. We used the galactic rotation curve published by Clemens (1985). Temperatures T were estimated by fitting a blackbody to the IRAS 100/60 μm ratio. The masses are listed in Table 1.

Mass estimates based on CS line data also appear in Table 1. They are calculated assuming the emission to be optically thin. An intermediate observational CS/H_2 abundance of 10^{-9} was assumed. We computed the optical depth for CS (2-1) based on the mm dust mass. Table 1 tells that τ is generally below 1. CS masses are generally much higher than dust masses, indicating that here we probe dense, high-temperature regions.

However, the assumption of optically thin emission breaks down earlier if the CS is not spread out over the total 1.3mm beam. This is at least strongly suggested in those cases where the dust emission seems to be strongly concentrated. In Fig.1, this is the case for I10286-5041, where the visual extinction ($\approx$ 2mag) derived from the mm continuum flux density is much lower than what is estimated from the H-K values (> 10mag). Also, the VLA sources in I17233-3606 detected by Hughes & MacLeod (see crosses in Fig.1 are not among our detected K' sources and presumably much more embedded.

3 LSA Prospects

In our survey, cluster star forming mode dominates isolated star formation. I17233-3606 shows even clustering of massive, ionizing stars, with relatively few near-IR sources representing the presumed low to intermediate mass cluster members. The extinguishing dust seems to be concentrated around individual objects separated by several arcseconds only. Similar conclusions can be drawn for other clusters' members by comparing their near-infrared reddening to the total dust mass per beam.

For these cold IRAS objects, a highly resolving large southern array would provide detailed maps of the dust emission, with the potential of assigning an evolutionary status to each member. Arcsecond resolution is not enough since the individual members are too close to each other.

Such highly resolved dust maps can also give in detail what was mentioned above for I10286-5041: Using [H-K],[J-H] diagrams and dust emission as two independent means of determining the circumstellar extinction will provide important clues on concentration and even geometrical distribution of the dust.

References

Clemens D.P. (1985): ApJ **295**, 422
McCutcheon W.H., Dewdney P.E., Purton C.R., Sato T. (1991): AJ **101**, 1435
Hughes V.A., McLeod G.C. (1993): AJ **105**, 1495
Ossenkopf V., Henning Th. (1994): AA **291**, 943
Wilking B.A., Mundy L.G., Blackwell J.H., Howe J.E. (1989): ApJ **345**, 257

Observations of Young Stellar Objects with Large Millimeter-wave Arrays

Steven V. W. Beckwith

Max-Planck-Institut für Astronomie, Heidelberg, Germany

Abstract. Observations of young stars with large millimeter-wave arrays should reveal the dynamics of the circumstellar disks, the rate of accretion and infall from the molecular clouds, the distribution of mass density, and the presence or absence of gaps cleared by large bodies such as planets condensing around the stars. Very few of these goals have been achieved with existing aperture synthesis arrays. The major advance offered by larger arrays is greatly improved spatial resolution and the ability to work at higher frequencies. One can expect specific structures to be revealed as the spatial resolution improves from a few seconds of arc (clouds, disks), to a few tens of milliseconds (formation of planets), to a few hundred microseconds (star/disk boundary layers, generation and collimation of jets). Hints of these structures are already seen through observations at many wavelengths.

1 Introduction

Observations of young stars have emerged as among the most exciting, and potentially most important, new applications of millimeter-wave astronomy with aperture synthesis arrays (Sargent 1996). The dense condensations around stars emit most of their energy at far infrared and mm-wavelengths over a large range of size scales from a few to a few thousand AU. Thermal continuum emission can be detected with large single telescopes, such as the 30 m at Pico Veleta, but it is almost always unresolved. Observations of molecular line emission, a potentially much richer source of information, can only be made with with interferometers. The higher spatial resolution of the interferometric images is necessary to distinguish the circumstellar structures from more extended molecular cloud material.

Circumstellar material resides in flattened disks, in compact spheroidal halos, and in fast, well collimated jets. The long wavelengths can reveal the material distribution and dynamics, essential to understand if we are to discover the evolution of the matter, whether it be to isolated stars, binaries, or planetary systems. The circumstellar disks are bright in molecular lines and thermal continuum. They are potential cradles for the birth of planets, a process of considerable interest to astronomy and one which may be directly observable at long wavelengths (Beckwith & Sargent 1993). The outflows may be an essential means by which stars rid themselves of excess angular momentum; the generation and collimation of the winds is one of the most important unsolved problems in the study of young stars.

On scales of a few thousand AU or more, the environment will be dominated by the molecular clouds: the velocities, densities, and temperatures should be characteristic of unbound cloud material. Understanding this region should give clues to the origin and extent of that portion of the star that collapsed to create the star and the rate at which residual infall feeds the central object. Over distances of a few hundred AU, we expect much of the material to be bound to the stars, with a small portion flowing out in an unbound wind. These regions contain the outer parts of the circumstellar disks and any residual spheroidal halos surrounding the star/disk systems.

On tens of AU scales, the disk is the most interesting component; the giant planets, Jupiter, Saturn, Uranus, and Neptune were created on these scales around the proto-Sun (e.g. Cameron 1988). Within a few AU, terrestrial-like planets condensed from the denser parts of the disk in the primitive Solar nebula. Within a few stellar radii - a few hundredths of an AU - the combination of accretion onto the star and mass loss through the creation of a wind, perhaps mediated by a magnetic field, will dominate the energy balance. This region is especially interesting, because the physical origin of the winds is a subject of great importance and considerable controversey (Shu et al. 1991). The various regions and corresponding phenomena are outlined in Table 1.

Table 1. Size Scales in Circumstellar Environments

Scale (AU)	θ @ 150 pc	T (K)	Phenomena
$\gtrsim 1500$	$\gtrsim 10''$	~ 20	Cloud structure, infall
150–1500	$1''$–$10''$	~ 20	Outer disks, infall, orbital motion
15–150	$0''\!.1$–$1''$	~ 50	Disk orbital motion, density waves, spheroidal halos (infall)
1–15	$0''\!.007$–$0''\!.1$	~ 200	Giant planet zone (gaps)
0.1–1	0.7–7 mas	$\gtrsim 300$	Terrestrial planet zone (gaps), holes, wind collimation
0.01–0.1	0.070–0.7 mas	$\gtrsim 1000$	accretion boundaries, magnetic channeling wind generation

The closest young stars are those in the dark clouds Taurus-Auriga, Ophiuchus, and Chameleon; these stars are almost all $1\,M_\odot$ or less, and the clouds are between 140 and 200 pc from the Sun. Higher mass stars are seen in Orion, three times more distant. Orion also has a very high density of low-mass stars (McCaughrean & Stauffer 1995), making high spatial resolution essential to distinguish the various objects. Table 1 also gives the resolution needed to study the different scales and outlines those which are most suited to mm-wave observations.

2   Molecular lines

The molecular line emission from young stars is often bright enough to observe with existing synthesis arrays(e.g. Sargent & Beckwith 1991; Hayashi, Ohashi, & Miyama 1993; Koerner, Sargent, & Beckwith 1993; Dutrey, Guilloteau, & Simon 1994; Ohashi & Hayashi 1995; Saito et al. 1995; Koerner & Sargent 1995; Dutrey et al. 1996). On the other hand, the available spatial resolution has been adequate to study only the outer parts of the circumstellar regions.

At these scales, one sees flattened structures in the emission lines from molecules such as CO and its isotopes (e.g. Sargent & Beckwith 1991) and CS (Ohashi et al. 1991, Ohashi et al. 1996), suggestive of centrifugally supported disks. Examples are shown in Figure 1. The velocity fields are consistent with this interpretation in several cases but not always: substantial radial infall or outflow can dominate rotation in the outer parts of the structures (Hayashi, Ohashi, & Miyama 1993, Saito et al. 1995). Individual structures, most likely disks, are seen around binary stars such as GG Tau (Dutrey, Guilloteau, & Simon 1994), when the separation is more than a few seconds of arc. Outflows can be seen in a few cases at velocities of a few $km\,s^{-1}$ (Sargent & Beckwith 1991; Keene & Masson 1990).

There is general agreement that observations of the gas with existing interferometers are probing regions dominated by the gravity of the stars rather than turbulence in the surrounding clouds, but there is continuing controversey about the detailed interpretation of the velocity fields and gas mass in individual objects. Stellar gravity is not so dominant on these scales to allow tests of accretion disk theory or to search for signatures of planet-building. Furthermore, the molecular lines used for these studies may well have high optical depths in interesting regions such as the disks, making an unambiguous interpretation of the observations difficult – compare, for example, Sargent & Beckwith 1991; Hayashi, Ohashi, & Miyama 1993; Omodaka, Kitamura & Kawazoe 1992; Dutrey, Guilloteau, & Simon 1994; and Kitamura, Kawabe, & Saito 1996). These problems have diminished the early hope that molecular emission would provide unequivocal proof that circumstellar disks are seen around many young stars.

Observations of the gas with spatial resolutions an order of magnitude greater than presently available – of order $0.''1$ – could overcome these ambiguities by resolving disks on scales dominated by orbital motion. Such observations could yield considerable information about the disk structures: temperatures, volume densities, and relative abundances are routinely inferred from millimeter-wave observations of molecular lines. The high spectral resolution of the heterdyne receivers means that line velocities and, hence, dynamical maps of the disks and spheroidal halos might be obtained.

A remaining uncertainty is if the heavy molecules usually observed to trace the molecular hydrogen - CO, CS, HCN, etc. - are in the gas phase or even present (Zuckerman, Forveille, & Kastener 1995). Many of these molecules freeze onto grain surfaces at the high volume densities and low temperatures thought to pertain throughout the disks. Additionally, the line widths will increase as approximately $r^{-\frac{1}{2}}$, where r is the distance from the star, owing simply to the higher keplerian speeds. This increase will place concomitant demands on the sensitivity.

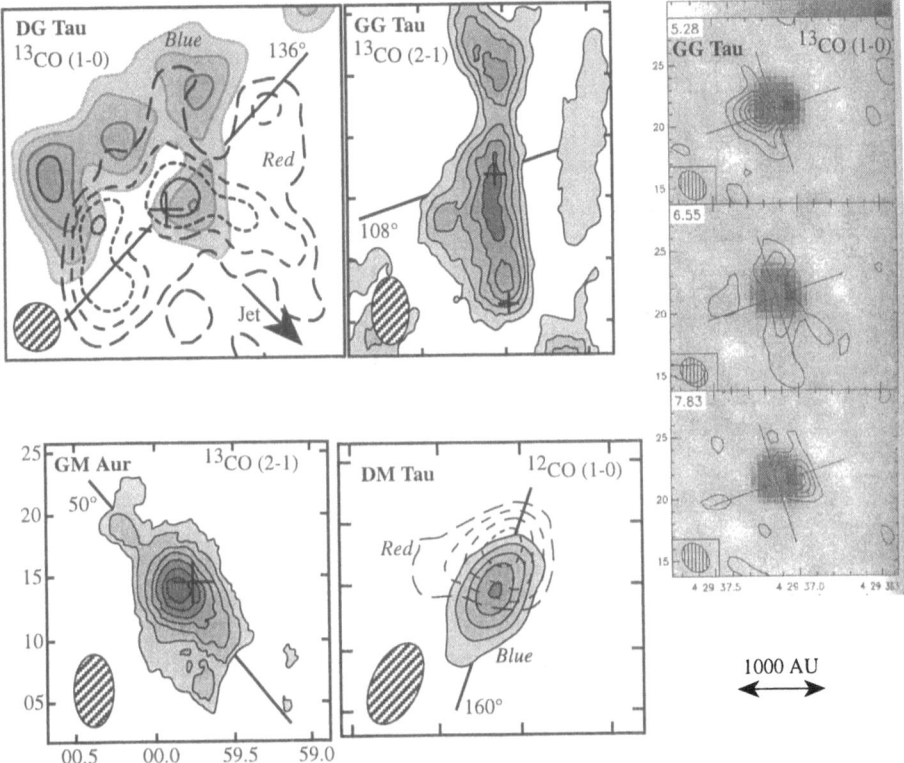

Fig. 1. DG Tau: Kitamura et al. (1996); GG Tau (center): Koerner et al. (1993); GG Tau (right): Dutrey et al. (1994); DM Tau: Saito et al. (1995); GM Aur: Koerner & Sargent (1995); all maps are to the same scale. The solid lines indicate the position angles of the major axes of the disks as derived by the authors from numerical modelling of the velocity distributions. Beamsizes are shown as hatched ellipses in the corners of the maps; red-shifted and blue-shifted velocity components are shown for several of the objects as well as the position of the CO at different velocities superimposed on the gray-scale continuum map for GG Tau (right). In several cases – DG Tau, DM Tau, and HL Tau (not shown) – the velocity distributions are consistent with disks only by assuming infall or outflow combined with rotation. The general complexity of the intensity and velocity distributions illustrates the difficulty of unambiguously interpreting these maps in terms of centrifugally supported disks.

It is still not known how the highly collimated jets from these young stars are created and focussed (Shu et al. 1991). An example of such a jet in HH 30 is seen in Figure 2. A substantial, perhaps dominant fraction of the mass outflow is neutral and dark at optical and near infrared wavelengths. Millimeter-wave observations could be the only means to observe this neutral component directly, making mm-wavelength interferometry a unique way to understand the generation and collimation of the winds. This very promising direction will be most demanding of spatial resolution, probably requiring baselines of 10 km or greater operating at a wavelength of 1 mm. Even VLBI experiments aimed at the innermost wind regions would be enormously beneficial, if sufficient sensitivity can be obtained by combining distant arrays of telescopes.

3 Continuum emission

Thermal emission from small particles - dust and dust aggregates - is commonly observed from young stars and is thought to originate in the circumstellar disks (Beckwith et. al. 1990). The emission is usually dominated by matter within 1 to 100 AU of the stars. The optical depths or column densities can be derived from mm-wave observations and provides one of the few means to assess the disk masses. Thermal emission is rarely seen in very young, embedded objects, often seen in classical T Tauri stars and Herbig Ae/Be stars, and rarely seen in weak-line or naked T Tauri stars (André & Montemerle 1994, Osterloh & Beckwith 1995). This progression may be indicative of an evolutionary sequence: disks develop slowly during the infall, or embedded phase, they are massive in the classical T Tauri phase, and they ebb during later stages as the particles clump together into larger bodies or are purged from the disk through accretion onto the stars (Ohashi et al. 1991, Ohashi et al. 1996) - the available evidence is against the common belief that a stellar wind expels the gas and dust from the system (Beckwith et. al. 1990).

There are at least three tests of our understanding of these regions which will be made with improved spatial resolution. The first is to measure the full width at half maximum of the thermal emission, thereby measuring the optical depth and column density of particles from which the mass may be inferred. Quite a few stars have spectral energy distributions consistent with blackbody emission between 0.3 and 2 mm (Beckwith & Sargent 1991, Mannings & Emerson 1994). If the emission is optically thin, e.g. gray, the particles themselves are black and, therefore, a few millimeters or more in size. Such an inference is an important step in deciding if the particles coagulate and grow from submicron size to rocks and ultimately to small planets.

For most stars, resolution of a few tenths of a second of arc requiring baselines of order 1 km will be needed to measure the sizes of the emission regions (Lay et al. 1994). In a few cases, the resolution attained by existing arrays - $\gtrsim 1''$- is adequate to measure the extent, in GG Tau, for example (Dutrey, Guilloteau, & Simon 1994), but these appear to be unusual. Direct images of disks in emission (O'Dell, Wen, & Hu 1993) and in silhouette with the Hubble Space Telescope (O'Dell & Wen 1994) give us a good idea of the distribution of disk sizes (McCaughrean & O'Dell 1996), and it is clear that subarcsecond resolution will be essential. A few examples of these images are shown in Figure 2: the HH 30 disk in Taurus subtends no more than $1''$ in silhouette (Burrows et al. 1996); the largest of the disks in Orion is about $1''$ (450 AU) across, and most are smaller.

The column density of dust seen in these images is very small, and the bulk of the optically thin emission at millimeter wavelengths will be dominated by higher column densities that subtend smaller angles.

The second test will require resolution of order 70 milliseconds of arc (mas) or better, baselines of several km operating at sub-mm wavelengths. At this resolution, one can map the optical depth as a function of distance from the star, thus getting a direct measurement of the radial surface density distribution. Knowledge of this distribution is an important constraint on theories of disk structure and evolution; if measured, it should give the value (or distribution) of α in the standard α-disk models; more generally, it should determine the viscosity in the disk, a parameter about which we can now only guess (Pringle 1981).

The third test requires another order of magnitude improvement to 10 mas or better. With this resolution, one can look for gaps in the material density indicative of clearing by large bodies. The prevailing theory of planet formation is that large bodies grow within a continuous disk, sweeping up material near their orbits as they evolve. This sweeping will leave large gaps in the particle distribution that might be detected directly with high enough resolution. There is already good indirect evidence for holes in the inner parts of disks, as well as the direct measurements of holes in the disks surrounding main sequence stars (Skrutskie et al. 1990; Lagage & Pantin 1994).

Binary and triple systems are at least as common among young stars as on main sequence; more than half of the young systems (including single stars) are multiple (Ghez, Neugebauer, & Matthews 1993, Leinert et al. 1993). The distribution of semi-major axes peaks below $1''$. Subsecond resolution will be needed simply to distinguish the individual components in most young stars. A striking example is shown in Figure 3. T Tau is a young binary with an apparent separation of $0''\!.7$ (100 AU) in the Taurus cloud (Dyck, Simon, & Zuckerman 1982). The northern component is a normal T Tauri star, whereas the southern component is a highly obscure, yet very active object which produces most of the luminosity of the combined system seen in the far infrared (Ghez et al. 1991). One anticipates that it is the southern component that will dominate observations at mm wavelengths, both because of its strong thermal infrared emission and because of its high luminosity.

The T Tau system is one of the brightest mm-continuum sources in Taurus (Beckwith et. al. 1990). It has unusually strong H_2 emission associated with shocks (Beckwith et al. 1978, Herbst et al. 1996), and its complex environment contains most of the phenomena one will see around many young stars. The figure reiterates the need for better spatial resolution to distinguish the various processes at play in this system. Observations at other wavelengths suggest we are looking at an almost face-on pair of stars, each of which has a circumstellar disk, with an extended circumbinary disk surrounding the entire system (Weintraub, Masson, & Zuckerman 1989). The large mass associated with the particles around these stars makes T Tau system a good candidate for the study of ongoing planetary formation.

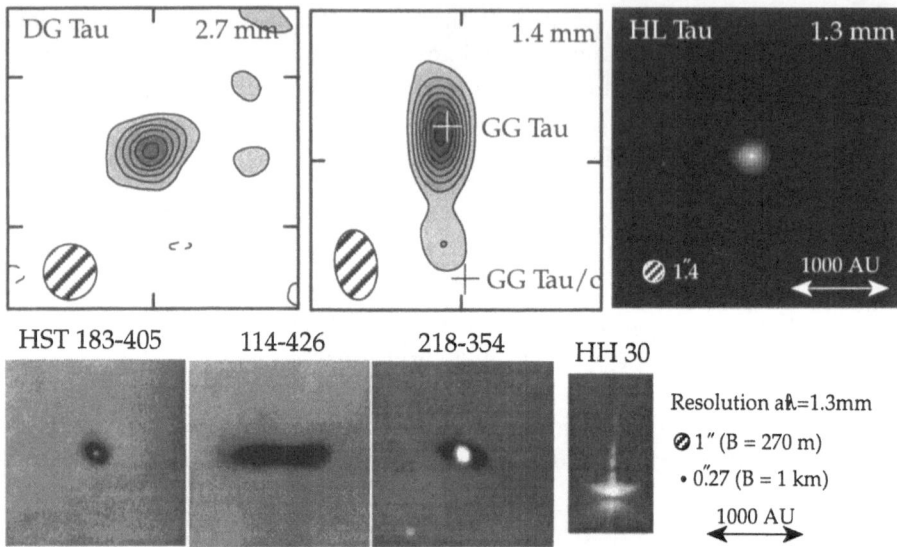

Fig. 2. The maps in the top row are of stars in Taurus ($D = 150\,\mathrm{pc}$) in boxes $20''$ (3000 AU) on a side: DG Tau (Saito et al. 1995), GG Tau (Koerner, Sargent, & Beckwith 1993), HL Tau (Koerner & Sargent 1995). The bottom row presents optical images with the Hubble Space Telescope: the three objects on the left are disks seen in silhouette against the Orion nebula by McCaughrean & O'Dell (1996); they have been scaled to appear as if they were at 150 pc for comparison with the top row. HH 30 in Taurus it displays a prominent jet (green), as well as a flared disk seen nearly edge-on. Also shown are beamsizes achieved with two interferometer baselines at 1.3 mm.

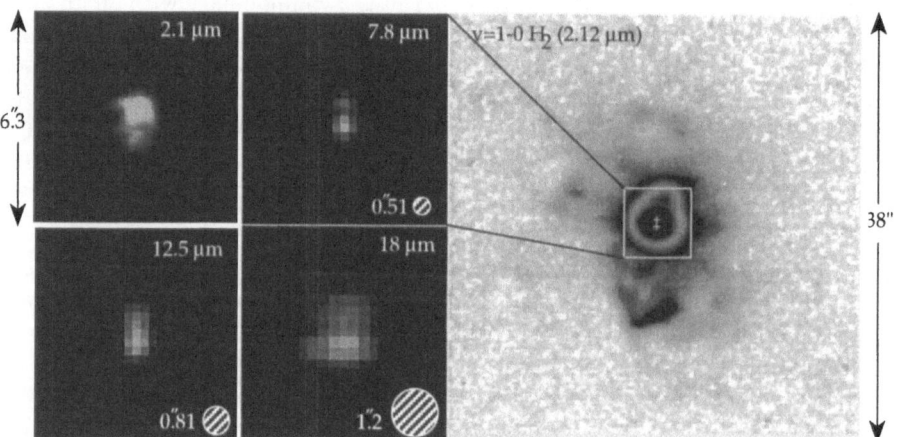

Fig. 3. The infrared images (Herbst et al. 1997) show clearly the T Tau binary pair and the increasing brightness of the T Tau South at longer wavelengths. The right-hand image is of the H_2 emission ($2.122\,\mu m$ $v = 1 - 0$ S(1)) by Herbst et al. (1996). It shows the rich structure of the wind-cloud-disk inteface which is shock heated.

A few observations using existing arrays on very long baselines will be quite useful in getting the flavour of what is to come but will certainly be inadequate for studies of large samples of stars that are necessary to understand the ubiquity of the phenomena discovered around the brightest objects. Adequate images will also require good u-v plane coverage, accomplished only when there is a correspondingly large number of telescopes to fill in and synthesize a complete aperture.

It has been only recently that astronomers can assess the issue of planets forming around young stars through direct observation, and even more recently, in which they can search indirectly for the planets themselves using observations of the reactions of the star. The announcements of planets around the pulsar PSR 1257 (Wolszczan & Frail 1992) and the stars 51 Pegasus (Mayor & Queloz 1995), 70 Virginis (Marcy & Butler 1996) and 47 Canis Majoris (Butler & Marcy 1996) underscore a growing interest in the circumstellar environments around young stars and a study of the conditions which give rise to other planetary systems. Millimeter wave observations have become increasingly important in the study of this evolution and will continue to be so in the future.

References

André, P. & Montemerle, T. 1994, ApJ, **420**, 837.

Beckwith, S., Gatley, I., Matthews, K., and Neugebauer, G. 1978, ApJ, **223**, L41.

Beckwith, S. V. W. and Sargent, A. I. 1993, in *Protostars and Planets III*, ed. G. Levy and J. Lunine (Tucson:U. Arizona Press), p. 521–541.

Beckwith, S. V. W., Sargent, A. I. 1991, ApJ, **381**, 250.

Beckwith, S. V. W., Sargent, A. I., Chini, R., and Güsten, R. 1990, AJ, **99**, 924.

Burrows, C., Stapelfeldt, K., et al. 1996, HST images available on the World Wide Web.

Butler, & Marcy, G. M. 1996, ApJ, in press.

Cameron, A. G. W. 1988, ARA&A, **26**, 441–472.

Dutrey, A., Guilloteau, S., Duvert, G., Prato, L., Simon, M., Schuster, K., & Menard, F. 1996, A&A, in press.

Dutrey, A., Guilloteau, S., & Simon, M. 1994, A&A, **286**, 149.

Dyck, H. M., Simon, T., and Zuckerman, B. 1982, ApJ, **255**, L103.

Ghez, A. M., Neugebauer, G., & Matthews, K. 1993, AJ, **106**, 2005.

Ghez, A. M., Neugebauer, G., Gorham, P. W., Haniff, C. A., Kulkarni, S. R., Matthews, K., Koresko, C., & Beckwith, S. 1991, AJ, **102**, 2066.

Hayashi, M., Hasegawa, T., Ohashi, N., & Sunada, K. 1994, ApJ, **426**, 234.

Hayashi, M., Ohashi, N., & Miyama, S. M., ApJ, **418**, L71.

Herbst, T. M., Beckwith, S. V. W., Glindemann, A., Tacconi-Garman, L. E., Kroker, H., & Krabbe, A. 1996, AJ, in press.

Herbst, T. M., Robberto, M., Beckwith, S. V. W., Bizenberger, P., & Birk, Ch. 1997, in preparation.

Jensen, E. L. N., Mathieu, R. D., & Fuller, G. A. 1994, ApJ, **429**, L29.

Keene, J. & Masson, C. R. 1990, ApJ, **355**, 635.

Kitamura, Y., Kawabe, R., & Saito, M. 1996, ApJ, **457**, 277.

Koener, D. W. & Sargent, A. I. 1995, AJ, **109**, 2128.

Koerner, D. W., Sargent, A. I., and Beckwith, S. V. W. 1993, Icarus, **106**, 2.

Lagage, P. O. & Pantin, E. 1994, Nat, **369**, 628.

Lay, O. P., Carlstrom, J. E., Hills, R. E., & Phillips, T. G. 1994, ApJ, **434**, L75.

Leinert, Ch., Zinnecker, H., Weitzel, N., Christou, J., Ridgway, S. T., Jameson, R., Haas, M., & Lenzen, R. 1993, A&A, **278**, 129.

Marcy, G. M. and Butler, 1996, ApJ, in press.

Mannings, V. & Emerson, J. P. 1994, MNRAS, **267**, 361.

Mayor, M. & Queloz, D. 1995, Nat, **378**, 355.

McCaughrean, M. J. & O'Dell, C. R. 1996, ApJ, in press (April 1996).

McCaughrean, M. J., & Stauffer, J. R. 1995, AJ, **108**, 1382.

Mundt, R., Ray, T. P., Bührke, T., Raga, A. C., and Solf, J. 1990, A&A, **232**, 37.

Natta, A. 1993, ApJ, **412**, 761.

O'Dell, C. R., & Wen, Z. 1994, ApJ, **436**, 134.

O'Dell, C. R., Wen, Z. and Hu, X. 1993, ApJ, **410**, 696.

Ohashi, N. & Hayashi, M. 1995, Ap&SS, **224**, 13.

Ohashi, N., Kawabe, R., Hayashi, M., & Ishiguro, M. 1991, AJ, **102**, 2054.

Ohashi, N., Hayashi, M., Kawabe, R., & Ishiguro, M. 1996, ApJ, **466**, in press.

Omodaka, T., Kitamura, Y., & Kawazoe, E. 1992, ApJ, **396**, L87.

Osterloh, M. & Beckwith, S. V. W. 1995, ApJ, **439**, 288.

Pringle, J. E. 1981, ARA&A, **19**, 137.

Saito, M., Kawabe, R., Ishiguro, M., Miyama, S. M., & Hayashi, M. 1995, ApJ, **453**, 384.

Sargent, A. I. 1996, in *Disks and Outflows from Young Stars*, eds. S. Beckwith, J. Staude, A. Quetz, & A. Natta, in press (Springer, Heidelberg).

Sargent, A. I. and Beckwith, S. V. W. 1991, ApJ, **382**, L31.

Shu, F. H., Ruden, S. P., Lada, C. J., and Lizano, S. 1991, ApJ, **370**, L31.

Skrutskie, M. F., Dutkevitch, D., Strom, S. E., Edwards, S., and Strom, K. M. 1990. AJ, **99**, 1187.

Smith, B. A., and Terrile, R. 1984. Sci, **226**, 1421.

Weintraub, D. A., Masson, C. R., and Zuckerman, B. 1989, ApJ, **344**, 915.

Wolszczan, A. & Frail, 1992, Nat, **355**, 145.

Zuckerman, Forveille, $ Kastner, J. H. 1995, Nat, **273**, 394.

Protoplanetary Disks

Anne Dutrey

Institut de RadioAstronomie Millimétrique, 300 rue de la Piscine, F-38406 Saint-Martin D'Hères, France

Abstract. Understanding the physics and the chemistry of the protoplanetary disks is one challenge of the next ten years mainly because we would like to analyse the physical processes leading to planet and life formation. Therefore it is interesting to focus on low mass PMS objects, namely TTauri stars, which are possible progenitors of solar-type stars.

For this issue, the LSA is really needed in order to get the sensitivity and the resolution (below 0.1″) which are necessary to map and resolve the disk structures both in molecular lines and in continuum.

1 Introduction

Recent NIR and bolometric mm surveys have shown that most low-mass stars are surrounded by extended dust disks (Beckwith et al., (1990)). Among these PMS stars, many TTauri stars have been found to be binary or multiple systems. Dust distributions around such systems are disturbed by tidal effects. Disks around single stars have typical radii of $\sim$ 500-1000 AU. These disks are principally heated by the central stars which have low luminosities, and about $\sim$90 % of the disks radiate in the mm range. Hence detailed studies of these objects at mm-wave are fundamental and can only be achieved by interferometers.

In the last few years, the current generation of arrays has observed at 3-1.3mm in continuum and in the first rotational lines of CO many TTauri stars in the Taurus-Auriga and ρ Oph molecular clouds. Detailed mapping leading to accurate modelling is always the exception because these instruments are still limited by the lack of sensitivity and resolving power.

2 Scientific Goals and Current Knowledge

The final goal of these studies is to analyse the physical processes which lead from a dusty and gaseous disk around a PMS star to the formation (or not) of planets. The links between disk and star properties are not well understood, and many points remain unclear:

- What are the morphology and the kinematics of dust and gas around a single new-born star, how do they evolve with the star?
- What does happen in case of multiple systems ?
- Among a given sample, how many systems have the mass required to form planets ?

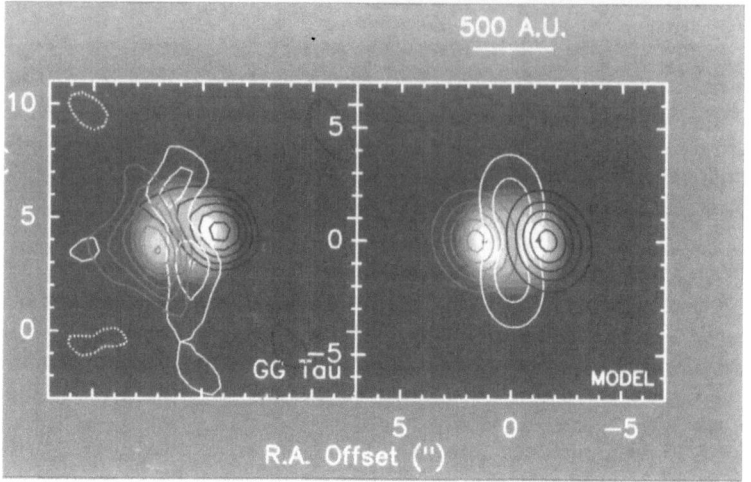

Fig. 1. Left: GG Tau observations at 2.7mm and in ^{13}CO J=1-0 using PdBI. Right: the best model. The dust emission is given in grey scale. Contours show three representative channels of ^{13}CO J=1-0 emission at blueshifted (grey), systemic (white) and redshifted (black) velocities.

– What are the chemical abundances of dust and gas components in disks? What are their evolution with time ? When do they start to form planetesimals ?

Up to now, we have only a few convincing cases of Keplerian motions inside disks of TTauri stars: GM Aur (Koerner et al. (1993)), GG Tau (Skrutskie et al., (1993), Dutrey et al., (1994)), DM Tau (Guilloteau & Dutrey (1994)). Good candidates are still rare. Moreover because of beam dilution and of a lack of sensitivity even CO and its isotopes, which are still the best kinematical tracers, remain difficult to detect.

Figures 1 (GG Tau) and 2 (DM Tau) illustrate these problems. Both stars, intensively studied with the IRAM instruments, are located in the Taurus cloud at 150 pc ($1'' = 150$ AU). GG Tau is a binary star with separation in the plane of sky $0.26''$ (40 AU) while DM Tau is a relatively old TTauri star with age $\sim 5 \cdot 10^6$ years. The left part of figure 1 shows the image of the binary GG Tau obtained with the IRAM interferometer (PdBI) in continuum at 2.7mm and 3 channels (blue, systemic and red velocities) of the ^{13}CO J $= 1 \rightarrow 0$ emission. Due to tidal effects related to the binarity, the inner part of the disk is cleared (hole size $\sim$ 360 AU). The right part shows the best Keplerian model of both dust and CO emissions (Dutrey et al., (1994)). Figure 2 is a montage of CO J $= 1 \rightarrow 0$ observations of DM Tau obtained recently with the PdBI (Guilloteau et al., (1996)). The gas disk is partially resolved (radius $\sim 5''$) and the velocity gradient is parallel to the major axis of the disk as expected in the case of Keplerian rotation (bottom model). For comparison, an infalling disk near free fall velocities is simulated on the top model: the velocity gradient is then parallel to the minor axis of the disk. In both cases, Keplerian and Infalling models, the integrated flux density profiles are double-peaked. To study the kinematics, it is therefore necessary to resolve the disk structure.

3 Limitations of Current Arrays

In the observations shown above, the resolution is still very limited compared to the scale of our solar system (Neptune is at $\sim$ 30 AU from the sun i.e. 0.2″ at the Taurus distance). Investigating physical processes inside the inner part of disks (<100-150 AU) is still impossible.

Taking into account parameters like a better site and better performances for the next generation of receivers, we can reasonably expect between the PdBI with 5 antennas and the LSA with 50 antennas a gain in sensitivity of about 40. This allows in many cases high resolution studies. These points are summarized by figures 3 and 4 where the limitations of current arrays are clearly shown using the DM Tau disk as an example.

Using the PdBI with 5 antennas, one can expect to map a line like ^{13}CO $J = 2 \rightarrow 1$ in DM Tau down to a resolution of about 1.5″. $H_2CO(3_{13}\text{-}2_{12})$ and $H^{13}CO^+$ $J = 1 \rightarrow 0$ recently detected using the 30-m radiotelescope (Dutrey et al., (1996)), should be detectable in a reasonable amount of time with the PdBI but impossible to map with a high enough resolution (< 1″) to allow a detailed modelling of the distribution and excitation conditions of the molecules.

For the LSA, on optically thick species like CO $J = 2 \rightarrow 1$ high resolution mapping can be achieved down to 0.05″, while mass distribution and kinematics on optically thin species like $J = 2 \rightarrow 1$ can be studied at a resolution of 0.1″ – 0.2″.

4 Perspective for the LSA

Therefore using the LSA, the physical and chemical properties of inner disks will be investigated. All these physical and chemical points should be studied in single and multiple systems.

4.1 Morphology and Kinematics of Disks

High resolution mapping is needed to constrain models of disk formation and evolution (Shu et al., (1987)). Due to the lack of resolution on current arrays, many points requiring high resolution are still speculative. These points would be addressed using "standard" gas tracers like CO and its isotope and continuum emission.

The accretion mechanism, its rate, its evolution with time, its link with Keplerian rotation have not yet been observationally investigated. The role of the jet and the stellar winds in the disruption of the disk near the central star is not yet understood. The links between the magnetic field, the jet and the accretion disk are only investigated from the theoretical point of view.

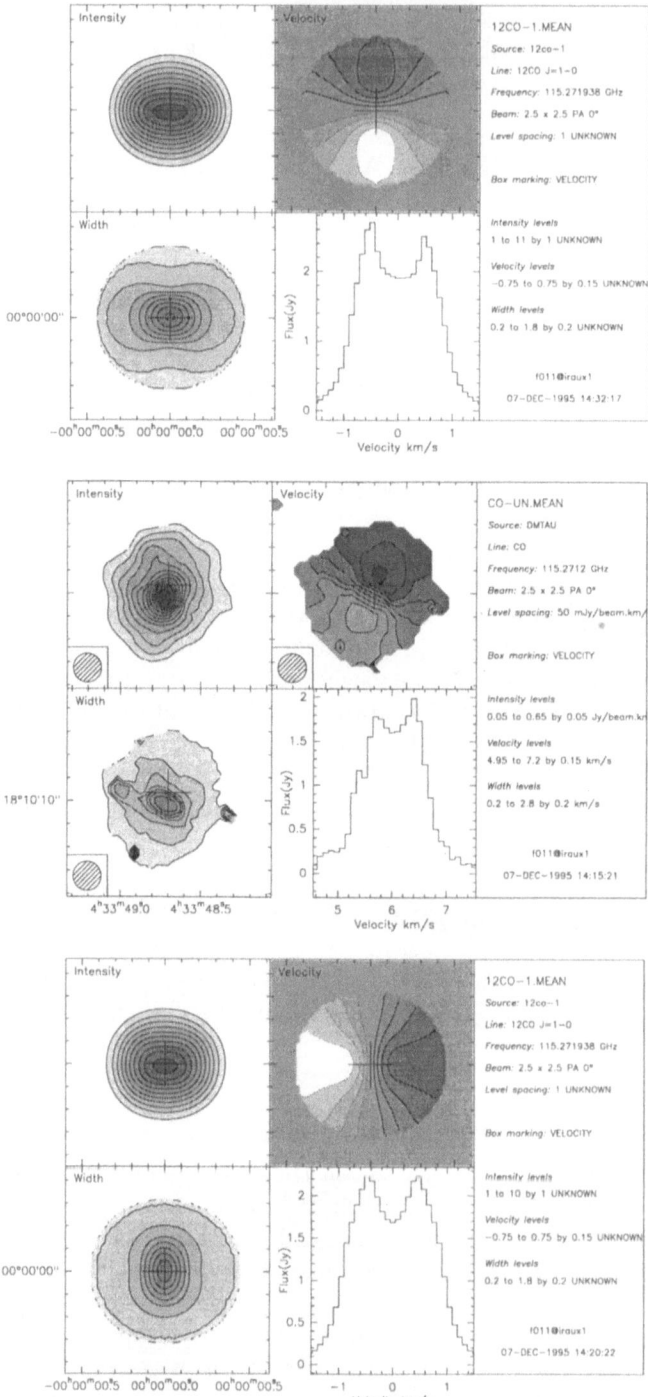

Fig. 2. Simulations of the ^{12}CO J=1-0 emission (LTE) from a disk seen inside an interferometric beam of 2.5″ and comparison with DM Tau data. Top panel: Infalling disk. Medium panel: DM Tau data from the PdBI. Bottom panel: Keplerian model.

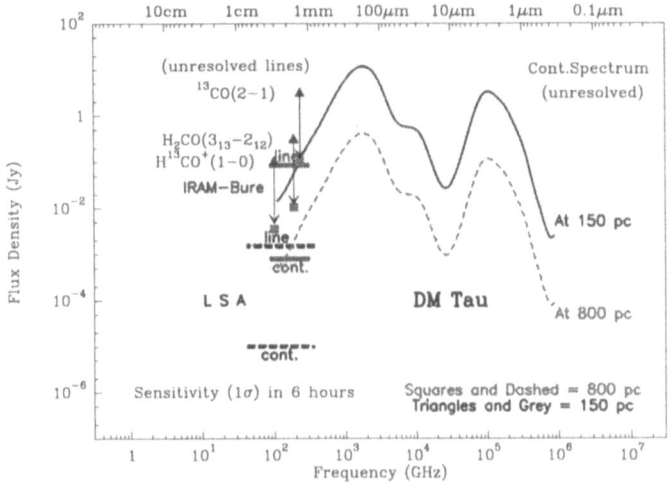

Fig. 3. Comparison between LSA and Bure. Sensitivity estimates have been done assuming 6 hours of integration time, spectral resolution of 100KHz for lines, 1GHz (Bure) and 2GHz (LSA) for continuum. The system temperature was 300 K for Bure and 150 K for LSA.

The LSA will also be a powerful instrument to find indirect evidence for planet formation. Around single stars, high resolution mapping allow to image gap clearing due to planetary formation: at 150 parsecs, a 10 AU central hole ($\sim$ Jupiter orbit) will be seen with a resolution of $\sim 0.1''$ like the GG Tau hole (see figure 1).

Moreover on multiple systems, which represent the majority of the PMS systems, surveys including relatively close binaries (separation $\leq 0.1''$) will allow to understand the role and the evolution of circumbinary disks in the formation of planets around binary systems, the link between circumbinary disks and circumstellar disks. The recent image of the circumbinary ring of GG Tau obtained in NIR adaptative optics by Roddier et al, (1996) in the J band (figure 5) is a beautiful extrapolation of possibilities offered by the LSA.

4.2 Continuum Properties of Disks

At mm wavelengths, most of the continuum arises from thermal dust emission. However in some cases free-free contributions should be present and even synchrotron variable emission has been observed (V773 Tau, Dutrey et al, (1996)). Moreover, the properties of the dust grains are still debated (Agladze et al., (1994)), their composition, their sizes, their spectral index, their evolution with time are not known.

With the LSA, multiple wavelength mapping in the range 3mm-0.8mm where the dust is mostly optically thin will provide the best way to determine the dust properties. These mappings require a high resolution ($\sim 0.5'' - 0.1''$) to

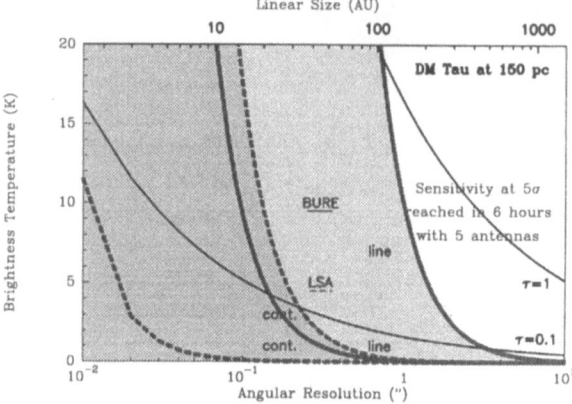

Fig. 4. Comparison between LSA and Bure. Sensitivity estimates assume same parameters than in figure 3. The brightness temperature $T_B(r) \propto r^{-0.5}$ is plotted in the optically thick ($T_B = T_k$) and optically thin ($T_B = 0.1\ T_k$) cases.

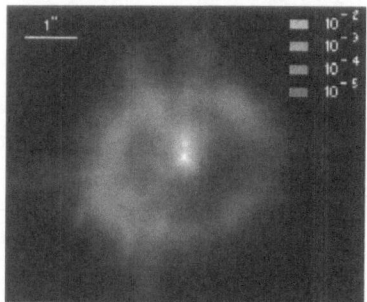

Fig. 5. GG Tau in the J band observed at the CFHT, from Roddier et al. 1996. The inner part of the circumbinary ring is seen in scattered light. The derived ring parameters are in excellent agreement with the Bure data (figure 1).

properly model the disk emissivity $F_\nu(r)$. Moreover, large surveys on objects from classes 0 to III are needed to trace the dust evolution from interstellar grains to planetesimals.

4.3 Chemistry and Physics of Gas and Dust Disks

Up to now, the gas-to-dust ratio and thus the mass distribution is poorly constrained in disks. When we compare the mass estimates derived from CO with the mass estimates derived from the dust, there are still discrepancies as high as $\sim$20-100. This point is of prime importance because it is actually impossible to determine how many TTauri stars have a disk with the minimum mass required to form planetary systems similar to our solar system.

Moreover until very recently CO was the only gas species detected inside disks. To understand the chemistry of protosolar-like nebulae, one need to study the dust properties *in relation* with the gas properties and their *related evolution*. In this field, the resolution is fundamental in order to avoid beam dilution of optically thick cores with typical radius ~30 AU (see Guilloteau & Dutrey (1994)). For example, figure 5 shows the first attempt for a chemical survey on the DM Tau disk using the 30-m radiotelescope (Dutrey et al., (1996), in prep.). Typical integration time is ~ 500 minutes per spectrum. High resolution and high sensitivity mapping is needed to map the molecule distributions and derive their excitation conditions and abundances as a function of the disk radius.

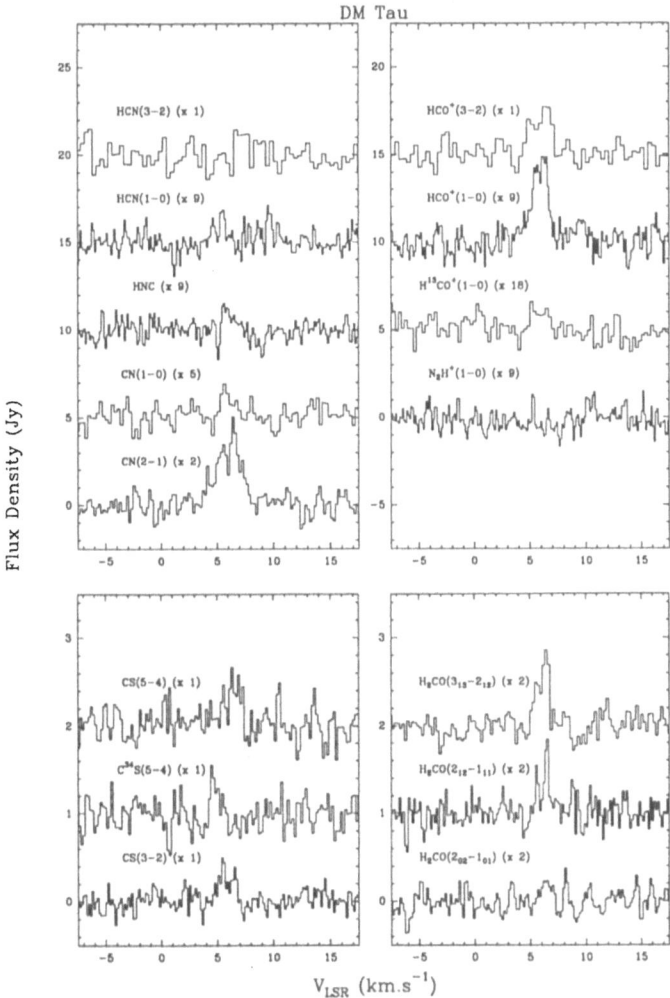

Fig. 6. The DM Tau disk seen in several molecules with the 30-m telescope. Typical integration time per spectrum is around 500 minutes.

These mappings are fundamental if we want to understand the origin of the disk mass discrepancies using the dust and several gases as mass tracers.

They would also provide the first analyses of the chemical evolution of protosolar-like nebulae. For this purpose, high sensitivity is needed to search for rarer optically thin species.

5   Other Objects

With the LSA, studying the gas and dust components of objects more evolved than TTauri stars will be feasible. For example, inside the β Pic disk, CO has only been detected in the UV bands (Vidal-Madjar et al., (1994)) and we do not know yet the large scale kinematics of the disk.

High sensitivity mapping will also allow to investigate the properties of disks at larger distances, particularly around high-mass stars (AeBe), or to study low-mass disks associated to high-mass star clusters like in the Orion region.

Finally for the first time, we will be able to statistically analyse the properties of fully resolved disks.

Acknowledgements

I would like to thank S.Guilloteau, T.Forveille, F.Malbet and A.M. Lagrange for many fruitful discussions and C.Roddier for the image of GG Tau in the J band.

References

Agladze, N.I., Sievers, A.J., Jones, S.A., Burlitch, J.M., Beckwith, S.V.W., (1994): Nat. **372**, 243

Beckwith, S.V.W., Sargent, A.I., Chini, R., Gusten, R., (1990): AJ **99**, 924

Dutrey, A., Guilloteau, S., Simon, M., (1994): A&A **286**, 149

Dutrey, A., Guilloteau, S., Duvert, G., et al, (1996): A&A, *in press*

Dutrey, A., Guilloteau, S., Guélin, M, (1996): *in prep.*

Guilloteau, S., Dutrey, A., (1994): A&A **291** L23

Guilloteau, S., Dutrey, A., Guélin, M, (1996): *in prep.*

Koerner, D.W., Sargent, A.I., Beckwith, S.V.W., (1993): *Icarus* **106** 2

Shu, F., Adams, F.C., Lizano, S., (1987): ApJ **25** 23

Roddier, C., Roddier, F., et al., (1996): ApJ *in press*

Skrutskie, M.F., Snell, R.L, Strom, K.M., et al, (1993): ApJ **409** 422

Vidal-Madjar, A., Lagrange-Henri, A-M., Feldman, P.D., et al., (1994): A&A **290** 245

Observations of bipolar molecular outflows with large millimeter arrays

Rafael Bachiller

Observatorio Astronómico Nacional (IGN)
Apartado 1143, E–28800 Alcalá de Henares, Madrid, Spain

Abstract. The ejection of a bipolar outflow is a central phenomenon in the processes by which a new star is formed, and the molecular component of such outflows is particularly interesting because of its large mass, momentum, and energy, and because it contains information about all the past mass-loss episodes from the central engine. This paper illustrates the fact that the research on bipolar molecular outflows has progressed thanks to the increase in angular resolution of the available instrumentation. I stress the importance of observing bipolar outflows with large millimeter arrays, and conclude that a millimeter-wave array providing an angular resolution of $0.1''$ will solve many of the challenging questions on outflows which remain open nowadays.

1 Introduction

Most young stellar objects (YSOs) are observed to eject violent outflows of high-velocity gas. The properties of bipolar outflows have been recently reviewed by Bachiller (1996). Such outflows are:

• *Bipolar.*- The outflow usually consists of two emission lobes placed at symmetrical positions with respect to the central YSO.

• *Spatially extended.*- The sizes of bipolar outflows range from less than one tenth to several pc.

• *Clumpy.*- Filling factors are often of the order of 0.1 at high-velocities (for observations with standard telescopes) but can approach unity at the lowest velocities. Individual clumps are resolved in the nearest outflows.

• *Very energetic.*- outflows involve enormous amounts of mass (up to $200\,M_\odot$) and energy (up to 10^{48} erg).

• *Time variable.*- over a wide range of time-scales (from $\sim 10^3$ to 10^5 yr).

It is widely accepted that the bipolar outflows from YSOs limit the mass of the star/disk system which is forming. Outflows are also important to carry away the excess of angular momentum from accretion disks. Finally, by dispersing the ambient molecular material, bipolar outflows can determine the evolution of the dense core where the new star is born. Because of all these reasons, there is a wide consensus that understanding star formation physics requires understanding the infall and outflow phenomena in a unified fashion.

Hence, the observations of outflows are essential for star formation studies. Bipolar outflows consist of at least three main components: high-excitation ionized gas, neutral atomic gas, and molecular low-excitation gas. The molecular component is the most massive (up to a few hundred solar masses). The large mass of this component implies that it is formed by swept-up ambient gas.

The outflowing molecular gas is probably pushed by a primary wind emerging from the central star/disk system. As a consequence, the molecular component of an outflow contains information about the entire history of the mass-loss activity in the central YSO. Observations of this component in the millimeter-wave CO lines, or in other weaker spectral lines, are of the highest interest to constrain theories on star formation.

This article stress that the progress achieved in the research on bipolar molecular outflows has been driven by the increase in the available angular resolution. The history of molecular outflow observations is quickly reviewed, emphasizing on the most recent interferometric observations. Finally, the results expected with a very large mm interferometer are extrapolated.

2   The role of angular resolution in outflow studies

The history of the research on molecular outflows can be divided in three periods following the increase in the angular resolution of the available instrumentation. The first period corresponds to the studies resulting from observations with "standard" single-dish radiotelescopes which provided an angular resolution in the range of 1 to 2 arcmin. A second period started with the put in operation of large single-dish telescopes with 10 to 30″ of angular resolution. Finally, a new era in the outflow research is opening thanks to the observations with large arrays which provide a resolution close to the arcsec at 1 mm of wavelength.

2.1   Observations with "standard" single-dish telescopes

Molecular outflows were discovered in the form of broad wings in the millimeter-wave CO profiles observed toward the Orion A molecular cloud in the mid 70's (Kwan & Scoville 1976; Zuckerman et al. 1976). High-velocity CO emission was soon detected toward other objects, and the structure of the outflowing material was found to be bipolar (Snell et al. 1980; Rodríguez et al. 1980). The first surveys revealed that these bipolar outflows are extraordinarily common around young stars (Bally & Lada 1983; Edwards & Snell 1982, 1983, 1984). Lada (1985) compiled a first catalogue containing 68 outflow sources. Further searches carried out with unbiased selection criteria led to the detection of many more outflows. Fukui et al. (1993) listed 157 outflows confirmed through complete or partial mapping. Observations since then have increased the number of presently known molecular outflows to nearly 200.

The observations with single-dish mm-wave radiotelescopes providing 1– 2′ of angular resolution revealed the importance of bipolar outflows in star formation (see the review paper by Lada 1985). Such observations made clear the ubiquity of outflows, and unveiled many of their surprising properties.

2.2   Observations with "large" single-dish radiotelescopes

The put in operation of large radiotelescopes by the mid 80's provinding a resolution of 10 to 30″ in the CO lines (the NRO 45-m telescope, the IRAM 30-m, the 15-m JCMT, and the 10-m CSO) instigated important progress in the study of outflows. By 1990, a new class of molecular outflows with very high collimation (> 10) were discovered. Good examples of such high-collimation CO outflows are L1448-mm (Bachiller et al. 1990), VLA1623 (André et al. 1990), and the monopolar outflow from NGC2024/FIR5 (Richer et al. 1992).

In some of these jet-like CO outflows, a weak component at extremely high velocities (EHV, i.e. velocities in excess of 50 km/s) was observed. Examples of EHV CO jets are HH7–11, IRAS03282 and L1448. The momentum in the EHV jet-like component was found to be large, generally enough that it could put into motion the standard high-velocity (SHV) bipolar outflows. In addition, in some particularly clear cases (such as IRAS03282 and L1448), the terminal velocity of the EHV jet is observed to decrease with distance from the outflow origin, while the terminal velocity of the standard high-velocity component is observed to increase. This behavior strongly suggests that the EHV jet-like component is injecting momentum into the ambient gas to produce the standard CO outflow.

The EHV component presents discrete peaks well defined in space and in velocity. Such peaks are referred to as "molecular bullets" (Bachiller et al. 1990). The SHV outflow is observed as extended lobes surrounding the EHV jet. This kind of molecular bullets are observed in the majority of highly collimated CO outflows. A remarkable example is that of the HH7–11 flow. The observed radial velocities of the bullets in this outflow exceed 100 $km\,s^{-1}$ with respect to the ambient cloud, and their CO linewidths are of about 20 $km\,s^{-1}$ (Bachiller and Cernicharo 1990; Masson et al. 1990). More recent examples of EHV outflows containing bullets include HH111 (Cernicharo & Reipurth 1996), and IRAS2005 (Bachiller et al. 1995a). The typical sizes of molecular bullets are a few 10^{-2} pc, and their masses are a few $10^{-4}\,M_\odot$. The kinematic timescales range from a few 10^2 to a few 10^3 yr. Thus, the masses and timescales of bullets are similar to those of the "optical" outburst observed in FU Ori stars. The FU Ori eruptions can be well explained by a large increase in the accretion rate through a circumstellar disk up to $10^{-4}\,M_\odot$/yr. For an average duration of about 100 yr, this yields a total accreted mass up to $10^{-2}\,M_\odot$. The masses of the molecular bullets would thus be consistent with a ratio of accretion rate to mass outflow in the range of $\sim$ 10 to 100.

In summary, the large telescopes mentioned above have allowed the detection and study of a new component in the CO outflows, the EHV component, well differentiated from the standard one (which essentially consists of swept-up gas). The EHV CO component, with its knotty structure, is reminiscent of the HH jets observed in the visible. One could ask whether these EHV CO jets are the neutral winds driving the standard CO outflows. Clearly it is very difficult to distinguish the primary wind material actually ejected from the central star/disk system from the high velocity gas accelerated and processed by shocks. It is not obvious if some amount of this primary material can remain unprocessed. What appears clear is that the EHV CO component is very intimately linked to the primary driving agent. It thus appears that there is at least an important class of CO outflows that are driven by jets. The presence of EHV jet-like molecular outflows is a clear step towards the "unification" of two phenomena, HH jets and CO outflows, which were initially considered to be distinct.

2.3   Interferometric observations

Interferometry at mm wavelengths makes now possible to image the CO outflows with a resolution close to the arcsec (Sargent & Welch 1993). This technique is providing crucial information on the structure of bipolar outflows. To illustrate this point, we will briefly discuss two of the most detailed observations available so far: those of the L 1448 and L 1157 outflows. Figure 1 shows a comparison of the H_2 image of L 1448 (from Bally et al. 1993b) with the images of the slow-moving CO obtained with the IRAM interferometer (Bachiller et al. 1995b). The H_2 images reveal a series of well-defined bow shocks in the blue-shifted lobe, whereas the slow-moving molecular gas traces the edges of a biconical limb-brightened cavity. The blue-shifted part of the cavity is also seen by the continuum near-infrared emission which is scattered at the cavity walls (Bally et al. 1993b). It is remarkable that the walls of the CO cavity seen at blue-shifted velocities (top left panel in Fig. 1) seem to be complementary to the bow shock H_2 structure which is the closest to the exciting source, L 1448-mm. The arc structure delineated by the H_2 emission seems to close the conical CO cavity. This configuration strongly suggests that the large opening angle of the SHV CO outflow results from the entrainment of ambient material through the large bow shocks traced by the H_2 line emission.

Another example in which interferometers have been very informative is that of L 1157 (Gueth et al. 1996). Figure 2 shows velocity channel CO images of the blueshifted lobe. The images reveal at least two prominent limb-brightened cavities which also seem to be created by the propagation of large bow shocks. These observations also exemplify the importance of combining single-dish data with the interferometric data. In fact, with only the purely interferometric images (top row of the figure), one could think that the cavities are empty structures. When one adds the zero-spacing information (middle row) it is shown that there is significant CO emission arising from the inner part of the cavities. The bow shocks at the head of the cavities are also well observed in NH_3 emission (Bachiller et al. 1993), and VLA images reveal a structure similar to that seen in CO (Tafalla & Bachiller 1995). Interestingly, the two cavities are not well aligned on a single line passing through the exciting source, L 1157-mm, as if the axis of the underlying jet had precessed from the first ejection event to the second one. A simple spatio-kinematic model in which the jet precesses on a narrow cone (of opening angle close to 6°) provides an accurate description of the observations (bottom row of the figure). Thus also in this case the large opening angle observed at the base of the CO outflow is very likely determined by the large size of the propagating bow shocks, rather than to the jet precession which happens in a very narrow cone.

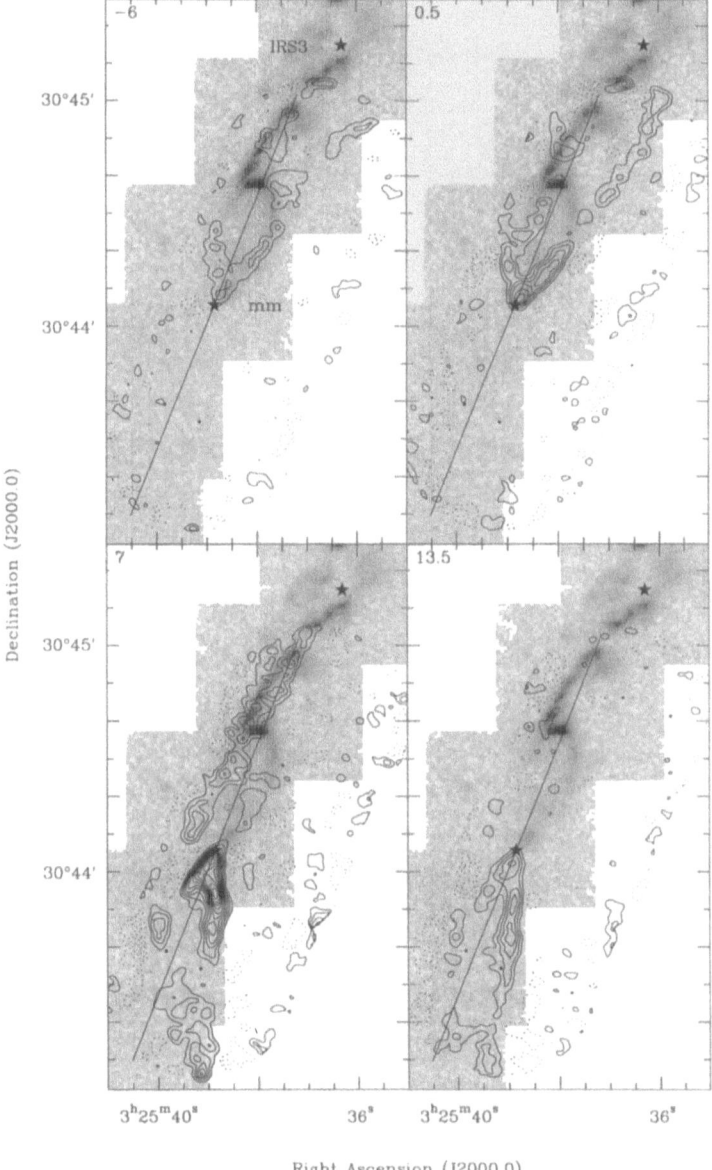

Fig. 1. Superposition of a grey-scaled H_2 image of the L1448 jet (from Bally et al. 1993) and the interferometric images of the CO 1–0 emission integrated over four intervals at low velocities (from Bachiller et al. 1995b). The central LSR velocity for each interval is given at the upper left corner of each panel. First contour and step are 0.92 K km s^{-1}. The jet direction, defined by SiO observations (see Dutrey et al. 1996), is indicated by the solid line. The positions of L1448-mm and IRS3 are marked with stars. A part of the CO outflow from IRS3 is visible in the 7 km s^{-1} panel. The H_2 emission traces large bow shocks, whereas the CO delineates the walls of a biconical cavity. Note that the walls of the CO cavity in the blueshifted lobe are complementary of the first H_2 bow shock. This morphology strongly suggests that the CO bipolar outflow results from entrainment of ambient material through the propagation of large bow shocks.

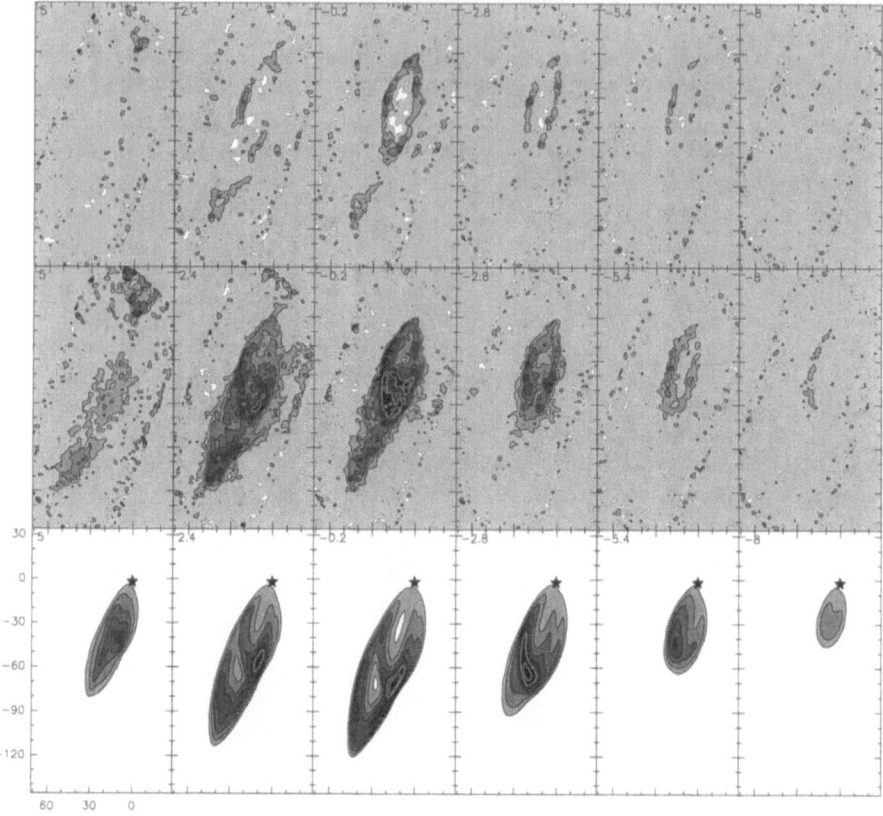

Fig. 2. CO emission from the blueshifted lobe of the L1157 outflow integrated over velocity intervals of 2.6 $km\,s^{-1}$ wide. The central LSR velocity for each interval is given at the upper left corner of each panel. The LSR velocity of the ambient gas is 2.75 $km\,s^{-1}$. Position offsets are in arcsec with respect to the central Class 0 source L1157-mm, whose position is indicated with a star. First contour and step are 155 mJy/beam (1.3 K). The beam size is $3.6'' \times 3''$ at P.A. 90°. Top row: CO 1–0 maps reconstructed from purely interferometric IRAM data. Middle row: images obtained after inclusion of the short spacing information obtained at the IRAM 30-m telescope. Bottom row: synthetic maps obtained with a precessing, episodic jet model smoothed to the resolution of the observations. Adapted from Gueth et al. (1996).

To summarize, the recent observations carried out with mm-wave interferometers are providing key information about the propagation of protostellar outflows. It results that the prompt entrainment is the dominant process, as corresponds to fast jets with Mach numbers >10. The formation of large bow shocks seems to be a rule, and other phenomena like turbulent entrainment seem to have a less important role in the formation of molecular outflows.

3 Open problems and perspectives

Despite the enormous progress achieved in outflow research after their discovery about twenty years ago, many important questions remain open.

The precise origin of the outflows is obviously one of the most important issues. Interferometric SiO observations carried out around L1448-mm show that the outflow is already well collimated within about 100 AU from the central star (Guilloteau et al. 1992), which is the approximate size of the disk around the protostar. Locating the position where the jet is launched would require to observe the base of the flow with a resolution of the order of 0.1″ to resolve out the disk.

The outflows from high-mass stars remain poorly studied, and this is because the regions with massive star formation are more distant than low-mass star forming regions. We know that there are also extremely high velocity wings in this kind of sources (e.g. Cep A) but to study such sources with a linear resolution similar to that achieved in Taurus-Perseus will require an angular resolution significantly higher than that available so far. An increase of one order of magnitude in the angular resolution (i.e. going from 1 to 0.1″) would allow to reach regions within 1.5 kpc with a linear resolution comparable to that presently available in Taurus. Many of the most important high-mass star forming regions are within this distance.

The chemistry associated with the bow shocks is known to be extremely rich (e.g.: van Dishoeck et al. 1995, Bachiller 1996). Interferometric observation of different chemical species have already given important results on some compact outflow sources such as OriA/IRc2 (Plambeck et al. 1982), Orion S, and IRAS05338-0624 (McMullin et al. 1993, 1994). A 0.1″ resolution mm array will be tremendously advantageous for such cases of study.

Such an interferometer would also be decisive in the study of large bow shocks. In fact, the physical and chemical structure of the shock front is expected to change steeply across the front, and a large interferometer will certainly allow to study in detail such complex chemically stratified regions. High sensitivity will permit the study of the differential distribution of all minor chemical components throughout the shock.

I thus conclude that a large millimeter array providing an angular resolution close to the tenth of the arcsec will likely solve out the challenging open questions raised here. And, most notably, such an instrument will certainly set up other even more fundamental questions that remain unsuspected today.

Acknowledgments

I acknowledge my colleagues at the Observatorio Astronómico Nacional of Spain for interesting discussions on outflows and interferometers. I also thank F. Colomer for a critical reading of the manuscript, and A. Dutrey and F. Gueth for providing some of the figures showed in the oral presentation. Funding support from Spanish DGICYT (through grant PB93–48) is gratefully acknowledged.

References

André P, Martín-Pintado J, Despois D, Montmerle T. 1990. A&A **236** 180–92

Bachiller R, 1996. ARA&A, in press.

Bachiller R, Cernicharo J. 1990. A&A **239**, 276–86

Bachiller R, Cernicharo J, Martín-Pintado J, Tafalla M, Lazareff B. 1990. A&A **231**, 174–86

Bachiller R, Fuente A, Tafalla M. 1995a. ApJ **445**, L51–54

Bachiller R, Guilloteau S, Dutrey A, Planesas P, Martín-Pintado J. 1995b. A&A **299**, 857–68

Bachiller R, Martín-Pintado J, Fuente A. 1993. ApJ **417**, L45–48

Bally J, Lada CJ. 1983. ApJ **265**, 824–47

Bally J, Lada CJ, Lane, AP. 1993. ApJ **418**, 322–27

Cernicharo J, Reipurth B. 1996. Preprint

Dutrey A, Guilloteau S, Bachiller R. 1996. Preprint

Edwards S, Snell RL. 1982. ApJ **261**, 151–60

Edwards S, Snell RL. 1983. ApJ **270**, 605–19

Edwards S, Snell RL. 1984. ApJ **281**, 237–49

Fukui Y, Iwata T, Mizuno A, Bally J, Lane AP. 1993. In *Protostars and Planets III*, ed. EH Levy, JI Lunine. Tucson: Univ. Ariz. Press

Gueth F, Guilloteau S, Bachiller R. 1996. A&A, in press

Guilloteau S, Bachiller R, Lucas R, Fuente A. 1992. A&A **265**, L49–52

Kwan J, Scoville N. 1976. ApJ **210**, L39–42

Lada CJ. 1985. ARA&A **23**, 267–317

Masson CR, Mundy LG, Keene J. 1990. ApJ **357**, L25–28

McMullin JP, Mundy LG, Blake GA. 1993. ApJ **405** 599–607

McMullin JP, Mundy LG, Blake GA. 1994. ApJ **437**, 305–16

Plambeck RL, Wright MCH, Welch, WJ, Bieging JH, Baud B, Ho PTP, Vogel SN. 1982. ApJ **259**, 617–24

Richer JS, Hills RE, Padman R. 1992. MNRAS **254**, 525–38

Rodríguez LF, Ho PTP, Moran JM. 1980. ApJ **240**, L149–52

Sargent AI, Welch WJ. 1993. ARA&A **31**, 297–343

Snell RL, Loren RB, Plambeck RL. 1980. ApJ **239**, L17–20

Tafalla M, Bachiller R. 1995. ApJ **443**, L37–40

van Dishoeck EF, Blake GA, Jansen DJ, Groesbeck TD. 1995. ApJ **447**, 760–82

Zuckerman B, Kuiper TBH, Kuiper ENR. 1976. ApJ **209**, L137–42

Accretion and outflow in a protostellar system in Corona Australis

Jorma Harju[1] and Iain Moray Anderson[1,2]

[1] Observatory, P.O. Box 14, FIN-00014 University of Helsinki, Finland
[2] Swedish-ESO Submillimetre Telescope, European Southern Observatory, Casilla 19001, Santiago 19, Chile

Abstract. We discuss the structure of one of the closest protostellar objects, located in the R Coronae Australis cloud core, in the light of recent radio, infrared and optical observations. The deeply embedded infrared object IRS 7 ($A_V \geq 35^m$), or its possibly undetected companion, is giving rise to a powerful outflow associated with HH-objects and strong thermal SiO emission. A large molecular disk or torus surrounds the central source(s) of the outflow. The object may represent a very early stage of evolution where outflow is accompanied by substantial accretion from the surrounding cloud. Possible indications of this process in the existing data are discussed.

1 Introduction

Our current understanding of true protostellar disks (i.e. those buried within molecular cloud cores) is still poor. This is in contrast to the recent rapid progress made in the study of T Tauri disks (see the reviews by Dutrey and Beckwith in this volume). Observations of protostellar disks are hampered by the surrounding whilst the theoretical framework for accretion disks around protostars is much less complete than that for T Tauri disks. The former disks are probably thick, self-gravitating and non-Keplerian, i.e. the assuptions of a thin disk have to be abandoned, and one has to resort to hydrodynamical simulations (e.g. Tscharnuter & Boss 1993).

In this paper we review observations towards a deeply embedded protostellar system in the nearby Corona Australis cloud, located at a distance of 130 pc. A 1.3 mm dust continuum survey conducted at the SEST by Henning et al. (1994) indicates that the strongest millimetre radiation source in the region is IRS 7 (R1 in Taylor & Storey 1984). The observations of Anderson *et al.* (1996) suggest the presence of a large molecular disk around IRS 7. In the light of these new data, the previous conclusions about the origin of the molecular outflow (Levreault 1988) and the numerous HH-objects in the region (Hartigan & Graham 1987, Graham 1993) need to be reconsidered.

Due to its southern location no interferometric molecular line data of the CrA cloud is available. It is conceivable that only interferometric studies towards IRS 7 can yield data from which the relationship between a possible inner circumstellar disk and the large molecular disk can be deduced. At the disk's distance, the planned maximum resolution of the LSA, $0''.1$, corresponds to 13 AU. With this resolution, the velocity, density and temperature distribution of the molecular disk could be determined with high accuracy. Such data would be of importance in checking the validity of hydrodynamical models of accretion disks.

2 New molecular line maps

Anderson *et al.* (1996) have mapped the environs of IRS 7 in several molecular tracers at 3 and 1.3 mm wavelengths using the SEST. The purpose of this study was to find the central source(s) of the R CrA outflow (Levreault 1988) and to determine whether a large accretion disk is present as suggested by the VLA 6 cm continuum observations of Brown (1987). The maps derived from the spectral line data reveal a complex system consisting of 1) a quiescent dense core, 2) a dense bipolar outflow originating somewhere close to IRS 7 and 3) a rotating molecular disk around IRS 7.

The outflow structure is exhibited by the $HCO^+(J = 1 - 0)$ channel maps presented in Fig. 1. The blueshifted lobe is peaked close to the HH-objects HH104 A and B. The approaching outflow collides with the outskirts of the foreground core causing a powerful shock which give rise to asymmetric SiO profiles. The dense outflow detected here almost certainly corresponds to the E-W oriented compact outflow detected by Levreault (1988). No trace of the extended NE-SW oriented outflow aligned with the well known string of HH-objects (Hartigan & Graham 1987) was seen.

The integrated line intensity map of the $HCO^+(J = 3 - 2)$ transition reveals an elongated condensation centred on IRS 7. In addition to the two lobes located on either side of IRS 7 there is a prominent extension south of the infrared source. The length of the condensation is in excess of some 6 000 AU, perhaps extending to over 8 000 AU. The condensation is aligned with the two compact continuum objects centred on IRS 7 detected by Brown (1987) with a separation of 14″ (1800 AU). Brown suggested that the compact sources represent thermal emission from the shock-excited inner boundaries of a thick accretion disk surrounding IRS 7.

A superposition of the K'-band image of the region around R CrA, taken from the survey of Hodapp (1994), and the $HCO^+(J = 3 - 2)$ integrated intensity map is presented in Fig. 2. The near–IR image is dominated by R CrA and the associated reflection nebulosity. At 2.11 μm the extinction is ∼ 10 times lower than at optical wavelengths. Therefore, the structures seen in the K'-image are comprised of relatively dense dust. IRS 7 appears as a point source at $(0,0)$ and to its NE there is an elongated, compact nebula some 7″, or 900 AU at 130 pc, in length. Two alternative interpretations of this feature can be considered: 1) It is a reflection nebula which is illuminated by radiation from IRS 7 escaping through the nothern polar cavity of the molecular disk; and 2) It is the inner circumstellar disk of another protostellar object which is located at its centre.

The $HCO^+(3 - 2)$ data exhibit deep absorption profiles caused by dense, material foreground which occur close to the cloud's average velocity $v_{LSR} = 5 \rightarrow 6 \, km \, s^{-1}$. This fact makes the data unsuitable for determining the velocity structure of the condensation close to the average of the cloud. Instead, the optically thin $C^{18}O(J = 2 - 1)$ line can be used for this purpose. The plot in Fig. 3 shows

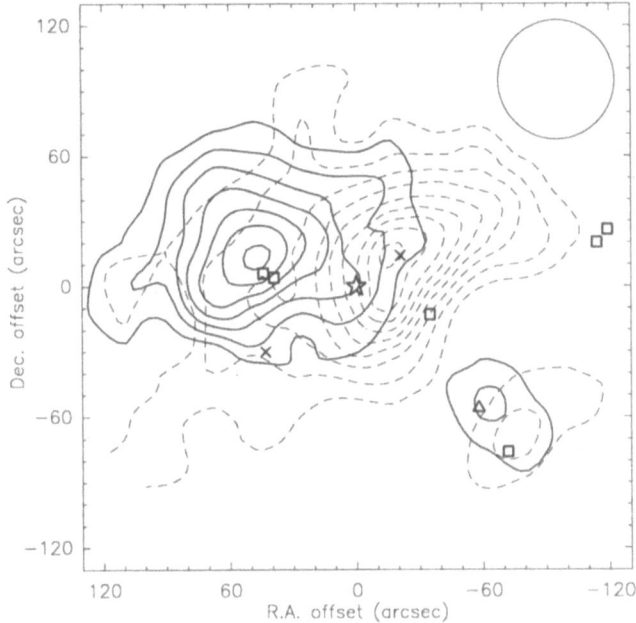

Fig. 1. The HCO$^+$($J = 1 - 0$) map of the blue (full line contours) and redshifted (dashed line contours) outflows adopted from Anderson *et al.* (1996). The velocity ranges are $-15 \to 0$ km s^{-1} (blue) and $10 \to 25$ km s^{-1} (red). The contour levels are from 0.5 K km s^{-1} in increments of 0.5 K km s^{-1}. The location of IRS 7 is denoted by a star, T CrA (left) and R CrA (right) are denoted by crosses, HH104 A and B (east), HH98 (centre) and HH104 C and D (west) and HH100 (south) are denoted by squares whilst the infrared source HH100–IRS is denoted by a triangle. The coordinate offsets are relative to IRS 7 ($\alpha_{1950.0} = 18^h58^m33.0^s$, $\delta_{1950.0} = -37°01'43''$).

the gaussian peak velocities of the C^{18}O lines along two strips with 12$''$ separation parallel to axis of the condensation. The data set form a Z–shaped pattern, the signature of differential rotation. Anderson *et al.* reproduced the observed pattern by an edge–on Keplerian disk. The best fit to the observed velocity pattern was obtained by setting the inner radius to 860 AU and and the central mass to 0.25 M$_\odot$. The predicted velocity distribution (smoothed to correspond to the 24$''$ resolution of the telescope) is shown as a solid curve.

3 Discussion

The observations of Anderson *et al.* (1996) suggest that the two outflows originating close to IRS 7 in CrA are of different ages, the E-W oriented outflow being the younger. This outflow collides with the foreground core material causing a shock which is manifest as HH 104A and B and the strong SiO emission

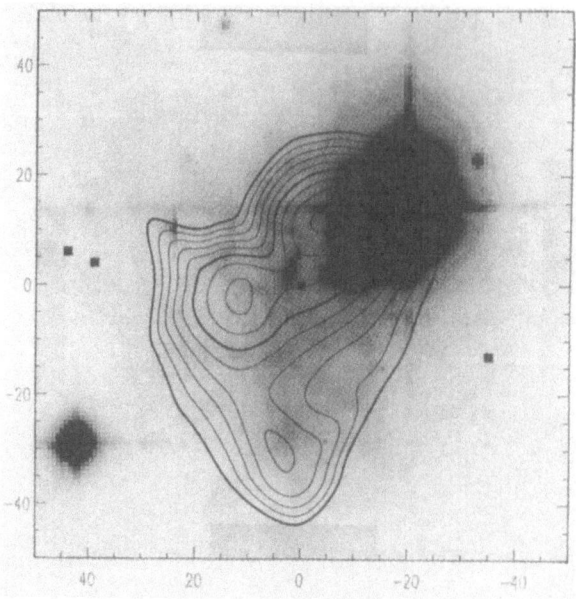

Fig. 2. The integrated intensity map of HCO$^+$($J = 3 - 2$) emission obtained by Anderson *et al.* (1996) superposed onto a K'-band image of the R CrA region. The K'-band image data is from the file RCRA.FIT in Hodapp (1994). The contour levels are from 1.0→18.0 K km s^{-1} in increments of 1.0 K km s^{-1}. IRS 7 is located at (0,0). The triangles to the E and W of IRS 7 mark the locations of HH104 A/B and HH98.

in their neighbourhood. Since SiO can exist only for a short period in the gas phase (see e.g. Martín-Pintado *et al.* 1992), its presence serves as an indicator of current activity. Contrary to the suggestion of Graham (1993), HH104A and B are probably not excited by R CrA.

The older, extended NE–SW oriented outflow is perpendicular to the major axis of the large molecular disk around IRS 7 and is likely associated with it. The dense E–W oriented outflow cannot be centrifugally driven from the molecular disk since the blueshifted flow is seen close to the redshifted side of the disk. Therefore, it is probably collimated by a circumstellar disk which has an orientation totally different from that of the molecular disk. Whether this circumstellar disk is anchored by a possible close companion of IRS 7 (as suggested in the case of the double outflow near NGC1333 IRAS 2 by Sandell *et al.* 1994), or if it represents the twisted inner regions of the detected large disk, cannot be ascertained from the existing data. More information about the nature of the compact elongated nebula in the K' image of Hodapp (1994) may clarify the situation.

The southern extension of the molecular disk in the HCO$^+$($J = 3 - 2$) map of Anderson *et al.* (1996), is slightly redshifted $(1 - 2.5$ km s$^{-1})$ with respect to the

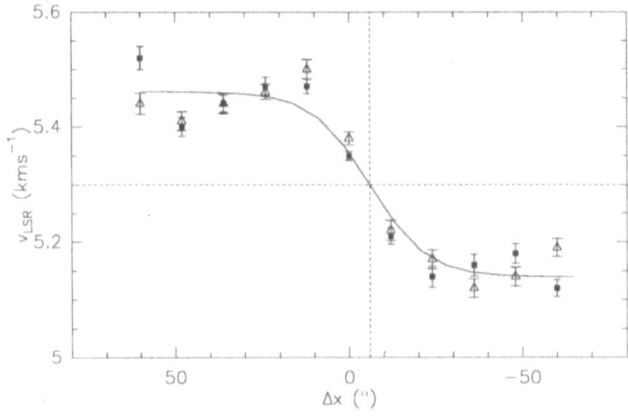

Fig. 3. The Gaussian peak velocities of the $C^{18}O(J = 2-1)$ transition as a function of offset along two lines passing through the elongated condensation shown in Fig. 2. The solid curve represents a Keplerian disk model.

central velocity of the disk and is seemingly connected with its receding side. The continuum map of Brown (1987) also shows evidence of this feature. The rather low velocity of the feature led Anderson *et al.* to suggest that it could represent material streaming to the disk from the dense foreground core. Such material exchange may be a natural development in the evolution of a rotating double core system. However, because of the complexity of the system, the evidence for an accretion stream in CrA is not conclusive and yet another outflow could provide a possible explanation for this observed feature.

References

Anderson I.M., Harju J., Knee L.B.G., Haikala L.K. 1996, *submitted to* A&A

Brown A. 1987, ApJ 322, L31

Graham J.A. 1993, PASP 105, 561

Hartigan P., Graham J.A., 1987, AJ 93, 913

Henning Th., Launhardt R., Steinacker J., Thamm E., 1994, A&A 291, 546

Hodapp K.-W., 1994, ApJS 94, 615 (AAS III CD-ROM)

Levreault R.M., 1988, ApJS 67, 283

Martín-Pintado J., Bachiller R., Fuente A. 1992, A&A 254, 315

Rodríguez L.F., Cantó J., Torrelles J.M., Ho P.T.P 1986, ApJ 301, L25

Sandell G., Knee L.B.G., Aspin C., Robson I.E., Russell A.P.G. 1994, A&A 285, L1

Taylor K.N.R., Storey J.W.V., 1984, MNRAS 209, 5P

Tscharnuter W.M., Boss A.P. 1993, in *Protostars and Planets III*, p. 921, eds. Levy E.H., Lunine J.I., The University of Arizona Press

Masers at Millimeter and Submillimeter Wavelengths

Karl M. Menten

Harvard-Smithsonian Center for Astrophysics, 60 Garden Street, Cambridge, MA 02138, USA

Abstract. Observations of interstellar and circumstellar maser emission at millimeter and submillimeter wavelengths can be used to probe the physical conditions and the chemistry in the emitting regions. Furthermore, because of their high brightnesses, many maser lines can be utilized as phase reference sources by future (sub)millimeter-wavelength interferometric arrays.

1 Introduction

Observations of cosmic masers have long been used to study a number of very interesting astrophysical environments, such as the immediate neighbourhoods of newly formed stars and the envelopes of evolved red giants and supergiants. In many cases, maser emission arises from very compact regions with sizes $\lesssim 1$ AU and has extremely high brightness temperatures ($> 10^{10}$ K). Masers can therefore be studied with very high (milliarcsecond) spatial resolution and many of the most interesting "applications" of maser observations involve their usage as high precision tools to probe the small scale structure and the dynamics of their emitting regions in great detail. For this reason, the development of spectral line radio interferometry went hand in hand with progress in maser research (see Moran 1993 for a historical account of "the early days"). In the following, we shortly summarize some of these applications in §2; for detailed discussions of the topics addressed we refer to the reviews by Reid & Moran (1988), Cohen (1989), Elitzur (1992), and Watson (1993),the book by Elitzur (1991), and the conference proceedings edited by Clegg & Nedoluha (1993).

Many maser molecules have most of their rotational transitions in the (sub)millimeter range. Therefore, multi-transition studies at these wavelengths have great potential for putting tight constraints on the physical conditions in the masing regions. We illustrate this in §3 with the example of submillimeter water masers. Finally, we discuss in §4 the use of masers as phase references to achieve high quality sub-arcsecond imaging at millimeter and submillimeter wavelengths and give examples, in §5, for some of the science profiting from this.

2 Masers as Astrophysical Tools

Almost always, masers are *signposts* of energetic phenomena in their emitting regions. This is because many prominent maser lines, such as all of the observed water (H_2O) maser transitions (see §3), are at high energies above the ground-state and, thus, need high temperatures for their excitation. Moreover, high temperatures are required to drive the *chemistry* that produces the high abundances of the masing molecules, which are inferred from the strong maser signals. Processes include endothermic neutral-neutral gas phase reactions in warm postshock gas in the case of interstellar water masers (Elitzur et al. 1989) or evaporation of icy dust grain mantles producing the high methanol (CH_3OH) abundances necessary to explain the strong masers observed from this species toward the warm molecular envelopes of newly formed massive stars (Hartquist et al. 1995).

Many maser lines can be observed with (sub)milliarcsecond resolution using VLBI techniques. This makes them excellent *kinematical probes*, allowing studies of velocity fields on AU-size scales for sources throughout the Milky Way and on sub-parsec scales in external galaxies. Repeated VLBI observations allow *proper motion studies* and, using the method of kinematic or statistical parallax, direct *distance measurements*, e.g. of the distance to the Galactic Center (Reid et al. 1988).

The Zeeman effect influences the spectral lines of some maser molecules in the presence of a magnetic field. For the paramagnetic OH radical, which shows strong maser action in several radio lines toward numerous ultracompact HII regions, this results in a line splitting, allowing determinations of *magnetic field strengths*.

Finally, because of their high intensities, maser lines may be used as tools in the basic sense of the word. Subarcsecond resolution interferometric observations, particularly at high frequencies, often only produce useful images if adaptive calibration techniques can be used to correct for the effects of short-term atmospheric fluctuations. However, frequently the continuum or spectral-line emission one wishes to observe will be too weak to serve as a reference signal. As described in §4, a strong maser line within the field of view of the interferometer may serve as such a reference signal: a *"maser guide star"*.

3 Masers in the (Sub)Millimeter-Wavelength Range

For a long time our understanding of masers was seriously limited by the fact that for some of the most interesting masing species only a few maser transitions were known. Observations of as many maser lines as possible toward the same region are highly desirable. Such multi-transition studies can be used to put tight constraints on maser excitation models, which have to explain the observed occurence (or absence) of maser action in the lines in question.

Some of the most prolific maser molecules, such as H_2O, CH_3OH, and SiO have most of their rotational transitions at (sub)millimeter and far-infrared wavelengths. Technical progress leading to sensitive receivers at ever higher frequencies and the advent of (sub)millimeter telescopes with large collecting areas at high, dry sites such as the IRAM 30-m telescope on Pico Veleta, Spain, and the JCMT and CSO on Mauna Kea, Hawaii, have opened most of the millimeter- and submillimeter-wavelength region for ground-based astronomy. Remarkably, this led to the discovery of a large number of new maser lines from the species mentioned above (e.g., Menten 1991, 1996; Slysh et al. 1995; Cernicharo & Bujarrabal 1992), as well as from HCN (Guilloteau et al. 1987; Lucas & Cernicharo 1989; Izumiura et al. 1995). Surprisingly, maser action was also observed in various millimeter- and submillimeter wavelength hydrogen recombination lines arising from the ionized stellar wind of the peculiar star MWC 349 (Martín-Pintado et al. 1989; Thum et al. 1994).

For water, ground-based observations of its lower excitation transitions are impossible because of absorption in the Earth's atmosphere. However, a number of its higher excitation lines can be observed from high altitude sites. In the following we use recent observations of submillimeter water masers in star-forming regions to illustrate how multi-transition maser studies can be used to constrain the physical conditions in the masing regions.

3.1 Submillimeter Water Masers as an Example

Maser emission in the 22 GHz radio transition of H_2O has been studied in hundreds of star-forming regions and circumstellar envelopes around oxygen-rich evolved stars. Milliarcsecond resolution, VLBI, proper-motion studies have firmly established that interstellar H_2O masers form in high velocity outflows driven by young stellar objects. For a long time, the excitation of water masers remained poorly understood, since observations of a single line are insufficient to meaningfully constrain excitation models.

Recently, the observational situation has improved with the detection of various new water transitions at submillimeter wavelengths (Menten et al. 1990a,b; Melnick et al. 1993) with the CSO. Also, high quality observations of the 183 GHz H_2O line toward various regions have been made with the IRAM 30-m telescope (Cernicharo et al. 1990). Maser emission is observed in all of these lines. Model calculations by Neufeld & Melnick (1991) show that these masers can be collisionally pumped in the warm ($T \gtrsim 500$ K) gas found in postshock regions. Melnick et al. (1993) use their multi-transition observations to put constraints on the physical conditions of the masing gas in several star-forming regions, W49N among them (see Fig. 1). They find that in all cases the kinetic temperature has to be of order 1000 K or higher to explain the observed line ratios. This seems to suggest that the masers are found in the postshock gas behind non-dissociative C-shocks, which can have higher temperatures than the gas behind dissociative J-shocks. Another H_2O maser model (Elitzur et al. 1989) had placed the masers in the warm (300 – 400 K) molecular reformation zone behind J-shocks.

While such multi-transition studies have great potential for providing interesting information on the maser regions, their usefulness is presently limited by the fact that no information is available on the emission distribution of the individual lines on scales smaller that the $\gtrsim 10''$ resolution obtainable with single dish telescopes. This problem will be resolved by synthesis imaging with future (sub)millimeter interferometers such as the Large Southern Array (LSA), which

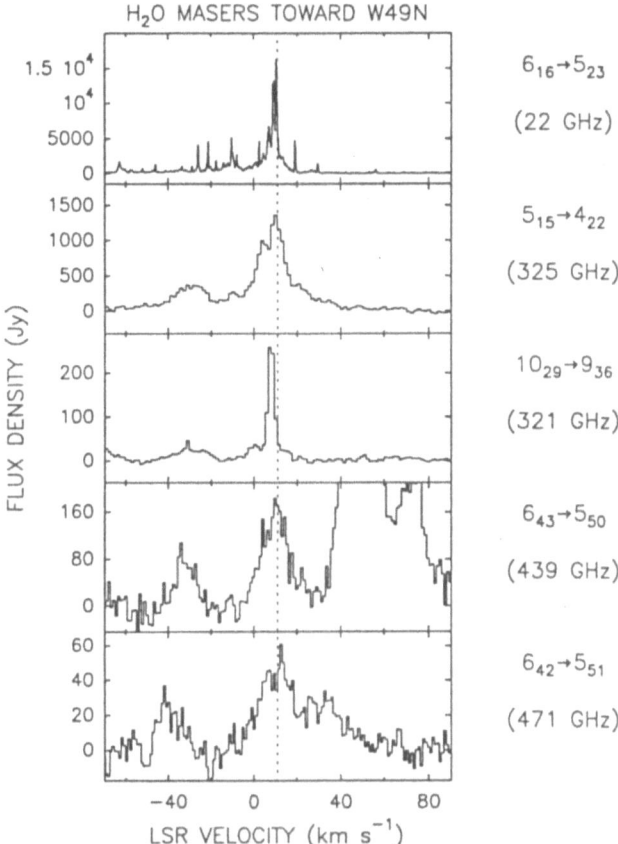

Fig. 1. Various H$_2$O maser transitions observed toward the W49 N star forming region

will produce maps with resolutions of a few tenths of an arcsecond, comparable to or surpassing the resolution attainable with the Very Large Array (VLA) for the 22 GHz line.

4 Masers as Phase Calibration Sources

An instrument such as the LSA with a maximum baseline length of 10 km will yield a minimum synthesized beamwidth of order $\theta_{min} = 0\rlap{.}''03 \times [230/\nu(\text{GHz})]$. It is unlikely that even under favorable atmospheric conditions "high quality", i.e. thermal-noise-limited, imaging will be possible with the LSA at its highest resolutions by using "traditional" millimeter-wave interferometer calibration techniques, i.e. switching between a phase calibrator and the program source with a cycle time of a few minutes. Other calibration techniques have to be used; a particularly promising method involves monitoring variations of the sky brightness temperature (Bremer 1995).

A more direct method, self-calibration, is used extensively at centimeter wavelengths to retrieve corrupted phase information. (see Thompson, Moran, & Swenson 1986 and references therein for a general description). For self-calibration to work, the source to be mapped has to be strong enough for a signal-to-noise ratio (SNR) of a few to be achieved on a single baseline within the coherence time of the interferometer. At (sub)millimeter wavelengths the brightness of many astrophysically important objects will be too low to achieve the SNR necessary for straightforward self-calibration to work. It is therefore interesting to examine whether a technique first used by Reid & Menten (1990) for VLA data might be extended to calibrate (sub)millimeter data in certain cases. This method achieves calibration of weak line or continuum data by using phase (and amplitude) corrections determined by self-calibration of a simultaneously observed strong maser source at a different frequency. If the LSA were to allow simultaneous observations with two different receivers, strong maser emission could be observed with one receiver tuned to the frequency of the maser line, ν_M, while the other receiver would be tuned to the target line or continuum frequency, ν_T, the observer is principially interested in. After self-calibrating the strong maser line, the phase corrections determined are scaled by the ν_T/ν_M frequency ratio and applied to the data taken at the target frequency.

Table 1. (Sub)Millimeter Maser Lines for Phase Referencing – Examples

| Species | Frequency (MHz) | T_{sys}^a (K) | ΔS^b (Jy) | Detected Sources | | | | |
| | | | | Evolved Stars | | Star Forming Regions | | |
				N known	Observed S	N known	Observed S	Ref.
H_2O	183310	20000	400	$\approx 10^c$	100 – 600 Jy	$\approx 10^{d,e}$	100–15000 Jy	1
	658006	4400	50	$\approx 10^c$	100 – 3000 Jy	none	–	2
CH_3OH	156828	70	1	none	–	4	10 – 60 Jy	3
SiO	86243	80	2	$> 100^c$	10 – 2000 Jy	1^f	≈ 1000 Jy	4
HCN	177283	220	5	7^g	$? – 400\ Jy^h$	none	–	5

Notes: (a) System temperature is calculated assuming that the single-sideband receiver temperature is equal to 5 times the "quantum limit" $h\nu/k$, 1 mm of precipitable water vapor above a site at an altitude of 4200 m, and observations at an elevation angle of 45°. (b) Rms flux density for a single baseline of two 10 m diameter antennas assuming the quoted T_{sys}, an integration time of 10 s, and a bandwidth corresponding to a 1 km s^{-1} velocity interval. (c) Detected in O-rich stars. (d) Detected in high-mass star-forming regions. (e) Detected in low-mass star-forming regions. (f) Orion-KL only. (g) Detected in C-rich stars. (h) Ref. 5 quotes flux density only for one object (IRC+10216)

References: (1) Cernicharo et al. 1990; (2) Menten & Young 1995; (3) Slysh et al. 1995; (4) e.g., Benson et al. 1990; (5) Lucas & Cernicharo 1989

In recent years, an appreciable number of maser lines has been detected at millimeter and submillimeter wavelengths in astrophysically interesting environments. In Table 1 we have examined for a few of these lines their usefulness as phase reference sources. For each line, the molecular species, frequency, system temperature, and rms flux density achievable in a 10 s interval are listed. Futhermore, we indicate whether the line in question has been detected toward evolved stars and/or star-forming regions and list approximate numbers of sources detected and typically observed flux densities in each category. It is clear from Table 1, that self-calibration using maser lines will be possible for a variety of scenarios and many different sources.

5 Sub-Arcsecond/(Sub)Millimeter-Wavelength Science

A number of very important scientific questions can be addressed most effectively with sub-arcsecond resolution observations. In the following we shall describe the impact that (sub)millimeter interferometry with resolutions $\lesssim 0\rlap{.}''1$ will have on our understanding of protostellar disks, the driving mechanism of outflows from young stellar objects, and dust formation and mass-loss processes of evolved stars. Toward many of the regions in question maser lines have already been detected.

5.1 Protostellar/Protoplanetary Disks

The inner parts of disks surrounding solar-mass and more massive stars give rise to dust continuum emission that can be readily studied at sub-arcsecond resolution. For example, a disk enclosing Pluto's orbit would have a diameter of $0\rlap{.}''2$ at the distance of the Orion molecular cloud and could thus be resolved with the LSA. While it is clear that a substantial fraction of pre-main sequence stars have circumstellar disks (Beckwith et al. 1990), many of these disks' characteristics, such as their temperature structure and mass, are poorly constrained by existing observations. With the LSA, dust emission from circumstellar disks can easily be imaged with sub-arcsecond resolution for all conceivable values of the dust temperature.

Little is known about the velocity structure of the inner disks on scales of a few AU and high resolution molecular line observations are necessary. Maser lines are particularly useful probes given their high brightness. Plambeck et al. (1990) modeled their 3 mm BIMA observations of the SiO maser in the Orion-KL region in terms of a rotating and expanding torus. New high quality VLA observations of this source in the 7 mm SiO lines reveal intricate fine structure in the velocity field and prove that the SiO emission distribution is centered on a radio continuum source that is located near the prominent infrared source IRc 2. The weak radio continuum emission was imaged with the VLA by using the SiO maser as a phase reference (Menten & Reid 1995).

5.2 What is Driving Protostellar Outflows?

While hundreds of outflows from young stellar objects have been mapped mostly in rotational lines of the CO molecule with resolutions ranging from $\approx 10''$ to arcminutes, the basic mechanism driving these outflows is still uncertain. Shu and collaborators (e.g., Shu et al. 1993) propose a mechanism that uses magnetocentrifugal acceleration of gas in the equatorial region of a protostar that rotates at breakup speed because it is being spun up by its accretion disk. Testing such a mechanism involves observations in the closest vicinity of the protostar. A suitable molecular probe for this region might be rotational emission from the SiO molecule, which chemical models predict to reach large abundances in Shu et al.'s wind scenario (Glassgold et al. 1989). Observations of maser as well as thermal emission from the innermost regions of these winds thus seem feasible and might reveal a characteristic spiral velocity structure. These winds should also mix with the surface regions of the circumstellar disks. One might speculate that the Orion-KL SiO emission mentioned above might arise from such an interaction region.

5.3 The Innermost Parts of Circumstellar Envelopes

Studies of the circumstellar envelopes of evolved stars are of great importance for many areas of astrophysics. Mass-loss from these objects is responsible for most of the mass return to the interstellar medium. Some of the basic processes involved in the mass-loss mechanism are presently poorly understood. For example, it is unclear how material can escape far enough from the stellar photosphere to condense into dust grains. Studies of elemental depletion would provide important information on grain formation and composition. Practically speaking, the latter process could be observationally studied by measuring radial abundance profiles of molecules undergoing substantial depletion, such as SiO. Recent measurements at $11 \mu m$ with the Berkeley Infrared Spatial Interferometer (ISI) have revealed that, surprisingly, dust formation commences at distances as close as 3 – 4 stellar radii to the star, where temperatures are of order 1200 K (Danchi et al. 1990). Therefore, observations with resolutions of order $0.''1$ or better are necessary to study molecular emission from within the dust formation zone of nearby ($D \approx 200$ pc) Mira variables. Imaging of rotational lines from many species seems feasible with the brightness sensitivities of instruments such as the LSA, since these lines are expected to have substantial optical depths and the emitting region is very hot (1000 – 2000 K). Direct imaging would lead to size and temperature estimates of the emitting region, and thus complement infrared absorption spectroscopy observations, which yields information averaged along the line-of-sight to the star only. Measurements of the abundances of refractory molecules as a function of distance from the star will yield unique information on depletion and dust grain formation. Most interestingly, many oxygen-rich stars have a variety of H_2O and SiO maser lines with frequencies throughout the (sub)millimeter range that will allow self calibration, while toward carbon-rich stars HCN masers may be used for that purpose (Table 1).

References

Beckwith, S. V. W., Sargent, A. I., Chini, R. S., & Güsten, R. (1990): AJ **99**, 924

Benson, P. J., et al. (1990): ApJS **74**, 911

Bremer, M. (1995): *IRAM Working Report No. 238*

Cernicharo, J., & Bujarrabal, V. (1992): ApJ **401**, L109

Cernicharo, J., Thum, C., Hein, H., John, D., Garcia, P., & Mattioco, F. (1990): A&A **231**, L15

Clegg, A. W., & Nedoluha, G. E., eds. (1993): *Astrophysical Masers*, (Springer, Berlin, Heidelberg)

Cohen, R. J. (1989): Rep. Prog. Phys. **52**, 881

Danchi, W. C., Bester, M., Degiacomi, C. G., McCullough, P. R., & Townes, C. H. (1990): ApJ **359**, L59

Elitzur, M. (1991): *Astronomical Masers*, (Kluwer, Dordrecht)

Elitzur, M. (1992): ARA&A **30**, 75

Elitzur, M., Hollenbach, D. J., & McKee, C. F. (1989): ApJ **346**, 983

Glassgold, A. E., Mamon, G. A., & Huggins, P. J. (1989): ApJ **336**, L29

Guilloteau, S., Omont, A., & Lucas, R. (1987): A&A **176**, L24

Hartquist, T. W., Menten, K. M., Lepp, S., & Dalgarno, A. (1995): MNRAS **272**, 184

Izumiura, H., Ukita, N., & Tsuji, T. (1995): ApJ **440**, 728

Lucas, R., & Cernicharo, J. (1989): A&A **218**, L20

Martín-Pintado, J., Bachiller, R., Thum, C., & Walmsley, M. (1989): A&A **215**, L13

Melnick, G. J., Menten, K. M., Phillips, T. G., & Hunter, T. (1993): ApJ **416**, L37

Menten, K. M. (1991): in *Skylines: Proc. Third Haystack Observatory Meeting*, eds. A. D. Haschick & P. T. P. Ho (ASP, San Francisco), 119

Menten, K. M. (1996): to appear in *Radio Emission from the Stars and the Sun*, eds. A. R. Taylor & J. M. Paredes (ASP, San Francisco)

Menten, K. M., Melnick, G. J., & Phillips, T. G. (1990a): ApJ **350**, L41

Menten, K. M., Melnick, G. J., Phillips, T. G., & Neufeld, D. A. (1990b): ApJ **363**, L27

Menten, K. M., & Reid, M. J. (1995): ApJ **445**, L157

Menten, K. M., & Young, K. (1995): ApJ **450**, L67

Moran, J. M. (1993): in *The Structure and Content of Molecular Clouds*, eds. T. L. Wilson & K. J. Johnston (Springer, Berlin, Heidelberg), 89

Neufeld, D. A., & Melnick, G. J. (1991): ApJ **368**, 215

Plambeck, R. L., Wright, M. C. H., & Carlstrom, J. E. (1990): ApJ **348**, L65

Reid, M. J., & Menten, K. M. (1990): ApJ **360**, L51

Reid, M. J., & Moran, J. M. (1988): in *Galactic and Extragalactic Radio Astronomy*, eds. G. Verschuur & K. I. Kellermann (Springer, Berlin, Heidelberg), 225

Reid, M. J., Schneps, M. H., Moran, J. M., Gwinn, C. R., Genzel, R., Downes, D., & Rönnäng, B. (1988): ApJ **330**, 809

Shu, F., Najita, J., Galli, D., Ostriker, E., & Lizano, S. (1993): in *Protostars and Planets III*, eds. E. H. Levy & J. I. Lunine (The University of Arizona Press, Tucson), 3

Slysh, V. I., Kalenski, S. V., & Val'tts, I. E. (1995): ApJ **442**, 668

Thompson, A. R, Moran, J. M., & Swenson, G. W. (1986): *Interferometry and Synthesis in Radio Astronomy* (John Wiley & Sons, New York)

Thum, C., Matthews, H. E., Martín-Pintado, J., Serabyn, E., Planesas, P., & Bachiller, R. (1994): A&A **283**, 582

Watson, W. D. (1993): in *The Structure and Content of Molecular Clouds*, eds. T. L. Wilson & K. J. Johnston (Springer, Berlin, Heidelberg), 109

Millimeter Recombination Lines from Dense Envelopes Around Young Stars

Clemens Thum and Jörn Wink

Institut de Radioastronomie Millimétrique, Domaine Universitaire de Grenoble, 300, rue de la piscine, F-38406 Saint–Martin–d'Hères, France

Abstract. We suggest that millimeter recombination lines are a powerful tool for studying the accretion disks around massive young stars. Specific targets are the disk surface which may be ionized by the diffuse Lyman radiation field, and the dense ionized flow leaving the disk. Subarcsec angular resolution and high sensitivity are required, such as available with the LSA.

1   Introduction

Studies of the evolution of massive young stellar objects face significantly more serious observational problems than investigations of their lower mass ($> 8\ M_\odot$) counterparts. To date firm evidence for accretion disk in massive YSO is available for only two sources: the star MWC349 and the source IRc2 in Orion (Sargent and Welch 1993). These problems are primarily due to the much shorter evolutionary time scale of these massive YSOs which reach the main sequence while still embedded into their residual infalling envelope and their parential cloud. Optical and, in most cases, near IR studies are thus impossible due to prohibitively large obscuration by dust. Studies at longer wavelengths need very high angular resolution, not in general available now, in order to discriminate the YSO from the dust or molecular emission of their surroundings.

The massive YSOs considered here will start to ionize the surrounding material whence they have reached the main sequence. We expect that the diffuse Lyman continuum radiation field will be sufficiently intense to ionize the dense surfaces of the disk where then an ionized wind may be set off as suggested for MWC349 (Thum and Martin–Pintado 1994; Hollenbach et al. 1995). The very high density of this medium renders its free–free continuum emission optically thick and thus bright. Short millimeter wavelengths may be necessary to penetrate through to those dense layers kinematically connected to the disk.

Their high line–to–continuum ratio makes recombination lines (RLs) at short millimeter wavelengths a powerful tool for the study of this dense medium around massive YSOs. Furthermore, RLs are easily excited and they are not subject to chemical processing like the dust or most of the abundant molecules. Guided by the RLs observed in MWC349 (Thum et al. 1995, and references therein) we expect these liness *(i)* to be broadened due disk rotation and/or outflowing motions, *(ii)* their center velocities to change systematically with position and to be generally different from that of any possible surrounding UCHII region, and *(iii)* the emission to be mildly amplified.

2 Angular resolution

The relevant parameter discribing the effective size of the disk around a massive YSO is the gravitational radius, R_g (Hollenbach et al. 1995). Beyond R_g a star of given mass cannot gravitationally bind ionized hydrogen whose electron temperature we adopt here as 6500 K as observed in MWC349. For the interesting range of stellar masses R_g ranges from 100 to 700 au. Outside R_g ionized gas will flow away from the disk rather than rotate with it as a stable "disk atmosphere". This range of R_g is indicated in Fig. 1 which shows the angular resolution for the two extreme configurations of the LSA. We note that MWC349–like disks can be resolved out to the galactic center. While the largest disks are resolved throughout the Galaxy, disks around the more numerous 10 $M_\odot$ stars can be resolved beyond the distance of W3.

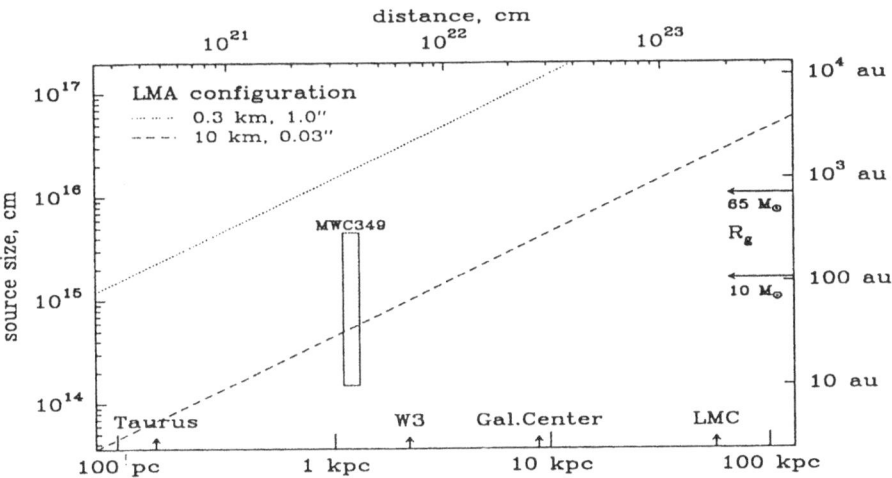

Fig. 1. Linear resolution of the LMA for two extreme configurations as a function of distance. The characteristic size, R_g, of an accretion disk around massive stars of 10 and 60 $M_\odot$ is indicated by arrows on the right–hand ordinate.

3 Sensitivity

We estimate the strength of the disk RLs from the expected brightness temperature T_C of the continuum combined with the calculated line–to–continuum ratio L/C. Following the formalism derived by Altenhoff et al. (1981) we get $L/C = 0.27$ for the H40α transition near 3.0 mm originating in an MWC349–like stellar wind source (electron temperature 6500 K, line FWHP of 100 km s^{-1}). Such a source is optically thick in the continuum over the wavelength range of interest. Following our arguments in section 2 we also assume that the source is resolved and adopt a continuum brightness temperature of $T_C = 2000$ K. Our model source has thus $T_L = 550$ K for RLs near 3mm.

From Table 1 of the LSA document we obtain a rms noise of 1 K for the 3.0 km configuration (0.1″ synthesized beam) and a spectral resolution of 2 km s^{-1}. We finally derived the signal–to–noise ratio as a function of the line width Δv (in km s^{-1}) and the angular resolution θ_b (in arc sec)

$$S/N = 550 \left(\frac{\Delta v}{100}\right)^{-\frac{1}{2}} \left(\frac{\theta_b}{0.1}\right)^2$$

We see that the short–mm RLs of the model source are indeed detectable, even with the highest LSA resolution, as long as the source is resolved and is optically thick in the continuum. A considerable margin exists for sources which are not generally optically thick, like UCHII–regions, or for unresolved sources.

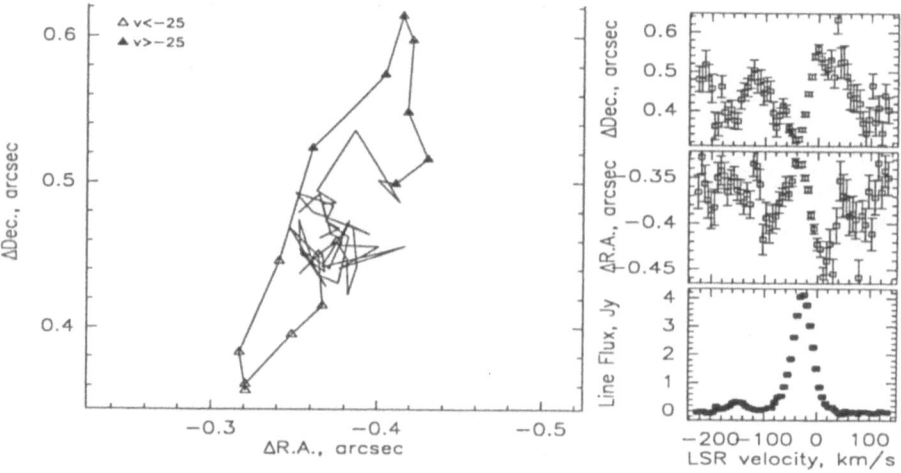

Fig. 2. Map of H40α recombination line emission of the ultracompact HII–region G70.29+1.60 obtained with the IRAM Plateau de Bure interferometer. The lower right panel shows the hydrogen and helium (near -150 km s^{-1}) line profiles. The other right–hand panels show the position offsets from our nominal phase tracking center. The left panel gives the locations of the velocity channels across the hydrogen line.

4 An example: the case of G70.29+1.60

We have used the IRAM interferometer on Plateau de Bure in summer 1995 to try some of the ideas described above (Thum and Wink 1996). Using maximum baselines of about 100m we looked at the H40α line from a sample of 10 UCHII (Wood and Churchwell 1989; Kurtz, Churchwell, and Wood 1993) distinguished by high continuum optical depth at cm wavelengths, their compact structure, and their relative isolation from confusing sources. Due to time and other limitations no full synthesis was obtained on any of our sources yet, but instead we looked for spatial variations across the H40α line profile by fitting each 3 km s^{-1} wide velocity channel in the uv–plane after subtraction of the continuum. According to our criteria (see end of section 1) such variations may represent the signature of a disk, either directly if we see rotation or indirectly if we see an outflow.

Several such sources were found of which we show G70.29+1.60 in Fig. 2. The source has an elongated shape (position angle $\sim$ 160 deg). While most velocity channels (average $v_c = -25$ km s^{-1}) cluster in a narrow central band roughly perpendicular to the elongation axis, the channels in the NW lobe have velocities systematically more positive, while those in the SE lobe are systematically negative with respect to v_c. Inspection of the velocity pattern suggests a decelerating, well collimated outflow whose projected velocity at the tips of the lobes is $\sim$ 15 km s^{-1}. The highest velocities, ±80 km s^{-1}, are detected in the central band whose characteristic size is no larger than 0.02″ or 200 au, while the outflow is detected over an extent of 2000 au. We are continuing to study this source at higher resolution.

5   Conclusion

The source G70.29+1.60 which may well be the first ionized outflow clearly identified and imaged in a UCHII has many of the features expected from our scenario of a massive YSO still surrounded by its accretion disk or parts thereof. Our preliminary 3mm investigation agrees in general well with the only previous high resolution study conducted at 2cm with the VLA (De Pree et al. 1994). A significant difference, however, consists in the size of the high velocity band which seems to be an order of magnitude larger at 2 cm. We attribute this striking difference to the fact that the mm RLs are weighted toward higher electron densities, likely to occur much closer to the origin of the outflow.

We thus think that the use of short mm wavelengths at the highest possible angular resolution is essential for the proposed research. The LSA combining these requirements is in our view ideally suited for detecting and investigating the hitherto elusive accretion disks of massive YSO.

References

Altenhoff, W.J., Wendker, H.J., Strittmatter, P. 1981, Astron. Astrophys. 93, 48

Hollenbach, D., Johnstone, D., Lizano, S., and F. Shu 1994, Ap.J., 428, 654

Kurtz, S., Churchwell, E., Wood, D.O.S. 1993, Ap.J.Suppl. 69, 831

De Pree, C.G., Goss, W.M., Palmer, P., and Rubin, R.H. 1994, Ap.J. 428, 670

Sargent, Welch 1993, A.R.A.A. 31, 297.

Thum, C., Martín–Pintado, J. 1994, ASP Conf. Ser., 62, 265

Thum, C., Wink, J. 1996, Astron. Astrophys., in preparation

Thum, C., Strelnitski, V.S., Martín–Pintado, J., Matthews, H.E., Smith, H.A. 1995, Astron. Astrophys., 300, 843

Wood, D.O.S., Churchwell, E. 1989, Ap.J.Suppl. 69, 831

Millimeter Continuum Observations of Stars

Roberto Pallavicini[1] and Stephen M. White[2]

[1] Osservatorio Astronomico, Palazzo dei Normanni, I-90134 Palermo, Italy
[2] Department of Astronomy, University of Maryland, College Park, MD 20742, USA

Abstract. We review the scientific advances to be expected from millimeter continuum observations of stars by a Large Southern Array (LSA). We discuss briefly several topics including millimeter observations of the Sun, winds of hot stars, circumstellar disks of pre-main sequence stars, circumstellar shells and mass loss in cool giants, symbiotic stars, and non-thermal emission in active binaries, flare stars and accretion powered X-ray sources. We show that even in the more limited area of continuum observations, the proposed LSA will allow major advances in virtually all fields of stellar astronomy.

1 Introduction

Various papers at this conference have discussed millimeter radio emission for specific classes of stars which are either very young (protostars) or are in very late stages of evolution (AGB and post-AGB stars). These are objects in which a rich spectrum of molecular lines is observed and which are sufficiently extended to be spatially resolved in some cases with the presently available instrumentation (see Sargent and Welch 1993 for a review). In addition to these objects, there are many other stars (including the Sun) for which continuum observations will be very fruitful. Although more limited in scope, this type of observation provides essential clues to understand the physics of various classes of stellar objects, and constitutes a powerful addition to current studies of stars at centimeter and decimeter wavelengths. We will try to demonstrate that even in this more limited field a large millimeter array will provide major advances of which the present data give only some tantalizing hints.

2 Millimeter Emission from the Sun

Before discussing stars, it may be useful to summarize briefly a few basic facts of solar millimeter radiation. Although the proposed Large Southern Array may not be suitable for solar observations, the Sun remains the prototypical stellar object which can be observed in far more detail than any other star. Whereas solar microwave emission consists of various components originating in the corona (a quiescent component due to thermal free–free emission, an active component associated with magnetic active regions which is due mostly to thermal gyro-resonance emission in strong magnetic fields, and a flare component which contains gyrosynchrotron emission at microwave frequencies and coherent plasma emissions at decimetre- and metre-wavelengths; see, e.g. Dulk 1985), observations of the Sun at millimeter wavelengths (such as those currently carried out with the BIMA array) provide information quite different from that obtained at lower frequencies. Thermal gyroresonance and plasma emission are not relevant for millimeter continuum emission. Since radiation of increasingly shorter wavelengths originates from deeper and deeper layers in the solar atmosphere, at $\lambda = 3$ mm the emission comes not from the corona but from an optically–thick layer in the chromosphere at a temperature of ≈ 7000 K. At $\lambda = 1$ mm most of the radiation comes from a somewhat deeper layer at ≈ 6000 K. Millimeter observations, therefore, provide a unique way to investigate the upper photosphere and low chromosphere of the Sun and stars; they have an advantage over UV and optical observations in that they measure the temperature of the ambient electrons directly. In flares, nonthermal gyrosynchrotron emission at millimeter wavelengths is produced by much higher energy electrons than are typically relevant at microwave frequencies. This is due to the fact that the peak frequency of the incoherent synchrotron emission scales as $\nu_{max} \sim \gamma^3 \, \nu_B$ where γ is the Lorentz factor and $\nu_B \propto B$ is the cyclotron frequency. Since typically $\nu_B \sim 1$ GHz ($\nu_B < 10$ GHz everywhere) in the solar corona, emission in the millimeter range requires $\nu_{max} \gg \nu_B$, i.e., $\gamma \gg 1$. Thus the millimeter emission is produced by MeV–energy electrons, whereas the microwave emission is typically produced by mildly relativistic electrons with energies of several hundred keV. Observations at millimeter wavelengths therefore provide a powerful diagnostic tool for high energy electrons in solar (and stellar) flares, complementary to that provided by γ-ray astronomy for nuclear particles. Extensive observations carried out at BIMA at 3.5mm (e.g. Lim et al. 1992, Kundu et al. 1994) have beautifully demonstrated this in the case of solar flares, and reveal differences in the production of the low and high energy electrons.

3 Millimeter Emission from Stars

We expect stars to produce millimeter continuum emission by three main mechanisms: thermal free-free, nonthermal gyrosynchrotron, and dust opacity. The most complete survey of millimeter emission from stars made up to now is that of Altenhoff, Thum and Wendker (1994) who have used the IRAM 30m single-dish telescope with the MPIfR bolometer at 250 GHz (1.2mm). 268 stars were surveyed, mostly with a flux at 5 GHz in excess of 1 mJy. 122 stars were detected at a limiting sensitivity of $\approx 10 - 30$ mJy. The spectral index from 5 to 250 GHz may be used to give some crude information on the emission mechanism. Altenhoff et al. have organized their sample in various subclasses which form a useful starting point, even though a given class may contain objects which differ drastically in their physical properties. They include:

- O - B stars with fluxes at 250 GHz from 10 to 150 mJy
- Wolf-Rayet stars, with similar fluxes (up to ≈ 380 mJy)
- hot stars with shells, a very heterogeneous class of objects with high millimeter fluxes, typically $\approx 100 - 500$ mJy (up to 1.6 Jy)

Table 1. Sample of millimeter continuum detections of stars

Star name	Stellar type	S_{1mm} (mJy)	Comments
P Cygni	LBV	146	Stellar wind source
η Carinae	LBV	16000	Wind + nebula; variable, extended
MWC 349	Bpe	1630	Bipolar outflow; extended
Cyg OB2-5	O	55	Stellar wind
WR 147	WR/O bin.	379	Somewhat peculiar binary system
WR 105	WR	48	Stellar wind
CRL 961	Massive YSO	465	Probably warm dust
MWC 957	Evolved	77	Flat spectrum; optically thin shell
VV Cep	M2Ia+B8V	45	Cool M-star wind ionized by B star
α Aur	G5III-G0III	23	Nearby binary; chromospheres?
α CMa	A1Vm	11	Nearby main-seq. A star; stellar disk
α CMi	F5IV	11	Nearby star; photosphere/chromosphere
Betelgeuse	M2I	351	Stellar disk
Mira	M7III	129	Warm dust in circumstellar shell
T Tauri	PMS	296	Dust in disk around low-mass PMS
HL Tau	PMS	961	Dust in disk around low-mass PMS
β Lyrae	B7Ve+?	56	Shell around binary system?
SS 433	XRB	112	Nonthermal synchrotron from jets?
Cyg X-3	XRB	20-260	Highly variable nonthermal source
AE Aqr	Mag. CV	24	Nonthermal; not well understood
UV Cet	dM5e	< 9	Flare star

- pre-main sequence (PMS) stars, including T-Tauri and post-T Tauri stars, FU Ori stars etc., with fluxes of $\approx 20 - 500$ mJy (up to 1.7 Jy).
- late-type giants and supergiants, including normal stellar photospheres as well as Miras and AGB stars with circumstellar envelopes (with typical fluxes of $\approx 10 - 100$ mJy, but up to 1.5 Jy in some cases).
- optically variable stars (another quite heterogeneous group of objects which includes flare stars, X-ray emitting accretion binaries, symbiotic stars, etc.) with fluxes in the range $\approx 20 - 300$ mJy.

Note that many of the 270 objects in this survey, from virtually all classes, were not detected at the sensitivity limit of $\approx 10 - 30$ mJy. In the following, we will discuss briefly these various types of objects, taking into account the Altenhoff et al. survey as well as the results of other millimeter observations obtained recently. Table 1 presents a sample list of 1 mm continuum detections of stars covering a wide range of stellar types (see also Bastian et al. 1995).

3.1 Winds from Hot Stars

The radio emission from hot stars (including both O - B and Wolf-Rayet stars) is dominated by optically–thick free–free emission from their massive radiatively driven winds. Observations by Leitherer and Robert (1991) with the SEST 15m single-dish and by Altenhoff et al. (1994) with the IRAM 30m telescope, show that hot stars typically have millimeter radio spectra of the form $S_\nu \sim \nu^{0.67}$ expected for optically-thick thermal free-free emission from an isothermal spherically symmetric wind expanding at constant speed. For such a wind, the radio flux $S_\nu \propto \dot{M}^{4/3} v_\infty{}^{-4/3} \nu^{2/3} d^{-2}$, where $\dot{M}$ is the mass loss rate, v_∞ is the wind terminal velocity, and d is the distance (e.g. Panagia and Felli 1975, Wright and Barlow 1975). The mass loss rates $\dot{M}$ derived at present from radio observations of hot stars are of the order of $\approx 10^{-5} M_\odot yr^{-1}$. However, since the mass loss rate scales as $\sim S_\nu^{3/4}$, the improved sensitivity of the LSA will extend the range of mass–loss rates amenable to study well below the current limit, which samples only the most extreme mass–loss rates in a given class of stars. Moreover, since the radial distance R_ν from which radiation of frequency ν originates scales as $\sim \nu^{-3/2}$, higher frequencies will sample regions closer to the stellar surface, allowing study of the region in which the wind is accelerated by optical/UV line emission.

The millimeter emission of Wolf-Rayet stars has properties similar to those of O - B stars, indicating that the emission mechanism is basically the same. However, deviations from the canonical $\sim \nu^{0.67}$ law are occasionally observed for Wolf-Rayet stars, implying that accurate determinations of the spectral slope from centimeter to millimeter wavelengths can provide important information on the structure of the wind and its ionization state.

Two stars in this class deserve specific mention: they are probably the brightest stellar millimeter radio continuum sources in the southern and northern skies, respectively. The massive southern "luminous blue variable" (LBV) η Carinae has a $\lambda = 1$ mm flux of up to 16 Jy (Cox et al. 1995), making it brighter than nearly all known calibrator sources. Further, the source is extended and the flux comes from a region of just a few arcseconds in extent, and is highly variable. The northern object MWC 349 is famous for its masing millimeter recombination lines (a property which it shares with η Car); at $\lambda = 3$ mm it has a flux of 1 Jy, and is also extended on a subarcsecond scale, showing a bipolar outflow and a circumstellar disk.

3.2 Be Stars and Other Hot Stars with Shells

Hot stars with shells (of which there is a variety of different types in the survey of Altenhoff et al.) are usually stronger at millimeter wavelengths than wind sources. The radio spectra show evidence for a turn-over around 5 -10 GHz from optically thick to optically thin emission. The spectrum at high frequencies is typically flat indicating optically thin shells. A contribution from warm dust may be present in some cases. An important class of hot stars with shells is represented by the Be stars for which millimeter observations have been obtained by Waters et al. (1989, 1991) at the JCMT in Hawaii. Observations of ψ Per and other Be stars show a steepening of the spectrum at 0.8 and 1.1 mm with respect to IRAS data. This has important implications for the radial density distribution of the ionized gas in the circumstellar disk.

3.3   Stellar Photospheres

In stars without massive ionized outflows, the millimeter radiation may originate close to the stellar photosphere and (for a uniformly illuminated disk) the flux at 250 GHz is given by:

$$S_{250} = 1.4175 \times 10^{-4} \ T_{disk} \ r_{mas}^2 \tag{1}$$

where T_{disk} is the temperature of the emitting layer and r_{mas} is its radius in milliarcsec. For $T_{disk} \approx T_{eff}$ and $r_{mas} \approx R_*/d$, a sizeable number of nearby stars (mostly giants and supergiants) are predicted to be detectable at millimeter wavelengths even at the present sensitivity levels. About 50 were in fact detected in the survey of Altenhoff et al. at flux levels > 10 mJy at 250 GHz, and $\approx$ 600 were predicted to have fluxes > 1 mJy at 230 GHz by the MMA working group (Dulk et al. 1989). Since the number of detectable sources increases as $\sim S_{lim}^{-3/2}$ for a homogeneous spherically symmetric distribution, thousands of sources will be detectable at the very low flux limits ($\approx$ 0.1 mJy) expected for the LSA, thus making the study of millimeter emission from normal stars a major field of investigation. The importance of millimeter observations in this field is that they are directly sensitive to the electron temperature in the atmosphere, and do not require assumptions such as LTE as does the analysis of optical and infrared spectra. For stars of known parallax and effective temperature, it will also be possible to measure stellar radii using (1) and the measured millimeter fluxes. Thus millimeter data can measure fundamental parameters for a much larger range of stars than presently possible.

3.4   Other Late-type Giants and Supergiants

Millimeter emission was detected from the photospheres of many late-type giants and supergiants in the survey of Altenhoff et al., but not all of their detected cool giants and supergiants are "normal" giants. This group of stars also contains various classes of objects with ionized winds, such as VV Cep stars, and circumstellar envelopes in which dust may make a significant contribution to the radio emission (Miras, and AGB and post-AGB stars). All these are potentially interesting objects for high-sensitivity millimeter wavelength observations.

For example, symbiotic stars (studied at millimeter wavelengths by Seaquist and Taylor 1992 and by Ivison et al. 1992) are interacting binaries with mass transfer from a late-type giant to a compact object. Their spectra, obtained at JCMT, show, in combination with VLA and IRAS data, a variety of behaviours ranging from optically thick free-free emission all the way from centimeter to IR wavelengths, to optically thick free-free emission plus a dust contribution at IR wavelengths, to a transition from optically-thick to optically thin emission plus a dust contribution at longer wavelengths. Millimeter observations are essential for studying these objects since they provide an important link between the centimeter data and the IR fluxes and allow the modelling of the ionized circumstellar enviroment. Similar considerations apply to VV Cep stars, which are main-sequence B stars orbiting a late-type supergiant. In this case, millimeter radiation originates from free-free emission in the massive wind of the cool supergiant which is ionized by UV radiation from the B star.

Miras and Semi-Regular (SR) variables have been studied at millimeter wavelengths by Walmsley et al. (1992) using the IRAM 30m single dish at Pico Veleta. The observed millimeter radiation is mostly consistent with photospheric emission. Excess millimeter emission (at levels as high as $\approx 500 - 1500$ mJy) is observed in AGB and post-AGB stars (including proto-planetary nebulae): this arises from warm durst in the circumstellar envelopes of these stars.

3.5 Pre-Main Sequence Stars

Pre-main sequence stars (including classical and weak-lined T-Tauri stars, FU Ori stars and Herbig Ae/Be stars) constitute an important class of millimeter continuum sources (Beckwith et al. 1990; Weintraub et al. 1991, Henning et al. 1994, Osterloh and Beckwidth 1995; Dutrey et al. 1995). PMS stars have a high rate of detection in the Altenhoff et al. survey, with very high fluxes in some cases ($\approx 100-1000$ mJy). Their spectra have a spectral index $\alpha > 2$ from 5 to 250 GHz. All this is consistent with thermal radiation from cold dust in circumstellar disks, as first discussed by Beckwith et al. (1986). Millimeter continuum observations allow the determination of the mass of the disk, and more sensitive measurements should allow the determinations of the mass of thinner disks, which is important for weak-lined T-Tauri stars and post-T Tauri stars. A recent survey by Osterloh and Beckwidth (1995) has shown that the rate of detection of classical T-Tauri stars (CTT) is indeed higher than for the weak-lined T-Tauri stars (WTT), likely because the former have more massive circumstellar disks. Post-T Tauri stars (i.e. PMS stars that are older and more evolved than T-Tauri stars) are believed to lack dense circumstellar disks (Skinner et al. 1992, Gahm et al. 1994); yet a few of them (e.g. HD 560, Altenhoff et al. 1994) have been detected at millimeter wavelengths. This rises the question whether the observed emission is thermal radiation from very thin disks or rather non-thermal emission from flares in young stars characterized by strong magnetic activity. Dusty disks have also been observed around some main-sequence stars like Vega (Chini et al. 1990).

3.6 Optically Variable Stars

The class of optically variable stars in the survey of Altenhoff et al. is very heterogeneous and shows a low detection rate at the present sensitivity levels (only 15 of the 76 surveyed stars were detected, including 5 symbiotics). This class includes RS CVn binaries, flare stars, accretion-powered X-ray sources, β Lyrae binaries, and novae. In many cases, the emission mechanism is likely to be non-thermal. This is the case for instance of flares on RS CVn binaries and flare stars where one expects, by analogy with the Sun, that millimeter radiation originates from synchrotron emission of high energy electrons (> 1 MeV). The emission is expected to be highly variable, and so although $\lambda = 3$ mm fluxes in excess of 1 Jy may be expected at times of outburst, only a few RS CVn's have been detected so far (e.g. HR 1099, UX Ari, UV Psc) and no undisputed detection of flare stars has yet been reported (Lim and White 1995). The higher sensitivity of the LSA should contribute substantially to this area thus allowing the study in other stars of flares similar to, but much more energetic, than solar flares.

In accretion-powered X-ray sources (like SS433, Cyg X-1, Cyg X-3, etc), non thermal emission is expected to arise from the accreting flow, while in β Lyr variables it is unclear whether we are observing non-thermal emission or thermal emission from ionized shells. The magnetic cataclysmic variable AE Aqr has been detected at $\lambda = 3$ mm (Abada-Simon et al. 1993), and while the emission is clearly nonthermal there is no accepted model as yet. Finally, in novae (such as Nova Cyg 1992 studied by Ivison et al. 1993), the radio spectrum is consistent with thermal free-free emission from an expanding shell of ionized gas. Observations at different epochs after the explosion (including crucial millimeter observations) allow one to follow the evolution of the phenomenon and the transition from optically thick to optically thin emission in the expanding shell.

4 Conclusion

From the survey above, it is clear that a large variety of different stellar objects can emit millimeter continuum radiation. About one hundred stars have already been detected at millimeter wavelengths at flux limits $> 10 - 30$ mJy. The detected stars cover the entire HR diagram, but with present instruments only a handful of sources is available within each class. With the large sensitivity of the proposed LSA, it will be possible to detect several thousands of them and obtain statistically significant samples for each class. In some cases it should even be possible to resolve the sources, although the majority of stellar sources will remain unresolved if the spatial resolution is not significantly better than 0.1 arcsec. Continuum observations to be carried out with the LSA will have a strong impact on many stellar problems, such as:

- the winds of hot stars
- the circumstellar disks of PMS stars and Vega-type objects
- the upper photopheres and low chromospheres of giants and supergiants
- the non-thermal emission of RS CVn binaries, flare stars and accretion powered X-ray sources
- the winds of symbiotic stars and VV Cep stars
- the circumstellar shells of Be stars, AGB and post-AGB stars, planetary nebulae, and novae

To conclude, it can reasonably be expected that the proposed LSA will produce for radio stars a quantum jump as large as or larger than the one produced in the early eighties at centimeter wavelengths by the opening of the VLA.

References

Abada-Simon, M., Lechacheux, A., Bastian, T. S., Bookbinder, J., and Dulk, G. A. (1993) *Ap. J.* **406**, 692

Altenhoff, W.J., Thum, C., and Wendker, H.J. (1994) *A&A* **281**, 161

Bastian, T., Gagné, M., Gary, D., Gordon, M., Hjellming, R., Hogg, D., Lindsey, C., Kundu, M., Rabin, D., and White, S. (1995) *Report of the Working Group on "The Sun and Stars"*, Second MMA Science Workshop

Bechwith, S.V.W., Sargent, A.I., Scoville, N.Z., Masson, C.R., Zuckerman, B., and Phillips, T.G. (1986) *Ap. J.* **309**, 755

Beckwith, S.V.W., Sargent, A.I., Chini, R.S., and Güsten, R. (1990) *A.J.* **99**, 924

Chini, R., Krügel, E., and Kreysa, E. (1990) *A&A* **227**, L5

Cox, P., Mezger, P. G., Sievers, A., Najarro, F., Bronfman, L., Kreysa, E., and Haslam, G. (1995) *A&A* **297**, 168

Dulk, G.A. (1985) *Ann. Rev. Astron. Ap.* **23**, 169

Dulk, G.A., Gary, D.E., Hjellming, R.M., Kundu, M.R., and Bastian, T.S. (1989) *Report of the Working Group on "The Sun and Stars"*, First MMA Science Workshop

Gahm, G.F., Zinnecker, H., Pallavicini, R., Pasquini, L. (1994) *A&A* **282**, 123

Dutrey, A., Guilloteau, S., Duvert, G., Prato, L., Simon, M., Schuster, K., and Ménard, E. (1995) *A&A*, in press

Henning, Th., Launhardt, R., Steinacker, J., and Thamm, E. (1994), in *ASP Conference Series* **62**, 171

Ivison, R.J., Hughes, D.H., and Bode, M.F. (1992) *MNRAS* **257**, 47

Ivison, R.J., Hughes, D.H., Lloyd, H.M., Bang, M.K., and Bode, M.F. (1993) *MNRAS* **263**, L43

Kundu, M.R., White, S.M., Gopalswamy, N., and Lim, J. (1994) *Ap. J. Suppl.* **90**, 599

Leitherer, C., and Robert, C. (1991) *Ap. J.* **377**, 629

Lim, J., and White, S.M. (1995) *Ap. J. Letters*, submitted

Lim, J., White, S.M., Kundu, M.R., and Gary, D.E. (1992) *Solar Phys.* **140**, 343

Osterloh, M., and Beckwith, S.V.W. (1995) *Ap. J.* **439**, 288

Panagia, N., and Felli, M. (1975), *A&A* **39**, 1

Sargent, A.I., and Welch, W.J. (1993) *Ann. Rev. Astron. Ap.* **31**, 297

Seaquist, E.R., and Taylor, A.R. (1992) *Ap. J.* **387**, 624

Skinner, S.L., Brown, A., Walter, F.M. (1992) *A.J.* **102**, 1742

Walmsley, M., Chini, R., Kreysa, E., Steppe, H., Forveille, T., and Omont, A. (1991) *A&A* **248**, 555

Waters, L.B.F.M., Boland, W., Taylor, A.R., van de Stadt, H., and Lamers, H.J.G.L.M. (1989) *A&A* **213**, L19

Waters, L.B.F.M., van der Veen, W.E.C.J., Taylor, A.R., Marlborough, J.M., and Dougherty, S.M. (1991) *A&A* **244**, 120

Weintraub, D.A., Sandell, G., and Duncan, W.D. (1991) *Ap. J.* **382**, 270

White, S.M., and Kundu, M.R. (1992) *Solar Phys.* **141**, 347

Wright, A.E., and Barlow, M.J. (1975) *MNRAS* **170**, 41

AGB star envelopes as probes of stellar evolution and time-dependent chemistry

M. Guélin, R. Lucas, R. Neri

Institut de Radioastronomie Millimétrique,
300 rue de la Piscine, F-38406 Saint Martin d'Hères, France

Abstract. Detailed mm-wave studies of circumstellar envelopes teach us how and at which rate AGB stars loose mass, and yield the isotopic composition of the stellar ejecta. They tell us which molecules are present and how fast these molecules are formed and destroyed.

The sensitivity and angular resolution of the present mm telescopes restrict these studies to a score of AGB stars and PPN, all located within 2 kpc from the Sun. The LSA will allow to map thousands of circumstellar envelopes and to reach the Galactic Center region.

1 Introduction

The envelopes expelled by AGB stars offer an unique opportunity to test stellar evolution and chemistry models.

Stellar evolution: the phase of high mass loss, one of the ultimate phases of stellar evolution, depends critically on the initial star parameters, as well as on all successive evolution steps. The prediction of composition and mass of the ejected matter raises a difficult challenge to the theoreticians.

Chemistry: while the matter is expelled from the stellar atmosphere and expands in outer space, it assumes a wide range of physical conditions. The density decreases by several orders of magnitude and the temperature drops from a few thousands to a few tens of kelvin. In the process, about every molecule formation/destruction mechanism is likely to occur: 3-body reactions in the hot atmosphere, grain condensation in the upper atmospheric layers, ion-molecule and radical-molecule reactions in the circumstellar envelope, and photodissociation in the envelope outer 'skin'. All of this makes an exceptionally rich chemistry, as attested by the more than 50 molecular species identified in the carbon star envelope IRC+10216; the latest of those, the C_8H radical, has a weight of 97 a.m.u. (Cernicharo & Guélin 1996).

Circumstellar chemistry is not only rich, it is comparatively simple. The chemical processes are largely driven by the physical conditions, which, due to the simple geometry, are relatively well known. The velocity field is also remarkably regular, making possible to date the different envelope layers. By studying the molecular composition of the layers, it is possible to test directly the time dependence of the chemistry.

Until recently, the large visual extinctions and small angular sizes have impeded the observation of the most interesting objects, those experiencing high mass loss. The development of interferometry has allowed substantial progresses in this domain. Molecules are now routinely mapped at arcsecond resolutions with the large mm interferometers, such as the IRAM interferometer on Plateau de Bure, providing a data basis to which the model predictions could be confronted. So far, however, only a small number of bright and relatively close stars are within the reach of the present instruments. The LSA, which will have an angular resolution and a sensitivity 10 times larger than the IRAM interferometer, will enlarge considerably the number and nature of observable stars, allowing to reach objects in the Galactic Center region.

We review below recent interferometric results obtained at mm wavelengths on AGB star envelopes and on their immediate offsprings, pre-planetary nebulae (PPN). The data are compared to model predictions, with emphasis on density/velocity structures and outer envelope chemistry. The question on the envelope isotopic composition is then briefly discussed and we underline the prospects open by the new generation interferometers.

2   Mass loss in AGB stars

The main component of the expelled matter, molecular hydrogen, is hard to detect. Envelope masses and mass loss rates are therefore derived primarily from the observation of the two most stable trace components: CO and dust.

The CO molecule is observed in emission at mm wavelengths, or in absorption in the IR. Because the central continuum source (which corresponds to the dust condensation region and has typically a radius of a few a.u.) is always much smaller than the telescope beam, mm emission and IR absorption observations are weighted differently: the IR spectra show mostly the gas close to the star and the mm spectra the gas in the mid & outer envelope. Ideally, the mass loss rate should be determined by combining both types of observations.

In order to derive the envelope mass from its CO mass, one has to model the CO chemistry. This has been done in particular by Mamon et al. (1988), who introduced laboratory data on CO photodissociation in their chemical network. In the case of cold red giants ($T_* < 3000$K), the far UV photons come from interstellar space; their energy distribution can be described in terms of an attenuated 'standard interstellar radiation field' (or 'ISRF'). The interstellar radiation is absorbed by the dust grains, as well as by the H_2 and CO molecules present in the outer envelope layers. In Mamon et al.'s model, efficient CO self-shielding and shielding by H_2 lines make that the CO fractional abundance $x(CO)$ relative to H_2 follows almost a step function: it raises from zero at large radii to a few $\times 10^{-4}$ at a critical radius R_{ph}. The measure of this radius and, if possible, of the CO column density from $R = 0$ to $R = R_{ph}$ provides the best measurement of the mass loss rate and of the local ISRF (see e.g. Olofsson et al. 1993, Loup et al. 1993).

Fig. 1 shows the ^{12}CO, J= 2–1 line profile along an EW cut passing through CW Leo/IRC+10216. CW Leo, the archetype of carbon stars, located only $D \sim$ 200 pc from the Sun, is believed to have reached the very end of the AGB

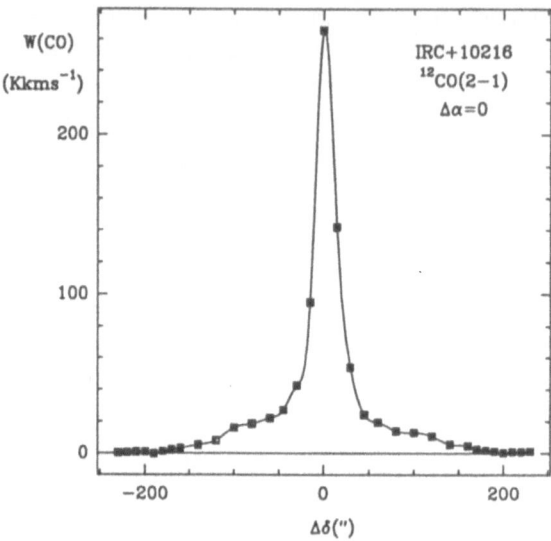

Fig. 1. ^{12}CO J= 2–1 line emission, integrated over the central 13 kms^{-1} velocity interval, along a strip at constant R.A. passing through the carbon star CW Leo/IRC+10216 (Guélin & Cernicharo *in preparation*). This profile, coupled to J= 1–0 and 3–2 CO profiles, allows to derive the mass loss rate of the star during the last 10^4 yr. The LSA will allow to study in this way thousands of stars with high mass losses.

phase and to be just on the verge of evolving toward the PN phase (Guélin et al. 1993 –hereafter GLC– Kastner et al. 1994). It is surrounded by a thick envelope of dust and gas. The CO profile shows a broad 'plateau' component whose intensity drops briskly at $R \simeq 120''$. The mass loss rate derived from this profile is $2\,10^{-5}M_\odot\mathrm{yr}^{-1}$ in the recently ejected 'core' and $4\,10^{-5}M_\odot\mathrm{yr}^{-1}$ in the older 'plateau': the mass loss rate seems thus to have decreased by a factor of 2 in the last 10^3 yr (Truong-Bach & Rieu 1991), which agrees with the picture of a star leaving the AGB phase.

The observation of the dust thermal emission offers another way of studying the stellar mass loss rates (see e.g Walmsley et al. 1991). Close to the star, where the dust grains are hot, the dust mass and the mass loss rate can be derived from IR interferometric observations; further out, one has to rely on sub-mm and mm wavelengths, the only ones sensitive to cold dust emission. Note that, contrary to molecules, the dust grains survive to the unattenuated ISRF. The sub/mm dust continuum emission provides the best way to study the envelopes beyond R_{phot}.

The number of stellar envelopes detected in the mid and far IR exceeds one hundred thousand. That of envelopes detected at mm/sub wavelengths, through the dust thermal emission (Walmsley et al. 1991) or through CO (see the compilation by Loup et al. 1993), amounts to a few hundreds. Of these, less than one hundred are bright and large enough to be mapped in CO or dust with the present mm arrays. A recent survey of the brightest objects, made by Neri et al. (1996) with the IRAM interferometer, resolved only two scores of envelopes, most of them located within 1 kpc of the Sun. With the LSA it will be possible to measure thousands of CO brightness profiles such as that of Fig. 1.

Mapping CO and dust in a much larger number of CSEs would make possible to study the mass loss rates and asymmetries as functions of the stellar type, local environment, galactocentric distance... Clear departures of the envelopes from spherical symmetry and high velocity bipolar flows are expected to mark the very end of the AGB phase; they have been so far observed in only a few envelopes which have already reached the PPN stage.

3   Pre-planetary nebulae

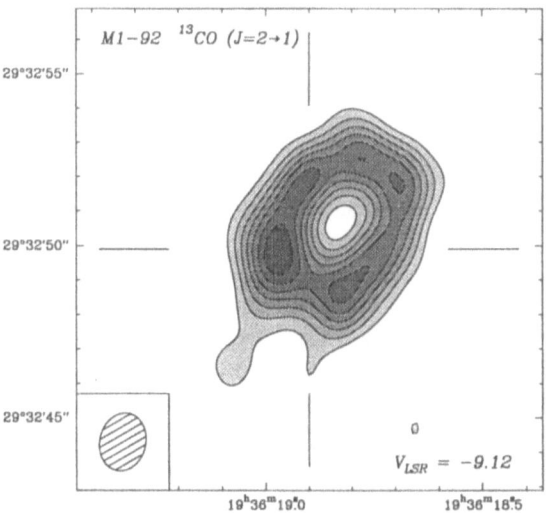

Fig. 2. Map of the ^{13}CO (2–1) line emission in the pre-planetary nebula M1-92, integrated over a 3 kms^{-1}-wide velocity interval centred on $V = -9.1$ kms^{-1}. The map was observed by Alcolea et al. (1996) with the IRAM Plateau de Bure mm array and has an angular resolution of $1.7'' \times 1.4''$. The central hole in the map shows that the envelope is already detached from the star.

Among the objects showing strong asymmetries, are the well known post-AGB stars M1-92 and CRL 618, which exhibit each a high velocity (HV) molecular outflow ($V_{exp} = 60$ kms^{-1} in M1-92 and > 150 kms^{-1} in CRL 618). These outflows and the central cores of the envelopes have been mapped in CO with a high sensitivity with the Plateau de Bure interferometer. Fig. 2 shows the ^{13}CO (2–1) line emission in M1-92, integrated over a 3 kms^{-1}-wide interval (Alcolea et al. 1996). The star is surrounded by a detached shell expanding at $\simeq 15$ kms^{-1}. A collimated bipolar flow, arising probably in the innermost regions, is dragging the dense gas from the envelope polar caps at P.A.= +125° and -35°. The HV flow itself, not visible in Fig. 2, is marginally resolved by the 1.4" (FWHP) synthetised beam. A similar situation was found in the case of CRL 618 (Neri et al. 1992, 1996, Yamamura et al., 1994, Hajian et al. 1995). A higher angular resolution would be needed to resolve the HV flows and extend this study to other PPNs.

The HV flow of CRL 618 is also detected in CN, HCN, HCO$^+$ and HC$_3$N. Since it is difficult to accelerate molecules to $V = 100$ kms^{-1}, it is likely that those were first dissociated, then reformed. A higher angular resolution would probably allow to separate the CO, CN, HCN, and HC$_3$N sources and measure the reformation times of the different species. Most likely, the HV jet changes its direction and velocity with time. A resolution of $\theta = 0.1"$, easily achievable with the LSA, may let observe significant changes if they occur over time scales of $\theta.D/V_{HV} \simeq 2 - 3$ yr (the distance to M1-92 and CRL 618 is 1–2 kpc).

4 Chemistry

We have stressed above the richness of the chemistry in AGB star envelopes and the possibility to follow in time the molecule formation/destruction by measuring the molecular abundances at different radii. Such studies, however, require good linear resolutions, which are now reachable only in the closest objects. IRC+10216, $\simeq 200$ pc away, is obviously a prime target.

The chemistry in the envelope is far from thermodynamic equilibrium, as illustrated by the large abundance of radicals (e.g. x(CP) > x(HCP)), and is highly time dependent. The short dynamical timescales (in IRC+10216, it takes 10^4 yr for the matter ejected by the central star to cross the entire CO envelope) imply that chemical processes must be fast. The formation of dust grains (which removes from the gas most elements heavier than carbon), photodissociation (which breaks out molecules into small radicals and atoms), photoionization and ionization by cosmic rays (which produce C$^+$, C$_2$H$_2^+$,...) and, finally, fast neutral and ion-molecule reactions shape out the molecular composition. It is also likely that grain chemistry plays a role in the final composition, all the more that the dust grains, while drifting outwards, are exposed to an increasingly large UV intensity.

The most recent models of gas phase chemistry in CSEs (with particular emphasis on IRC+10216) are those of Glassgold and co-workers (see Cherncheff et al. 1993 and references therein) and Millar & Herbst (1994). In addition to the photodissociation and ion-molecule reactions already present in previous models, they include neutral reactions, such as radical-radical reactions. Although neutral reactions have smaller rate coefficients than the fast ion-molecule reactions, they turn out to be much faster in the dense envelope cores where the density of ions is particularly low. Gas phase models predict centrally peaked radial distributions for SiS, CO, C$_2$H$_2$ and HCN, which come from the stellar atmosphere, and hollow shell distributions for C_2H, CN, and their products C_3H, C_4H, C_3N, HC_3N, etc..., which are formed in the envelope.

Fig. 3 shows some examples of maps observed with the Plateau de Bure array at a resolution of $\sim 3''$. The emission is integrated over a 5 kms^{-1}-wide interval, centered on the source systemic velocity ($v_{sys} = -26.4$ kms^{-1}). The 5 kms^{-1} interval was chosen narrow enough that the contours represent roughly the line brightness distribution in the meridian plane perpendicular to the line of sight. The CN map results from a mosaic of 7 interferometer fields spaced by 20$''$ steps; the other maps are derived from single field observations and are not corrected for the primary beam attenuation (HPBW$\simeq 50''$). Some of the maps (CN, C_3H, C_4H, CS, SiC_2 and MgNC) include short spacing data, obtained by mapping the Bure field of view with the 30-m telescope, the others only the central 30-m spectrum (HPBW 23-27$''$). The continuum emission (essentially, a point like source of flux 60 mJy) has been removed.

The maps of Fig. 3 confirm that the radicals and the cyanopolyynes have hollow-shell distributions (i.e. ring-like distributions in the velocity interval plotted in the figure) and that the small stable molecules are centrally peaked, as predicted by the gas-phase models. They show much more, however. First, the rings and the central sources show a rich azimuthal structure, in the form of thin arcs and clumps, with a crude axis of symmetry (see e.g. the C_2H and SiS maps). Second, the rings show a radial structure suggesting the presence of double or even triple shells (e.g. the CN and C_2H maps). Third, some maps show at the same time a central source and a ring (CS and SiC_2). Fourth, the rings are off-centered by $\sim 2''$ to the E of the central star.

According to gas phase models, the species in the envelope must peak at different radii, since they are formed from different parent molecules through reactions with different time scales. For example, in Cherchneff et al.'s model (1993) C_4H is mainly formed by the neutral reaction $C_2H + C_2H$, whose characteristic time is $t = (k_n x(C_2H)n_{H_2})^{-1} \sim 2\,10^3$ yr, whereas C_3H is formed from C_2H, but by radiative association with C, a reaction at least 5 times slower. Neutral carbon, moreover, is abundant only in the outer envelope, further out than C_2H, so that the C_3H abundance is predicted to peak at a radius at least twice larger than C_4H. C_4H, itself, is predicted to peak at a radius ~ 1.5 times larger than C_2H (A. Glassgold, private communication).

This is not observed: the bright blobs in the C_2H, C_3H and C_4H maps (as well as in the maps of several species, such as MgNC–see GLC) coincide within 2$''$. This implies a time delay for the formation of C_3H and C_4H from $C_2H \leq 150$ yr, or a few percent of the characteristic reaction times! Other examples of discrepancies between observations and predictions are the similarities between the CN and HNC distributions (Fig. 3) and the relatively small radius of the CN shell. Clearly, the models, which assume spherical symmetry and a smooth density distribution, are much too crude in view of the complexity of the new data and attempts to modelize line excitation and gas phase chemistry in clumpy, asymmetrical media are in the way. Nonetheless, the above discrepancies probably ask for a reconsideration of the chemical networks.

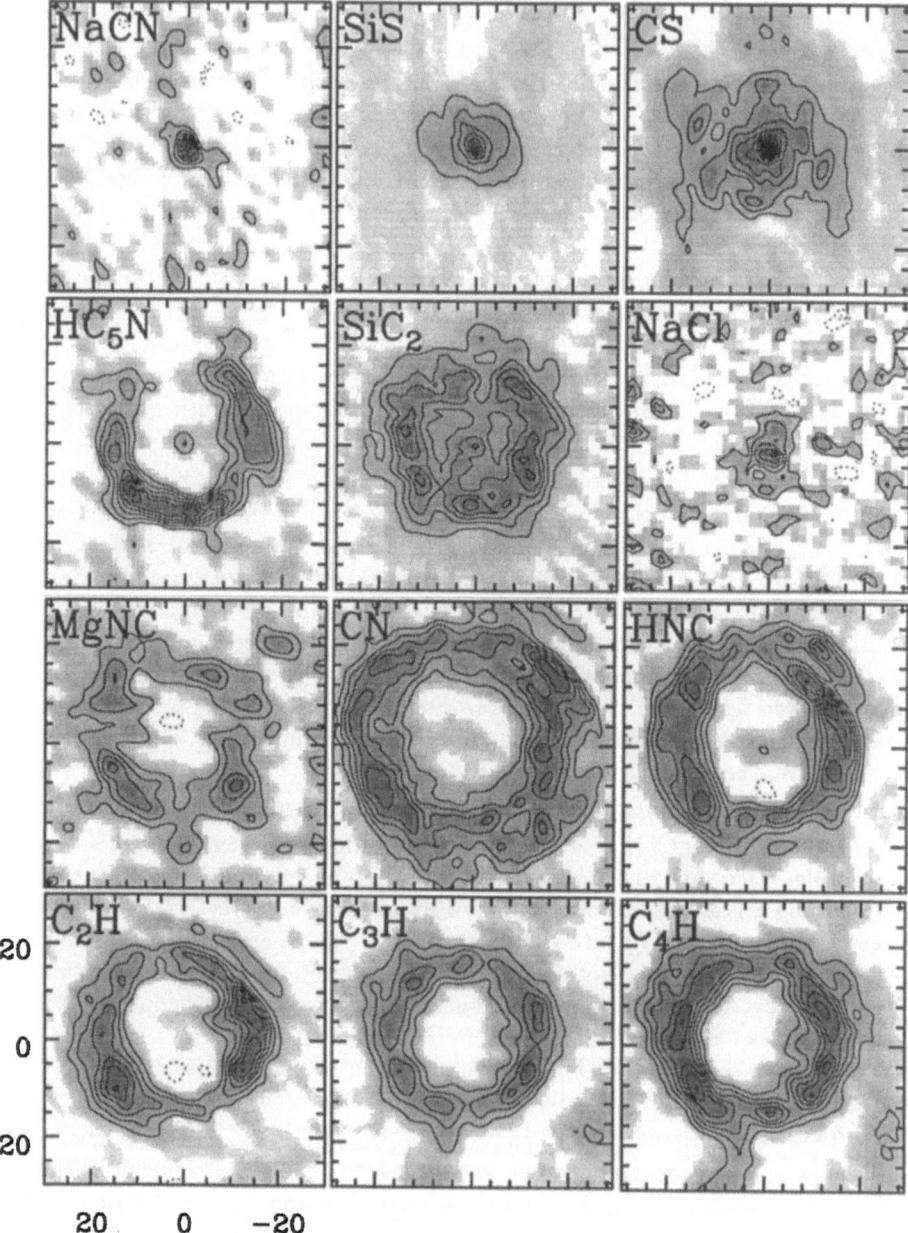

Fig. 3. The IRC+10216 envelope, observed with an angular resolution of $\simeq 3''$ (FWHP) in the 3 mm lines of different molecules. The line intensities are integrated over the central 7 kms^{-1} interval (IRAM interferometer).

It is tempting to see in the remarkable similarities between different species a common formation mechanism, capable of building large molecular abundances in timescales of few 10^2 yr. Photon-related processes, such as photodissociation and photodesorption from dust grains, on the one hand, and shock desorption of molecules weakly attached to grains, on the other hand, can be fast enough. What role they actually play in IRC+10216's complex chemistry remains however to be seen.

5 Elemental isotopic ratios

The matter expelled by AGB stars has been processed in the stellar interiors and its composition reflects only remotely that of the surrounding interstellar medium. Depending on the initial stellar mass and on the evolution stage, the abundances of C, N, O, and even of heavier elements such as Mg, Al, Si and S may be significantly altered.

Model calculations allow to follow the evolution of stars with masses of few $M_\odot$ from the Hayashi contraction phase to the end of the thermally pulsating AGB phase (see e.g. Forestini & Charbonnel 1996). They include nucleosynthesis (core and shell burning, hot bottom burning,...) as well as the mixing of the synthesis products with the envelope matter and their transportation to the surface. They yield detailed predictions of the elemental isotopic abundance ratios. Some ratios (e.g. $^{12}C/^{13}C$, $^{14}N/^{15}N$, $^{16}O/^{17}O$), which depend sensitively on the stellar parameters, may be used as indicators of the star masses and ages. The isotopic abundances vary more rapidly at the end of the AGB phase, when the convective envelope is thin, hence easily enriched in freshly synthesized elements. This is illustrated by Fig. 4 (from Forestini and Charbonnel 1996), which represents the evolution of the C, N, O ratios in a 5 $M_\odot$ star, throughout the TP-AGB phase. One see that it may be possible to observe large abundance variations in just a few 10^3 yr, provided the star has reached the end of the TP phase. Such variarions would be visible across the envelope.

The rotational lines of molecules are certainly the best tools for the investigation of isotopic enrichment (at least for the heavy elements $\geq$ C): the rotational isotopic shifts are always much larger than the line widths and line confusion is less severe at mm than at IR wavelengths. The large number of molecular species which can be studied in the mm/submm domain allows to check for isotope fractionation effects. Twenty four different isotopes pertaining to 8 different elements have been observed to date in IRC+10216 and the corresponding isotopic ratios accurately measured. The measured ratios, after comparison with model predictions, have allowed to derive the mass of the central star and confirmed that it has reached the end of the AGB phase (Guélin et al. 1995). No isotopic variation with radius has been found.

So far, such a detailed study of molecular and isotopic composition is possible for only one star; it is based essentially on 30-m telescope observations. This star, IRC+10216, is in no way unique; it is just the closest C-star at the end of the AGB phase. The above measurements could be extended to a wide range of

$M_{ini} = 5M_\odot$; $Z = 0.02$

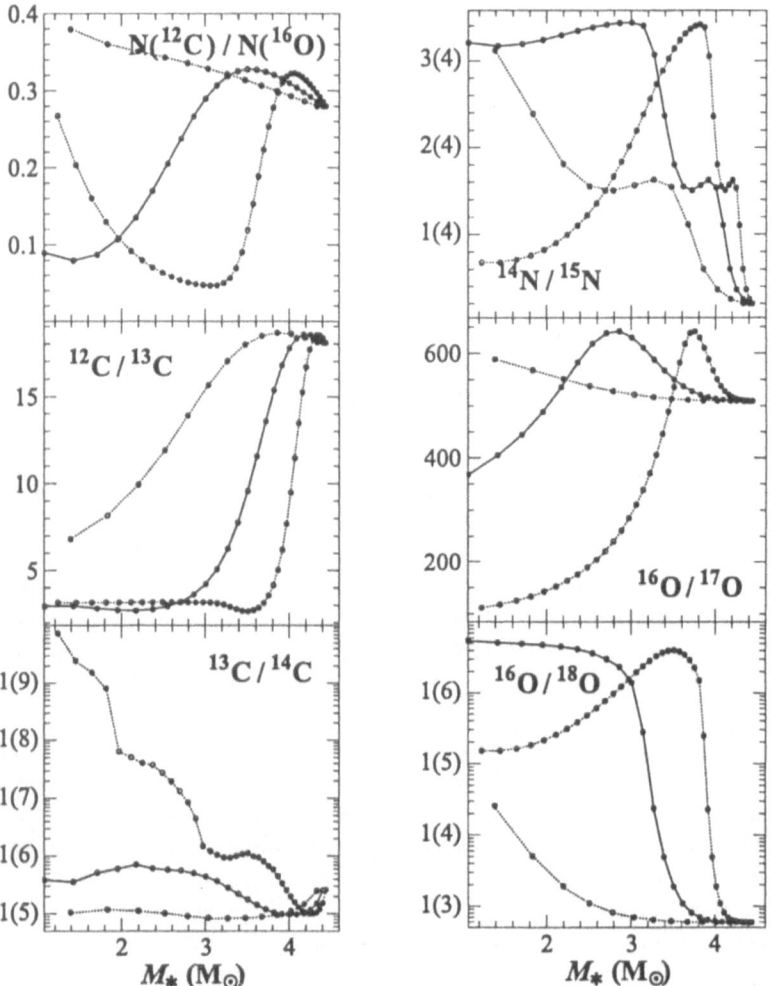

Fig. 4. The behaviour of the C/N abundance ratio and of the C, N and O isotopic ratios in a 5 $M_\odot$ star during its evolution through the TP-AGB phase. The data points (black dots) are calculated after each thermal pulse and are separated in time 3700 yr. The time increases from right to left. The total mass ($M_* =$core+convective envelope) remaining in the star is indicated on the bottom scale; it decreases from $\simeq 4.5M_\odot$ to $1M_\odot$ as the star expells its convective envelope. The central curve (thick line) corresponds to the mass loss rate predicted by the model; the other curves to mass loss rates reduced and enhanced, respectively, by a factor of 2 (from model calculations of Forestini & Chardonnel 1996)

stellar types and ages, if one could reach a similar sensitivity in a circle of 1 kpc radius.

Gains in sensitivity can be reached by increasing the telescope surface and, for the light molecular species, by observing at higher frequencies. For the heavy molecules, which constitute the most interesting component from the viewpoint of chemistry, they require a larger surface. A telescope with an area 25 times larger than the effective area of the 30-m telescope (410 m^2 at 100 GHz and 250 m^2 at 230 GHz) would be needed to reach objects like IRC+10216 at 1 kpc, and like M1-92 at $\simeq$ 8 kpc. Such a large instrument, for practical reasons, can only be an interferometer.

Aknowledgements: We thank Drs. M. Forestini and A.E. Glassgold for helpful discussions and Drs. J. Alcolea and V. Bujarrabal for communicating results prior to publication.

References

Alcolea, J., Bujarrabal, V., Neri, R. 1996, *in preparation*
Cernicharo, J., Guélin, M. 1996, A&A 309, L27
Cherchneff, I., Glassgold, A.E., Mamon, G.A. 1993, Ap. J. 410, 188 A&A 280, L19 (GLC)
Guélin, M., Forestini, M., Valiron, P., Ziurys, L.M., Anderson, M.A., Cernicharo, J., Kahane, C. 1995, A&A 297, 183
Forestini, M., Charbonnel, C 1996, A&A *submitted*
Kastner, J.H., Weintraub, D.A. 1994, Ap. J. 434, 719
Loup, C., Forveille, T., Omont, A. and Paul, J.F. 1993, A&A Supp. Ser. 99, 377
Mamon, G., Glassgold, A.E., Huggins, P.J. 1988, Ap. J. 328, 797
Hajian, A., Phillips, J.A., Terzian, Y. 1995, *preprint*
Millar, T. J., Herbst, E. 1994, A&A 288, 561
Neri, R. et al. 1992, A&A 262, 544
Neri, R., Kahane, C., Lucas, R., Bujarrabal, V. 1996 A&A *submitted*
Olofsson , H., Eriksson, K., Gustafsson, B., Carlstrom, U. 1993, Ap. J. *Supl* 87, 267
Truong-Bach, Morris, D., Nguyen-Q-Rieu 1991, A&A 249, 435
Walmsley, M. et al. 1991, 248, 555
Yamamura, I., Shibata, K., Kasuga, T., Deguchi, S. 1995, Ap. J. 427, 406.

CO observations of short period Miras

Martin Groenewegen[1], F. Baas[2], J. Blommaert[3], E. Josselin[4], R.P.J. Tilanus[2]

[1] Max-Planck Institut für Astrophysik, Karl-Schwarzschild Straße 1, D-85740 Garching, Germany
[2] Joint Astronomy Centre, 660 N. A'ohoku Place, University Park, Hilo, HI 96720, U.S.A
[3] ISO Science Operations Centre, Astrophysics Division of ESA, Villafranca, Spain
[4] Institut d'Astrophysique de Paris, CNRS, 98 bis Boulevard Arago, F-75014 Paris

Abstract. CO observations are reported of Miras with periods shorter than 400 days.

1 Introduction

In two previous contributions (Groenewegen 1995a, b) it was found from modeling the spectral energy distributions (SEDs) of carbon-rich Mira variables with a dust radiative transfer code (Groenewegen 1993) that there exists a tight relation between mass loss rate (MLR) and pulsation period for periods $\gtrsim$ 400 days, but that the MLRs appear to be significantly lower for periods $\lesssim$ 400 days (see Fig. 1, crosses). This was interpreted as an observational signature that radiation pressure on dust is less effective at these relatively low luminosities, in qualitative agreement with model calculations by e.g. Habing et al. (1994).

More observations were needed (in particular CO) to confirm this trend at short periods, and, at the same time, the question was raised if a similar trend could be observed among oxygen-rich Miras. To this end we have made CO observations of a sample of C- and O-rich Miras with periods below 400 days.

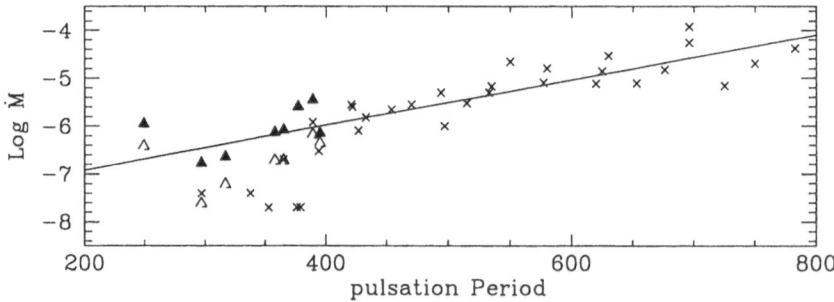

Fig. 1. $\dot{M} - P$ relation for carbon Miras derived from modeling the dust shells (crosses). The line is a fit to the crosses wich represent periods longer than 400 days. The new CO data is indicated (filled $\triangle$ = 2-1, open $\triangle$ = 1-0 transition)

2 Observations

Observations were performed in Dec. 1994 (IRAM, CO J=1-0, 2-1, observers MG & JB), Jan. 1995 (IRAM, J=1-0, 2-1, 3-2, by MG), Aug. 1995 (IRAM, J=1-0, 2-1, by EJ) and Aug. 1995 (JCMT, J= 2-1, 3-2, partly ^{13}CO as well, by FB & RT). Among carbon Miras, 19 were observed and 10 detected; among O-rich Miras, 24 were observed and 19 detected. Many are new detections. The different detection rates are simply due to the available numbers in the General Catalog of Variable Stars (GCVS). Among carbon stars, ALL stars north of δ = $-30°$ (and P < 400 days) were observed, while there are so many such O-rich Miras in the GCVS that we have picked the brightest in IRAS, to increase the detection probability.

Two things are remarkable. First, some of the stars have very small expansion velocities, down to 6 km s^{-1}. Often these stars also have the lowest MLRs. This could indicate that radiation pressure on dust in stars with short periods (= low luminosities) can not drive the outflow to the typical outflow velocities of 10-15 km s^{-1} observed in most AGB stars. Second, the ratio between the 2-1 and 1-0 peak temperatures are sometimes very large. In 'normal' AGB stars with 'high' MLRs and expansion velocities near 10-15 km s^{-1} this ratio is about 2. In the present sample, we find ratios which are almost always larger than 2, even up to 10. This could indicate that the excitation of CO may be very different. Possibly, radiative excitation in these very optically thin dust shells is much more important than collisional excitation. An example of a star with such a high ratio and low expansion velocity is the 356 day period O-Mira R Ser (Fig. 2).

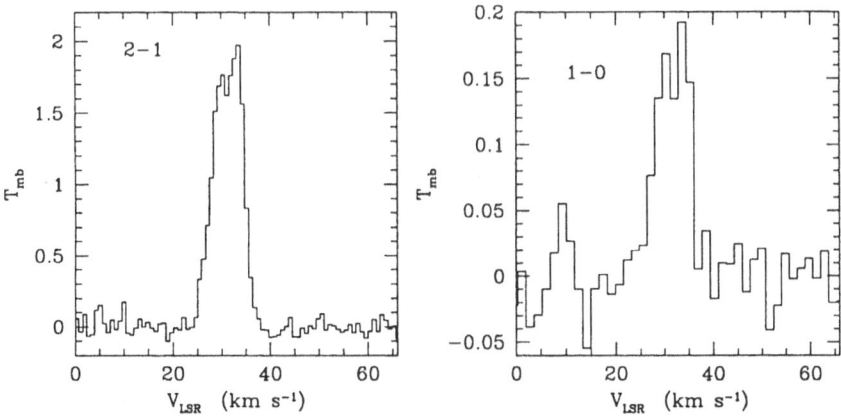

Fig. 2. R Ser

3 Analysis

Luminosities are derived from the period-luminosity relation of Feast et al. (1989) for O-rich Miras, and a newly derived one (Groenewegen, MNRAS submitted) for C-rich Miras. Distances are estimated from the IRAS 12 and 25 μm flux-densities using a pre-calculated grid of dust models, basically giving the bolometric correction at 25 μm as a function of S_{25}/S_{12}. In the future it is planned to model the SEDs of individual stars to accurately determine the dust MLR and the distance.

The (gas) MLRs are calculated from the standard Knapp & Morris (1985; KM85) formula (also see Van der Veen & Oloffson 1990) relating MLR to peak temperature, distance, expansion velocity, beam width and transition. Especially in view of the a-normal line ratios it could very well be that the simple KM85 formula does not apply. The MLRs based on the 2-1 line are systematically higher than those based on the 1-0 line. In the future, it is planned to model the CO line profiles in detail using the model of Groenewegen (1994). For a discussion on the applicability of the KM85 formula when there is strong chromospheric or interstellar UV radiation, see Josselin et al. (1996).

The lowest MLR among the C-stars is 1.6 10^{-7} $M_\odot$ yr^{-1} in V Oph which has the lowest period as well, and an expansion velocity of 7.0 km s^{-1}. Among O-stars the "record holder" is R Aqr with a MLR of 1.5 10^{-8} $M_\odot$ yr^{-1}.

Final conclusions regarding the MLR must await detailed modeling. However, the first impression of the new CO data for carbon stars with periods below 400 days is that there is better agreement with the extension of the $\dot{M}$-P relation than with the dust MLRs (see Fig. 1). For the moment we list the preliminary results in Table 1, where the name, period, expansion velocity and the (uncertain) mass loss rate based on the 2-1 and 1-0 line are given. Observed but not detected are the carbon Miras UV Aur, R CMi, V390 Cas, PT Cas, CN Per, GY Per, VX Gem, AO Lac, LS Cas, and the oxygen-rich Miras BD Eri, RS Eri, T Lep, R Peg, S Scl.

References

Feast M.W., Glass I.S., Whitelock P.A., Catchpole R.M., 1989, MNRAS 241, 375

Groenewegen M.A.T., 1993, Ph.D. thesis, Chapter 5, University of Amsterdam

Groenewegen M.A.T., 1994, A&A 290, 531

Groenewegen M.A.T., 1995a, in: "Circumstellar matter", eds. G.D. Watt & P.M. Williams, p. 321

Groenewegen M.A.T., 1995b, in: "Astrophysical applications of stellar pulsation", eds. R.S. Stobie & P.A. Whitelock, p. 141

Habing H.J., Tignon J., Tielens A.G.G.M., 1994, A&A 286, 523

Josselin E., Loup C., Omont A., Barnbaum C., Nyman L.A., 1996, A&A Letters, submitted

Knapp G.R., Morris M., 1985, ApJ 292, 640 (KM85)

Van der Veen W.E.C.J., Olofsson H., 1990, in: "From Miras to Planetary Nebulae", eds. M.O. Mennessier, A. Omont, editions Frontieres, Gif-sur-Yvette, p. 139

Table 1. Results

Name	Period (days)	v_{exp} (km s^{-1})	$\dot{M}$(2-1) (10^{-7} M$_\odot$ yr^{-1})	$\dot{M}$(1-0) (10^{-7} M$_\odot$ yr^{-1})
		carbon Miras		
Y Per	249	7.2	11	3.8
V Oph	297	7.0	1.6	0.24
ZZ Gem	317	6.8	2.3	0.61
R Cap	345	10.5	76.	–
V Aur	353	20(:)	15.	<6
V Crb	358	8.2	7.3	1.9
R Pyx	365	8.5	8.2	1.9
R Ori	377	10(:)	25.	<8
R For	389	16.9	35.	6.6
AX Cep	395	13.2	7.1	4.4
		oxygen Miras		
R Cet	166	9.0	14.	–
Z Cyg	264	7.5	41.	–
X Hya	301	6.8	2.2	–
R Leo	310	7.3	0.49	0.05
R Tau	321	6.0	11.	–
TU Peg	322	9.0	15.	–
RT Aqr	327	6.0	25.	–
X Oph	329	6.3	1.2	–
W Peg	345	11.0	19.	–
RU Cap	347	7.5	48.	–
RS Vir	354	6.2	57.	4.6
R Ser	356	6.4	7.9	0.56
U Ori	368	7.5	1.5	0.10
RT Eri	372	8.5	5.5	–
R LMi	372	8.0	6.3	0.74
W Eri	377	10.0	23.	–
R Peg	378	6.3	4.7	0.44
R Aqr	387	16.7	0.15	–
BQ Cyg	420	7.5	50.	–

Planetary Nebulae

Pierre Cox[1,2]

[1] Institut d'Astrophysique Spatiale, Bât. 121
   Université de Paris XI, 91405 Orsay Cedex, France
[2] Institut d'Astrophysique de Paris
   92b. bd Arago, 75014 Paris, France

Abstract. We describe recent results obtained at high spatial resolution on post-AGB sources and planetary nebulae which trace the molecular envelopes of these evolved stars of a few $M_\odot$. Possible directions in the study of these molecular envelopes are sketched based on the sensitivity and spatial resolution expected from the future Large Southern Array.

1 Introduction

Recent observations have demonstrated that molecular gas is an important component of the planetary nebulae (PNe). Surveys have shown that the near-infrared 1−0 S(1) line of H_2 (Kastner et al. 1996) and the millimeter lines of CO (Huggins et al. 1996) are readily detected in a large number (about 40) of PNe, OH maser emission has been found in a dozen of PNe (Zijlstra et al. 1989), and several other species (CN, HCN, HCO^+) have been detected and mapped in PNe including evolved ones (Cox et al. 1992, Bachiller et al. 1996). The molecular observations reveal that many PNe are surrounded by massive envelopes of neutral gas, and they provide a wealth of information on the physical state and chemistry of the gas, as well as on the overall kinematics of the envelope. The presence of the neutral envelopes in PNe firmly establishes continuity with mass-loss during the precursor AGB and post-AGB phases, and provides a framework for understanding the formation and evolution of the ionized nebulae (Huggins 1993).

Observations done at high spatial resolution using near-infrared facilities or millimeter arrays allowed significant progress to be made in the study of the small scale structure of the envelopes around PNe. Rather than being homogeneous, PNe envelopes consist of a collection of numerous, dense and small globules expanding together with the ionized gas in which they are immersed. In this paper we review some recent advances in our understanding of the morphology, physics and chemistry characterizing the molecular envelopes of post-AGB objects and PNe. The aim is to illustrate the fact that PNe are potential targets for further investigations with the planned millimeter and/or submillimeter instrumentation at higher spatial resolution and increased sensitivity, and in particular with the Large Southern Array.

2   Results of CO and H_2 surveys

Huggins et al. (1996) reported a comprehensive survey of the millimeter CO emission in 91 planetary nebulae using the IRAM 30 m and SEST 15 m telescopes. These observations provide new detections and improved data for 23 nebulae in the CO(2−1) line, and sensitive limits for those not seen in CO. Including all these detections with additional detections reported in the literature (see Huggins et al. 1996 and references therein) 44 PNe have now been detected in the CO(2−1) line as compared to 80 PNe for which upper limits or non detections are reported. The PNe which are detected are characterized by massive (10^{-2} to a few $M_{\odot}$) envelopes of molecular gas. These nebulae typically have abundance ratios of N/O $\gtrsim$ 0.3 and bipolar morphologies indicative of a young disk population. Similar findings have been reported by Kastner et al. (1996): from the imaging survey of the 2.12 μm H_2 emission from $\sim$ 60 PNe, they concluded that all the PNe which are detected also possess bipolar morphology. The column density through the envelopes and their mass relative to the mass of ionized gas show dramatic decreases with increasing nebular size, documenting the expansion of the envelopes and the growth of the optical nebulae at the expense of the molecular gas. The molecular envelopes remain a major mass component in the objects until the nebulae reach a radius of about 0.1 pc. The nebulae not detected in CO have little or no molecular gas ($10^{-2} - 10^{-3} M_{\odot}$), and their envelopes must be rapidly photo-dissociated before or during the compact phase.

3   Chemistry

To survive the harsh conditions (high temperatures and shock velocities) characterizing the transition from the AGB to the PNe phase, the neutral envelopes of PNe must have peculiar properties over the dynamical lifetime of the nebulae (about 10,000 years). From detailed mapping results, it is now known that clumpiness is a fundamental property of the PNe envelopes and that the densities in the inner most exposed regions are high ($10^5 - 10^6 \, cm^{-3}$). The neutral envelopes are thus extreme cases of dense and warm photon-dominated regions (PDRs)

Besides the physical conditions which are found to be extreme, recent studies of PNe envelopes have demonstrated that the chemical composition of its molecular gas is radically different from that in the interstellar medium and the circumstellar envelopes around AGB stars (see M. Guélin, this volume). The chemistry starts to be dominated by ion-molecular reactions, species such as CN, HCN, HCO^+ increase dramatically (Cox et al. 1992, Bachiller et al. 1996) and molecular ions such as CO^+ (Latter et al. 1994) and N_2H^+ (Cox et al. 1993) have been detected in the case of NGC 7027. Such molecular species clearly point to an on-going chemistry which is likely driven by photodissociation, ion-molecule reactions and shocks (see e.g., Sternberg & Dalgarno 1995).

From the recent study by Bachiller et al. (1996) the following trends have been observed for carbon-rich objects: the abundances of CN, HCN, and HCO^+

first increase as the AGB star evolved to a proto-PN, and then to a bona fide PN; the chemical composition of the dense clumps in the PN envelope is then fixed at the beginning of the PN phase and will not change much during the dynamical lifetime as long as molecular gas is present. In evolved objects, CN is an order of magnitude more abundant than HCN, HNC, and HCO$^+$, which have all similar abundances: on average, CN/HCN = 9.0, HNC/HCN = 0.45, and HCO$^+$/HCN = 0.54. These ratios are respectively one, two, and three orders of magnitude higher than in the prototypical AGB envelope IRC+10 216. Besides chemical abundances, the observation of molecular species other than CO provides useful constraints on the physical conditions inside the envelopes of PNe. Most notably, Bachiller et al. (1996) find that the volume density of the molecular clumps does not change significantly with nebular age. Dense clumps can thus survive for much of the nebular evolution with essentially constant density. Clumpines is thus an essential ingredient of the PN envelopes that determines their physical and chemical evolution throughout their lifetime.

4 Clumps: the Helix nebula

The best case study of clumpiness inside the envelope of a PN is the Helix nebula one of the most familiar planetary nebulae, often considered as the prototype of evolved PNe. At a distance of 130 pc, it is the closest PN (in fact the closest PDR) and is far larger in angular size than any other nebula. The Helix nebula thus affords the opportunity to explore its large and small scale stucture in great detail. The Helix nebula contains at least 0.03 M$_\odot$ of molecular gas, or 25% of the total nebular mass (Huggins & Healy 1986). The molecular gas forms a large incomplete shell structure which breaks down in numerous sub-structures unresolved with the 10$''$ beam of the 30 m telescope (Forveille & Huggins 1991). Subsequent observations (Huggins et al. 1992) have shown that the molecular gas has managed to survive in the cometary globules which are embedded within the ionized cavity. These small ($\sim$ 0.001 pc) globules consist of dense condensations of molecular gas with typical masses ($\gtrsim 5\,10^{-6}$ M$_\odot$) and densities ($\gtrsim 10^5$cm^{-3}). Similar numbers have recently been derived for these globules from the spectacular *HST* imaging data published by O'Dell & Handron (1996).

5 High spatial resolution near-infrared observations

Observations at near-infrared wavelengths, which give access to both the atomic and the molecular hydrogen ro-vibrational lines, have recently proven to be remarkably successfull in probing the morphology and the physical conditions in the fast evolving environments of proto-PNe and young PNe. Such observations are complementary to the millimeter measurements described above. In particular, a new instrument called BEAR coupling a 256x256 pixel NICMOS camera to the Fourier Transform Spectrometer (FTS) has now been used extensively at the 3.60 m CFHT at Mauna Kea (Hawaii) to explore the most obscured inner regions of transition objects at both high spatial ($0.4''$) and spectral ($\sim 10 - 20\mathrm{kms}^{-1}$) resolution. A detailed description of BEAR can be found in Maillard (1995). We will briefly describe the morphology of two small ($\lesssim 10''$) prototypical sources (a proto-PN, AFGL 2688 and a young PN, NGC 7027) which illustrates the gains obtained by combining high sensitivity and spatial resolution when studying molecular gas in PNe.

5.1 AFGL 2688

AFGL 2688 is one of the rare sources known to be in the rapidly evolving transition from AGB to the PN phase and has been the subject of numerous observational and theoretical investigations (see Cox et al. 1996 and references therein). It is a bright infrared source with a bipolar optical and near-infrared nebula scattering the light of the cool (supergiant F5, $T_{eff} \sim 6500\,\mathrm{K}$) central star. Photon scattering models convincingly reproduce the optical and the near-infrared appearance of AFGL 2688 with the equatorial torus lying in the east-west direction and the bipolar axis aligned north-south along the reflection nebula. Single-dish measurements in millimeter lines reveal a nearly circular, cold and dense component tracing the slow expanding AGB envelope, which is shocked by a warm, optically thin, fast moving wind. Figure 1 displays the BEAR image of the H_2 1–0 S(1) line at an unprecendented level of details. The molecular hydrogen emission shows four bright clumps forming a remarkable cross-like pattern. The four clumps are at about the same distance (5 $''$) from the central (hidden) star. No continuum emission of line emission other than H_2 is associated with these structures. The velocity field of the H_2 indicates a significant velocity dispersion in the four clumps and second, that clumps S and W are red-shifted whereas clumps N and E are more detected at blue-shifted velocities (the velocity difference is of the order of 50 km s^{-1}). This velocity field is consistent with gas outflowing in a small-opening cone along the polar regions, the bipolar axis being inclined (i $\sim 16°$) to the plane of the sky with the northern lobe towards the observer. The interpretation of the velocity dispersion in the equatorial plane is less straightforward. Possibly, the detailed structure of the inner regions may allow the fast wind to break through near the equatorial plane in two opposite directions, forming the two main regions of interaction where the strongest H_2 emission is seen. Both from the morphology (bow-like) and the H_2 line ratios the molecular hydrogen in AFGL 2688 traces the outer, shocked regions of the tilted expanding, dense equatorial torus. AFGL 2688 is one of the very few regions wherein the onset of the fast winds and their interaction with the slowly expanding envelope expelled during the the AGB phase can be studied in great detail. Future millimeter observations at high spatial resolution are needed to further investigate the relation in the equatorial plane between the dense, central torus and the high-velocity molecular gas traced in H_2.

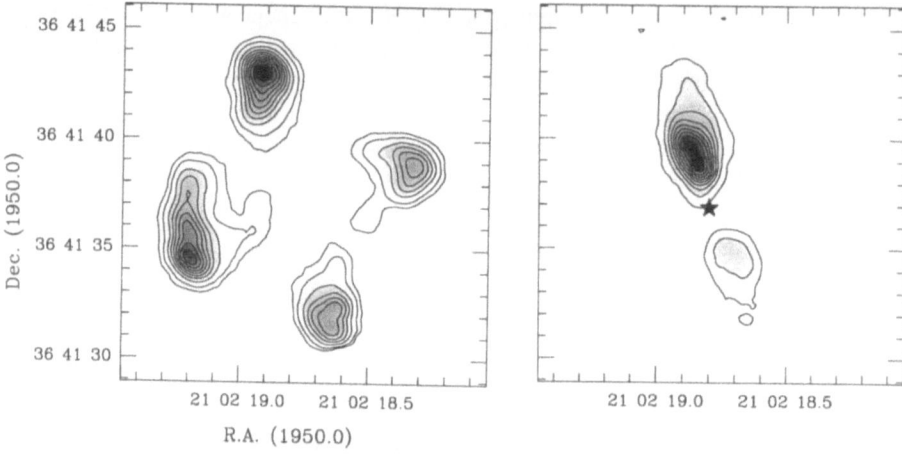

Fig. 1. Image of AFGL 2688 in the 1–0 S(1) transition of H_2 (left panel) and in the underlying continuum at 2.12 μm (right)

5.2 NGC 7027

NGC 7027 is a key object in the study of PNe because it is relatively young, with strong line and continuum emission. The dense inner regions of the extended circumstellar envelope have been ionized by the intense radiation field of the hot central star (T $\sim$ 200,000 K), whereas farther out the envelope is still molecular. The general properties of the neutral envelope have been investigated in detail by many authors and the inner parts of NGC 7027 including the ionized region and its immediate periphery abutting the molecular envelope have been studied at arcsec and subarcsec resolutions in the near-infrared and in the radio (e.g., Cox et al. 1996 and references therein).

Images of NGC 7027 obtained with BEAR in the 1−0 S(1) transition of H_2 and in the continuum are displayed in Figure 2. The molecular hydrogen is distributed along a symmetrical structure showing four lobes encompassing the inner equatorial torus. The inner regions are dominated by the continuum and line emission from the ionized gas. The velocity field of the H_2 shows that at blue-shifted velocities the brightets emission is mainly seen along the southern lobe and the northern segment of the inner equatorial torus, whereas at red-shifted velocities the brightest H_2 emission is observed along the northern lobe and the southern segment of the torus. Around the systemic velocity, the H_2 emission shows an almost complete ring-like structure tracing the equatorial torus and the two lobes around the central ionized cavity (Fig. 2).

The kinematics of the H_2 emission in the outer loops match the velocity fields

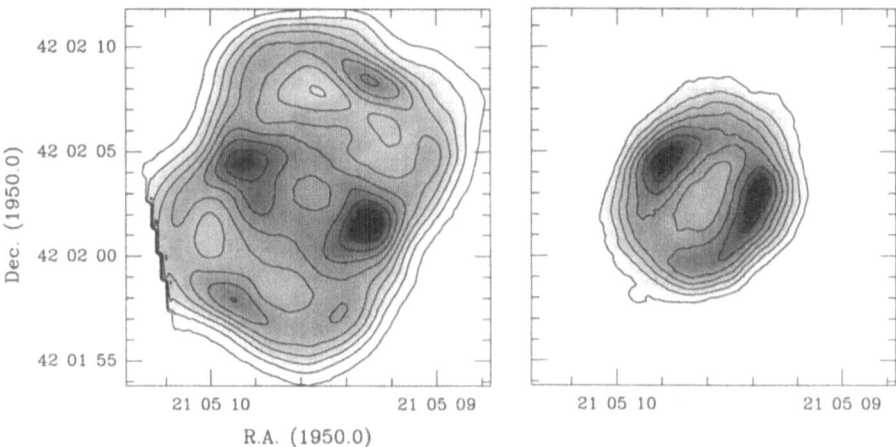

Fig. 2. Image of NGC 7027 in the 1–0 S(1) transition of H_2 (left panel) and in the underlying continuum at 2.12 μm (right)

derived from millimeter line studies in CO and HCO^+ (Graham et al. 1993, Cox et al. in preparation). The near-infrared data thus confirm the general dynamical model derived from millimeter observations where the overall kinematics are explained by an inclined thin, dense, expanding torus with the outer loops towards the poles, tacing the warm, high-density gas of the PDR, which lies along the inner surface of the molecular envelope.

6 Evolution of nebulae

The PNe originate from precursors stars in the AGB and proto-PN phases, surrounded by a massive (0.01 - a few $M_\odot$) envelope of molecular gas. First, the fast winds emitted by the central star compresss, fragment and accelerate the molecular material. Then the central star rapidly evolves across the H-R diagram to higher temperatures, it becomes hot enough to begin ionizing the core of the envelope so that the latter does not change rapidly. However, as the envelope expands, the radial CO column density decreases and the dissociating and ionizing UV radiation consumes more and more of the molecular gas. The ionized mass continues to grow, and becomes comparable to the mass of molecular gas at a radius of 0.1 - 0.2 pc, and the major mass component at still larger radii. The mass of the ionized nebula will level off when the bulk of the neutral gas in the vicinity of the star has been ionized, although given the complex geometry of these PNe, the transition from an ionization-bounded to a density bounded nebula is a not a simple matter.

The high spatial resolution measurements of now many PNe either in the millimeter or in the near-infrared add important details to the above picture. In the earliest stages the molecular envelopes are typically extended like those of the precursor AGB envelopes (though possibly with a good deal of structure - see M. Guélin in this volume), but they rapidly become etched by the penetration of ionized radiation into the least dense gas, leaving the density structures to survive the longest. These structures are typicaly partial shells around elliptical PNe or equatorial rings around waists of bow-tie PNe (e.g., NGC 7027), and are probably composed of smaller condensations like those seen in the Helix nebula. The localized, vestigial molecular gas found in the largest PNe resides in the final stages of these structures.

The destruction of the molecular envelopes in these PNe is expected to proceed in PDRs which are formed by the penetration of stellar radiation beyond the ionization fronts into the neutral gas. In comparison to the interstellar medium which is relatively well understood, the theoretical understanding of the physical conditions and evolution of PDRs associated with PNe is still in the making (Tielens 1993, Hollenbach & Natta 1996, Bertoldi & Draine 1996). One feature expected of these interface regions is emission in the near-infrared lines of H_2, due to either fluorescence or thermal excitation in the warm gas. The results outlined above and the recent paper by Kastner et al. (1996) strongly support this picture of the evolution of PNe. It should be noted that the warm H_2 is only a minor constituent, consistent with surface emission from PDRs (such as in NGC 7027), or possibly from localized shocks (e.g., AFGL 2688). A second important feature of the PDRs is that the stellar radiation field causes a transition from CO to its constituents atoms in the form of O I, C I, and C II. Depending on the physical conditions and the degree of evolution, the PDR can act as important reservoirs of neutral gas. For instance, in the case of the Ring nebula Bachiller et al. (1994) found that the C I mass exceeds the molecular mass by a factor of about 10. One might expect that the relative importance of the PDR gas increases as the envelopes expand and fragment because of the increase in exposed surface area (see Dinerstein et al. 1995). Clearly the data now collected by the Infrared Space Observatory are essential for the study of PDRs in PNe although ground-based studies of the submillimeter lines of C I will be very useful in this respect as well.

7 Conclusions

The above results demonstrate that the recent increase in sensitivity and spatial resolution both in the millimeter and near-infrared have been essential in the study of the physical and chemical properties of the molecular envelopes of PNe throughout their evolution. In particular, the gain in angular resolution has proven to be a key factor in our understanding of the inhomogeneous morphology of these envelopes and the actual sensitivities achieved with single-dish and interferometric instuments such as the IRAM instruments have been essential in probing for the first time the chemical conditions. It is clear that the prospects offered by the Large Southern Array with its vast collecting area, expected improvement in angular resolution and sensitivity of about an order of magnitude over the present instruments will provide a major advance in the study of the latest stages of evolution of stars of a few solar masses. It is also clear that such an instrument will allow definite investigations of the changes occuring in the morphological, physical and chemical conditions and their links with the rapid and still poorly understood evolution from the AGB to the PN phase.

It is a great pleasure to thank colleagues Drs. P.J. Huggins, R. Bachiller, T. Forveille, S. Guilloteau and J.P. Maillard: the results of our collaborative programs form the basis of this paper.

References

Bachiller, R., Forveille, T., Huggins, P.J., Cox, P., Omont, A. (1996): AA, submitted

Bachiller, R., Huggins, P.J., Cox, P., Forveille, T. (1994): AA 281, L93

Bertoldi, F., Draine, B.T. (1996): ApJ 458, 222

Cox, P., Omont, A., Huggins, P.J., Bachiller, R., Forveille, T. (1992): AA 266, 420

Cox, P. et al. (1993): in Planetary Nebulae, Weinberger R. & Acker A. (eds.) Kluwer, Dordrecht, p. 227

Cox, P., Maillard, J.-P., Huggins, P.J., Forveille, T. et al. (1996): AA, in press

Dinerstein, H., Haas, M., Erickson, E.F., Werner, M.W. (1995): in Airborne Astronomy Symposium on the Galactic Ecosystem M.R. Haas, J.A. Davidson, and E.F. Erickson (eds.) ASP Conference Series, vol. 73, p. 387

Forveille, T., Huggins, P.J. (1991): AA 248, 599

Graham, J.R. et al. (1993): ApJ 408, L105

Huggins, P.J., Healy, A.P. (1986): ApJ 305 L29

Huggins, P.J., Bachiller, R., Cox, P., Forveille, T. (1992): ApJ 401, L43

Huggins, P.J. (1993): in Planetary Nebulae, Weinberger R. & Acker A. (eds.) Kluwer, Dordrecht, p. 147

Huggins, P.J., Bachiller, R., Cox, P., Forveille, T. (1996): AA, in press

Hollenbach, D.J., Natta, A. (1996): ApJ 455, 133

Kastner, J.H., Weintraub, D.A., Gatley, I., Merrill, K.M., Probst, R.G. (1996): ApJ 462, 777

Latter, W.B., Walker, C.K., Maloney, P.R. (1994): ApJ 419, L97

Maillard, J.P. (1995): in 3-D Opical Spectroscopy methods in astronomy IAU Colloq. n. 149, Comte G. & Marcellin M. (eds) ASP Conf. Series 71, p. 316

O'Dell, C.R., Handron, K.D. (1996): AJ 111, 1630

Sternberg, A., Dalgarno, A. (1995): ApJSS 99, 565

Tielens, A.G.G.M. (1993): in Planetary Nebulae, Weinberger R. & Acker A. (eds.) Kluwer, Dordrecht, p. 155

Zijlstra, A.A., Pottasch, S.R., Bignell, C. (1989): AAS 79, 329

Supernovae with the Large Southern Array

Claes Fransson

Stockholm Observatory, S - 133 36 Saltsjöbaden, Sweden

Abstract. The relevance of the LSA for radio observations of supernovae and supernova remnants are discussed. The possibility of observing supernovae at mm wavelengths offers great advantages in terms of early detection and better spectral coverage. The increased resolution for especially LMC remnants can provide us with imaging at the same resolution as galactic remnants today. Finally, the possibility of detecting the synchrotron nebula around a putative pulsar in SN 1987A, as well as the collision of the ring and the ejecta are discussed.

1 Introduction

Observations of supernovae in the radio domain are important for understanding both the objects themselves, and physical processes of interest for a larger class of objects. Usually one distinguishes between relatively young supernovae and supernova remnants. As a somewhat arbitrary definition we define radio-supernovae as supernovae younger than 100 years, observable at radio wavelengths, and supernova remnants as remnants older than this.

A list of areas where radio observations can improve our understanding includes:

- Circumstellar environment
- Progenitor evolution, mass loss history
- Shock structure. Collisionless processes
- Particle acceleration in a time-dependent environment
- Transition from circumstellar to interstellar interaction
- Synchrotron nebulae around newborn pulsars
- Radio supernovae as distance indicators.

In this review we expand on these points, and first discuss the relevance of the LSA to radio supernovae, then some comments on remnants, and finally we discuss some possible applications for SN 1987A.

2 Young Radio Supernovae

2.1 Background

With the advent of the VLA radio supernovae are now regularly observed, and a large number of objects have been monitored with varying degree of coverage in time and wavelength. A review of the current observational situation can be found in Weiler *et. al.* (1996), while theoretical models are discussed in Chevalier (1990) and Fransson (1994). In general, for well observed supernovae the radio emission can be characterized as a gradual rise of the emission on a time scale of days to months up to a constant or slowly decreasing level. This turn-on occurs first at short wavelengths and later at longer wavelengths (Fig. 1.). After the initial phase the spectrum is a non-thermal power law, with spectral index $\alpha = 0.7 - 1.1$, showing that the emission is basically synchrotron radiation.

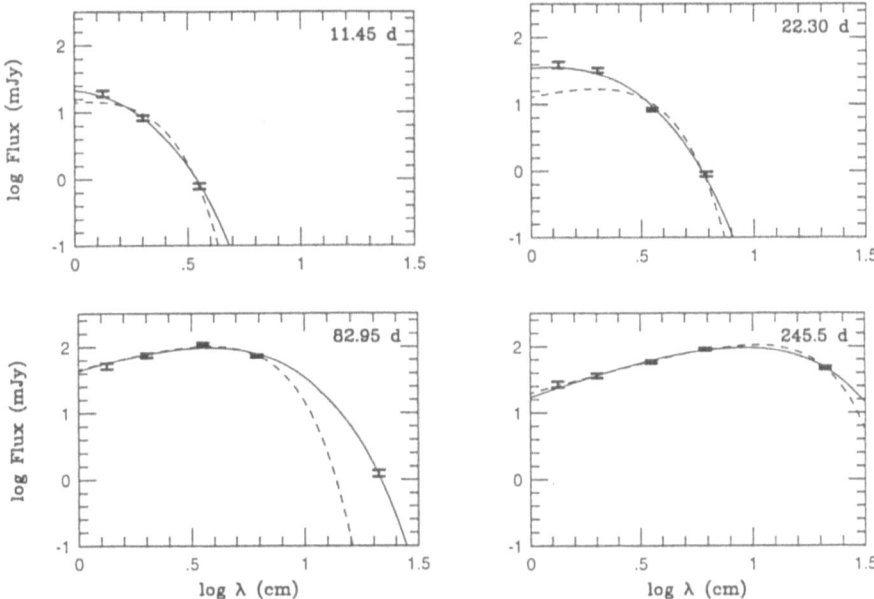

Fig. 1. VLA spectra of SN 1993J at four different epochs from Van Dyk *et. al.* (1994). Note the progression of the cut-off to longer wavelengths. At 245 days the spectrum is close to the unsupressed power-law synchrotron spectrum. Observations at mm wavelengths would show this at a much earlier epoch. The solid line shows a fit to the spectrum based on the Razin effect, while the dashed line gives a similar fit with external free-free absorption.

Compared to the strongest galactic remnant the radio luminosities for 'normal' Type II's and Ib/c's are in the range $10 - 300$ times Cas Á. The most extreme examples are the narrow line Type II's SN 1986J and SN 1988Z, with luminosities of ~ 2000 times Cas A (Rupen *et. al.* 1987, Weiler, Panagia, & Sramek 1990, Van Dyk *et. al.* 1993). These two supernovae are most likely rare objects, visible at large distances only due to their high radio luminosities. The high flux may persist for several decades. There is also a strong correlation with X-ray emission and broad optical and UV emission lines. These are all a result of the circumstellar interaction, and provide together with the radio observations a consistent picture of the interaction (see Fransson 1994). Up to date, supernovae of Types Ib, Ic, and II, but *not* Type Ia's, have been observed as radio emitters. Even upper limits are, however, of interest for the understanding of the Type Ia progenitors (Boffi & Branch 1995, Eck *et. al.* 1995, Cumming *et. al.* 1996).

Based on these observations a standard model has emerged, first proposed by Chevalier (1982), where the emission arises as a result of the interaction of the supernova with its circumstellar medium. Most likely the latter is the result of mass loss, either in the form of a stellar wind, or, in a binary system, as Roche lobe overflow. The interaction gives rise to one shock wave with velocity $10,000 - 20,000$ km s^{-1}, propagating into the circumstellar medium, and one slow shock with velocity $500 - 1,000$ km s^{-1}, propagating into the ejecta. Electrons are either injected close to the circumstellar shock, and are there accelerated across the shock by first order Fermi acceleration, or in the turbulent magnetic field close to the Rayleigh-Taylor unstable contact discontinuity between the circumstellar and reverse shocks. The radio emission arises as a result of synchrotron emission in the compressed magnetic field.

The wavelength-dependent rise of the radio emission is explained in the original model as a result of free-free absorption by the external circumstellar medium. Assuming a fully ionized wind with constant mass loss rate and velocity, $\rho = \dot{M}/4\pi u_w r^2$, the free-free optical depth at wavelength λ is

$$\tau_{ff}(\lambda) \approx 7.1 \times 10^2 \lambda^2 \left(\frac{\dot{M}_{-5}}{u_{w1}}\right)^2 T_5^{-3/2} V_4^{-3} t_{days}^{-3} \tag{1}$$

where $\dot{M}_{-5}$ is the mass loss rate in units of 10^{-5} M$_\odot$ yr^{-1}, u_{w1} the wind velocity in units of 10 km s^{-1}, V_4 the maximum expansion velocity in 10^4 km s^{-1}, and T_5 the temperature of the circumstellar gas in 10^5 K. The expansion velocity can be obtained from the line widths in the optical spectrum. However, the heating of the gas by the radiation from the supernova has to be estimated from models of the circumstellar interaction, making the temperature in the gas somewhat uncertain. Lundqvist & Fransson (1988) find that initially the radiation heats the gas to $10^5 - 10^6$ K. The temperature then decreases with time, and after a year $T_e \approx (1.5 - 3) \times 10^4$ K. In addition, the medium may recombine, which further decreases the free-free absorption. From $t(\tau(\lambda)_{ff}) = 1$ the ratio of $\dot{M}/u_w$ can be calculated. A caution is that clumping of the medium may increase the absorption and lead to an overestimate of $\dot{M}/u_w$.

The mass loss rates for 'normal' Type II supernovae obtained in this way are between 3×10^{-5} and 2×10^{-4} M$_\odot$ yr^{-1}. These are at the high end of those determined for red supergiants, but given their sensitivity to the temperature and clumping of the medium, it is premature to draw any strong conclusions from this.

The effects of the circumstellar medium are not only observable in the radio. The interaction also gives rise to a very large X-ray flux from both the reverse and circumstellar shocks. This X-ray flux may be directly observable, as for SN 1993J, or absorbed by the ejecta and cool gas between the two shocks. This absorbed fraction is there re-emitted as optical and UV radiation. These aspects are discussed more fully in Chevalier & Fransson (1994) and Fransson, Lundqvist, & Chevalier (1996).

While 'classical' type II supernovae, like SN 1979C and SN 1980K, are well modeled with external free-free absorption (e.g., Chevalier 1984, Weiler *et. al.* 1986, Lundqvist & Fransson 1988), the best observed radio supernova, SN 1993J, did not fit into this picture (van Dyk *et. al.* 1994, Fransson *et. al.* 1996). As was first proposed by Lundqvist (1994), a possible way out could be to modify the density law from the standard $\rho \propto r^{-2}$ to a flatter distribution like $\rho \propto r^{-1.5}$ (see also van Dyk *et. al.* 1994). There is, however, no obvious explanation for this, although variable or non-sherically symmetric mass loss have been suggested. In addition to the problematic fits of the light curves, neither did the spectrum agree with that of free-free absorption, $F_\nu \propto \lambda^\alpha \exp(-\tau_{ff})$, where $\tau_{ff} \propto \lambda^2$. Especially at long wavelengths the cutoff was too steep.

An alternative view of the spectrum formation has recently been proposed by Fransson (1996). In the case of SN 1993J it is found that an essentially perfect fit can be obtained with a synchrotron spectrum, suppressed by the Razin effect (e.g., Pacholczyk 1970). The Razin effect arises in a synchrotron emitting plasma of finite density, changing the refractive index to less than one. At frequencies less than $\gamma \nu_p$, where ν_p is the plasma frequency, the effective Lorentz factor is decreased. The beaming of the radiation will then be less pronounced, and as a result the electron will radiate less efficiently. In the Razin case the spectrum is given by $F_\nu \propto \lambda^\alpha \exp(-\lambda/\lambda_R)$, where $\lambda_R = 20\,B/n_e$ cm. The spectrum therefore less steep at long wavelengths, compared to the free-free case (Fig. 1).

For each epoch λ_R can be determined, and one finds that $\lambda_R \propto t^{0.84}$. With $n_e \propto r^{-2}$ and $r \propto t^m$, where $m \approx 0.9$ (Chevalier 1982), one finds for SN 1993J that $B \propto r^{-1}$, as expected for a 'Parker-type' spiral magnetic field in the wind (Weber & Davies 1967). X-ray observations allow a determination of the mass loss rate to $\dot{M} \approx 5 \times 10^{-5}$ M$_\odot$ yr^{-1} (Fransson *et. al.* 1996), and consequently the absolute value of the magnetic field in the emitting region can be determined. Finally, the normalization of each radio spectrum gives the number of relativistic electrons as function of time, and therefore also the injection rate of relativistic particles.

Both the injection mechanism and the acceleration process are largely unknown. An increased understanding of these are therefore very important, both for understanding radio supernovae, but perhaps most important, for a general understanding of particle acceleration. The fact that one is here studying a time-dependent process means that, in contrast to e.g. older supernova remnants and extragalactic jets, one has considerably more constraints. Of particular interest for the acceleration process is the work by Achterberg & Ball (1994) on quasi-perpendicular shock acceleration in SN 1978K and on the radio lightcurve of SN 1986 J (Ball & Kirk 1995). This is further discussed in connection with SN 1987A in §4.1.

This discussion shows that in both the external free-free absorption case and in the Razin case important information about both the circumstellar medium and the particle acceleration can be obtained.

2.2 The Contribution of the LSA

With the VLA observations between $1 - 90$ cm can be obtained. The possibility of observing radio supernovae at shorter wavelengths is, as the above discussion shows, important for several reasons. Both in the case of free-free absorption and the Razin effect the radio emission first becomes observable at short wavelengths. In the free-free case $t(\tau_{ff} = 1) \propto \lambda^{2/3}$, and in the Razin case $t(\lambda = \lambda_R) \propto \lambda$. Observations at mm wavelengths therefore makes it possible to observe the radio supernova sooner after explosion by a factor of about ten compared to cm observations. A fexible scheduling making a prompt reaction possible is cruical. For example, SN 1979C was only detected after about one year, so this can have dramatic effects. Constraining the spectrum at shorter wavelengths is also valuable, since this helps defining the spectral index better. In addition, the large collecting area makes it possible to observe supernovae as distant as $\gtrsim 30$ Mpc with the same or better quality as has been possible for SN 1993J with IRAM and OWRI.

Direct imaging of radio supernovae is hardly realistic, because of the small angular diameter, corresponding to only

$$\theta \approx \frac{2Vt}{D} \approx 1 \times 10^{-3} \left(\frac{t}{1 \text{ year}} \right) \left(\frac{D}{10 \text{ Mpc}} \right)^{-1} \text{ milliarcsec.} \qquad (2)$$

VLBI technique is therefore necessary, as demonstrated very nicely for SN 1993J by Marcaide *et. al.* (1995). However, as a component in a VLBI network the LSA could be very valuable.

3 Supernova Remnants

The increased spatial resolution of the LSA gives new possibilities to study the fine structure close to the shock and the time evolution of the radio emission. Both these aspects are important for understanding the acceleration process at the shock. Achterberg, Blandford, & Reynolds (1994) discuss limits on the diffusion length of the electrons from the spatial correlation of the X-ray and radio emission at the shock. While the radio resolution will be increased by a factor of about ten, improved modelling will, unfortunately, probably be limited by the resolution of the X-ray observations.

Observations of temporal changes in small scale structures are useful for understanding the acceleration of the particles. Examples of this are the observations of Cas A by Arendt & Dickel (1987), showing changes of compact knots with size of ~ 0.3" on a time scale as short as ~ 14 years. This decay rate, $\sim 7\%$ yr^{-1}, is much faster than that of the remnant as a whole, $\sim 0.76\%$ yr^{-1}. Similar studies for southern remnants, like Pup A (Dubner *et. al.* 1991), with the LSA would be of great interest.

A set of supernova remnants where the LSA can advance our understanding considerably are the remnants in the LMC and SMC (e.g., Dickel & Milne 1994). The known distances and moderate reddening in the optical and UV make these objects of special interest. The diameters of these remnants are between $\sim 20 - 200$". Especially for the smaller of these, it is obvious that the increased resolution possible with the LSA is very important.

One of the most interesting objects in this respect is SNR 0540-693, often referred to as the 'Twin of the Crab'. This remnant is estimated to be only 762 ± 50 years old (Kirshner *et. al.* 1989), and has central pulsar with a period of 50.3 ms. The total luminosity and energy loss are a factor ~ 3 less than the Crab. Optically, this remnant is seen as a roughly spherical bubble, emitting both lines and a synchrotron continuum (Kirshner *et. al.* 1989, Caraveo *et. al.* 1992), with the highest velocities corresponding to a velocity of ~ 2735 km s^{-1}. Observations with the Australia Telescope Compact Array at 1.5 and 5 GHz show the same component, with a diameter of ~ 8", is seen, but in addition an outer shell with diameter of 65 ", or velocity $\sim 10,800$ km s^{-1}, is also observed (Manchester, Staveley-Smith, & Kesteven 1993). This contrasts to the Crab, where the absence of a high velocity shell is somewhat of a mystery. The central pulsar has a very steep spectrum, $F_\nu \propto \nu^{-2.5}$, and is unlikely to be observable at mm wavelengths even with the LSA. In addition, the 0540-693 shows considerable structure both in the radio and the optical domain. The resolution of the 5 GHz ATCA observations is ~ 2.7". With the LSA the resolution is ~ 0.1", and with the large collecting area one can obtain a map with a reasonable integration time (a few hours). The linear resolution will be similar to the best maps of the Crab nebula we have today. Proper motion studies of various features 0540-693 would e.g. be possible.

Another interesting remnant is the oxygen rich N132D.This is relatively young, $2350 - 3150$ yr (Sutherland & Dopita 1995; Morse, Winkler & Kirshner 1995), with a diameter of ~ 100". The characteristics of this are similar to Cas A in terms of kinematics with slow moving knots, mixed with fast moving, oxygen rich knots. The radio image indicates a central hole, indicating a shell geometry. A map with a spatial resolution comparable to Cas A would be of great help for the understanding of the expansion and structure.

4 SN 1987A

4.1 A Central Pulsar?

Millimeter observations of SN 1987A are of interest for both the ejecta and the circumstellar ring. Starting with the ejecta, the possibility of detecting a pulsar in the center is of obvious importance. As is demonstrated by the pulsar in SNR 0540-693, direct observations of the pulsar in the mm-domain is probably very difficult, because of the steep spectrum. SNR 0540-693, however, also shows that if a typical pulsar is present, the synchrotron nebula around it may be observable. The main problem for a detection of any plerion is the free-free absorption by the ejecta. The free electrons responsible for this are produced either as a result of the ionization of the ejecta by the radioactive decay, or as a result of the photoionization by the synchrotron radiation from the plerion.

The former has been estimated by Fransson (1987) and McCray (1993), and one finds that the ejecta may be optically thin already a few years after explosion. A severe problem for any predictions is that the ionization and free-free absorption are both sensitive to clumping.

The dynamics and ionization structure of the synchrotron nebula have been discussed by Chevalier & Fransson (1992). They found that a slowly expanding bubble of relativistic particles and magnetic fields will be formed in the center. This is separated to the ejecta by a thin, dense shell of swept up ejecta gas, which is probably dynamically unstable. In terms of velocity one estimates a radius of $\sim 490\, L_{38}^{1/5}$ km s^{-1} for the shell, where $10^{38}\, L_{38}$ is the total energy loss from the pulsar. The electron density and free-free optical depth of the ionized gas close to the bubble depends on both the degree of clumping and the chemical composition. Given these uncertainties, one can in this case estimate that if radioactivity dominates the ionization, the ejecta should already be optically thin to free-free absorption at cm wavelengths. In the case of ionization by the plerion Fransson (1987) estimates that $t(\tau_{ff}) \approx 2.5\, \lambda^{5/6}\, L_{38}^{1/4}$ years, if the ionizing luminosity is similar to the total energy loss of the pulsar. Again, this estimate is sensitive to clumping and composition. Because $\tau_{ff} \propto \lambda^2$, observations at mm wavelengths are more favorable than at longer wavelengths, in spite of a lower flux. The advantage of the LSA is here obvious.

4.2 The Circumstellar Medium and the Ring

Radio emission from the interaction of the ejecta and the circumstellar medium was observed at $0.843 - 8.41$ GHz shortly after the explosion (Turtle et. al. 1987). This decayed quickly and became unobservable after a few months. Compared to ordinary Type II supernovae the radio luminosity was very small, $\sim 5\%$ of that of Cas A. This is not surprising in view of the low circumstellar density expected outside a blue supergiant, with a wind speed of ~ 500 km s^{-1} (Chevalier & Fransson 1987).

In 1990 July (day 1200), however, a revival of the radio emission occurred (Staveley-Smith et. al. 1992, 1993, Ball et. al. 1995). During this phase the emission has increased steadily. The flux from the remnant is now ~ 7 mJy at 8.6 GHz, with a spectrum $F_\nu \propto \nu^{-1}$. Accurate astrometry shows that the center of the radio emission coincides with the center of the optical ring (Reynolds et. al. 1995). Ball & Kirk (1992) and Duffy, Ball, & Kirk (1995) have, using a simple model for the interaction, modeled the acceleration of the relativistic electrons in considerable detail. In particular, from the fact that in a first order Fermi mechanism high energy electrons take longer time to accelerate they explain the observed delay of the emission at 0.843 GHz compared to 4.8 GHz. From the acceleration time of the electrons the mean free path of the collision-less scattering, thought to be responsible for the Fermi acceleration, can be estimated. Although the observational uncertainties are large, this example shows the potential of time dependent observations for the understanding of the acceleration process.

The ejecta expands with a velocity of $(10 - 30) \times 10^3$ km s^{-1}. The exact value is sensitive to the circumstellar density around the ejecta. Earlier models for the early radio burst from SN 1987A, based on free-free absorption of the radio emission, indicated a $\dot{M} \approx 6 \times 10^{-6}$ M$_\odot$ yr^{-1} circumstellar medium (Chevalier & Fransson 1987). This would give a lower maximum ejecta velocity. This model, however, has a serious problem in that the expansion velocity of the ring, resulting from the ram-pressure of this wind, would be considerably higher than observed (Blondin & Lundqvist 1993). A recent re-evaluation by Chevalier & Dwarkadas (1995) of the observations, however, interprets the early absorption as a result of synchrotron self-absorption, as was suggested earlier by Storey & Manchester (1987). This requires a much smaller wind density, $\dot{M} \approx 7.5 \times 10^{-8}$ M$_\odot$ yr^{-1} for a wind velocity of 450 km s^{-1}. To explain the steady increase in the flux at both the radio and X-ray wavelengths they propose that the supernova is now interacting not with the shocked blue wind, as in previous models, but with the denser H II region gas of the progenitor. This increases the emission at both radio and X-ray energies, in accordance with the observations. One important consequence of this model, especially for the LSA, is that because of the dense H II region, the collision with the ring occurs considerably later than previously assumed. While the earlier estimates when this would happen was 1999 ± 3 (Luo, McCray, & Slavin 1994), Chevalier & Dwarkadas estimate 2005 ± 3.

The radio emission from the collision of the ejecta and the ring has been modeled by Luo & McCray (1991a, 1991b). They find that the flux should increase on a time scale of several years, and then reach a fairly constant, or slowly decreasing level. The rise is determined by light travel effects, so that one first observes the front side of the ring and later the back. This simple picture may, however, be complicated by several effects. First, the geometry is highly non-spherical, causing the ejecta to expand faster in the polar regions. Secondly, the turn-on could be strongly affected by any non-azimuthally symmetric density distribution inside the ring. The collision with the ring will then first take place at the point where the column density from the supernova is lowest. This may dominate the light travel effects. Finally, the estimate of the level of the radio emission is very sensitive to the assumed efficiency factor for the acceleration of the relativistic electrons. Luo & McCray (1991b) estimate that $F_\nu \approx 30 \, (f/1 \times 10^{-2})^2$ mJy at ~ 1 GHz, where the efficiency factor, f, may be between $10^{-3} - 10^{-2}$. At 99 GHz the flux will be a factor $50 - 100$ lower. From models of normal radio supernovae, Chevalier (1982) estimates $f \approx 10^{-2}$. This flux estimate is very uncertain (R. McCray, priv. comm.), and has in fact already been superseded at 8.6 GHz, and it is likely that the collision should be observable with the LSA. An interesting possibility is to study the temporal evolution of different features in the ring at different wavelengths. In the same way as discussed above, one expects a correlation of wavelength and time, allowing an estimate of the acceleration time scale. The combination of similar spatial resolution with the HST, VLT and LSA offers interesting possibilities to study this event in the whole UV to radio range.

5 Executive Summary

Thanks to its higher resolution, sensitivity and frequency range the LSA can improve the observational situation both for the understanding of radio supernovae, remnants, and the general physics of relativistic particle acceleration. For convenience we here summarize some of the main areas where the LSA can make an important contribution.

Radio supernovae

- Early detection of radio supernovae, important for probing the circumstellar medium close to the supernova
- Better defined spectrum at mm wavelengths, where free-free absorption and the Razin effect are unimportant
- Factor of $\sim 10^3$ increased detection volume at mm wavelengths compared to present day mm arrays

Supernova remnants

- Small-scale structure and temporal evolution in galactic remnants
- Resolution of LMC and SMC remnants comparable to todays galactic remnants

SN 1987A

- Most promising range for detection of the synchrotron nebula surrounding a putative pulsar
- Observation of the ejecta – ring collision with the same resolution as with HST

General

- Possibility to study the relativistic particle injection and acceleration in a time-dependent environment.

Aknowledgement. I am grateful to Robert Cumming, Peter Lundqvist and Dick McCray for comments.

References

Achterberg, A. & Ball, L. 1994, A & A, 285, 687
Achterberg, A. and Blandford, R. D. & Reynolds, S. P. 1994, A & A, 281, 220
Arendt, R.G. & Dickel, J.R. 1987, ApJ, 315, 567
Ball, L. & Kirk, J. G. 1992, ApJ, 396, L 39
Ball, L. & Kirk, J. G. 1995, A & A, 303, L 57
Ball, L., Campbell-Wilson, D., Crawford, D. F. & Turtle, A. J. 1995, ApJ, 453, 864
Blondin, F., & Lundqvist, P.. 1993, ApJ, 405, 337
Boffi, F., & Branch, D. 1995, PASP, 107, 347
Caraveo, P.A., Bignami, G.F., Mereghetti, S., & Mombelli, M. 1992, ApJ, 395, L103
Chevalier, R. A. 1982, ApJ, 259, 302

Chevalier, R. A. 1984, Ann. N. Y. Acad. Sci., 422, 215

Chevalier, R. A. 1990, in *Supernovae*, ed. A. G. Petschek, Berlin, Springer, 91

Chevalier, R. A & Dwarkadas, V. V. 1995, ApJ, 452, L45

Chevalier, R. A., & Fransson, C. 1992, ApJ, 395, 540

Chevalier, R. A., & Fransson, C. 1994, ApJ, 420, 268

Cumming, R. J., Lundqvist, P., Smith, L. J., Pettini, M., King, D. L. 1996, MNRAS, submitted

Dubner, G.M., Braun, R., Winkler, P.F., & Goss, W.M. 1991, AJ, 101, 1466

Dickel, J.R. & Milne, D.K. 1994, Proc. ASA, 11, 99

Duffy, P., Ball, L. & Kirk, J. G. 1995, ApJ, 447, 364

Eck, C. R., Cowan, J. J., Roberts, D. A., Boffi, F. R., & Branch, D. 1995, ApJ451, L53

Fransson, C. 1994, in *Circumstellar Media in the Late Stages of Stellar Evolution*, eds. R.E.S. Clegg, I.R. Stevens & W.P.S. Meikle, Cambridge University Press, Cambridge, p. 120

Fransson, C. 1996, in preparation

Fransson, C., Lundqvist, P. & Chevalier, R.A. 1996, ApJ, 461, 993

Kirshner, R. P., Morse, J. A., Winkler, P. F., & Blair, W. P. 1989, ApJ, 342, 260

Lundqvist, P. 1994, in *Circumstellar Media in the Late Stages of Stellar Evolution*, eds. R. E. S. Clegg, W. P. S. Meikle, & I. R. Stevens, CUP, Cambridge, p. 213

Lundqvist, P., & Fransson, C., 1988 A & A, 192, 221

Luo, D. & McCray, R. 1991a, ApJ, 372, 194

Luo, D. & McCray, R. 1991b, ApJ, 379, 659

Luo, D., McCray, R., & Slavin, J. 1994, ApJ, 430, 264

Manchester, R. N., Staveley-Smith, L., & Kesteven, M. J. 1993, ApJ, 411, 756

Marcaide, J. M., Alberdi, A., Ros, E., Diamond, P., Shapiro, I. I., Guirado, J. C., Jones, D. L., Krichbaum, T. P., Mantovani, F., Preston, R.A., Rius, A., Schilizzi, R. T., Trigilio, C., Whitney, A. R., & Witzel, A. 1995, Science, 270, 1475

Morse, J. A., Winkler, P. F., & Kirshner, R. P. 1995, AJ, 109, 2104

Pacholczyk, A.G. 1970, *Radio Astrophysics*, Freeman

Reynolds, J. E., *et. al.* 1995, A & A, 304, 116

Rupen, M. P., Van Gorkom, J. H., Knapp, G. R., Gunn, J. E., & Schneider, D. P. 1987, AJ, 94, 61

Staveley-Smith, L., Manchester, R. N., Kesteven, M. J., Reynolds, J. E., Tzioumis, A. K, Killeen, N. E. B., Jauncey, D. L., Campbell-Wilson, D., Crawford, D. F., & Turtle, A. J. 1992, Nature, 355, 147

Staveley -Smith, L., Briggs, D. S., Rowe, A. C. H., Manchester, R. N., & Reynolds, J. E. 1993, Nature, 366, 136

Storey, M.C. & Manchester, R.N. 1987, Nature, 329, 421

Turtle, A. J., Campbell-Wilson, D., Bunton, J. D., Jauncey, D. L., & Kesteven, M. J. 1987, Nature, 327, 38

Van Dyk, S. D., Weiler, K. W., Sramek, R. A., & Panagia, N. 1993, ApJ419, L69

Van Dyk, S. D., Weiler, K. W., Sramek, R. A., Rupen, M. P., & Panagia, N. 1994, ApJ, 432, L115

Weber, E. J. & Davis, L., Jr. 1967, ApJ, 148, 217

Weiler, K. W., Sramek, R. A., Panagia, N., van der Hulst, J. M., & Salvati, M. 1986, ApJ, 301, 790

Weiler, K. W., Panagia, N., & Sramek, R. A. 1990, ApJ, 364, 611

Weiler, K.W., Van Dyk, S.D., Sramek, R.A., & Panagia, N. 1996, in *IAU. Coll. No. 145, Supernovae and Supernova Remnants*, eds. R. McCray & Z. Wang , Cambridge University Press, p. 283

Study of planetary atmospheres

André Marten

Observatoire de Paris–Meudon, F–92195 Meudon, France

Abstract. As a necessary background for discussion of future prospects this review attempts to summarize current knowledge on observational data relating to the properties of planets at radio wavelengths. According to Whitcomb *et al.* (Icarus, 1979, **38**, 75), a consensus has emerged among the observers that "measurements of planetary brightness temperatures at wavelengths between 200 μm and 2 mm are of practical value in submillimeter astronomy since the planets are used as calibration sources in studies of other objects. In addition, such measurements can provide tests for theories of planetary atmospheres". Although the former topic deserves great consideration, only the latter is developed here with special emphasis on the most recent observations of planets in the millimeter and submillimeter range. I briefly outline some major observational perspectives which embody very attractive goals of future planetary research.

1 Introduction: A picture of the solar system objects

A classical approach to reviewing the solar system objects is to mention the traditional division between the terrestrial planets (Mercury, Venus, Mars) and the giant planets (from Jupiter to Pluto). This argument has been used in the past on the basis of considerations of radius, mass, mean density and distance from the Sun. Another basic way is to introduce some pertinent aspects on the differences of chemical composition in order to construct models of planet formation and evolution of the solar system throughout the Galaxy. Finally, one may easily envisage that the most significant clues pertaining to the earliest history of the solar system are observable in the primitive bodies like comets and outer planets which preserved the elemental abundances of the primordial solar nebula and thus some records of the early processes of planetary formation. In contrast to the large planets, the inner planets and Earth have lost all memory of the earliest epochs due to evolutionary processes affecting their initial compositions. Notwithstanding the fact that this general outlook is quite schematic, we will use it to highlight the astrophysical problems that are addressed with the planetary observations.

2 Exploration of the solar system

In the last three decades much research was devoted to the investigation of planets. The field of planetary astronomy has been revolutionized by results from space missions coupled with advances in technology that produced large ground–based telescopes in different domains of wavelengths from the visible to the microwave range.

2.1 Space missions

In the framework of the solar system exploration more than two dozen spacecrafts made close encounters with most of the planets. The entry probes like those of the Veneras collected a wealth of data on the chemical composition, the thermal structure, the nature of the clouds and the surface characteristics of Venus. The last mission to Venus, Magellan, was able to acquire radar images of most of the solid surface at a resolution ranging from 100 to 250 meters. Several flyby missions and orbiters - the Mariner, Viking spacecrafts - provided almost all our in-depth knowledge of the Mars atmosphere and made high resolution images of the entire surface. After Pionner 10 and 11 initiated the exploration of Jupiter and Saturn, the Voyager 1 and 2 spacecrafts accumulated major discoveries during their tour of the outer solar system. Important steps forward concern now Jupiter and its satellites with the Galileo spacecraft composed of an orbiter and an atmospheric entry probe, but also Saturn and its largest moon Titan with the Cassini mission scheduled for launch in 1997.

2.2 Ground–based observations

Even in the era of *in-situ* spacecraft missions, the planetary astronomy has a striking history of continuing productive research conducted from Earth–orbital, airborne, and ground–based telescopes. Since many extensive reviews exist to date, we will focus here on topics related to the compositional aspects that we can retrieve from spectroscopic measurements at long wavelengths. Note first the important role played by the infrared instruments which provide high sensitivity at relatively high spectral resolution. At radio wavelengths observations of the thermal emission permit us to probe the deep atmospheres of the giant planets or the surface layers of the terrestrial planets. In the centimeter range interferometric techniques have produced excellent radio images of most of the planets. Of particular interest are the extensive observations performed with the VLA. A detailed review of this subject was given by de Pater (1990).

Heterodyne spectroscopy in the millimeter and submillimeter range allows the detection of narrow spectral lines and thus to infer quite small abundances of molecular compounds. We note, however, that broad lines (~ 1 GHz) originating from dense atmospheres are extremely difficult to measure. In fact, the heterodyne detection of non–equilibrium species predicted to be present in the upper tropospheres of giant planets has not been productive so far. In contrast, by the early 1990s advances in millimeter technology began to open up the field of search for weak, narrow lines. According to Encrenaz *et al.* (1995), Table 1 gives a somewhat exhaustive list of mm/submm lines observed in planetary atmospheres by means of heterodyne techniques. While all the submillimeter lines were detected using the JCMT and CSO, most of the recent millimeter measurements were performed either at the NRAO and IRAM–30m telescopes or at the OVRO and IRAM millimeter interferometers.

In the next sections many of the observational aspects are examined and a few relevant analyses are reviewed briefly to shed more light on future development of comparative planetology.

Table 1.

Molecular lines observed in planetary atmospheres by heterodyne techniques

Frequency (GHz)	Species	solar system object
88.63	HCN	Jupiter-SL9, Titan
115.27	CO	Venus, Mars, Jupiter-SL9, Neptune, Titan
143.06	SO_2	Io
145.56	HC_3N	Titan
146.97	CS	Jupiter-SL9
183.31	H_2O	Venus
218.32	HC_3N	Titan
218.90	OCS	Jupiter-SL9
219.56	$C^{18}O$	Mars
220.40	^{13}CO	Venus, Mars, Titan
220.70	CH_3CN	Titan
221.97	SO_2	Io
225.90	HDO	Venus
230.54	CO	Venus, Mars, Jupiter-SL9, Neptune, Titan
244.94	CS	Jupiter-SL9
265.89	HCN	Jupiter-SL9, Neptune, Titan
345.34	$H^{13}CN$	Titan
345.80	CO	Neptune, Titan
354.51	HCN	Jupiter-SL9, Neptune
461.04	CO	Neptune

3 Inner planets

3.1 Venus

Microwave spectral line observations of Venus have been performed in the first transitions of ^{12}CO, ^{13}CO and $C^{18}O$. Clancy and Muhleman (1991) presented a thorough analysis of the interannual changes in the mesosphere (above 75 km) from the retrieval of (whole–disk) CO distributions and vertical temperature profiles. Disk–resolved spectra taken with the OVRO interferometer have recently been used by Gurwell et al. (1995) to derive the spatial distribution of CO. Some significant variations have been seen between night- and day- sides of the disk. Other studies by Lellouch et al. (1994) were concerned with the mesospheric circulation and the dynamics. They showed that the Doppler shift measurements of CO lines performed at very high spectral resolution on various planetary regions provided an excellent picture of the wind velocities near 100 km. Their results suggest that similar observations should permit a global study of the climatology of Venus' mesosphere wind system.

In complement to that long–term project a search for trace species not yet detected in the stratosphere appears as an attractive goal. Although photochemical models predict a large number of molecular species, attempts to detect them in the microwave range were not successful due to the lack of sensitivity. Since sulfur is a key–element governing a complex atmospheric chemistry on Venus, a systematic search for the most important S–bearing compounds expected above the cloud regions is certainly the main priority of future investigations. Species like SO_2, H_2S, SO and OCS which exhibit relatively strong transitions at 1.3 mm are thus good candidates for pertinent observations.

3.2 Mars

Temperature profiles and CO distributions in the martian atmosphere were inferred from CO millimeter spectra recorded at single–dish telescopes, (Clancy *et al.* 1990, Lellouch *et al.* 1991). However, in contrast to Venus, the spatial coverage of Mars is relatively poor due to the smaller size of the planet and needs to be improved. The current mm measurements probe the atmosphere from the surface up to $\sim$ 70 km with a vertical resolution of 10 km. Clancy *et al.* found that the 1982–1989 microwave data indicated a colder, cloudier and less dusty atmosphere than that observed during the 1976-1978 period of Viking measurements. We can thus emphasize the remarkable role played by the millimeter observations in the temporal monitoring of the atmosphere. Maybe more important is the compositional aspect of the atmosphere in relation with the seasonal change and the climate.

A search for trace species was motivated by the suggestion of Korablev *et al.* (1993) that organic compounds could be present in significant amounts. In spite of strong efforts to detect minor organic and hydrogeneous species at the IRAM–30m telescope during the last three years, we only derived disk–averaged upper limits on abundances in apparent conflict with a refined dynamical and chemical two–dimensional model initially proposed by Moreau *et al.* (1991). Further detailed analysis using an updated version of the Moreau model allowed us to deduce that strong compositional variations must exist both spatially and annually. A confirmation of our predictions requires disk–resolved observations that a next–generation mm interferometer could provide and also a temporal monitoring of emission lines associated with molecular compounds formed in the lower atmosphere.

4 Outer planets

4.1 Jupiter, Saturn

No heterodyne detection of millimeter transitions has been performed on Saturn hitherto. For Jupiter the same situation prevailed until July 1994. At this time the collision of comet Shoemaker–Levy 9 (SL9) with Jupiter caused cometary debris and many new constituents to be introduced in the jovian atmosphere. Quite strong narrow lines of CO, CS, OCS and HCN were observed in emission at various impact locations using the IRAM-30m telescope and the JCMT. These surprising results were indicating that strong local modifications of composition and temperature occurred in the stratosphere, (Lellouch *et al.* 1995,

Marten *et al.* 1995). Subsequently, the emission faded and absorption features were detected. Many months after the infall of comet SL9, signatures of CO, CS and HCN are still present on Jupiter spectra taken on the southern hemisphere. Of particular interest are the observations at the planetary limbs. The spectral line shapes which exhibit a mix of absorption and emission yield coherent information on both stratospheric temperature and abundances (Matthews *et al.* 1995, Moreno *et al.* 1995). From data taken at the IRAM-30m telescope in May 1995 when Jupiter had a maximum size, significant intensity variations in longitude were seen on the impact regions, providing evidence in favor of a slow angular extent of newly created species. Finally, there is no doubt that this remarkable study of Jupiter gives some clues on unusual photochemistry processes which produced long–lived species not discovered before in planetary atmospheres. But a double key question emerges for future work in the longer term: how rare can be the captures of comets by large planets and then, how exceptional the collisions are?

4.2 Uranus, Neptune

Since the pionnering discoveries of the Voyager 2 spacecraft the most innovative observations on Neptune were concerned with the detections of CO and HCN performed in the mm and submm range at the JCMT and CSO in 1991, and subsequently at the IRAM-30m telescope (Marten *et al.* 1993, Rosenqvist *et al.* 1992). The IRAM interferometer measurement of the CO(1–0) line (Guilloteau *et al.* 1993)[1] confirmed the existence of carbon monoxide in the troposphere of Neptune and hence allowed us to definitely conclude to an internal origin of this compound.

Since current thermochemical and photochemical models did not predict correct profiles of CO and HCN much attention has been paid in the retrieval of the HCN vertical distribution from complementary submm observations and also in the development of best–suited models. It is worth noting that both species have not yet been detected on Uranus. Satisfactory explanations exist now in the literature but, nevertheless, our global understanding of the formation mechanisms that are effective in the two planets appears severely limited. The following questions remain: *i.e.* existence of more complex nitriles and sulfur compounds on Neptune, detection of trace species on Uranus, dynamics on both planets. One may anticipate that such searches will deserve some growing interest when further progress in sensitivity will be achieved by the large millimeter interferometers.

---

[1] Until now, this continuum measurement is the only published observation of a broad line (> 2 GHz) detected in absorption in a planetary atmosphere using a millimeter interferometer.

5 Planetary satellites

5.1 Io

The innermost Galilean satellite of Jupiter is known to maintain a tenuous neutral atmosphere. Following the first observation provided in 1973 by the Pionner 10 spacecraft a great deal has been learned about the atmospheric composition and the surface from the disk–resolved images and infrared spectra taken by Voyager 1. Many active volcanoes and traces of SO_2 were discovered on localized regions at this time. Subsequently, using the IRAM–30m telescope Lellouch et al. (1992) observed several rotational transitions of SO_2 at mm wavelengths. Their measurements have represented a profound advance in general understanding of the composition and thermal structure of Io's atmosphere. Nevertheless, a search for other minor compounds such as SO and H_2S was not successful and certainly requires more investigations. Since the molecular lines like those of SO_2 presumably originate from small regions on the disk the retrieval of the space distribution of trace species demands that a better sensitivity and high spectral resolution (< 100 kHz) be simultaneously reached. Such a mapping program that a large mm array only can ensure is surely a promising goal. Incidentally, we do not know whether the other Galilean satellites possess tenuous atmospheres or exospheres and thorough investigations are needed in this respect.

5.2 Titan

Current knowledge about atmospheric properties of the largest moon of Saturn is essentially based on intensive Voyager 1 observations made in 1980. Particularly important are the infrared data recorded by the IRIS spectrometer which examined Titan's disk at various locations during the closest approach of the spacecraft. Several minor species such as hydrocarbons and nitriles were detected in the lower stratosphere. In addition, the stratospheric thermal structure below 300 km was inferred at the same time. According to the most recent analysis of IRIS spectra presented by Coustenis and Bézard (1995) we note that substantial latitudinal variations in mixing ratios for all hydrocarbons and nitriles were found with striking enhancement of nitriles abundances retrieved in the northern polar regions. Recent observations in the mm and submm range permit not only firm identifications of CO and HCN and determinations of abundances but also the discovery of complex nitriles such as HC_3N and CH_3CN. While the former was only detected at the north pole of Titan, the latter was not identified in the Voyager data (Bézard et al. , 1993). Though the derived abundances correspond to disk–averaged values, they provide a valuable insight into the photochemistry occurring in Titan's atmosphere. At present it is clear that the mapping of Titan's disk (~ 1 arcsec) is an impossible task. So we emphasize that a next-generation millimeter array being used in the sub-arcsecond domain at 1.3 mm should make considerable advance in the compositional study and ultimately in the development of appropriate two–dimensional photochemical models of Titan.

6   Concluding remarks and outlook

Much of planetary research has largely been motivated by the *in situ* investigations of planetary systems. It is worth–while to remark that complementary information has still come out from ground–based observations. In this respect large telescopes contribute to extend our in–depth knowledge partly because of their renewed instrumentations (not appropriate for space applications) and partly because of high possibilities of follow-ups. As discussed above, many original discoveries were made in the field of radioastronomy: the Jupiter observations in the aftermath of the SL9 collision constitute excellent examples of unexpected results that radiotelescopes (IRAM and JCMT) are able to obtain.

In the next decades we can obviously expect that the planetary research will pursue a very productive phase of observational studies by using either improved heterodyne receivers or bolometer array instrumentations. One can envision observational studies of the atmospheres of Pluto or Triton if both better sensitivity and high angular resolution are achieved.

However, as a final conclusion, I want to highlight a major goal for future prospects that deserves great attention. It concerns the search for other planetary systems and the identification of these objects through detection of eventual atmospheres. Since, at present, we know the existence of terrestrial–mass planets orbiting two radio pulsars and the presence of one Jupiter–mass companion discovered around three solar–type stars, we are aware that this field of research will constitute a long-term challenge of various investigations. Therefore one may wonder whether the 21st–century radioastronomy will contribute to this fascinating research.

Acknowledgments. I wish to thank Y. Biraud and T. Hidayat for valuable comments on the manuscript.

References

Bézard, B., Marten, A., Paubert, G. (1993): Detection of acetonitrile on Titan. Bull. Am. Astron. Soc. **25**, 1100

Clancy, R.T., Muhleman, D.O. (1990): Long–term (1979–1990) changes in the thermal, dynamical, and compositional structure of the Venus mesosphere as inferred from microwave spectral line observations of ^{12}CO, ^{13}CO and $C^{18}O$. Icarus **89**, 129–146

Clancy, R.T., Muhleman, D.O., Berge, G.L. (1990): Global changes in the 0–70 km thermal structure of the Mars atmosphere derived from 1975–1989 CO spectra. J. Geophys. Res. **95**, 14543–14554

Coustenis, A., Bézard, B. (1995): Titan's atmosphere from Voyager infrared observations. IV. Latitudinal variations of temperature and composition. Icarus **115**, 126–140

de Pater, I. (1990): Radio images of the planets. Ann. Rev. Astron. Astrophys. **28**, 347–399

Encrenaz, T., Bézard, B., Crovisier, J., Coustenis, A., Lellouch, E., Gulkis, S., Atreya, S.K. (1995): Detectability of molecular species in planetary and satellite atmospheres from their rotational transitions. Planet. Space Sci. **43**, 1485–1516

Guilloteau, S., Dutrey, A., Marten, A., Gautier, D. (1993): CO in the troposphere of Neptune. Detection of the J=1–0 line in absorption. Astron. Astrophys. **279**, 661–667

Gurwell, M.A., Muhleman, D.O., Pierce Shah, K., Berge, G.L., Rudy, D.J., Grossman, A.W. (1995): Observations of the CO bulge on Venus and implications for mesospheric winds. Icarus **115**, 141–158

Korablev, O.I., Ackerman, M., Krasnopolski, V.A., Moroz, V.I., Muller, C., Rodin, A.V., Atreya, S.K. (1993): Tentative identification of formaldehyde in the Martian atmosphere. Planet. Space Sci. **41**, 441–451

Lellouch, E., Paubert, G., Encrenaz, T. (1991): Mapping of CO millimeter–wave lines in Mars' atmosphere: the spatial variation of carbon monoxide on Mars. Planet. Space Sci. **39**, 219–224

Lellouch, E., Belton, M., de Pater, I., Paubert, G., Gulkis, S., Encrenaz, T. (1992): The structure, stability, and global distribution of Io's atmosphere. Icarus **98**, 271–295

Lellouch, E., Goldstein, J.J., Rosenqvist, J., Bougher, S.W., Paubert, G. (1994): Global circulation, thermal structure, and carbon monoxide distribution in Venus' mesosphere in 1991. Icarus **110**, 315–339

Lellouch, E., Paubert, G., Moreno, R., Festou, M.C., Bézard, B., Bockelée-Morvan, D., Colom, P., Crovisier, J., Encrenaz, T., Gautier, D., Marten, A., Despois, D., Strobel, D.F., Sievers, A. (1995): Chemical and thermal response of Jupiter's atmosphere following the impacts of comet Shoemaker–Levy . Nature **373**, 592–595.

Marten, A., Gautier, D., Owen, T., Sanders, D.B., Matthews, H.E., Atreya, S.K., Tilanus, R.P.J., Deane, J. (1993): First observations of CO and HCN on Neptune and Uranus at millimeter wavelengths and their implications for atmospheric chemistry. Astrophys. J. **406**, 285–297

Marten, A., Gautier, D., Griffin, M.J., Matthews, H.E., Naylor, D.A., Davis, G.R., Owen, T., Orton, G., Bockelée-Morvan, D., Colom, P., Crovisier, J., Lellouch, E., de Pater, I., Atreya, S.K., Strobel, D.F., Han, B., Sanders, D.B. (1995): The collision of the comet Shoemaker–Levy 9 with Jupiter: Detection and evolution of HCN in the stratosphere of the planet. Geophys. Res. Lett. **22**, 1589–1592

Matthews, H.E., Marten, A., Griffin, M.J., Owen, T., Gautier, D. (1995): JCMT observations of long–lived molecules on Jupiter in the aftermath of the comet Shoemaker–Levy 9 collision. Bull. Am. Astron. Soc. **27**, 67

Moreau, D., Esposito, L.W., Brasseur, G. (1991): The chemical composition of the dust–free Martian atmosphere: Preliminary results of a two–dimensional model. J. Geophys. Res. **96**, 7933–7945

Moreno, R., Marten, A., Lellouch, E., Paubert, G., Wild, W. (1995): Long–term evolution of CO and CS in the Jupiter stratosphere after the comet Shoemaker–Levy 9 collision: millimeter observations with the IRAM–30m Telescope. Bull. Am. Astron. Soc. **27**, 75

Rosenqvist, J., Lellouch, E., Romani, P.N., Paubert, G., Encrenaz, T. (1992): Millimeter–wave observations of Saturn, Uranus and Neptune: CO and HCN on Neptune. Astrophys. J. **392**, L99–L102

Asteroids and Comets: The Prospect for Observations with a Large Millimetre Array

Jacques Crovisier

Observatoire de Paris-Meudon, F-92195 Meudon, France

Abstract. Radio continuum observations of asteroids provides an appreciable complement to the radiometric observations at other wavelengths, contributing to the determinations of their size and albedo. In addition, it offers the unique possibility to sample
their subsurface temperature and to study their thermal properties. Radio continuum
observations of comets reveal the large-size particles of their dust coma rather than
their nucleus. Radio spectroscopic observations allow us to determine the molecular
composition of the nucleus ices sublimating into the coma, to study the kinematics
of the cometary atmosphere and to investigate its physical conditions. In addition to
an increased sensitivity, a large millimetre array would permit, by imaging comets,
to study the structure and evolution of the gas and dust jets, in connection with the
outgassing processes of the cometary nuclei.

1 Introduction

The study of small Solar System bodies with radio astronomical techniques is
relatively recent, due to the weakness of the signals emitted by these objects and
the demand for highly sensitive instruments. We will review here the current
status and future prospect of radio astronomical observations of asteroids and
comets at millimetre and submillimetre wavelengths with a special emphasis on
imaging observations with a large millimetre array.

2 Asteroids

Imaging asteroids — if one excepts the beautiful pictures taken by *Galileo* during
the Gaspra and Ida flybies — is still very difficult at distance, because of their
small sizes. Even the largest ones are barely resolved: Ceres has at most 0.85"
angular diameter (for a 930 km diameter and a closest approach of 1.50 AU).
Attempts have been made to image such objects with adaptative optics (Saint-
Pé *et al.* 1993) or with the *Hubble Space Telescope*. However, one may have the
luck to observe a *Near Earth Asteroid*, like recently Toutatis ($\simeq$ 4.6 km size and
a closest approach of 0.024 AU, making an angular diameter of 0.26") with more
conventional instrumentation.

Reviews on radio astronomical observations of asteroids were given by Altenhoff (1985) and Webster & Johnston (1989). See also Ostro (1989, 1993) for
reviews on radar observations, to be complemented by recent observations of
Toutatis (Ostro *et al.* 1995; Hudson & Ostro 1995; Zaytsev *et al.* 1993). Tens of

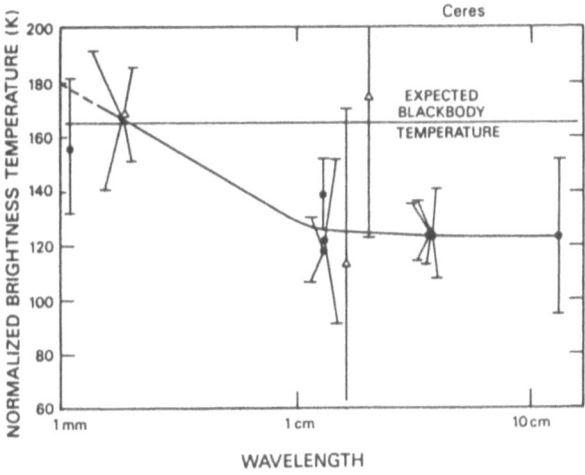

Fig. 1. The radio spectrum of Ceres (from Webster *et al.* 1988).

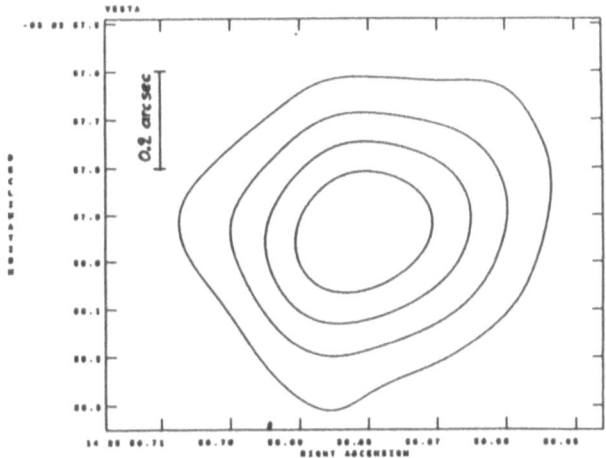

Fig. 2. The map of Vesta obtained at the VLA (from Johnston *et al.* 1989).

asteroids are accessible to existing large millimetre telescopes equipped with bolometers (Altenhoff *et al.* 1994; Altenhoff & Stumpff 1995; Jewitt & Luu 1992b; Redman et al. 1995). It should be noted that the largest asteroids are commonly used as calibrators for millimetre telescopes (c.f. Webster *et al.* 1988; Altenhoff *et al.* 1996; Fig. 1).

Asteroids (and cometary nuclei when they can can be separated from the contamination of emission by the dust coma — see next section) are thermal emitters at radio and infrared wavelengths. This emission is expected to come from material at a depth of several wavelengths below the surface. Therefore, comparative mid-infrared, submillimetric, millimetric, and centimetric observations probe the temperature at various depths and provide us with a very important clue to the outer structure of these bodies.

Radio observations of asteroids allow a reliable determination of their diameter with minimal model assumption. Combined with observations in the visible, they also provide a determination of their albedo. It was thus possible to determine the diameter (168 ± 20 km) and albedo (0.13) of Chiron with the 30-m IRAM telescope (Altenhoff & Stumpff 1995); these determinations were in close agreement with those from independent measurements in the thermal infrared (Campins *et al.* 1994).

Up to now, the VLA has the unique ability to image the surfaces of the largest asteroids; images of Ceres and Vesta observed at 2 cm with a resolution of 0.1" are reproduced in Webster & Johnston (1989) (see also Fig. 2. A large millimetre array would have the necessary sensitivity and spatial resolution for extending such studies to millimetre wavelengths.

3 Comets

Comets are among the most popular objects in the sky and their imaging has been practised for a long time. Besides naked eye observations of the large scale structure of spectacular comets at historical times, the short-scale structure of the inner coma has been imaged using telescopes and was recorded first by hand drawing, then by photography and now with the CCD technique. Imaging comets at radio wavelengths, however, is still very poorly developed because of the weakness of the signal, either continuum or spectroscopic.

3.1 Continuum Observations

After a succession of pioneering observations with doubtful results (see the review by Crovisier & Schloerb 1991), positive results were obtained at millimetre and submillimetre wavelengths on a handful of comets. This was mainly due to the efforts of two teams. W. Altenhoff and his collaborators at the IRAM 30-m detected continuum emission at 1.2 mm wavelength in comets P/Halley (Altenhoff *et al.* 1989), P/Brorsen-Metcalf and Austin 1990 V. Jewitt and Luu reported observations of P/Brorsen-Metcalf and several other comets at the JCMT at 0.8 mm wavelengths (Jewitt & Luu 1990, 1992a; Jewitt 1996).

The continuum signal detected from these observations appears to be too strong to be due to the cometary nucleus alone. It is rather interpreted as the emission of large-size dust particles. Since photometry in the visible or near and mid-infrared is rather sensitive to micrometre- or submicrometre-size particles, radiometry at millimetre wavelengths provides unique complementary information on the dust content of the coma. A theoretical basis for such an interpretation has been provided by Jewitt & Luu (1992a) (see also Walmsley 1985). Continuum emission can also be related to the hypothetical *icy grain halo* which is likely to have a transient existence at a specific heliocentric distance.

A tightly related topic is that of the radar observations of comets at centimetric wavelengths. An early review of the subject is given by Kamoun *et al.*

Table 1. A list of the molecular radio lines observed in comets (adapted from Crovisier, 1995, where references can be found).

	frequency MHz	species	transition	comet a)	comments
decimetric	1612	OH	ground F1-2	M, L	
	1665	OH	ground F1-1	K...	more than 40 comets
	1667	OH	ground F2-2	K...	"
	1721	OH	ground F2-1	M, L	
	3335	CH	ground F1-1	K	not confirmed
centimetric	4830	H_2CO	1_{10}-1_{11}	P/H	marginal
	8188	?		K	marginal, unidentified
	22235	H_2O	6_{16}-5_{23}	B, IAA	not confirmed
	23870	NH_3	(3,3)-(3,3)	IAA	not confirmed
millimetric	89088	HCN	1-0	K, P/H...	
	967xx	CH_3OH	(2,K)-(1,K) s	A, L	
	110700	CH_3CN	6-5 $\nu_8 = 1$	K	not confirmed
	115271	CO	1-0	P/SW1, H-B	
	1451xx	CH_3OH	(3,K)-(2,K) s	A, L...	
	165xxx	CH_3OH	(J,K)-(J,K-1) s	L, P/ST	
	168763	H_2S	1_{10}-1_{01}	A...	
	216710	H_2S	2_{20}-2_{11}	L	
	218440	CH_3OH	(4,2)-(3,1)E	L	
	225698	H_2CO	3_{12}-2_{11}	P/BM...	
	230538	CO	2-1	P/SW1, H-B	
	265886	HCN	3-2	P/BM, A...	
submillimetric	338xxx	CH_3OH	(7,K)-(6,K) s	P/ST	
	341416	CH_3OH	(7,1)-(6,1)A-	P/ST	
	345796	CO	3-2	P/SW1, H-B	
	351767	H_2CO	5_{15}-4_{14}	L, P/ST	
	354505	HCN	4-3	L, P/ST	
	2509949	OH	5/2-3/2	P/H	*KAO* with Fabry-Pérot

a) Full designation of comets: A: Austin 1990 V; B: Bradfield 1974 III; P/BM: P/Brorsen-Metcalf 1986 IX; P/H: P/Halley 1986 III; H-B: C/1995 O1 Hale-Bopp; IAA: IRAS-Araki-Alcock 1983 VII; K: Kohoutek 1973 XII; L: Levy 1990 XX; M: Meier 1978 XXI; P/SW1: P/Schwassmann-Wachmann 1 1989 XV; P/ST: P/Swift-Tuttle 1990 XXVIII.

s) Several lines of methanol are observed around the same frequency.

(1982), which has to be updated by the more recent observations of IRAS-Araki-Alcock at its close approach to the Earth (Goldstein *et al.* 1984; Harmon *et al.* 1989) and of P/Halley (Campbell *et al.* 1989). These observations demonstrate that, in addition to the signal of the nucleus, an echo is coming from large-size particles.

3.2 Spectroscopic Observations

The goals of spectroscopic observations of comets are the following:

- *Investigating the composition of comets.* Spectroscopy at radio wavelengths is peculiarly adequate for identifying the *parent molecules* directly sublimed from the nucleus ices (e.g. Crovisier 1994). This technique was thus able to identify without ambiguity, and for the first time, HCN, H_2CO, CH_3OH and H_2S. It is especially sensitive, since it is able to observe molecules such as

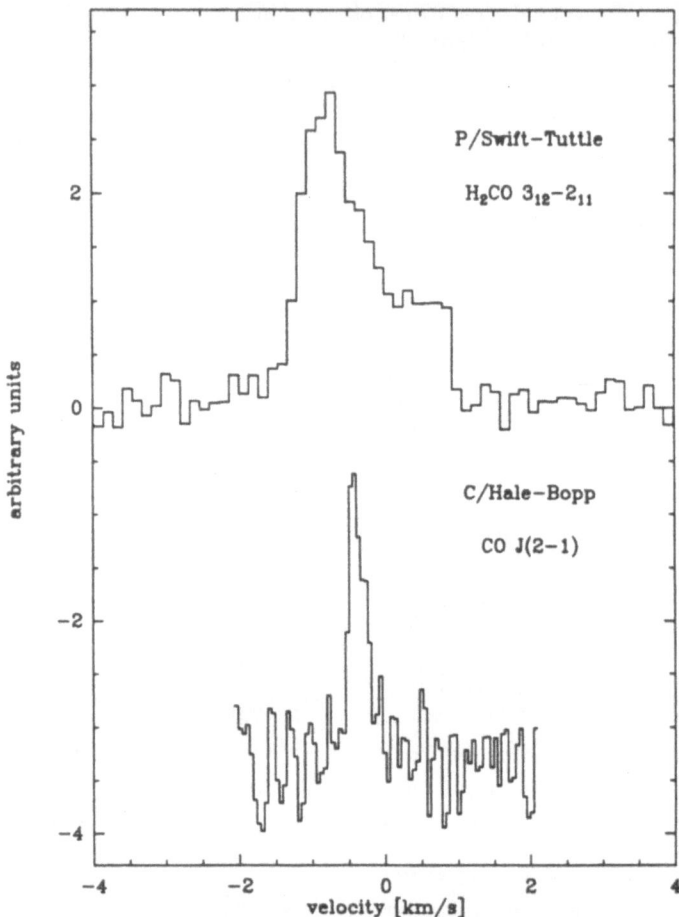

Fig. 3. Millimetre lines observed in comets with the IRAM 30-m telescope. Top: the formaldehyde line at 227 GHz observed in comet P/Swift-Tuttle 1992 XXVIII (Despois *et al.* 1996). Bottom: the CO line at 230 GHz observed in comet C/1995 O1 (Hale-Bopp) in Sept.–Oct. 1995, when it was at 6.5 AU from the Sun (Biver *et al.* 1996).

HCN and H₂S which have abundances of only 10^{-3} relative to water, the major gaseous output of comets. It was also able to observe the CO production of comets far from the Sun, where spectroscopy at other wavelengths (infrared, visible, ultraviolet), which observes *fluorescence* emission excited by the solar radiation, is inefficient: comet P/Schwassmann-Wachmann 1 at 6 AU from the Sun (Senay & Jewitt 1994; Crovisier *et al.* 1995) and quite recently comet C/1995 O1 (Hale-Bopp) at nearly 7 AU (Matthews *et al.* 1995; Rauer *et al.* 1995; Biver *et al.* 1996; Fig. 3).

– *Probing the kinematics of cometary atmospheres by observing molecular line shapes.* The unique spectral resolution at radio wavelengths allows to resolve the molecular lines, which have Doppler profiles (Fig. 3). From the line widths, one can thus determine the expansion velocity of the cometary atmospheres (Bockelée-Morvan *et al.* 1990). From the line centre, one can evaluate the bulk motion of the atmosphere with respect to the cometary rest of frame, which is due to the anisotropic sublimation of cometary ices, preferentially occurring from the nucleus hemisphere heated by the Sun, and which is related to the non-gravitational forces which affect cometary orbits (Colom *et al.* 1996).

– *Probing the excitation conditions within cometary atmospheres.* This can be done by observing the relative intensities of molecular rotational lines and retrieving the rotational distribution and the coma temperature. This possibility was especially applied to the methanol lines (Bockelée-Morvan *et al.* 1994).

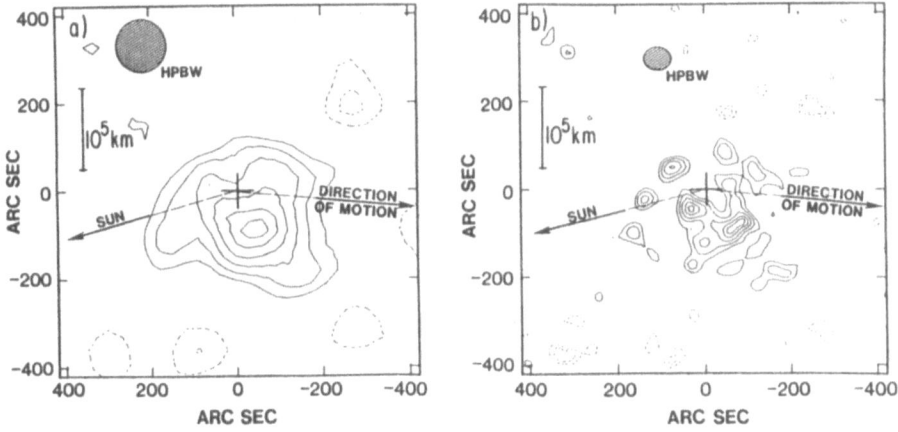

Fig. 4. Imaging of the 18-cm OH lines in P/Halley with the Very Large Array. The maps are averages of 13–16 November 1985, convolved to two different spatial resolutions. The highest contour levels are 19 and 10 mJy/beam for maps a) and b), respectively (from de Pater *et al.* 1991).

Observations with an Interferometer. Imaging observations of comets with radio interferometers are reviewed in detail by de Pater *et al.* (1991). Positive results were obtained at the VLA on the 18-cm OH lines of comets P/Halley and Liller 1988 V (Fig. 4). The maps are hampered by the limited signal-to-noise ratio in a synthesized beam and the lack of zero-spacing observations. The maps show the presence of small OH *clumps* (Fig. 4) which are difficult to understand in the context of the smooth spatial distribution expected for the cometary OH radical. It would be desirable to repeat such observations on future bright comets. De Pater *et al.* also claimed, from their VLA observations of comet Liller 1988 V, the existence of short-term variations of the OH line profiles, but it seems difficult to follow their conclusions due to the limited signal-to-noise ratio and spectral resolution of their data.

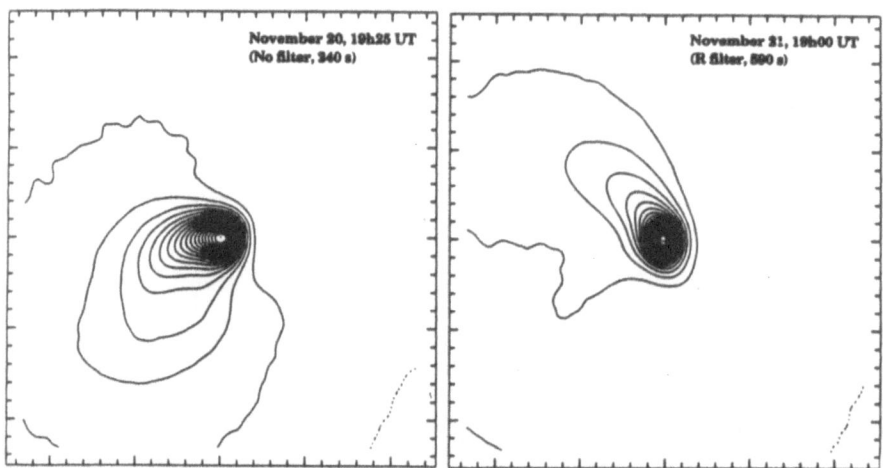

Fig. 5. Visible images of P/Swift-Tuttle obtained at the Pic-du-Midi Observatory showing the time evolution of the dust jets (from Jorda *et al.* 1994).

Prospects for a Large Millimetre Array. Using such an instrument as a phased array, we benefit from the increase in sensitivity due to the large collecting area. Tracking a comet with a radio telescope is often a problem, since one has to rely on ephemeris extrapolated from past astrometric positions. The advantage of the phased array versus a single dish with the same collecting area is that the tracking accuracy has to be of the order of the primary antenna beam only (on the other hand, its disadvantage is the difficulty to obtain accurate phase corrections). One should note the idiosyncrasy of the brightness distribution of a comet which is neither that of a point source nor that of a uniform extended source: the column density of cometary dust or of a parent molecule scales as $1/\rho$, ρ being the distance to the nucleus centre (in a spherical symmetry approximation). So, the average brightness in a beam of radius ρ scales as $1/\rho$. The *size* of a cometary coma has thus no meaning. In order to retrieve correctly such a brightness distribution, zero-spacing observation are a requisite (this coud be done either with the same instrument in the phased-array mode or independently with a larger single-dish).

Comets are variable objects. There are long-term variations due to the varying distance to the Sun. There are also important short-term variations are due to nucleus rotation and to outburst processes (Fig. 5). The time scales are of the order of a fraction of day. Therefore, it is not possible to synthesize images by grouping observations made on different periods, e.g. for different positions of movable antennas. It is even not desirable to use the Earth rotation to cover the uv plane. Images of comets will have to be obtained as *snapshots* of at most a few hours.

Mapping the space distribution of cometary molecules with a large millimetre array permits to investigate which part of these molecules are directly sublimed from the nucleus, or coming from distributed sources: icy grains, CHON grains, or decaying complex molecules. Molecules such as H_2CO and CO are known to come — at least partly — from yet unidentified distributed sources. This may be the case of several other *parent* molecules.

Mapping the space distribution and the velocity field of parent molecules (such as HCN, CH_3OH or CO) in the inner coma gives also the possibility to investigate the relations between dust jets (as imaged in the visible; Fig. 5) and the outgassing from sublimation of the nucleus ices and the icy grains of the dust coma.

In order to investigate the small features of the velocity field (anisotropic outgassing yields velocity offsets that may be as small as 0.1–0.2 km s^{-1}) and the profiles of the cometary lines, which may vary from lines as narrow as 0.15 km s^{-1} to structured lines about 2 km s^{-1} wide (see Fig. 3), a very good velocity resolution is required — of the order of 0.1 km s^{-1}.

4 Conclusion

Radio astronomical observations, and especially imaging with a large millimetre array, of the small bodies of the Solar System — asteroids and comets — are not only a useful and necessary complement to the observations at other wavelengths. They provide unique data on the thermal properties of the nucleus of these small bodies, on the large-size particles of cometary dust, on the composition, distribution and kinematics of the cometary atmospheres. Informations of paramount importance will also be provided from the space exploration of these objects. It can be noted that a microwave experiment has been proposed for the orbiter of the *Rosetta International Mission*, in order to map the nucleus of comet Wirtanen and rotational lines of selected molecular species in its coma at a very short distance (Gulkis *et al.* 1995). However, space explorations will be limited to a very small number of objects, whereas there is a large diversity in the Solar System objects to be studied. This could only be achieved from a long-term Earth-based investigation.

ACKNOWLEDGEMENTS. I am grateful to W. Altenhoff and E. Gérard for their useful comments on the manuscript and to P. Encrenaz for presenting this paper.

NOTE ADDED IN PROOF. Quite recently, comet C/1996 B2 (Hyakutake) was an outstanding opportunity for observations when it passed at only 0.10 AU from the Earth at the end of March 1996. Several new cometary molecules were detected at radio wavelengths (NH$_3$, OCS, HNC, CS, CH$_3$CN, HDO) through efforts at various telescopes. Successful observations were conducted with the IRAM interferometer at the Plateau-de-Bure where the emissions of the lines of CO $J(2-1)$ and HCN $J(1-0)$ were mapped with resolutions of 1.5 and 3 arcsec, respectively.

References

Altenhoff W.J. (1985): The Solar System: (sub)mm continuum observations. *ESO-IRAM-Onsala Workshop on (Sub)Millimeter Astronomy*, P.A. Shaver and K. Kjär edts, ESO Conference and Workshop Proceedings No 22, 591–601.

Altenhoff W.J., Baars J.M.W., Schraml J.B., Stumpff P. and von Kap-herr A. (1996): Precise flux density determination of 1 Ceres with the Heinrich-Hertz-Telescope at 250 GHz. A&A (in press).

Altenhoff W.J., Huchtmeier W.K., Kreysa E. et al. (1989): Radio continuum observations of comet Halley at 250 GHz. A&A **222**, 323–328.

Altenhoff W.J., Johnston K.J., Stumpff P. and Webster W.J. (1994): Millimeter-wavelength observations of minor planets. A&A **287**, 641–646.

Altenhoff W.J. and Stumpff P. (1995): Size estimate of "asteroid" 2060 Chiron from 250 GHz measurements. A&A **293**, L41–L42.

Biver N., Rauer H., Despois D. et al. (1996). Substantial outgassing of CO from comet Hale-Bopp at large heliocentric distance. Nature **380**, 137–139.

Bockelée-Morvan D., Crovisier J., Colom P. and Despois D. (1994): The rotational lines of methanol in comets Austin 1990 V and Levy 1990 XX. A&A **287**, 647–665.

Bockelée-Morvan D., Crovisier J. and Gérard E. (1990): Retrieving the coma gas expansion velocity in P/Halley, Wilson (1987 VII) and several other comets from the 18-cm OH line shapes. A&A **238**, 382–400.

Campbell D.B., Harmon J.K. and Shapiro I.I. (1989): Radar observations of comet Halley. ApJ **338**, 1094–1105.

Campins H., Telesco C.M., Osip D.J. et al. (1994): The color temperature of (2060) Chiron: a warm and small nucleus. AJ **108**, 2318–2322.

Colom P., Bockelée-Morvan D., Crovisier J. et al. (1996): Radio observations of anisotropic outgassing in comets. *Preprint*.

Crovisier J. (1994): Molecular abundances in comets. In *Asteroids, Comets, Meteors 1993*, A. Milani, M. Di Martino and A. Cellino edts (Kluwer Academic Publishers, Dordrecht) 313–326.

Crovisier J. (1985): Spectra of comets: infrared and radio regions. In *Laboratory and Astronomical High Resolution Spectra*, A.J. Sauval, R. Blomme and N. Grevesse edts (ASP Conference Series, Vol. 81) 383–395.

Crovisier J., Biver N., Bockelée-Morvan D. et al. (1995): Carbon monoxide outgassing from comet P/Schwassmann-Wachmann 1. Icarus **115**, 213–216.

Crovisier J. and Schloerb F.P. (1991): The study of comets at radio wavelengths. In *Comets in the Post-Halley Era*, R.L. Newburn, Jr, M. Neugebauer and J. Rahe edts (Kluwer Academic Publishers, Dordrecht) 149–173.

de Pater I., Palmer P. and Snyder L.E. (1991): A review of radio interferometric imaging of comets. In *Comets in the Post-Halley Era*, R.L. Newburn, Jr, M. Neugebauer and J. Rahe edts (Kluwer Academic Publishers, Dordrecht) 175–207.

Despois D., Biver N., Bockelée-Morvan D. et al. (1996). Radio line observations of comet 109P/Swift-Tuttle at IRAM. Planet. Space Scie. (in press).

Goldstein R.M., Jurgens R.F. and Sekanina Z. (1984): A radar study of comet IRAS-Araki-Alcock 1983d. AJ **89**, 1745–1754.

Gulkis S. et al. (1995): Microwave Instrument for Rosetta Orbiter (MIRO).

Harmon J.K., Campbell D.B., Hine A.A., Shapiro I.I. and Marsden B.G. (1989): Radar observations of comet IRAS-Araki-Alcock 1983d. ApJ **338**, 1071–1093.

Hudson R.S. and Ostro S.J. (1995): Shape and non-principal axis spin state of Asteroid 4179 Toutatis. Sci **270**, 84–86.

Jewitt D.C. (1996): Debris from comet P/Swift-Tuttle. AJ (in press).

Jewitt D. and Luu J. (1990): The submillimeter radio continuum of comet P/Brorsen-Metcalf. ApJ **365**, 738–747.

Jewitt D. and Luu J. (1992a): Submillimeter continuum emission from comets. Icarus **100**, 187–196.

Jewitt D. and Luu J. (1992b): Submillimeter continuum observations of 2060 Chiron. AJ **104**, 398–404.

Jonhston K.J., Lamphear E.J., Webster W.J. et al. (1989): The microwave spectra of the asteroids Pallas, Vesta, and Hygiea. AJ **98** 335–340.

Jorda L., Colas F. and Lecacheux J. (1994): The dust jets of P/Swift-Tuttle 1992t. Planet. Space Sci. **42** 699–704.

Kamoun P.G., Pettengill G.H. and Shapiro I.I. (1982): Radar detectability of comets. In *Comets*, L.L. Wilkening edt. (University of Arizona Press, Tucson) 288–296.

Matthews H.R., Jewitt D. and Senay M.C. (1995): Comet C/1995 O1 (Hale-Bopp). IAU Circ 6234.

Ostro S.J. (1989): Radar observations of asteroids. In *Asteroids II*, R.P. Binzel, T. Gehrels and M.S. Matthews edts (University of Arizona Press, Tucson) 192–212.

Ostro S.J. (1993): Planetary radar astronomy. Rev. Mod. Phys. **65**, 1235–1279.

Ostro S.J., Hudson R.S., Jurgens R.F. et al. (1995): Radar images of asteroid 4179 Toutatis. Sci **270**, 80–83.

Rauer H., Despois D., Moreno R. et al. (1995): Comet C/1995 O1 (Hale-Bopp). IAU Circ 6236.

Redman R.O., Feldman P.A., Pollanen M.D., Balam D.D. and Tatum J.B. (1995): Flux density estimates at millimeter wavelengths of asteroids near opposition from 1996 to 2005. AJ **109**, 2869–2879.

Saint-Pé O., Combes M. and Rigaut F. (1993): Ceres surface properties by high-resolution imaging. Icarus **105**, 271–281.

Senay M.C. and Jewitt D. (1994): Coma formation driven by carbon monoxide release from comet Schwassmann-Wachmann 1. Nat **371**, 229–231.

Walmsley C.M. (1985): The interpretation of the radio continuum emission from comet IRAS-Araki-Alcock (1983d). A&A **142**, 437–440.

Webster W.B., Jr and Johnston K.J. (1989): Passive microwave observations of asteroids. In *Asteroids II*, R.P. Binzel, T. Gehrels and M.S. Matthews edts (University of Arizona Press, Tucson) 213–227.

Webster W.B., Johnston K.J., Hobbs R.W. et al. (1988): The microwave spectrum of asteroid Ceres. AJ **95**, 1263–1268.

Zaytsev A.L., Sokolsky A.G., Wielibinski R. et al. (1993): 6-cm radar observation of (4179) Toutatis. In *Asteroids, Comets, Meteors 1993* (LPI, Houston) 325.

An extension of the European Millimetre VLBI Network: Plans for a mm radiotelescope in Greece

John H. Seiradakis

University of Thessaloniki, Department of Physics, Section of Astrophysics, Astronomy and Mechanics, GR-54006 Thessaloniki, Greece

Abstract. We present preliminary plans for establishing a Radio Astronomy Institute in Greece which will support and maintain a 15-m radio telescope capable of operating at up to 300 GHz. The Institute will cover most branches of research in Radio Astronomy and will serve at the forefront of technological developments in Greece and abroad. Finally, it will transform the *one dimensional* structure of the mm European VLBI Network into a useful *two dimensional* network. Some tentative suggestions for the location of the telescope are presented.

1   Introduction

During the last decade, observations carried out with radio telescopes have been taking more than one third of the observational papers published in scientific journals, revealing a wealth of unsuspected and fundamental new processes in the universe. As technological developments are rapidly advancing, a rich harvest of new results is shedding new light to our understanding of nature and the world we live in. In particular, a new generation of high performance radio telescopes and recent developments in mm receivers have allowed detailed investigations of quasar emission mechanisms (e.g. Barvainis et al. 1994), star forming regions and molecular clouds chemistry (Dutry et al. 1994), stellar evolution, outflow jets, the structures in the centre of our Galaxy, proto-planetary disks, earth-like planets and a thorough examination of the constituents of the solar system (Bézard et al. 1994).

Construction of mm radio telescope is not a trivial project. At present, there are about 20 mm radio telescopes located in several sites in three continents (America, Asia, Europe) providing exciting new results to the astronomical community. Surface irregularities of the telescope's panels should not exceed 5% of the observing wavelength, which means that if the shortest wavelength to be used is 1 mm, then the surface accuracy should not exceed $25\,\mu m$. Telescopes with even better surface accuracy are already operational, working at submm wavelengths. The angular resolution that has already been achieved by existing mm radio telescopes is of the order of a few tens of arcseconds. Radio interferometers have pushed this limit down by a factor of ten and Very Long Baseline Interferometry (VLBI) techniques, linking radio telescopes across the globe have reached angular resolutions of the order of 10 milli arcseconds.

2   European mm astronomy

At the moment European mm radio telescopes play a leading role in astronomical research. The 30 m single dish at Pico Veleta, Spain built by IRAM and the mm interferometer (4 × 15 m array, soon to become 5 × 15 m) have been used by astronomers for several years. The 15 m SEST telescope at La Silla (Chile), the 15 m JCMT in Hawaii and the recent 10 m HHT submm telescope on Mt. Graham (Arizona) are providing first class results. They all belong to the most recent generation of radio telescopes and are placed on the best observing sites in the world. A 21st century mm array with a collecting area of about $10000\,m^2$ (Large Southern Array - - LSA-) is under consideration for construction high at the Andes in Chile.

All the instruments mentioned in the previous paragraph can be linked with the VLBI technique, resulting in a very powerful instrument with large collecting area, high frequency capabilities and very high angular resolution.

3   Astronomy in Greece

Greece played a very important role in the birth of Astronomy a subject which still fascinates and intrigues a fairly large number of scientists in the country. The Hellenic Astronomical Society (HEL.A.S.) numbers about 185 members, all professional astronomers mostly working in educational or research institutions. Their research activities cover a wide range of subjects including radio astronomy. At the moment, observing possibilities in Greece are restricted, although, recently, with the installation of the new 1.3 m reflector on Mt. Skinakas (Crete) and the upgrading of the 1.2 m telescope at Kryonerion and the 30 inch telescope at Stephanion (both in Peloponnese) optical observations have been elevated.

4   A new mm radio telescope in Greece

Taking the above into consideration, the natural question whether new radio telescopes are desirable on the European continent comes as a pragmatic concern for those involved in promoting Astronomy on a world-wide base. A quick look in the map of Fig. 1 immediately reveals a void in "telescopes' density" in the south eastern corner of Europe. Taking into account the observational advantages offered by sites close to the equator, it is obvious that the construction of a radio telescope in Greece would serve not only the Hellenic community but it could play a leading role in European astronomy. The geographical latitute of Greece ranges between $\phi = 35°$ and $\phi = 41°$. One is surprised to note that even northern Greece lies further south than areas in southern Italy. For example Thessaloniki in northern Greece, ($\phi = 40.5°$) lies further south than Naples ($\phi = 40.8°$) in southern Italy. The galactic centre is readily available for several hours from such locations together with several low latitude parts of the sky.

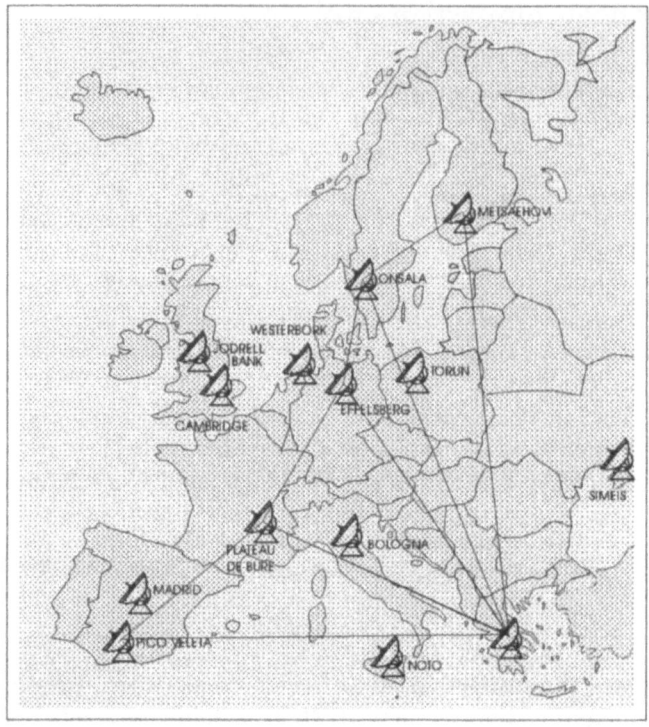

Fig. 1. Major radio telescopes in Europe. The telescopes capable of carrying out observations at mm wavelengths are located at *Pico Veleta, Plateau de Bure, Effelsberg, Onsala* and *Metsaehovi*.

A closer look in Fig. 1 reveals that all European radio telescopes capable of undertaking observations at mm wavelengths, seem to lie on a straight line starting from the 30 m telescope at Pico Veleta, Spain to the 14 m dish at Metsaehovi, Finland. They form an *one-dimensional distribution* on the European continent. The inclusion in the map of a mm radio telescope on the Greek peninsular would immediately transform this *one-dimensional distribution* into a *two-dimensional distribution*. This shall have a large impact on any future VLBI projects in Europe as the coverage in the UV-plane will be vastly improved. At the same time the resolution of such an instrument will also increase by almost a factor of 2.

Furthermore, Greece lies at the rim of two tectonic plates, making it a geologicaly active country. Movements of the plates have been monitored for several years (Billiris et al., 1991) revealing high activity in the south, which is accompanied by earthquakes. The island of Crete, in particular, is upraised at the place where the *African Plate* sinks under the *Eurasian Plate*. Using VLBI techniques, it is possible to monitor plate tectonic movements and assist geodetical and seismological investigations. A radio telescope in Greece, connected to the European VLBI network will be a prime instrument for such surveys (this technique is already in use in central Europe, Campbell 1991).

Here we propose the construction of a 15 m, fixed, IRAM-type mm radiotelescope in Greece, built and operated as a national facility from a *Radio astronomy institute* managed by a consortium of Greek Universities and Research Centres. A zero-order cost estimate of such an instrument is about 4 MECU, including a first generation of receivers and correlators, operating buildings and computer facilities.

If this project is actively supported by the geodetic community or the seismologists, then an obvious site for the location of a radio telescope in Greece would be the island of Crete. There are three mountain peaks in Crete, exceeding 2000 m height. There are several elevated plateau's in these mountains which would be ideal for installing such a telescope. If the project becomes a wholly astronomical project, then there is a larger choice in the high plateau's of continental Greece.

Is Greece capable of building and supporting such an instrument? We envisage that the major part of the building costs can be supported by national funds. However, taking into account the international character of such a facility, a small part of the cost could be taken by the European Community, within the frame of projects aimed to improve the technological differences between the richest and poorest countries in Europe and the integration of advanced networks. Concerning managing staff, as mentioned earlier, there exists a number of radio astronomers in Greece and an even larger number of Ph.D. students will be finishing their studies abroad within the next few years. These people could be involved in the project and promote its aims during construction but also after finishing.

5 Conclusions

The establishment of a medium size mm radio telescope in Greece could be built as a national facility. It could become a front-line instrument in astronomical research and a very useful tool in geodesy and seismology. Such a telescope will transform the present *one-dimensional* configuration of the European mm VLBI Network into a *two-dimensional* one.

References

Barvainis, R., Tacconi, L., Antonucci, R., Alloin, D. and Coleman, P. 1994, Nature, 371, 586

Bézard, B., Marten, A. and Paubert, G. 1994, Bull. Am. Astron. Soc., 25, 1100

Campbell J. 1991, The European Geodetic VLBI Network – Status Report; Proceedings of the 8th Working Meeting on European VLBI for Geodesy and Astrometry; Dwingeloo, The Netherlands

Dutry, A., Guilloteau, S. and Simon, M. 1994, A&A, 286, 149

Billiris H., Paradissis D., Veis G., et al. 1991, Nature, 350, 124

Science with Large Millimetre Telescopes

W. F. Wall and Luis Carrasco

Instituto Nacional de Astrofísica, Óptica, y Electrónica, C.P. 72000, Puebla, Puebla, México

Abstract. The sensitivity and resolution of large single-dish millimetre-wave telescopes can provide observations that complement those of millimetre-wave interferometers, such as the Large Southern Array. We present a discussion of some of the scientific objectives for the Large Millimeter-wave Telescope, a 50-metre aperture adaptive surface single-dish millimetre-wave telescope, whose construction will be completed within the next few years. The Large Millimeter-wave Telescope will combine the spatial resolution of a few arc-seconds with high sensitivity, making it one of the essential instruments for millimetre-wave observations in the 21st century.

1 Introduction

The upcoming Large Millimeter-wave Telescope (LMT) will fill an important niche in observational millimetre-wave astronomy. The principal institutions in the collaboration to build the LMT are the University of Massachusetts in Amherst, MA, USA and the Instituto Nacional de Astrofísica, Óptica, y Electrónica in México. The LMT will be a single-dish antenna with an aperture of 50 metres and an adaptive surface to maintain good aperture efficiency (see the specifications listed below). Hence, the LMT will combine good spatial resolution (5″ at $\lambda = 1\,\mathrm{mm}$) with high sensitivity ($\Delta S_{rms} \simeq 20\,\mu\mathrm{Jy/beam}$ in the continuum after 8 hours integration), allowing observations of extended low-surface brightness objects. This will complement observations by millimetre-wave interferometers such as the the proposed Large Southern Array (LSA), because the LMT data, which will contain information on large-scale structures resolved out by interferometers, can be combined with LSA data, which will contain information on the smallest spatial scales. In addition, the mapping speed of the LMT can be comparable to that of the LSA if the LMT is equipped with array receivers. Single-dishes are especially sensitive in continuum observations because they can make use of the widest possible bandwidth. The collecting area of the LMT, $\sim 2000\,m^2$ will be as great or greater than those of existing and proposed interferometers, except for the LSA. Nevertheless, the continuum sensitivity of the LMT will be comparable to that of the LSA because of the large bandwidth of the LMT bolometers. For years to come, the LMT will be the largest single-dish radiotelescope that will operate at wavelengths as short as 1 mm.

Specifications of the LMT and a description of two areas out of the many scientific objectives follow. Many of these objectives are similar to those of the LSA.

Fig. 1. An Artist's Conception of the LMT

SPECIFICATIONS:

Surface
- 50 metre aperture, radome enclosed
- active with 126 segments, 5 metres each
- better than 70 μm (rms) surface accuracy

Aperture Efficiency — 0.4 to 0.5

Nominal Wavelength Regime — 1 mm to 4 mm

Beamsize — 6.5″ at $\lambda = 1.3$ mm

Pointing Accuracy — $\simeq 1''$

Sensitivity — 3 Jy/K

Instruments — heterodyne array, bolometer array, radar

2   LMT SCIENTIFIC POTENTIAL

Of the many scientific goals of the LMT, we will concentrate on just two areas: high-redshift galaxies and star formation.

2.1   Galaxies at High Redshifts

The analysis of the CO luminosity function by Verter (1992) suggests that the LMT will be able to detect over one million galaxies. LMT's sensitivity and the spatial coverage provided by focal plane arrays of receivers will allow mapping of entire clusters of galaxies at high redshifts.

LMT will also have the sensitivity to detect primeval galaxies, determining their dust and molecular gas content. Deep continuum surveys with a bolometer array at 230 GHz will provide one of the most sensitive methods for searching for primeval galaxies. The 230 GHz observations would detect the red-shifted infrared emission from the warm dust in these high-redshift galaxies. The expected flux density for high-redshift objects will overcome the inverse-square law for sufficiently high redshifts, permitting observations of the far-infrared peak shifted to millimetre wavelengths. The 230 GHz band provides the best sensitivity for $z < 20$.

In addition to continuum emission, detection of the line emission in primeval galaxies is also possible. Searching for line emission from these high-redshift objects is complicated by the narrow bandpass of millimetre-wave spectrometers; one must know beforehand at which redshifts to observe. The search can be narrowed by observing a position near a quasar of known redshift. A 1 GHz bandpass would cover a few thousand $km \cdot s^{-1}$ in the rest frame of the $CO\ J = 3 \to 2$ line at redshifts of 3 to 4.

A simple comparison can nicely demonstrate the relative sensitivity of the LMT to high-redshift galaxies. The extremely luminous infrared galaxy IRAS F10214+4724 at redshift of 2.3 was detected in the $CO\ J = 3 \to 2$ line with 23 hours of integration with the NRAO 12-m telescope (Brown and Vanden Bout 1991). With the factor of 17 increase in collecting area, the LMT would have detected IRAS F10214+4724 in less than 20 minutes. IRAS F10214+4724 is probably not representative of primeval galaxies because it is believed to be a gravitationally-lensed object (e.g., see Trentham 1995). Nevertheless, IRAS F10214+4724 serves as an excellent concrete example in illustrating the advantages of large collecting area for detecting high-redshift galaxies.

2.2 Star Formation

Since stars form in molecular clouds, the physical conditions in molecular clouds are crucial to the understanding of star formation. Details of how molecular clouds collapse and fragment to form stars are still very uncertain. Molecular clouds possess structure on size scales from giant cloud complexes hundreds of parsecs across down to the densest cores thousands of AU in size. The LMT would be able to trace structure at the smallest scales, 700 AU, in the nearest clouds like those in Taurus. At these scales, extended circumstellar disks would be detectable.

The rotational lines of different molecular species can provide clues to the temperature, density, and kinematic structure in star-forming molecular cloud cores. To disentangle radiative transfer effects from other physical effects, it is necessary to observe the optically thin – and, hence, less abundant – isotopic variants of molecules such as CO, CS, OCS, etc. The LMT will have the sensitivity for mapping clouds in the spectral lines of the low abundance species, enabling us to determine the physical and kinematic properties of cloud cores with sufficient reliability for detailed comparisons with star formation models.

The continuum sensitivity of the LMT permits observations of areas of molecular clouds that would be otherwise difficult or impossible to observe. For example, at the edges of molecular clouds the molecular abundances drop drastically, making molecular line observations unfeasable. However, dust grains are present in sufficient quantity to allow the LMT to map the edges of molecular clouds. (The LMT can detect emission from a line of sight having $A_v = 1\ mag$ in ten minutes at $\lambda = 1.3\ mm$.) Important physical processes, such as photoelectric heating and the input of Alfvénic wave energy, occur on molecular cloud edges.

Understanding these processes will provide insights into the environments in which star-forming cores exist.

The LMT can also provide insights into the role played by magnetic fields in star formation. Magnetic fields, which affect the dynamical collapse of the star-forming cloud cores, align dust grains so that their continuum emission is partially polarized. Consequently, a map of the continuum polarization is an important tool in understanding magnetic fields in star-forming regions. The high sensitivity of the LMT make it a powerful instrument for making polarization maps of star-forming clouds. A 500 mJy source with a 1% polarization could be measured with a signal to noise ratio of 10 to 1 in 100 seconds at $\lambda = 1.3$ mm. Since the total flux per beam ranges from few $\times 0.1$ to a few $\times 10$ Jy, a two thousand pixel polarization map of a $5' \times 5'$ region of a molecular cloud would only require a few hours. The LMT equipped with a bolometer array and a polarimeter could obtain high resolution images of magnetic field geometry throughout a molecular cloud.

The early stages of star formation exhibit both infall and outflow of gas (Strom, Strom, and Edwards 1987). While the circumstellar disk is accreting material, there are also energetic outflows of material along the axis of the disk. These outflows strongly influence the evolution of the proto-star and the surrounding molecular cloud because of the momentum and magnetic flux that they carry. Therefore high-sensitivity maps of these outflow regions, like those that can be acquired with the LMT, are crucial for understanding the processes of star formation and its effect on the environment of the surrounding molecular material.

The LMT would also help address important questions involving the circumstellar disks themselves, including how these disks evolve. As the proto-stars approach the main-sequence the spectral signatures of the circumstellar disks disappear, but a sample of solar-type pre-main sequence stars will clear the dust from only the inner 1 AU or so, while maintaining a longer-lived outer disk that extends out to about 100 AU (Beckwith *et al.* 1990). The LMT will have a spatial resolution of a few hundred AU for pre-main sequence stars out to about 100 pc. The LMT will have sufficient sensitivity to carry out an extensive survey of these circumstellar disks, to compile the necessary statistics for determining disk accretion time scales and the time scales for these disks to form planetary systems.

References

Beckwith, S., Sargent, A., Chini, R. S., & Güsten, R. 1990, ApJ, 99, 924.

Brown, R. L. & Vanden Bout, P. A. 1991, AJ, 102, 1956.

Downes, D. 1995, LSA: Large Southern Array, Garching: ESO.

Schloerb, F. P. 1995, The Large Millimeter Telescope: LMT Science Rationale, Amherst: University of Massachusetts.

Strom, S. E., Strom, K. M., & Edwards, S. 1987, in NATO Advanced Study Institute: Galactic and Extragalactic Star Formation, ed. R. Pudritz and M. Foch (Dordrecht: D. Reidel)

Trentham, N. 1995, MNRAS, 277, 616.

Verter, F. 1992, ApJ, 386, 398.

Part 3

A Large Millimetre Array
in the Southern Hemisphere

The Synergy Between a Large Southern Array and the Very Large Telescope of ESO

James Lequeux[1]

Observatoire de Paris, 61 Av. de l'Observatoire, 75014 Paris, France

Abstract. This paper summarizes the main programmes that can be foreseen to be conducted in synergy between the Very Large Telescope of ESO and a future Large Southern array

1 Introduction

If a Large Southern Array (LSA) is built not far from the Very Large Telescope (VLT), two major instruments for astronomy will exist close to each other, each of which might be the largest of its class. It is of interest to try to foresee what kind of science might arise from the combination of these facilities.

First of all, it is remarkable that the angular resolution will be very high and similar for both instruments. The goal for the LSA is 0.1", similar to the angular resolution of individual telescopes of the VLT with adaptive optics in the near-IR. At 10 mm, the 8-m telescopes will have a slightly lower resolution of 0.3", but this resolution will of course be better in the VLTI interferometric mode, and could reach 0.01".

Next to the VLT, there should certainly be a 3-4 m class auxiliary telescope allowing wide-field imaging in the visible and in the near-IR with mosaïcs of receivers (CCD, NICMOS etc.). This telescope is for me an integral part of the VLT and will play an important role in the synergy I am going to describe.

2 Programmes concerning individual objects

It is difficult to foresee what will be astronomy in the next two or three decades, when both instruments will be operational. The only reasonable thing to do is to outline what could be the programmes performed by the instruments if they were both available now. This can only give a flavor of the future, keeping in mind that the most interesting discoveries will be the unexpected ones and that any novel large instrument as will be the LSA and VLT has a large potential for such discoveries. Table 1 gives an idea of possible programmes taking advantage of the complementarity of the two instruments. The left column indicates the scientific topic, and the second and third ones the relevant observations with the LSA and the VLT respectively. The problem to be solved is indicated in italics. This table is self-explanatory and does not need further comments.

Table 1. Complementary programs with the LSA and the VLT: individual objects

Field	LSA	VLT
SOLAR SYSTEM	Continuum	Near- and mid-IR continuum
	Albedo and size of asteroïds	
	Molecular lines	Lines (mainly IR)
	Atmospheres of planets, their dynamics	
HOT STARS	free-free continuum	Line shapes
		VLTI
	Mass loss [f(Z) in Magellanic Clouds]	
SUPERNOVAE	Continuum f(time)	Hα, [OII], rings
	History of progenitor mass loss	
AGB, post-AGB STARS	Molecular lines	IR molecular lines, dust emission
	Evolution and chemistry	
PLANETARY NEBULAE	Molecular lines	H$_2$ v>0, 10 µm imaging
	Neutral envelopes	
DISKS	Continuum and line mapping	Visible, IR scattering, 10 µm
		VLTI
	Dust and gas disks, kinematics	
OUTFLOWS	Molecular lines	Hα, [SII], [OI], H$_2$ v>0 etc.
	Nature and origin of outflows	
ISM: CLOUDS	Molecular lines	IR absorption lines (CH$_4$..)
	Physics and chemistry	
	Outflow detection and mapping	IR imaging
	Star formation	
DIFFUSE ISM	Absorption lines	CH$^+$, C2, CH abs. lines on same object (ex. BL Lac)
	Physics and chemistry	
NORMAL GALAXIES	Molecular lines, continuum	Continuum, line imaging
	Structure, ISM, kinematics and dynamics	
NUCLEI OF GALAXIES, QSOs	Continuum, molecules	Near IR: stars, ionized gas mid IR: dust, gas from [NeII]
		VLTI
	Physics, central monster	

3 Cosmology

The LSA will be able to observe galaxies to the highest redshifts at which they have formed heavy elements and dust. The millimeter surveys will of course be biased in favor of star-forming galaxies rich in dust and in molecules. The selection effects will be very different in the visible, where the surveys will be biased against dust-rich objects and strongly affected by extinction by intervening galaxies. Finally, a survey in the redshifted [CII]$\lambda158\mu m$ line with the LSA will offer an unbiased view of star formation at cosmological distances, as this line is one of the major cooling agents of interstellar matter heated by star formation. The VLT and of the LSA will thus be very complementary and their combination will have a tremendous power for solving cosmology problems. Table 2 summarizes the kind of programmes which could be conducted within their synergy, at least as we can foresee them now.

Table 2. Cosmology with the LSA and the VLT

OPTICAL SURVEYS AND FOLLOW-UPs
- VLT: - detection and classification of distant galaxies (in particular wide-field imaging)
       - redshift, luminosity, morphology, star formation rate, abundances
- LSA: - dust, molecular gas (content and imaging), kinematics.

MILLIMETER SURVEYS AND FOLLOW-UPs
- LSA: - surveys in the continuum (dust), CO lines, CI lines and C^+ line
- VLT: - redshifts for galaxies detected in the mm continuum
       - luminosity, morphology, star formation rate, abundances

GRAVITATIONAL LENSING
- VLT: - search for lensing around strong, high-redshift sources found with the LSA
- LSA: - systematic study of known gravitational lenses found with the VLT
       - gravitational mapping through gravitational shear
- Wide-field telescope and VLT:
       - clusters, gravitational mapping through gravitational shear

4 Conclusions

It is clear that the main scientific drivers for *both* the LSA and the VLT, at least as we can foresee them today, are cosmology and star formation. These are also the domains where their synergy will be at its best. This is a major argument for building the LSA in a site not too remote from the VLT and to have close ties between ESO and the organism which will responsible for the LSA.

A European Study Project for a Southern Millimetre Array

R.S. Booth

Onsala Space Observatory, S-439 92 Onsala, Sweden

1 Introduction

For several years, European millimetre astronomers have been discussing a large millimetre-wavelength array for the southern hemisphere. The instrument we have defined will be a true 'next generation' millimetre telescope, capable of opening up new directions in millimetre science, as well as extending the exciting recent developments in the field, most of which we have heard about in this meeting. The main goals of the array should be sensitivity (an order of magnitude improvement in sensitivity over today's major instruments) and resolution (0.1 arcsec at a wavelength of 3 mm). In order to refine this concept, to review possible sites and to investigate the technology required, the Institute de Radio Astronomie Millimétrique (IRAM), the European Southern Observatory (ESO), Onsala Space Observatory (OSO) and the Netherlands Foundation for Research in Astronomy (NFRA) have recently agreed to pool their resources in a joint European study which has just begun. This short paper will introduce the goals of the study and subsequent papers will deal in more detail with certain key areas.

Arguments for a large millimetre array may be found in papers by Booth (1994), Ishiguro et al. (1994), Downes (1994), Booth (1996) and, in a earlier context, in the MMA Proposal of NRAO (1990).

2 Basic concept of the array

We have defined the main goals of the array in a new document compiled by Dennis Downes and entitled 'LSA: Large Southern Array'. These are:

1. High sensitivity - to be achieved through a collecting area of about 10,000 m². This implies 50×16 m telescopes or 100×11 m telescopes.
2. High resolution - 0.1 arcsec at a wavelength of 3 mm - implying baselines between 5 and 10 km.
3. High quality southern hemisphere site - the array should be built on a high, dry site (altitude above 3000 m), preferably in Northern Chile.
4. Frequency range - the array should be dedicated to observations in the frequency range, 43 GHz to 350 GHz (wavelength range 7 mm to 0.8 mm).

3 Background to the LSA

Before describing the study, I want to put the array in context both in terms of the strong millimetre/submillimetre astronomy background in Europe and in terms of the other projects being discussed.

3.1 European millimetre facilities

Any new development in European millimetre astronomy builds on a strong base, established over the past 20 years through a number of important instruments described below.

The Swedish 20 m diameter, radome enclosed telescope at Onsala was completed in the mid 70's and has recently been resurfaced and generally upgraded to give an aperture efficiency of nearly 50% at 115 GHz. In collaboration with ESO, Sweden also operates the 15 m diameter Swedish-ESO Submillimetre Telescope, SEST, on La Silla. However, the major millimetre facilities inside Europe have been built, and are operated by, the French-German-Spanish institute, IRAM. They consist of a 30 m diameter antenna located on Pico Veleta in Spain and a 4, soon to be 5 (even 6), ×15 m element interferometer on Plateau de Bure in the French Alps. (The design of SEST is based on that of the IRAM interferometer telescopes). Finally, we have the 15 m James Clerk Maxwell submillimetre telescope on Mauna Kea in Hawaii, operated by the UK, in collaboration with the Netherlands and Canada, and also the Heinrich Hertz 10 m submillimetre antenna in Arizona, a collaboration between the Max-Planck Foundation and the Steward Observatory.

3.2 Non-European projects

Two other array projects are being discussed in the field of millimetre/submillimetre astronomy: the Millimetre Array (MMA), a US project proposed by the National Radio Astronomy Observatory, NRAO and a Japanese project proposed by the Nobeyama Radio Observatory.

The MMA is a long standing project to build an array of 40×8 m telescopes (collecting area $2000 \, m^2$). Although they have considered several sites in the United States, NRAO is now enthusiastically promoting a site in Northern Chile - Cerro Chajnantor, near the village of San Pedro. The proposed frequency range for this array is the same as for the European LSA but the collecting area is only twice that of the full IRAM interferometer.

The Japanese project is more ambitious and calls for 50×8 m antennae working to a frequency of 890 GHz. Sites under discussion for this array are Hawaii or Northern Chile. The collecting area is not significantly larger than the MMA but the extension into the submillimetre wavelength regime is an exciting prospect.

Notwithstanding the relative merits of these projects, our European deliberations have identified high sensitivity and resolution in the millimetre wavelength range as the most important drivers for a next generation instrument. This is what we propose. The enthusiastic scientific discussion at this meeting would seem to support our plan.

4 The European study

The detailed study for the LSA will address several important design aspects. In general, to minimise the need for interventions on a spread-out, remote, desert site, a major design goal should be reliability of all components. To keep the cost to a minimum, components should be as simple as possible and capable of easy replication. The areas to be addressed in this first study will be discussed below.

4.1 Site quality

Some of the world's best millimetre sites lie in the Atacama desert, located on the coastal side of the Andes in Northern Chile, where there are many large flat tracts of land at altitudes even as high as 5000 m. Several apparently excellent, accessible sites have been identified recently by Japanese and Swedish teams, and 22 potential locations were found by Raffin & Kusunoki (1992) in a search for sub-mm interferometer sites. These sites are extremely dry, with precipitable water vapour around 1 to 1.5 mm for large fractions of the time. The expanse, flatness, altitude, and low water vapour content of many of these sites confirms that Northern Chile would be an ideal location for the LSA.

Extensive water vapour measurements have been made on Cerro Paranal (altitude 2650 m), the site for the ESO VLT. These data show the precipitable water vapour content is below 1.5 mm for at least 30% of the April-November period, and below 3 mm for 60% of this period (Sarazin 1990; Martin 1991). Paranal is only 15 km from the ocean; the higher inland sites probably have a lower water vapour content. We want to investigate such sites and an early investigation by Lars Bååth and Angel Otarola has led to the selection of the site of Pampa El Chino at an altitude of 3300 m for further study.

Site testing equipment is now being prepared by Swedish and ESO (Onsala/SEST) scientists, as part of the combined study. We will set up a simple interferometer to monitor atmospheric phase fluctuations by observing a geostationary satellite as described by Ishiguro et al. (1990), and will add a 220 GHz receiver to monitor the atmospheric opacity by measuring total power in the wing of the 180 GHz water line. It is our intention to compare opacities for all the Chilean sites being studied (see the paper by Lars Bååth) with other meteorological data, particularly wind speed, and so to identify the optimal site for the array.

The optimisation process will involve not only the details of atmospheric opacity and phase stability but such questions as wind speed, which has important consequences for the antenna design, and considerations of the altitude and how it affects the ability of engineers to work efficiently, which has important consequences for servicing the array as well as general operational costs.

4.2 Antennae

The individual telescopes should be simple and robust; designed to maintain their mm-quality surface (total precision 50 microns r.m.s.) in reasonably windy conditions (up to 15 ms^{-1} as a goal). Since we need to have antennae which can survive in the outdoors, without astrodomes, particular attention must be paid to their pointing accuracy and thermally stable structures must be designed. For a sufficiently large array, it might be possible to have most of the antennas on fixed stations, avoiding the need for rails and transporters, although at present, this seems unlikely.

The antenna study is being led by IRAM with the possibility of industrial involvement at a later stage.

4.3 Receivers

For ease of maintenance receivers should be modular and of robust construction. They should, ideally, be tunerless systems with broad instantaneous bandwidth i.e. they should be designed to avoid complex tuning procedures. Closed-cycle cryogenic systems must be developed to be maintenance-free for long periods and the need for 4K systems should be avoided by using HEMT technology where feasible, although this is unlikely at wavelengths below 3 mm.

Most of the institutes involved in the study will contribute to the receiver discussion, with additional input from the Max-Planck Institut für Radioastronomie, Bonn.

4.4 Intermediate frequency system

Broad band IF systems are required for sensitivity reasons and early digitisation of the IF signal should be considered. Fibre optic transport of the signals (IF and local oscillator) will be studied. NFRA and IRAM are most experienced in this area.

4.5 Correlator

The correlator is a major item in the study. With 50 antennae there are 1225 baselines per polarization per resolution interval (1000 resolution channels over a bandwidth of 1 GHz is typical). The new fast correlator chips coming on to the market will be evaluated and manufacturers identified. The need for a special chip development and development of fast samplers must also be considered. A comparison of correlators with more units per receiver band, using fast samplers and slower chips in parallel will be carried out. Finally, the question of separate line and continuum correlators will be addressed.

The correlator study will be conducted by IRAM and NFRA.

4.6   Atmospheric phase compensation

A crucial aspect of the study will be the development of a method to correct the interferometer phases for atmospheric fluctuations. Several systems have been suggested but the most promising results have been recorded at IRAM, using an idea, originally put forward by Wright and Welch (1990). This is to monitor the variation in sky brightness temperature in the direction of observation, preferably with a total power channel in parallel with the receiver used for observations. The brightness temperature variation is proportional to the water vapour fluctuation in that antenna beam and can be used to correct the phase. The method requires high receiver stability (1 part in 10^5), which is not usually achieved in today's millimetre wave receivers. This, and other phase compensation techniques will be studied using the IRAM interferometer.

5   The future

The current study is a two year project with very limited funds. The next phase, a real design study, will involve industrial contracts and will therefore cost considerably more – 10 MECU is a realistic estimate. However, it is clear already that the LSA is feasible. But it will be an expensive instrument, with a cost comparable to some of the current optical telescope projects, or a small scientific satellite. Nevertheless, the amazing breadth of scientific return, from cosmology to planetary science, revealed from the presentations in this meeting and the great enthusiasm for the project shown by the participants must make it a strong contender for future European funding.

References

Booth, R.S., 1994, in *Astronomy with Millimetre and Submillimetre Wave Interferometry*, eds. M. Ishiguro & W.J. Welch. Astronomical Society of the Pacific, p.413.

Booth, R.S., 1996, in *25 years of CO*, ed S. Radford, in press.

Downes, D., 1994, in *Frontiers of Space and Ground-Based Astronomy*, eds. W. Wamsteker, M. Longair, Y. Kondo, Kluwer, Dordrecht, p.133.

Ishiguro M., Kanzawa T., Kasuga T., 1990, in *Radio Astronomical Seeing*, eds. J.E. Baldwin & Wang Shouguan, Pergamon, Oxford, p 54.

Ishiguro, M., Kawabe, R., Nakai, N., Morita, K.-I., Okumara, S.K., Ohashi, N., 1994, in *Astronomy with Millimetre and Submillimetre wave Interferometry*, eds. M. Ishiguro & W.J. Welch. Astronomical Society of the Pacific, p.405.

Martin, R.N. 1990. *SMT Technical Memo* UA-90-2 (U. Arizona, Tucson).

National Radio Astronomy Observatory, 1990, *Proposal for a Millimetre Array*, (NRAO, Charlottesville, Va.)

Raffin, P., Kusunoki, A. 1992, *SAO Submillimetre Array Tech. Memo*, No 59. (Smithsonian Astrophysical Observatory).

Sarazin, M. 1990, *ESO VLT Site selection Working Group*, Final Rep. No. 62.

Wright, M.C.H. and Welch, W.J. 1990, in *Radio Astronomical Seeing*, eds. J.E. Baldwin & Wang Shouguan, Pergamon, Oxford, p.71.

Large Southern Array: Feasibility

Stéphane Guilloteau

Institut de RadioAstronomie Millimétrique, 300 rue de la Piscine, F-38406 Saint-Martin D'Hères, France

Abstract. The LSA should provide high sensitivity for detection experiments and very high angular resolution in the millimeter range. The goal is a resolution of $0.1''$ at 3mm. In this document, I examine the feasibility of the two major objectives and the resulting constraints on the array design and site.

The angular resolution can be reached on dry, cold sites at 3mm wavelengths under good weather circumstances, and at shorter wavelengths using active phase correction techniques under clear sky conditions. Sensitivity constraints deriving from spectral line imaging of narrow lines from relatively cold regions imply a total collecting area of order $10\,000$ m^2. This can be achieved using 50 to 60 15-m antennas. The antennas must be moveable to provide useful sensitivity for detection projects and long baselines for high angular resolution. The site should be dry and presumably cold, but extremely high altitudes are not *a priori* required.

1 Sensitivity

The first goal of the LSA is to provide high sensitivity. This is required both for detection projects and for high angular resolution. For observations of continuum sources, the sensivity can be increased by increasing the observing bandwidth. However, for spectral line observations, the bandwidth is specified by the Doppler width of the line. Spectral resolutions as high as 0.1 km.s^{-1} (35 kHz at 3mm) are required in some circumstances such as searching for infall motions in star forming regions.

Ultimately, the sensitivity is thus dependent only on the system temperature of the detectors and on the total collecting area.

1.1 Atmospheric Opacity

Figure 1 presents the atmospheric opacity of the two major constituents of the atmosphere: O_2 and H_2O. The curves have been computed for parameters relevant to Cerro Chajnantol, where extensive site survey has been performed for the MMA. The opacity distribution measured at 225 GHz during the site survey is consistent with precipitable H_2O content of less than 0.55 mm for 25% of the time, less than 0.85 mm for 50% of the time and less than 1.4 mm for 75% of the time.

An essential point which stands out from Figure 1 is that, for water vapor content around 1 mm, the opacity is totally dominated by Oxygen at frequencies

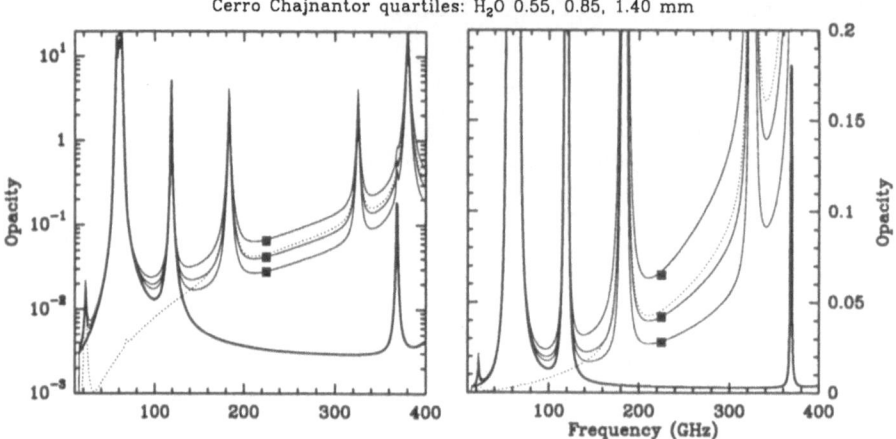

Cerro Chajnantor quartiles: H_2O 0.55, 0.85, 1.40 mm

Fig. 1. Dotted line: zenith opacity due to 1mm of precipitable water vapor. Thick line: zenith opacity due to O_2 alone at 5500 m altitude. Thin lines: total atmospheric opacity for 0.55, 0.85 and 1.4 mm of precipitable water vapor respectively. The black squares indicate the 25%, 50% and 75% quartiles of *measured* opacities at 225 GHz from a 6 month period of site survey.

below $\simeq$ 130 GHz. Moreover, because of the strong line near 60 GHz, this opacity does not decrease with decreasing frequency like the water vapor contribution.

Assuming the scale height of the water vapor is about 2 km, a site at 3500 m elevation would be expected to have a mean water vapor content of 1.7 mm, with less than 1 mm 25 % of the time. However, it has not yet been demonstrated that a scale height distribution is applicable. Since the general air flow runs from West to East over the Chilean Andes, we might expect that the same air layers smoothly drift across the Andes with an almost frozen water vapor content because it hardly ever rains in this area. Accordingly, the above estimate may be pessimistic.

1.2 System Temperature

The above values for the water vapor content can be used to estimate the expected system temperature for the LSA array.

Figure 2 presents the influence of some basic parameters on the system temperature, computed with the ATM model from J.Cernicharo. The behaviour can be understood as follows. A "standard" curve was computed with the following mean parameter values: local temperature 0°C, site elevation 3500 m, precipitable water 1mm, forward efficiency of the antenna 0.95, receiver temperature $2h\nu/k$, and image sideband rejection by 20 dB. A single parameter has been varied each time. The system temperature is given for SSB observations, as relevant for spectroscopy, and for a source elevation of 45°.

Temperature 0°C, Altitude 3.5 km, H_2O 1 mm
F_{eff} 0.95, T_{rec} 2hν/K, Rejection 20 dB

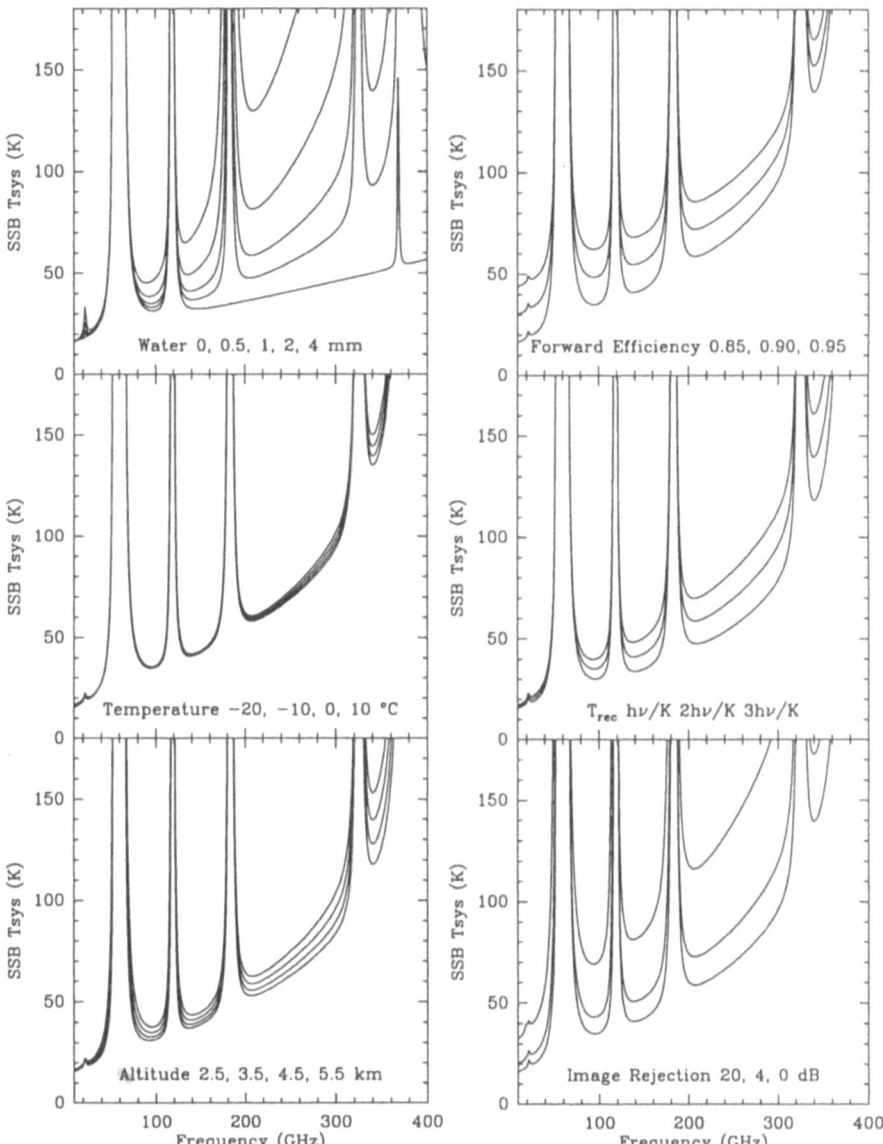

Fig. 2. Influence of various parameters on the Single-Sideband system temperature curves as a function of observing frequency. Top left: influence of precipitable water vapor content. Middle left: influence of the outside temperature Bottom left: influence of the site altitude. Top right: influence of the forward efficiency. Middle right: influence of the receiver temperature (including optics). Bottom right: influence of the image rejection.

- Precipitable water vapor

  Above 130 GHz, the effect is quite significant and a drier site offers a significant advantage in the 1 to 2 mm window. However, below 130 GHz, the system temperature becomes practically independent of the water vapor content, provided this content is less than 2 mm.
- Dependence on outside temperature

  The dependence comes essentially from the emissivity of the lower layers of the atmosphere, and is negligible.
- Dependence on Altitude

  The amount of precipitable water was fixed to 1 mm. The altitude effect is thus essentially due to the pressure difference. It can be seen that this is a relatively minor effect. Significant differences only appear above 320 GHz.
- Forward Efficiency

  In a good site, optical losses contribute quite significantly to the system temperature. Minimizing losses is thus absolutely essential. This must be kept as a major goal in the antenna design.
- Receiver Temperature

  A goal of $2h\nu/k$ seems realistic. Note that below 120 GHz, the dependence is relatively weak.
- Image Rejection

  Sideband rejection can lower the system temperature by a factor $\simeq 2$; The curve was computed for an IF of 1.5 GHz and should be redone with the effective value for better accuracy. This will in general degrade further the DSB case. The displayed improvement may be optimistic because SSB mixers usually have poorer noise performance than DSB mixers. In principle, high rejections are not mandatory in an interferometer because sideband separation (of the signal, but not of the noise) is possible in the correlator system. Accordingly, a compromise between the rejection and the receiver temperature is acceptable.

Except in frequency ranges close to some O_2 or H_2O line, the typical Tsys obtained for the "mean" set of parameters is well estimated by

$$T_{sys}(K) = 15 + 25 \times (\nu/(100 GHz) \tag{1}$$

i.e. typically 40 K at 100 GHz and 75 K at 230 GHz. This is about 2 to 3 times better than the currently obtained system temperatures with the IRAM interferometer. Furthermore, it should be stressed that below 130 GHz, such system temperatures **cannot be improved** by a better site and most likely represent the ultimate performances.

1.3 Collecting Area and Number of Antennas

The brightness sensitivity of the array is related to this system temperature by

$$\Delta T = T_{sys}(\theta_p/\theta_s)^2/\sqrt{N_b \delta \nu t}/B_{eff} \tag{2}$$

where N_b is the number of baselines, $\delta\nu$ the spectral resolution, t the integration time, θ_s the synthesized beam, θ_p the primary beam of the antennas and

B_{eff} the main beam efficiency. This rms noise is inversely proportional to the total effective collecting area (i.e. including antenna efficiency) through the term $\theta_p^2/\sqrt{N_b}/B_{eff}$.

To illustrate the minimum required sensitivity, let us consider the study of the gas (column) density distribution and bulk kinematics of a protoplanetary disk around a low mass star. An optically thin molecular tracer must be selected to sample the whole gas, and the spectral resolution should be significantly higher than the Keplerian speed. Inserting parameters appropriate for high angular resolution at 100 GHz, i.e. $\delta\nu = 100$ kHz (0.3 km.s^{-1}), $\theta_s = 0.1''$, $\theta_p = 50''$ (15-m dish), and using 50 antennas with $B_{eff} = 0.7$ during 24 hours, the obtained rms noise is 3 K. As the expected temperature is a few times 10 K in these disks, the instrument is **just sensitive enough** to image marginally optically thin lines ($\tau \simeq 0.1$) at such angular resolution, and hence to study the density distribution.

All other astronomical cases result in either less stringent requirements because of broader lines (galaxies, circumstellar envelopes, more massive stars, etc...), or in *ad-hoc* brightness temperature threshold levels since the line opacity can be arbitrarily low. Accordingly, the above results show that a collecting area of order 10 000 m^2 is the minimum required to provide 0.1'' resolution in astrophysically interesting cases.

Such a large collecting area cannot be done with small antennas, both for cost and feasibility reasons. Concerning the cost, below a minimum diameter, the infrastructure costs (receivers, stations and cables, specially if moveable antennas are required, etc...) become an incompressible part of the antenna price. Moreover, the correlator cost basically grows as the number of baselines. Very large antennas may indeed always provide the lower total cost.

However, antennas should not be too large because of the field of view constraints. A compromise around 50 antennas of 15-m diameter is reasonable. Slightly smaller dishes, e.g. 12-m, may be more convenient both for the field of view argument and because of better pointing performances under wind loading. However, even with 12-m dishes, the number of *receivers* may be unrealistically large (above 80), and the correlator too complex.

Eq. 2 shows (as expected) that higher angular resolution is easier at shorter wavelengths *provided the system temperature does not increase faster than $\nu^{2.5}$*. This assumption is valid in the atmospheric windows between 70 and 350 GHz, as shown by Eq. 1. However, because the atmospheric opacity dramatically increases at still higher frequencies, this is no longer true at sub-mm wavelengths. Hence, even without considering the antenna efficiency loss, from pure sensitivity considerations, the mm domain is the optimum for high angular resolution observations. Seeing limitations may also argue in the same direction.

2 Array Layout: Moveable Antennas

Although fixed antenna are cheaper and more reliable, the LSA antennas must be moveable. This derives from the need to fulfill two contradictory goals: high sensitivity for high angular resolution, and high sensitivity for detection experiments. In the first case, a uniform coverage of the UV plane up to the longest baselines is desired. In the second case, it is important to avoid resolving the objects in order to maximize the detected flux density. Since the typical size of objects which may be searched for is about 1-2″, baselines longer than 300 m provide no useful sensitivity and should be avoided for detections.

Simulations show that even with the best compromise, a single configuration offering adequate mapping capabilities for high angular resolution result in a sensitivity loss of about 3 for detection experiments of objects with typical size 1″.

Accordingly, the LSA should offer at least two configurations. The total number of configurations may be larger, in particular if some sort of "scaled array" is desired to provide similar angular resolutions at various wavelengths. Table 1

Maximum Baseline	300 m	900 m	2.7 km	8 km
Frequency	Angular resolution			
100 GHz	2"	0.7"	0.2"	0.07"
230 GHz	0.9"	0.3"	0.1"	0.03"

provides an example with the angular resolution varying by factors 3. Finally, reconfiguring the array may be a lengthy process. Based on the VLA experience, it is unlikely that the LSA could be reconfigured more than twice a year.

3 Phase Calibration and Seeing Effects

0.1″ angular resolution is a legitimate goal, but is it achievable, either by using the best seeing conditions or by appropriate calibration schemes?

Recent experiments carried on the Plateau de Bure with the IRAM provide a new and quite promising view on this topic. These experiences are based on the monitoring of water vapor content through the total power output of the 1.3 mm receivers. Because water vapor is by large the most important time variable dispersive constituent of the atmosphere, monitoring the fluctuations of the sky emissivity provides an almost direct measurement of the pathlength variations (see the article by Bremer et al. in these proceedings).

Observations with the IRAM interferometer have demonstrated the ability of such techniques to bring the atmospheric phase noise from more than 270° down to about 20° at 1.3mm, using *predictive* models, thereby demonstrating that the model accuracy is better than about 8%.

Furthermore, the same technique has allowed to determine the phase noise *at the outer scale of the turbulence*, just by monitoring the total power of one receiver. Three such measurements (expressed in terms of rms phase at 230 GHz) are displayed on Figure 3, in respectively excellent, average and poor conditions. Note that the measurement in excellent weather conditions is an upper limit to

Fig. 3. Total tropospheric phase excursion above one antenna derived from the total power fluctuations. Top: excellent conditions, Middle: normal weather, Bottom: poor weather. In all cases the outside temperature was below 0°C

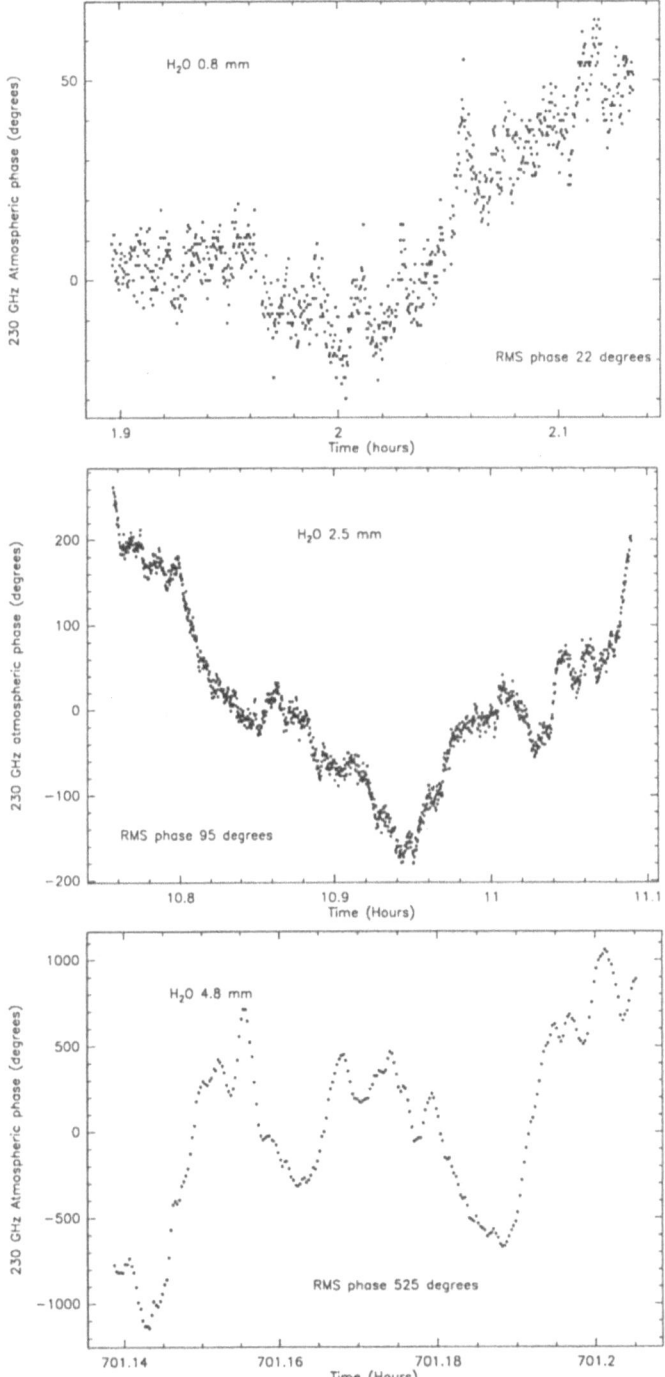

the tropospheric phase noise, because the measurement accuracy is limited by the intrinsic stability of the detection system.

Phase noise below 30° at 230 GHz on arbitrarily long baselines are thus possible on excellent days. This is sufficient to provide good quality images. Note that 3mm observations would have a lower phase noise, and hence comparatively better imaging quality.

On more typical days, the atmospheric phase noise is still low enough to be corrected by the total power monitoring techniques, even without further progress in the atmospheric modeling. It is only on "poor days" that the model accuracy may be unsufficient. However, such days should be extremely rare on the LSA site, as illustrated by the MMA site study at Cerro Chajnantor (even accounting for a possible degradation due to a lower altitude).

The phase compensation currently used on the IRAM interferometer is unable to properly ensure phase continuity between the source and the calibrator. It rather acts as a *decorrelation compensation*, allowing to recover the full amplitude but not the true phase. This is inherent to the current design in which standard receivers moving in elevation are used: receiver gain variations during the slewing between source and calibrator observations result in a lack of continuous phase prediction at this time. These gain variations are due to the antenna acceleration during the slewing motion, which is high because calibrators are often at least 10 to 15 degrees away from the source.

A more appropriate design for a dedicated phase monitoring receiver could in principle overcome this limitation. Moreover, an alternative solution to this problem consists in "semi-fast" calibration on a very nearby quasar. With the LSA sensitivity, the probability to find with 1 degree of the source a quasar sufficiently strong to allow calibration after 30 sec integration is extremely high. By observing such a quasar every 5 to 10 minutes, the proper phase continuity between the source and the quasar can be recovered.

4 Conclusions

Recent observations have demonstrated that seeing limitations may **not** be important at 3 mm wavelengths, and can be overcomed at shorter wavelengths using active phase correction techniques currently under tests at various interferometers. Hence, **high angular resolution is possible provided sufficient sensitivity is available.**

An estimate of the expected system temperatures shows that the required sensitivity is reached with a total collecting area of order $10\,000$ m^2. A dry site offering baselines up to 6-8 km is required. However, a high altitude site is better **only if** it is drier. The altitude may rather be specified from the average temperature, because the atmosphere is much more stable when the outside temperature is below 0°C. These two aspects of the site (water vapor, mean temperature) should be addressed very carefully during the site study.

Finally, antennas should be moveable, and a minimum number of 4 different configurations seems appropriate.

On the Imaging Efficiency, Speed, and Sensitivity of Millimetre Arrays

Alain Baudry

Observatoire de l'Université de Bordeaux, BP 89,
F-33270 Floirac, France

Abstract. We compare the imaging efficiency (defined from closure relations), speed, and sensitivity of existing and planned mm-wave arrays, and conclude that the Large Southern array (LSA) would be unsurpassed in the 1-3 mm band. We point out that the IRAM array expanded to 6-8 antennas would achieve sensitivity limits in the 3 mm-band nearly comparable to those of the MMA. Finally, we suggest that the LSA could consist of sub-arrays containing 6-8 large antennas.

1 Introduction

Advances in millimetre astronomy largely depend on improved sensitivity and spatial resolution performances of mm radio telescopes. Therefore, achieving a high sensitivity and mapping speed is always essential in designing or expanding new or existing mm arrays. Ignoring cost constraints and interferometer site differences (as well as techniques diminishing atmospheric phase fluctuations), we discuss how the European long term project for a Large Southern Array (LSA) compares with existing or planned mm arrays. Our comparison includes the IRAM array, the Berkeley-Illinois-Maryland-Association Array (BIMA), the Owens Valley Radio Observatory array (OVRO), the Sub-Millimeter Array (SMA under construction) of the Smithsonian-Harvard Center for Astrophysics, and the future arrays: the NRAO Millimeter Array (MMA), and the Large Millimeter and Submillimeter Array (LMSA) under discussion in Japan. In addition, for the IRAM interferometer on Plateau de Bure we contemplate an intermediate term plan adding 1 to 3 more antennas to the existing 5-element array. This plan could be conceived as a preparation to the LSA project, especially if it attracts new European partners.

Using closure relations we first define an imaging efficiency, then we compare relative detection speeds for sources smaller or wider than the primary beam in the simple (though fictitious) case of arrays with similar system temperatures and antenna efficiencies. Finally, we derive and briefly discuss the flux density and brightness sensitivities of the next generation mm arrays and of the IRAM array.

2 Closure Relations and Imaging Efficiency

Optimization of the algorithms used to constrain the brightness distribution of sources observed with synthesis radio telescopes depends on the number of independent closure relations. We may thus crudely estimate the "instantaneous" imaging efficiency of an array by the ratio of the number of closure phases and amplitudes to the total number of unknowns (antenna complex gains, and source amplitude and phases). From the maximum number of baselines, or the maximum number of independent correlators per polarization

$$N = \frac{n(n-1)}{2} \tag{1}$$

where n is the number of antennas, we deduce the number of independent phase and amplitude closure relations $Np = \frac{(n-1)(n-2)}{2}$ and $Na = \frac{n(n-3)}{2}$, and we define an imaging efficiency by

$$IE = \frac{Np + Na}{n^2 + n} . \tag{2}$$

This efficiency cannot obviously account for the actual mapping capability of an array and rather applies to snapshot observations. Table 1 compares N, Np,

Table 1. Closure relations and imaging efficiency of existing and planned (sub)millimetre arrays. (IRAM array: expansion beyond $n = 5$ is under discussion. Possible upgrades of BIMA, OVRO and SMA are not included.)

Array	Country/Organization	Antenna number n	Baseline number N	Closure phase number Np	Imaging efficiency IE
IRAM	France, Germany, Spain	5	10	6	0.37
		6	15	10	0.45
		8	28	21	0.57
BIMA	US	9	36	28	0.61
OVRO	US/CalTech	6	15	10	0.45
NMA	Japan				
SMA	US/CfA				
MMA	US/NRAO	40	780	741	0.90
LSA	Europe	50	1225	1176	0.92
LMSA	Japan				

and IE, and shows that there is a rapid increase of these quantities beyond $n = 5$. It is important to note that by increasing the number of closure relations one makes a better use of the longest baselines and one improves the antenna gain calibration made at the highest frequencies. The future MMA, LSA or LMSA significantly enhance the imaging efficiency. This will be achieved only if new generation correlators can be designed since processing around 1000 or more independent baselines in a flexible way is undoubtedly a technical challenge. We thus believe that the LSA will meet the goal of 10000 m^2 collecting area with a reasonable number of large individual antennas (of the order of 50).

3   Detection and Mapping Speeds of Synthesis Arrays

The r.m.s. flux sensitivity reached in one polarization for a point source or a compact source mapped with equally weighted uv cells is

$$\Delta S = \frac{2kTsys}{A\eta_a\eta \ (n(n-1)\Delta\nu\Delta\tau)^{\frac{1}{2}}} \ , \tag{3}$$

where Tsys is the system temperature outside the atmosphere, A the collecting area of a single antenna, η_a the aperture efficiency, η the product of the correlator efficiency and of the atmospheric and instrumental decorrelation factors, $\Delta\nu$ the bandwidth, and $\Delta\tau$ the integration time. For arrays with comparable instrumental performances and atmospheric conditions the time needed to reach ΔS is proportional to the inverse of $n(n-1)D^4$ where D is the antenna diameter, and we call detection speed the quantity DS

$$DS = ND^4 \ . \tag{4}$$

Table 2 gives the collecting area and DS for the arrays of Table 1. We arbitrarily

Table 2. Collecting area and speeds of existing and planned (sub)millimetre arrays assuming similar sky opacities and instrumental performances

Array	n D(m)	Collecting area (m^2)	Mapping speed $N = \frac{n(n-1)}{2}$	Relative detection speed ND^4	ND^2
IRAM	5x15	885	10	1.0	1.0
	6x15	1060	15	1.5	1.5
	8x15	1415	28	2.8	2.8
BIMA	9x6	255	36	0.09	0.6
OVRO	6x10.4	510	15	0.35	0.7
NMA	6x10	470	15	0.30	0.7
SMA	6x6	170	15	0.04	0.2
MMA	40x8	2010	780	6.3	22.0
LSA	50x16	10055	1225	159.0	139.0
LMSA	50x10	3930	1225	24.0	54.0

set the speed of the 5x15 m IRAM interferometer to 1 and compare all other speeds to this value. Although we do not account for differing site qualities, some conclusions may be drawn from Table 2. Augmenting the IRAM array to 6 or 8 antennas would significantly improve its detection speed and would make it much faster than all other existing arrays. In addition, the collecting area of a 6x15 m array is already half that of the planned MMA. The large arrays of the future offer very high detection speeds compared to all existing interferometers and would be extremely useful to make quick surveys of compact sources. For a source size Ω wider than the primary beam the total synthesis time scales with $\Omega \Delta \tau (\frac{D}{\lambda})^2$ where $\Delta \tau$ is now the time required to map the primary beam. Ignoring the extra time needed to fully calibrate the mosaic field, the detection speed to reach ΔS is

$$DSE = ND^2 \ . \tag{5}$$

For sources wider than the primary beam the IRAM array is still faster than all existing arrays in their present configuration (column 6 of Table 2). However, it will become much slower than the next generation arrays.

For extended sources and for line observations, the relevant sensitivity is the brightness temperature detectable over the synthesized beam $(\frac{\lambda}{B})^2$ where B is the maximum baseline of the array. To compare different interferometers with similar instrumental performances and sky opacities we may define a brightness detection speed which scales as ND^4B^{-4} and ND^2B^{-4} for compact sources or sources wider than the primary beam, respectively. For comparable baseline extents these speeds are given in the last two columns of Table 2.

Finally, one may easily compare the mapping speed of various arrays by using equation (1). The LSA would be roughly 100 times faster than the IRAM array, but adding a sixth antenna to the latter would increase its speed by 50 percent, and would make array reconfigurations much less frequent.

4 The Impact of Future Mm-Wave Arrays

Table 3 compares the r.m.s. flux density and brightness temperature ΔS and ΔT_b achieved at 90 and 230 GHz after 8 hours observing with future arrays and the present or expanded IRAM array. We estimate ΔT_b from $\Delta T_b = 1.2 \ 10^6 \nu^{-2} \theta^{-2} \Delta S$ where the observing frequency ν is in GHz, and the synthesized beam θ is in arc second. We restrict ourselves to a continuum bandwidth of 2 GHz which is appropriate for distant objects or faint galactic sources, and to 1 kms^{-1} resolutions with $\theta = 1$" and 0.5" which are adequate in several molecular studies of circumstellar envelopes and star-forming regions. For the next generation arrays (MMA, LSA, LMSA) we take $Tsys = 80$ and $100K$, $\eta_a = 0.85$ and 0.80 at 90 and 230 GHz, respectively, and $\eta = 0.80$ assuming that the antennas are located on the best sites. For the IRAM array we take $Tsys = 120$ and $400K$, $\eta_a = 0.6$ and 0.4, and, assuming good weather conditions, $\eta = 0.8$ and 0.6 at 90 and 230 GHz, respectively.

Table 3 shows that: (i) Future mm arrays, and especially the LSA, are suited to detect weak continuum sources and to map weak optically thin lines with high spatial resolution; (ii) Around 90 GHz the sensitivity of the expanded IRAM interferometer would not be greatly different from that of the MMA; (iii) For a given value of the synthesized beam, the brightness sensitivity is always better at 230 GHz than at 90 GHz (unless the site properties degrade much the 230 GHz sky opacity).

Table 3. 1σ sensitivity after 8 hours integration for 2 GHz broad continuum observations and for 1 kms^{-1} line channel observations

Array	n D(m)	Flux density (1 σ, mJy) 2 GHz bandwidth		Brightness temperature (1 σ, K) Line observations		
		90 GHz	230 GHz	90 GHz $\theta = 1$"	230 GHz $\theta = 1$"	230 GHz $\theta = 0.5$"
IRAM	5x15	0.12	1.6	1.4	0.9	3.6
	8x15	0.07	0.5	0.8	0.55	2.2
MMA	40x8	0.022	0.03	0.3	0.03	0.13
LSA	50x16	0.004	0.006	0.055	0.007	0.03
LMSA	50x10	0.011	0.015	0.14	0.017	0.07

5 Conclusions

The sensitivity and mapping speed of the Large Southern Array (LSA) would be unsurpassed in the 1-3 mm band. The LSA would be a powerful instrument to make quick surveys of compact sources, to detect sources as faint as a few μJy, and to map weak, complex continuum and line sources. However, the LSA will require new technological (hardware *and* software) developments among which one must note processing a huge number of baselines with a flexible correlator. We find that expanding the IRAM array to 6-8 elements would make it nearly comparable to the MMA in the 3 mm band, and we suggest that the LSA could be made of about 7 sub-arrays each one containing 6-8 large, high quality antennas capable of quick phase calibration.

Acknowledgements. I thank F. Viallefond for stimulating discussions.

Site Survey for a Large Southern Array

A. Otárola[1], G. Delgado[1], L. Bååth[2]

[1] SEST project, European Southern Observatory, Casilla 19001, Santiago 19, Chile
[2] Centre for Imaging Sciences and Technologies, Halmstad University,
Box 823, S-301 18 Halmstad, Sweden

Abstract. In this paper we describe ongoing site testing activities in the Atacama desert of northern Chile, including relevant geomorphologic and meteorological information on the sites under test as well as proposed new sites and on the area in general. Based on the collected information and the availability of fairly good infrastructure support, we conclude that the north of Chile is potentially a very good region to install a (sub-) millimetre array of the size and sensitivity of the proposed Large Southern Array.

1 Introduction

A consortium of observatories has been formed in Europe to project the building of a large mm- wavelength array in the southern hemisphere.

Among the most important characteristics that a site must meet to fulfill the requirements of the European mm-array are: high altitude to minimise the oxygen contribution to opacity; low water vapour content, with a maximum desirable of 1 mm of precipitable water vapour; flat and with an area of at least 10x10 km to allow long baselines; access to good roads for transport of heavy equipment; it is also desirable that the site lies close to power lines and, even better, near a large city.

2 General Description of Northern Chile

The site survey activities are concentrated in the II Region of Chile, which extends between 21.5 to 26 degrees South and 67 to 70.5 degrees west.

The very dry atmosphere in the Atacama desert, together with the availability of many extended flat areas at a range of elevations up to 6.000 metre, makes northern Chile an exceptional place for high frequency in radio astronomy [1, 2].

2.1 Geomorphology

From the geomorphologic point of view the II Region of Chile is dominated by three mountain ranges extending north-south: the Domeyko mountain range at mid-longitudes between the Coastal and Andes mountains, with elevations as high as 5.000 metre. The Andes mountains have some very high peaks, up to 6.200 metre. Some of them are volcanoes which have been inactive for centuries, the exception being is the volcano Lascar located 32 km south of the village San Pedro de Atacama (23.2 S 67.75 W, 2.400 m), characterised by persistent fume emissions and by occasional large plumes of visible white vapour, with no significant eruption recorded [3, 4, 5].

2.2 Meteorology

Airflow over the South Pacific area is dominated by the jet stream, which flows almost continuously from west to east. The dryness of the Atacama desert is explained by the fact that usually a very stable high pressure system in the area deflects the jet stream to the south [6] which, together with the high altitude of the Coastal and Andes mountain ranges, makes the humid air masses coming from the Pacific and Atlantic oceans condense before reaching the central part of the Atacama region.

The exception to the rule occurs in the period December to February when the probability of getting rain due to the altiplanic winter is higher. This altiplanic winter, also called the "Bolivian Winter", is due to a tropical high humidity air mass that reaches northern Chile from southern Bolivia.

2.3 Infrastructure

Cities Antofagasta, administrative Head of the region, is the largest city in the area, with more than 200.000 inhabitants. Other cities close to Antofagasta are Calama with about 110.000 inhabitants and Chuquicamata with a population of about 20.000.

Mining is the most important economic activity in the region, with Antofagasta having the lowest unemployment rate in Chile. These cities are large business centres for the northern area of the country with representatives of all government departments and agencies.

Roads The main road in the area is the Pan-American Highway which in the north of Chile does not go through the main cities. Access to them is through paved roads which, in general, are in fairly good condition.

More important for the mm-array project is the secondary system of roads crossing through the Atacama desert. Some of these roads are paved, but most are dirt roads, maintained in good shape by mining companies.

The two sites which are currently under study by the National Radio Astronomy Observatory (NRAO) and the Nobeyama Radio Observatory (NRO) have good access roads. In the case of NRAO, the access road to the site is paved almost all the way from Antofagasta.

Airports Antofagasta and Calama both have airports, suitable for the landing of big aeroplanes. Several daily flights are available from Santiago to Antofagasta and viceversa, usually serviced by Boeing 737s. At least two daily flights are available to and from Calama.

San Pedro de Atacama, a village 100 km east from Calama, has become the operations centre for the exploration and site survey of NRAO and NRO. San Pedro has a landing strip suitable for small aeroplanes.

Other landing strips, belonging to private mining companies, are distributed around the area, with one of these no more than 60 km from a good candidate site for the European mm-wavelength array.

Railroads An important railroad company in the area is the "Ferrocarril Antofagasta-Bolivia". Its main activities are transports of Blewster copper bars from Chuquicamata to Antofagasta and copper ore from other mining companies. This railroad also connects Chile with Argentina and Bolivia.

Many sites suitable for the installation of a big radio interferometer are close or even adjacent to the railroad, therefore making it feasible to transport heavy equipment during the construction and operation of the mm- wavelength array.

Harbours Mining being the main activity of this region it has very busy harbours: Antofagasta is the second biggest port in the country. The infrastructure for the harbour of Antofagasta is complete, with direct access to main roads and railroads. All the VLT components (including the 8 m mirrors) will go through this port. The port not only serves the mining industry, but also it is the port through which Bolivia moves all its sea freight.

Tocopilla, to the north of Antofagasta also is a big port, but utilised almost solely for the shipping of mineral ore.

3 Sites for a Radio Interferometer

3.1 Sites under test

Three sites are now under test with instrumentation already installed.

Paranal (base camp), 2.400 m This site is located 120 km south of Antofagasta and about 12 km east from the coast in a high peak on the Coastal mountains and is the site of the ESO Very Large Telescope [7].

The Vega Valley, a small flat area close to Paranal at about 2.200 m, has been monitored for more than one year by NRO with support from the SEST staff. A 220 GHz tipping radiometer, a radio seeing monitor and a portable weather station are in place. This area, depending on the size of the mm-array, can be a potential site. However, as pointed out by Martin [1], Vega Valley experiences only on average 20 % of the year with PWV less than 1.5 mm, with an exception in September where the PWV peaks at around 40 %.

Martin concludes that Paranal is an attractive site for millimetre wavelength observations, but its merit as a potential submillimetre site is not outstanding.

Río Frío, 4.200 m This is the main target of the NRO site testing project. Data on the opacity at 220 GHz, phase stability, and meteorological conditions have been collected since June 1994.

This site is a 3x5 km plateau at about 4.200 m, 350 km south-east from Antofagasta. It is almost at the same latitude as Paranal but on the east slope of the Domeyko mountain range.

Data collected during July and August 1995 show that the opacity at 220 GHz was lower than 0.066 nepers for 75 % of the time.

Chajnantor, 5.000 m Under investigation by NRAO since April 1995. It is located about 60 km from the village of San Pedro de Atacama. Access to the site is on paved roads all the way from Antofagasta to San Pedro de Atacama, from where the international road Paso de Jama (to Argentina) connects to the site.

This site is a lovely high plateau overlooking San Pedro de Atacama. It is very flat and the size is about 4x4 km. Some of the lowest water vapour contents ever recorded at any site on Earth have been measured here with opacities at 220 GHz as low as 0.017 nepers, this is the expected opacity due to oxygen alone (the collected data is available at http://www.tuc.nrao.edu/mma/mma.html).

The instrumentation deployed at the site consists of a 220 GHz tipping radiometer, a 300 m baseline interferometer for phase stability measurements, and a weather station.

3.2 Possible sites for the European millimetre wavelength array

The Onsala/ESO site testing team have explored some additional areas in order to find a site large enough to hold a big array with long baselines as required by the European concept.

A few sites has been visited during 1995 such as Pampa Loyoques (4.350 m) and Pampa Herrera (4.500 m). These site share extended flat areas of igneous rocks, with sparse vegetation. No further reasearch has been conducted at these sites because they are situated too far inside the mountain range.

Of the explored sites the two most promising ones are situated side-by-side at the southern edge of Salar de Punta Negra and are described in the following lines.

Pampa El Chino, 3.300 m Pampa el Chino is situated on the east side of the Domeyko mountain range, about 80 km of the Escondida mine, next to the southern end of the Salar de Punta Negra.

The area is 15x15 km, flat but slanting east to west with an altitude of 3.134 m at the eastern border and 3.550 m at the western border.

Pampa San Eulogio, 3.750 m It is an exceptionally flat area of 20x20 km. it lies close to several volcanoes (Llullaillaco, De la Pena, Aguas Calientes). No water erosion could be noticed (indicating little impact from the "Bolivian Winter".

The sites share characteristics of accesibility and geomorphology:

i) Accessibility The sites are located about 300 km east of Antofagasta through 240 km of good paved road, part of it built by the Escondida mine. Two lines of 80 MW each are already installed to provide the Zaldivar and Escondida mines with electrical power.

Since these mines are fairly large, with a combined workforce of over 4.000 workers, there is access to all reasonable logistic support: communication, lodging, medical care, supplies etc., as well as to landing strips.

A railroad is running about 80 km west of the sites. This railroad is mainly being used to transport the copper ore from the Zaldivar mine down to the port of Antofagasta.

ii) Geomorphology The Andes are 40-50 km to the east of the sites and the highest peak in the area is the inactive volcano Llullaillaco. The Domeyko mountain range protects the area from the predominant north-west winds.

4   Conclusions

Considering the magnitude of the planned European array every effort should be made in order to guarantee that the selected site is the best possible one.

Due to the experience and data gathered by the site survey efforts of NRO and NRAO, we can assure that the north of Chile shows exceptionally good atmospheric conditions for high frequency radio astronomy.

Despite the fact that the northern desert in Chile is one of the least populated areas of the country, the big mining activities in the area give access to good roads, power lines, means of transport (railway and airfields), and fairly large cities.

Of the sites currently under test, Chajnator is clearly superior if only the atmospheric conditions are considered. However, the platform is too small for

10x10 km baselines and a large part of the array would have to be located on other, nearby plateau's. This is possible, but would result in great difficulties in changing the array configuration with big antennas being moved over and around mountains at the very large altitude of 5000-5500 m. In addition we regard the altitude of 5300 m to be too high for the continuous and extensive labour that has to be done at the site to set-up new configurations, service the antennas, and install and test new hardware. The Chajnantor site seems an excellent choice for a smaller size array, such as the one planned by NRAO.

Río Frío is larger, but is also somewhat too small to fit a 10x10 km array. The site is also far away from existing infrastructure.

Paranal has the advantage that the infrastructure for VLT will be there. However, this site has a low altitude and the existing data confirm that this results in a significantly higher opacity at 220 GHz. We do not consider Paranal as a viable alternative site.

Because of the accessibility, geomorphology and excellent meteorological conditions, it is our opinion that the Pampa el Chino-Pampa San Eulogio complex area is one of the best candidates for the installation of a very large radio interferometer useful even at submillimetre wavelengths, following the guidelines of the present European Large Southern Array concept.

5 Recommendations

Based on the requirements of the Large Southern Array we make the following recommendations:

For a smaller array than the present concept and aimed at shorter, submillimetre wavelengths, we recommend the Chajnantor site. This is clearly the best site for very high frequency operations and can be extended with baselines up to 5-7 km. For a larger array, as the presently proposed 50 x 15m array, working at mm wavelengths, we recommend the area formed by Pampa el Chino and Pampa San Eulogio. The altitude of this place is such that hard labour can be done there and work on the antennas, construction and maintenance, can be performed at the site with standard vehicles and machinery. Power lines, water supply, infrastructure are available at the nearby Escondida mining site.

For the immediate future:

1. We recommend a full site test at the Pampa el Chino and Pampa Eulogio complex.

2. We also recommend that smaller scale tests are carried out at Pampa Loyoques. This area is 1 km higher than Pampa el Chino and therefore is expected to have proportionally lower water vapour content. It is also close to the proposed MMA site making collaborating projects natural for the future.

6 Acknowledgements

The authors want to thank the Environmental Department of the Escondida mine for access to their weather data and logistic support. Also the Civil Protection Department at ONEMI has collaborated by giving us access to their documentation.

References

[1] R. Martin, Preliminary Summary of the 22 GHz PWV Data from Paranal, SMT Technical Memorandum UA- 90-2, 1990

[2] K. Kono, R. Kawabe, M. Ishiguro, T. Kato, A. Otárola, R. Booth, and L. Bronfman, Preliminary result of Site Testing in Northern Chile with a Portable 220 GHz Radiometer, NRO Technical Report No. 42, 1995

[3] L. Glaze, P. Francis, S. Self, and D. Rothery, The 16 September 1986 eruption of Lascar volcano, north Chile: satellite investigations, Bulletin of Volcanology, Vol. No. 51, pp. 149-160, 1989

[4] M. Gardeveg, S. Sparks, S. Mathews, and P. McLeod, Noveno Informe Sobre el Comportamiento del Volcán Lascar, II Región, Febrero-Marzo de 1995, ONEMI Internal Report, 1995

[5] M. Gardeveg, E. Medina, M. Murillo, and A. Espinoza, La Erupción del 19-20 de Abril de 1993; Sexto Informe Sobre el Comportamiento del Volcán Lascar, ONEMI Internal Report, 1989

[6] M. Mornhinweg, Make your own weather forecast, Internal Memo from La Silla Observatory, 1994

[7] R. West, The VLT Goes to Paranal!, The Messenger, Vol. No. 62, pp. 1-2, 1990

Correction of Atmospheric Phase Fluctuations

Richard Hills

MRAO, Cavendish Laboratory, Cambridge, CB3 0HE, England

Abstract. Large millimetre-wave arrays will only achieve their expected angular resolution if steps are taken to remove the "seeing" introduced by the atmosphere. A range of techniques are available for doing this including self-calibration, frequent observations of calibration sources and the measurement of emission from the atmospheric disturbances that cause the problem. Although the last of these methods is not yet established as a routine procedure, initial results are very encouraging and it is likely that it will prove to be the most efficient and least constraining approach.

1 Introduction

All astronomers are familiar with the problem that turbulence in the atmosphere introduces distortions into the wavefronts of signals arriving from astronomical sources. This distortion sets a limit to the angular resolution that can be obtained by ground-based telescopes unless active steps are taken to correct for it. The amount of distortion can be characterised by the rms path difference Δp between two points in the telescope aperture separated by a distance r. It is found that Δp rises with increasing separation r and the limiting angular resolution at a particular wavelength λ is of order λ/r_0, where r_0 is the separation at which Δp reaches $\approx \lambda/2.4$. It is also well known that, because the increase in Δp with r is slower than linear (it goes as the 5/6th power in the case of Kolmogorov turbulence), the limiting angular resolution improves as one goes towards longer wavelengths, albeit only slowly.

At optical and near-IR wavelengths this "seeing" is primarily caused by the refractive index fluctuations associated with the density variations caused by temperature differences between the turbulent cells. The sizes of the turbulent structures that are important at those wavelengths range from centimetres to a few metres. The fluctuations in this "dry" component of the atmosphere are in fact not well characterised over the substantially larger scales of hundreds metres to several kilometres that will be important for the large millimetre-wave arrays. In general though, they are not likely to be significant under good observing conditions (although it is worth noting that the pressure gradients associated with weather systems do give rise to refraction effects which need to be corrected for accurate astrometry). In passing from the IR to the sub-millimetre wavebands, however, we have moved from frequencies above the main rotational transitions of water vapour to frequencies below them. This means that, because of the substantial dipole moment of the water molecules, the effective refractive index is greatly increased. In fact each additional of 1 mm of precipitable water vapour in the line of sight gives rise to more than 6 mm of additional path delay in the millimetre-wave and radio windows. Since water is not well mixed in the atmosphere (as a result of its tendency to condense out when the air is cooled and then to be reinjected into the atmosphere by evaporation at ground level) this creates an additional strong but highly variable source of "seeing" not present in the optical and IR wavebands. See Masson (1994a) for a detailed discussion of the relationship between optical and radio seeing.

These phase variations due to water vapour have been known to radio astronomers for many years, e.g. Hinder and Ryle, 1971 and corrections are often made by using one or more of the methods described below. At millimetre wavelengths, however, the effects are both stronger and harder to correct because turbulence with smaller physical scales becomes more important and this produces fluctuations with shorter timescales.

Since the amount of water vapour in the atmosphere and the extent to which it is well-mixed both depend on complicated physical, meterological and even biological processes, it is likely to vary strongly with site location, season, time of day and so on. In testing possible sites it is therefore important to measure the path fluctuations over substantial periods of time. This is most conveniently done by building a small radio frequency interferometer and observing the beacons or TV signals broadcast from a geostationary satellite. This approach was pioneered by Ishiguro et al. (1990) at Nobeyama and systems have since been installed on a number of sites.

Masson (1994b) has used this technique to measure the atmospheric phase fluctuations on Mauna Kea. He reports that without correction the limiting angular resolution at 345 GHz would be poorer than 1 arc second for rather more than half of the time. It is important to note that the strong diurnal effect on Mauna Kea means that the middle part of the day is often much worse than this (see Church and Hills, 1990), but the nights can be much better. This is confirmed by experience with the JCMT-CSO interferometer. The initial reports from the high altitude sites in Chile indicate that they are substantially better than Mauna Kea, but show that for much of the time the atmospheric fluctuations will still prevent observations at wavelengths of around a millimetre with the resolutions of 0.1 arc second or even lower which are demanded by many of the astronomical programmes that we are discussing here.

2 Methods of Correction

Under reasonably good conditions the variable refraction at millimetre and submillimetre wavelengths is weak, in the sense that the amplitude of the wave is not significantly changed, just the phase. For small antennas the slope of the wavefront across the individual apertures will also generally be small. This means that in order to correct the data from a synthesis array we only need to know the phase shifts introduced by the atmosphere into the paths between the source and each antenna. These must be measured more frequently than the time over which the corrections change as a result of the movement of the turbulent cells through the beams of the telescopes.

Note that the refraction due to water vapour is very nearly non-dispersive at frequencies below 1 THz, so that the phase shift is just linearly proportional to frequency. (The small deviations from linearity at frequencies near strong absorption lines could be calculated and applied if necessary.) In practice the corrections can usually be applied as a phase shift to the whole band of frequencies being observed, e.g. by rotating the local oscillator phase, but this is not strictly accurate and for the very large bandwidths now being considered it would be better to apply it as a correction to the delay.

I will now outline three different approaches to obtaining the information necessary to make these corrections.

2.1 Self-calibration

This is the method that is most often used on centimetre-wave aperture synthesis telescopes, see Thompson, Moran and Swenson (1986). The necessary phase information is obtained from the astronomical signals themselves, which is possible because there is redundant information in the correlations measured for real sources. The method relies on there being enough bright objects in the field of view to yield a good signal-to-noise ratio in a time short compared to the period of the atmospheric fluctuations. The related technique of measuring "closure" phases has the same limitation. These methods are roughly analogous to adaptive optics in the optical-IR bands using a "natural guide-star". Note, however, that because one is applying this correction in the data processing and constructing a model of the source brightness distribution, it is not necessary for there to be a point source in the field, but it is essential to have adequate fringe visibility above the noise on most of the samples. At millimetre wavelengths most of the objects we wish to observe have intrinsically much lower surface brightness than those at centimetre wavelengths and we are also likely to be working with higher system noise temperature, smaller dishes and shorter coherence times. For these reasons self-calibration will not solve the atmospheric phase problem for millimetre arrays anything like as well as it does on centimetre-wave synthesis telescopes.

2.2 Frequent Observations of Calibration Sources

It is in any case necessary to measure calibration sources (preferably bright unresolved objects of known position) when making aperture synthesis observations so that one can correct for instrumental phase and amplitude variations, obtain baseline parameters, and so on. If these measurements are made sufficiently frequently and the calibration sources are sufficiently close in the sky to the objects being observed, then the atmospheric phase variations can also be obtained in this way. The problem is that at millimetre and especially submillimetre wavelengths it will be difficult to meet these requirements.

Our submillimetre measurements on Mauna Kea show that the fluctuations can already be significant on timescales as short as a few seconds, so to correct these it would be necessary to build antennas capable of switching to calibration sources every few seconds. Sources bright enough to give an accurate phase measurement in such a short time are rare at high frequencies, so one would generally have to switch through a significant angle to find one. This reduces the accuracy of the correction since the signal is no longer passing through the same piece of atmosphere. Various strategies can be employed to improve the situation (see the memos by Holdaway and colleagues in the MMA series): one can make the phase calibration measurements at a different frequency – either at a lower one where the synchrotron sources are stronger, or perhaps at one where there are maser sources available; with a large array one could have some of the antennas tracking calibration sources continuously. It may even be possible to use a number of antennas switching rapidly between several bright calibrators to construct a model of the slowly-evolving phase screen presented by the atmosphere as it moves across the site. Any such methods are however likely to be expensive in terms of the amount of resources – antennas, receivers, computing and observing time – that have to be devoted to them.

2.3 Measurement of Millimetre-wave Emission from the Water

This third approach has long been advocated by Welch and was first demonstrated by him on the BIMA array (Welch, 1994). It makes use of the fact that the water molecules which cause the path delay also produce emission at millimetre wavelengths. Any variation in the water vapour content of the path to an antenna will therefore show up as a change in the total power output from the receiver on that antenna. Careful design of the receiver systems is necessary to ensure that instrumental instabilities do not mask the atmospheric fluctuations, but recent results from IRAM (Guilloteau, 1996) have shown that this method can be made to work very well. In particular total power measurements at 230 GHz, where the emission from the water vapour is relatively strong, provide very precise correction of the phase fluctuations at 100 GHz.

.There a number a factors which may limit the performance of systems which use the astronomical receivers themselves to measure the water vapour emission. First of all the emission is a strong function of frequency and a certain amount of work will be needed to calibrate the system at each observing frequency. Secondly the measurements may be confused by the signals from the astronomical sources, especially when one is observing a bright calibration source. Thirdly the emission in the atmospheric "windows" is proportional to the water vapour content of the path, w, but also depends on the pressure at the altitude where the water is, since the emission is due to pressure broadening of the far-IR lines. The path delay is also proportional to w but is essentially independent of pressure. The relationship between the emission measured and the phase shift to be applied therefore depends on the altitude of the water, which can only be estimated indirectly. There is also a possibility that the emission in the "windows" contains a component which depends on collisions between pairs of water molecules, which would depend on the square of the local water vapour density.

For these reasons a number of groups are developing special-purpose radiometers which will measure the water vapour along the line of sight to each antenna at a fixed frequency. These can be simple uncooled systems designed for high stability and with appropriate switching arrangements to provide frequent calibration (which would be inefficient in the astronomical receivers). Hall (these proceedings) is building a set of 220 GHz radiometers to provide accurate phase correction at centimetre wavelengths on the Australia Telescope. By observing at carefully chosen frequencies near water vapour emission lines the problem of pressure dependence can be also avoided. Woody is developing 22 GHz radiometers for the OVRO array which exploit this.

For the JCMT-CSO interferometer we are building radiometers which measure the 183 GHz emission line. This has the advantage that the signals are strong so that very small fluctuations can be measured in a short time: we expect to be able to correct the path delay to better than 25 microns with a 1 second time constant. The complication is that the optical depth near the centre of the line will be greater than one under most conditions so the relationship between emission and water vapour content is non-linear. To combat this we are using a broad-band mixer with 3 separate IF channels, spaced 1.4, 4.2 and 7.8 GHz away from the line centre. Using these three channels together should enable us to obtain an accurate value for the path delay over the range of conditions normally experienced on Mauna Kea. It should also be possible to detect the broad-band emission from water droplets which would confuse a single-frequency radiometer. The radiometers will be mounted near the secondary foci of each telescopes at positions which are slightly offset from the astronomical receivers. This means that the beam of the radiometer is offset from the astronomical one by a few arc minutes on the sky, so it will not normally be confused by the astronomical signals, but that there is an almost complete overlap between the beams as they pass through the lower part of the atmosphere.

As as final comment on this type of system, I note that such a radiometer could also be used to measure the gradient of water vapour emission across the aperture of a single antenna. This could for example be done by illuminating say quarter of the aperture at a time and using a scanning mirror to switch between sectors. Synchronous detection would provide a signal proportional to the emission gradient which could be used to correct the pointing of the antenna in such a way that it followed the variations in the apparent direction of the source caused by the refraction due to the gradient in the water vapour. This could be applied either to large single-dish telescopes, to make it possible to observe in the afternoons on Mauna Kea for example, or to the antennas of a synthesis array for those measurements, like "mosaicing", which require especially accurate pointing of the individual elements.

3 Conclusions

If nothing were done about them, atmospheric phase fluctuations would seriously limit the performance of large millimetre-wave arrays. Fortunately there are relatively simple techniques available for correcting them which should work well. The results from the systems now under development will show which of these is the best in practice and provide proof that the high angular resolution which has been set as a goal for the LSA can be achieved.

References

Church, S. E. and Hills, R.E. (1990), in URSI/IAU Symp. on *Radio Astronomical Seeing*, J E Baldwin and Wang Shougan (eds), IAP, Beijing, 75–80

Guilloteau, S. (1996), these Proceedings, see also recent issues of the IRAM Newsletter

Hinder, Richard and Ryle, Martin (1971), MNRAS, **154**, 229

Ishiguro, M., Kanazawa, T., and Kasugo, T. (1990), in URSI/IAU Symp. on *Radio Astronomical Seeing*, J E Baldwin and Wang Shougan (eds), IAP, Beijing, 60–63

Masson, C. R. (1994a), in IAU Symp 158, *Very High Angular Resolution Imaging*, J G Robertson and W J Tango (eds), Kluwer, Dordrecht, 1–9

Masson, C. R. (1994b), in: *Astronomy with Millimeter and Submillimeter Interferometry*, M Ishiguro and Wm J Welch (eds), ASP Conference Series **59**, 87–95

Thompson, A. Richard, Moran, James M., and Swenson, George W., Jr. (1986), *Interferometry and Synthesis in Radio Astronomy*, Wiley-Interscience, John Wiley & sons, New York

Welch, Wm. J. (1994), in: *Astronomy with Millimeter and Submillimeter Interferometry*, M Ishiguro and Wm J Welch (eds), ASP Conference Series **59**, 1–9

Atmospheric Phase Correction Based on Sky Emission in the 210-248 GHz Band

Michael Bremer, Stéphane Guilloteau, and Robert Lucas

Institut de Radio Astronomie Millimétrique, Domaine Universitaire de Grenoble, 300 rue de la Piscine, F-38406 St. Martin d'Hères, France

Abstract. For millimeter radio astronomy, the instrumental performance is nearly always limited by the amount of fluctuating tropospheric water vapor in the line of sight. Several phase correction schemes for interferometers have been proposed, both for current observatories and the large arrays planned for the next century.

We present some examples from the now operational real time phase correction system at the Plateau de Bure Interferometer (PdBI), which employs the stability of its dual frequency SIS receivers to monitor the atmosphere on timescales from one second to two minutes along the line of sight. The gain in coherent integration time was found to be considerable, and the phase noise became effectively independent of baseline length (with an optical path r.m.s. of 90 μm over one minute). However, the system works only during clear sky conditions and doesn't allow to track the phase over source changes.

Studies of the modelled phase show that on occasion (ground wind near 10 m/s), the atmospheric phase screen can move the phase at 230 GHz by 100° on timescales of two seconds (average over one hour, baseline 262 m) for an amount of precipitable water of 3 mm. The PdBI monitoring technique manages to transform this essentially random phase to an r.m.s. of about 25-35 degrees over one minute.

1 Introduction

Millimeter astronomy operates at the high frequency edge of the atmosphere's radio transmission window, a border that is ever shifting due to the amount of atmospheric water vapor. Because the average scale height of water is just two kilometers, many observatories have been constructed on dry sites at high altitudes to get above the greater part of it.

But even if the altitude is high enough for good transmission, the residual water can induce considerable phase shifts. This is because the refractive index of water vapor increases from optical to radio wavelengths due to infrared resonances of the water molecule. Furthermore, vapor mixes badly with dry air. Under the influence of atmospheric turbulence it tends to form bubbles on scale sizes between some centimeters and several kilometers. The phase shifts induced by these bubbles can therefore be correlated over distances up to several kilometers. Being lighter than dry air but rather close to its transitions towards liquid and solid, water vapor is one of the most unstable tropospheric constituents. Ground measured humidity values correlate only with the averages over several days of the vertical profiles (see e.g. Staelin (1966)), which makes remote sounding necessary for a real time phase correction.

1.1 Remote sounding

Vertical water distributions are also of interest for other disciplines, e.g. meteorology, aviation and Earth-space communication links, which have developed various monitoring techniques.

The phase monitoring at the PdBI uses simply the total power output of the new dual frequency SIS receivers at the chosen observing frequency in the 210-248 GHz band (for technical details about their design see Blondel et al. (1995)). Typical relative receiver stability over 10 minutes is $2 \cdot 10^{-4}$, which corresponds to a path r.m.s. of $18 \mu m$ ($5°$ phase noise per antenna at 230 GHz) for good atmospheric conditions.

Due to the distance to the 183.3 GHz water resonance, sensitivity to water is reduced, but also the risk of saturation during the summer months when atmospheric turbulence is worst. Even so, it can be shown (Bremer (1994)) that the sensitivity to water vapor in the 84-116 GHz band is about the same as at 22.2 GHz, and that the sensitvity at 210 to 248 GHz is by factor 5 to 6 higher (Fig. 1).

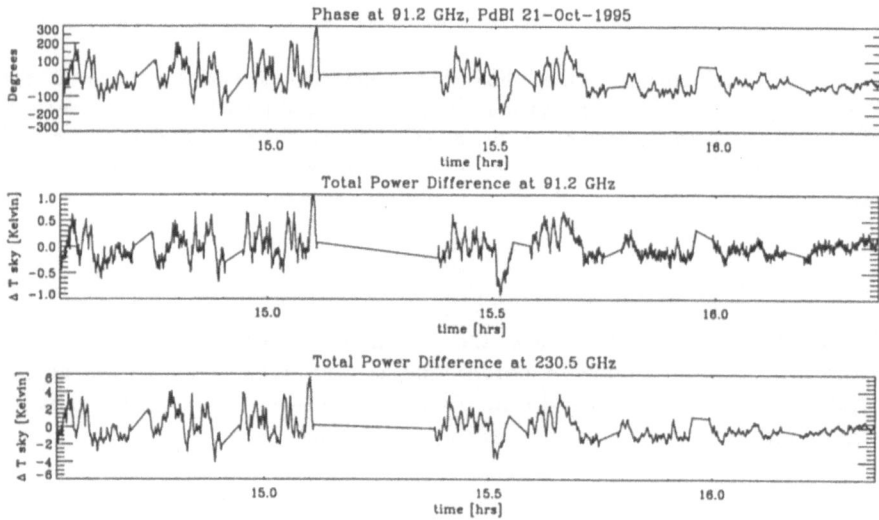

Fig. 1. Observation on the strong quasar B1749+096 during unstable atmospheric conditions after the passage of clouds. Time resolution is one second

1.2 The model

After converting the total power towards the antenna temperature scale via a standard load calibration, the atmospheric model by Cernicharo (1985) is used to calculate the corresponding opacity based on kinetic line profiles and the

U.S. 1962 45° northern latitude standard atmosphere (15 layers). A vapor scale height of 2 km is assumed. With the wavelength dependent refractive index by Hill and Clifford (1981), the optical path for each antenna is calculated to obtain the gradient $\Delta\text{path}/\Delta T_{sky}$, which is used until the next calibration to transform the fluctuations of atmospheric emissivity into phase shifts.

This technique seems to work *always* during clear sky conditions, as shown by tests in summer and autumn 1995 (Bremer (1995), Bremer, Guilloteau and Lucas (1995)). The main limitations were noise and time resolution in the total power measurement. In the presence of clouds, however, applying the correction *degrades* the signal. Hence, to avoid loosing valuable observations, two independent data sets are stored, one with the phase correction applied and the other without. The user can switch freely between both during data reduction.

It has been tried to include a cloud correction by combining the simultaneous total powers in the 84-116 GHz and 210-248 GHz bands, but up to now the results are not encouraging.

2 Examples

Because the observed phase is defined modulo 360°, the maximum phasenoise φ_{max} is $180°/\sqrt{3}$. No such restriction exists for the phase derived from the predicted pathlength (Tab. 1). Phase noise histograms for the raw and corrected observations on strong point sources allow to study the effect of the model on the phase stability. From the experiments, the predicted phase shifts were found to be more accurate than 20%.

Table 1. Phase shifts of the 230 GHz model used to correct the data on October 21 and 24 on the longest and shortest baselines, averaged over one hour. The amount of precipitable water was about 3 mm on both days, but the phase moved twice as fast on October 24 (compare the bold faced numbers).

Date	Baselines	Δ time [sec]	1	2	4	6	8
Oct. 21	64.0 m	$\Delta\varphi$ [°]	28	48	82	110	134
	176.0 m	$\Delta\varphi$ [°]	28	48	**84**	114	**140**
Oct. 24	48.0 m	$\Delta\varphi$ [°]	47	78	119	136	142
	187.2 m	$\Delta\varphi$ [°]	**49**	**85**	141	185	221
	262.4 m	$\Delta\varphi$ [°]	57	100	169	222	267

Detailed histograms for 115.1 and 230.5 GHz phase for one hour on October 24 for all baselines b are given in Fig. 2. The effect of the correction is such that the phase r.m.s. on $b = 262$ m for 230.5 GHz becomes better than the uncorrected 115 GHz phase at $b = 48$ m.

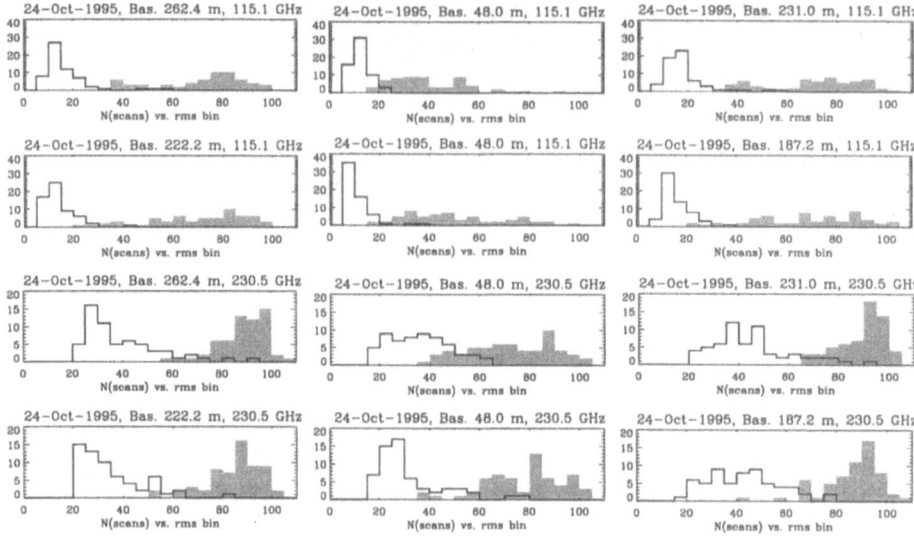

Fig. 2. Phase noise histograms for raw (shaded) and corrected (drawn) data at 115.1 (upper six plots) and 230.5 GHz (lower six plots)

3 Conclusions

Down to an optical path r.m.s. of about $90\mu m$ per minute, clear sky atmospheric phase correction is mainly limited by instrumental effects. Model assumptions can be quite basic for this precision. The behaviour of the phase shifts in time does not seem dominated by the amount of precipitable water, but rather by turbulence and wind speed. For observatories at high altitudes, this calls for phase correction with high temporal resolution.

It is likely that the atmospheric model must be improved to achieve a correction of higher precision. However, our current model requires no initial "training" on a phase calibrator which allows to use it even for VLBI.

References

Blondel, J., Carter, M., Karpov, A., Lazareff, B., Mattiocco, F. (1995), *Dual-Channel SIS Receivers for the Plateau de Bure Interferometer*, Digest of the 20th Int. Conf. on Infrared and Millimeter Waves, Florida, USA

Bremer, M., (1994): *The Phase Project: First Results*, IRAM Internal Report

Bremer, M., (1995): *The Phase Project: Observations on Quasars*, IRAM Working Report N°**238**

Bremer, M., Guilloteau, S., and Lucas, R. (1995), IRAM Newsletter, 24. Nov. pp 6,8

Cernicharo, J., (1985): *ATM: A program to compute theoretical atmospheric opacity for frequencies < 1000 GHz*, IRAM Internal Report

Hill, R.J., Clifford, S.F. (1981), Radio Science Vol. **16** No. 1, 77–82

Staelin, D.H., (1966) J. Geophys. Res. **71**, 2875–2881

Phase Correction Strategies for the Australia Telescope Compact Array at Short Wavelengths

Peter J. Hall

Australia Telescope National Facility, PO Box 276, Parkes, NSW 2870, Australia

Abstract. The Australia Telescope Compact Array (ATCA) is likely to be the first operational mm-wave interferometer in the Southern Hemisphere. This paper examines some possible atmospheric phase correction strategies for the array, notes the applicability of these to future large mm-wave arrays, and points to the usefulness of the ATCA in assessing the various approaches.

1 Introduction

The ATCA is an array of six 22 m dishes arranged in a linear east-west configuration and located near Narrabri, in northern NSW. Five antennas are movable on a 3 km rail track; the sixth is located a further 3 km west. Available baselines range from 6 km to 30 m. At frequencies below 50 GHz the full 22 m aperture is available on each antenna, while a precision 15 m inner section forms the effective reflector at frequencies up to 115 GHz. The array currently operates in a variety of modes (including mosaicing and real-time, one-dimensional imaging) in four broad observing bands below 9.2 GHz. Antenna and system tests show that the ATCA will function effectively at its highest design frequency of 115 GHz, making it viable as the first-generation southern millimetre array. This paper outlines possible solutions to the greatest challenge facing ATCA mm-wave observers: phase fluctuation induced by atmospheric water vapour instabilities. Some of the approaches are likely to be applicable in future large mm and sub-mm wave arrays.

The ATCA site is only 200 m above sea level but, notwithstanding the low altitude, the Narrabri climate is an inland continental one characterized by a clear and still winter season, with night-time temperatures often reaching freezing point. First site evaluation results, based on 30 GHz water vapour radiometry opacity measurements and 9 GHz interferometric stability tests, indicate that approximately 60 winter nights per year will support 3 mm band synthesis, assuming that only existing observing and calibration techniques are used. Precipitable water vapour under such conditions is often < 4 mm. As with other mm-wave interferometers, the challenge is to increase the array coherent integration time to the point where the resulting signal-to-noise ratio allows an armoury of imaging tools (e.g. self-calibration) to be used. At Narrabri the search is particularly tantalizing in that most winter (and many late autumn and early spring) days are cloud-free. A phase fluctuation correction scheme which accounts for atmospheric water vapour instability under such conditions promises substantial return, possibly more than doubling the available mm-wave observing time.

In common with other groups, the ATNF has been examining ways of correcting phase fluctuations induced by the motion of tropospheric water vapour inhomogeneities passing the array. At present, four techniques are being considered:

- Water vapour sounding using dedicated radiometers atop each antenna
- Water vapor sounding using a stable continuum channel in future 3 mm band receivers
- Phase correction at $\lambda = 7$ or 3 mm using simultaneous calibrator source observations at $\lambda = 12$ mm
- Phase correction at short wavelengths based on interspersed cm-wave calibrator observations.

This paper discusses early progress in the water vapour sounding program and mentions briefly the other schemes in the ATCA context.

2 Water Vapour Sounding

The basis of this technique is the use of the emission brightness of tropospheric water vapour as a measure of the integrated vapour column encountered by cosmic radio waves propagating towards an astronomy antenna. The amount of WV encountered determines the so-called "wet delay" and, in large measure, the electrical path (or phase) defining the transmission path. By observing the water vapour seen by each antenna in a synthesis array, it should be possible to make differential phase corrections to account for inhomogeneities in the vapour distribution above the array. Frequencies near atmospheric WV resonance lines are obvious choices in the design of sounding equipment. Practical considerations relating to brightness saturation and pressure broadening effects make it desirable to operate a little away from the lines, at least with robust phase correction algorithms based on straightforward atmospheric models.

After reviewing earlier (disappointing) attempts at phase correction based on 22 GHz sounding and considering the relative sensitivity of emission brightness to water vapour content, the wide range of precipitable water vapour encountered at Narrabri, the required degree of insensitivity to fluctuations in other tropospheric constituents, and available stable receiver technology, the ATNF chose 225 GHz as an operating frequency for its water vapour radiometers (WVRs). A subsequent collaboration with IRAM (Bremer 1994) confirmed, on the basis of more sophisticated atmospheric modelling, that 220-230 GHz operation was indeed a good choice.

While the results of Welch (1994) taken with the BIMA array provided good initial impetus for the renewed interest in water vapour sounding, recent results from IRAM (Bremer 1995) demonstrating the efficacy of the technique on the Plateau de Bure Interferometer (PdBI) have shown beyond doubt the merit of the approach. The IRAM-ATNF collaboration continues to explore the subject, including parallel investigations of astronomy antenna and dedicated WVR sounding. Specifically, the ATNF programs are designed to implement a pilot scheme which involves placing two precision 225 GHz WVRs on elements of the ATCA as well as investigating the merit of WV sounding using a stable continuum channel in future 3 mm band receivers. At present, one prototype 225 GHz WVR has been built and tested, and initial PdBI 110 GHz data, together with that of Welch (1994), encourages further the pursuit of sounding investigations at that frequency.

3   The ATNF Water Vapour Radiometer

The design philosophy adopted for this 225 GHz instrument was aimed at achieving the required sub-1 K sensitivity and 1-2 parts in 10^4 stability over tens of minutes (Hall and Abbott 1993). Figure 1 is a block diagram of the instrument.

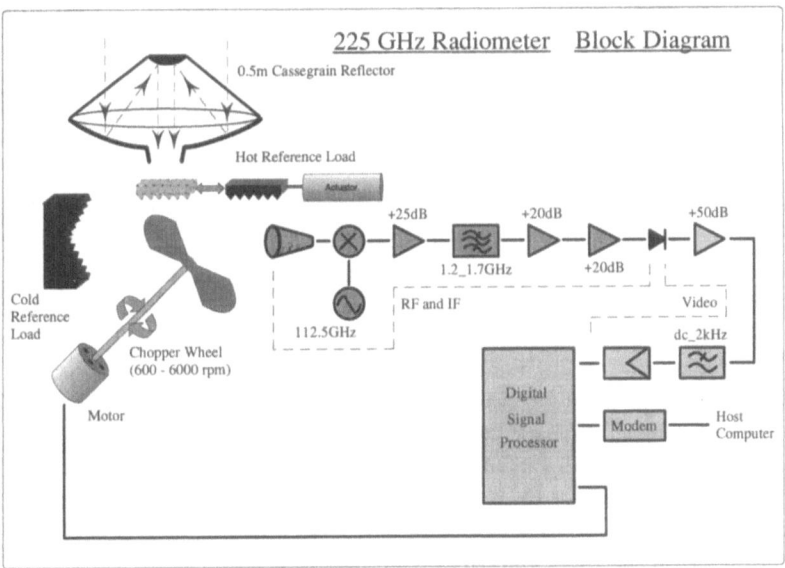

Fig. 1. Block diagram of the ATNF WVR

Table 1 lists the WVR major specifications and measured performance. The WVR is fairly conventional in design, with the front-end components being supplied by Millitech Inc. The Dicke reference load is cooled, rather than heated, to bring the physical temperature closer to typical sky brightness temperatures, thereby permitting more effective gain fluctuation cancellation in the quasi-optic Dicke switching process. An ambient load can be positioned in the sky path, allowing load-load switching and explicit gain and T_{sys} calibration. This load also serves as a fail-safe shutter, protecting the internal optical path from residual solar radiation in the event of a power failure. Within the optics box, mountings with high thermal mass ensure that critical components experience a minimum of short-term temperature variation.

Table 1. ATNF Water Vapour Radiometer Specifications and Performance

Operating frequency	225 GHz
Architecture	DSB superheterodyne, un-cooled Schottky mixer
Optics	Classical cassegrain
Primary reflector aperture	0.5 m
Beamwidth (-3 dB)	11°
Aperture efficiency	50%
System temperature	1600 K
Sensitivity	0.08 K (1 sec. integration)
Stability	0.08 K over 10 sec. (total power mode)
	0.15 K thereafter (switched mode)
Control/Data acquistion system	TMS320C25 DSP μP based
I/O system	Serial data, control and monitor via fibre-optic link

The digital signal processing backend, together with hardware controllers designed around a field programmable gate array, give great flexibility in data acquisition, on-the-fly calibration, and WVR control and monitoring. The fibre-optic serial data link allows a galvanically isolated connection to the remote package. Figure 2 is a photograph showing the prototype WVR mounted on the ATNF Mopra 22 m telescope, a similar antenna to those of the ATCA. Note the Goretex radome covering the optical elements. The T_{sys} penalty is $\lesssim$20K and the cover blocks 70% of incident solar radiation at infra-red wavelengths.

Fig. 2. The prototype WVR mounted on the Mopra 22 m telescope

The performance of the prototype WVR is encouraging and the next stage of the project is to produce a second unit so that interferometric phase correction trials can commence.

4 Other ATNF Phase Correction Schemes

The nature of the ATCA optics, involving an on-axis receiving package selected via an off-axis rotary turret, lends itself well to experimenting with other atmospheric phase correction and calibration techniques. For example, rapid alternation between, say, 9 GHz and higher frequency receivers is possible, allowing calibration to be done using the much larger number of strong sources observable at the lower frequency. It should of course be recognized that since phase scales directly with frequency, any uncertainty in estimating the low-frequency phase error will scale by the same factor. Even so, the source number-brightness statistics at the lower frequency make the technique attractive in principle.

As well as receiver alternation, the next-generation ATCA packages will include a 12 mm band (<18 - 25 GHz) and a shorter-wavelength receiver (either $\lambda = 7$ or 3 mm) within the same dewar. Thus, simultaneous 12 mm and shorter-wavelength observations will be possible, conferring dual-band calibration advantages without the time penalty involved in turret rotation. A particularly interesting possibility arises in the observation of (e.g.) CO sources that have H_2O masers located in the 22 GHz field. In these cases, ATCA observations with the longest baselines will be possible as the phase correction process can be essentially perfect. True sub-arcsecond angular resolution should therefore be obtainable.

With regard to the use of a stable 3 mm continuum channel to obtain phase correction data, it is worth noting that ATCA 3 mm receivers will use HEMT amplifiers cooled to 20 K (in contrast with the 4 K SIS systems in use elsewhere), either in discrete form or integrated into MMIC packages. A project to demonstrate this technology is already under way and it remains to be seen whether expected gains in receiver stability can be realised in practice.

5 Conclusion

With its location and existing infrastructure, the ATCA is well placed to be the first operational Southern Hemisphere mm-wave array. The challenge of obtaining usable performance at a difficult site, together with unique capabilities which allow development of techniques relying on wideband and frequency-agile operation, make it an ideal test bed for some aspects of future arrays.

References

Bremer, M. (1994): The Phase Project - First Results, IRAM Technical Note, April 1994.

Bremer, M. (1995): IRAM Newsletter, November 1995.

Hall, P., Abbott, D. (1993): Interferometer Phase Correction Using Millimetre Wave Water Vapour Radiometry, ATNF Technical Note, AT/31.6.7/018.

Welch, W. (1994): in *Astronomy with Millimetre & Submillimetre Interferometry*, eds. Ishiguro M. & Welch W., Astron. Soc. Pac. Conf. Series, **59**.

Innovative Telescope Designs

Dietmar Plathner

I.R.A.M., 300, rue de la Piscine, F-38406 St. Martin d'Hères, France

Abstract. Some innovative telescope designs of the last 50 years are presented showing the development to structures with higher stiffness and less weight of modern
approaches. A slant axis 15-m radiotelescope is detailed here as it seems to be the best
compromise when taking into account the costs as a criteria for an innovative telescope
design.

1 Introduction

The consortium of observatories preparing the construction of the Large Southern Array (LSA) has given the study of a proposal for the telescopes to IRAM.

Innovative telescope designs have considerably changed the instrument structures which hold the optics with growing accuracy and decreasing costs. In this
paper, it is tried to stress on requirements which can lead to a design of a compact telescope structure of excellent astronomical performance with reasonable
financial implications.

2 Innovative Telescope Designs

In telescope structures stiffness requirements are probably the most dominant
parameters and all efforts to obtain a good astronomical performance are governed by them. High stiffness is obtained by optimally passing the loads through
the structural elements, i.e. in a push-pull bar the whole cross-section is exposed
to the same load while in a beam under bending with a linear stress distribution
over the cross-section, the maximum load permitted is limited by the stresses in
the area of the outer fibres.

A very innovative design some 50 years ago was the 200 inch Mount Palomar
optical telescope (Fassaro, J.S., 1947) (Fig. 1) where for control reasons, the
equatorial mount had to be chosen with two bearings at the ends of the long
polar axis operating in bending mode and two bearings for the elevation axis
with its structure again exposed to bending.

Already much more compact and with less bending in the telescope mounts
are instruments with an alt-az configuration. A typical example is the 15-m
SEST (Delannoy, J., 1985), where three bearings are sufficient as the vertical
(azimuth) axis is compressed into one bearing with high stiffness in tilt.

Only two bearings and thus an even more compact design are possible with
mounts having the optical axis inclined against the elevation axis and the latter
not perpendicular to the azimuth axis. A proposal for an optical interferometer

Fig. 1. 200" Mount Palomar Optical Telescope

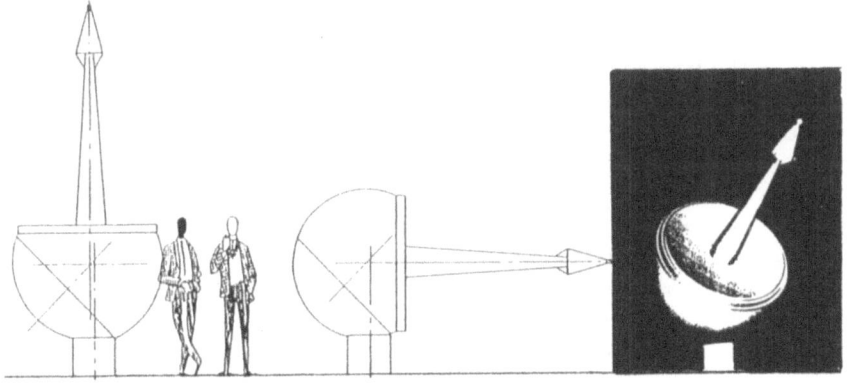

Fig. 2. Optical Telescope with Inclined Elevation Axis

was prepared at IRAM in 1987 for INSU based on a structure as shown in Fig. 2 (Plathner, D., et al. 1987). And it is also possible to design telescopes with only one (spherical) bearing as proposed by A. Labeyrie (Labeyrie, A.) (Fig. 3). Finally, neither mount nor bearings in the proper sense are needed for telescopes where the reflector is directly supported by a number of push-pull bars with variable length as indicated in Fig. 4.

The above-mentioned telescope mounts are taken as examples for innovative designs and illustrate the progress in weight reduction accompanied by an increase in stiffness which leads to natural frequencies of typically 10 Hz and higher for large telescopes.

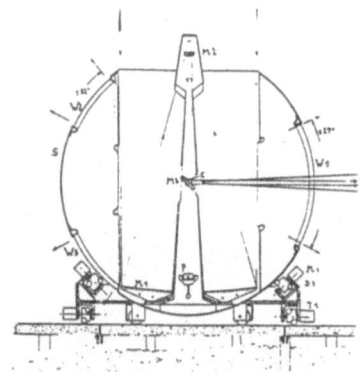

Fig. 3. Single Bearing Telescope (A. Labeyrie)

3 Design Goals

Of the five mounts presented in section 2, we have chosen here the one with the inclined optical and elevation axes, also known as slant axis mount. This mount seems to be the best compromise in view of observational requirements, weight reduction, increase of stiffness, and low technological complexity. The following design goals for the 50 to 60 telescopes shall be detailed:

- slant axis mount,
- CFRP axisymmetric reflector back-up structure,
- aluminium panels thermally stabilized on five support points,
- gravity deformations below 0.5λ for a fixed focus,
- displaceability and maintenance,
- costs down to 12.5 MFF/15-m unit.

3.1 Slant Axis Mount

The proposed design is shown in Fig. 5. Bearings and drives are concentrated in a bloc with minimum bending loads in the structures. This bloc connects to

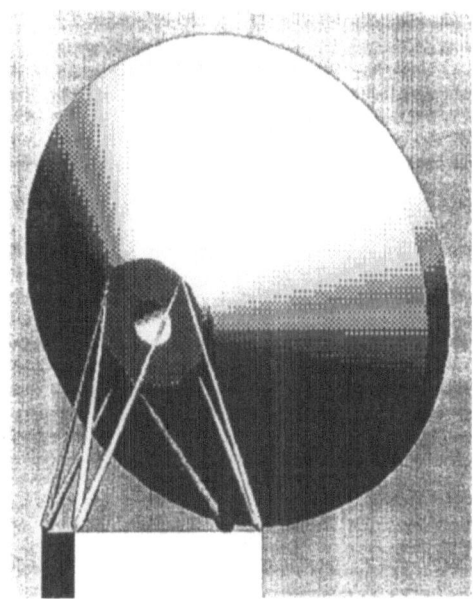

Fig. 4. Radio Telescope on Push-Pull Actuators

the reflector and a counterweight at one side and to a cylindrical tower on the other. The tower is made from prestressed concrete of high thermal inertia; the outside being covered with a naturally ventilated cladding to keep the surface of the concrete cool. Also the mechanics bloc is protected by a ventilated thermal shielding. The mechanics consist of classical backlash-free drives with bull gear and double pinion and also classical, preloaded 3-roller bearings (one per axis). The bearings are arranged in a way that the axes of the reflector optics and the vertical (azimuth) axis cross in one point. The mechanics bloc gives at its inside enough room to arrange reflecting mirrors and to pass the beam.

The arrangement has the following advantages:

− compact structure
− high stiffness
− reduced weight
− reduced thermal deformations
− easy balancing
− state of the art (low cost) technology

The reflector can reach the horizon but might have speed problems for astronomical observations at this elevation.

3.2 CFRP Axisymmetric Reflector Back-up Structure

Starting as early as 1980 with studies on Carbon Fibre Reinforced Plastic (CFRP) trusses for the 15-m Plateau de Bure dishes, IRAM has developed technologies

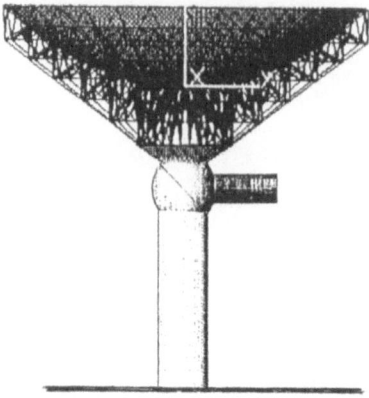

Fig. 5. Slant Axis Radio Telescope

which render CFRP reflector structures pricewise comparable to steel versions, but with the important advantages of reducing considerably the weight, stabilizing the thermal performance to qualities known from large optical mirrors and having "steel"-stiffness.

An axisymmetric reflector will certainly increase the stiffness value due to its rotationally symmetric back-up structure and the possibility to integrate the secondary mirror support into the load carrying system which is favoured by its central position and the short vertex separation between the primary and secondary mirrors.

The high number of identical structural elements of an axisymmetric design (typically 3500 pieces of each strut type) will also have a positive influence on the costs.

3.3 Aluminium Panels

Aluminium panels are used by all new projects and IRAM will have the fifth antenna equipped with this type of panels. They can be thermally stabilized by an appropriate 5-point support system and can be made light weight with masses as low as 13 kg/m^2. They have excellent long-term stability and need not much power for deicing. Due to the axisymmetric reflector design, there will be nearly 2000 identical panels per ring which permits a real series production with adequate low cost production means.

3.4 Gravity Deformations Below 0.5λ

Modern radiotelescopes are optimised in their astronomical performance by correcting certain surface errors on the primary through an active secondary mirror with focus, tilt and lateral motorized displacement possibilities. The motorization is in general complex and difficult to maintain.

It is therefore one of the design goals to find a back-up structure which keeps the surface deviations in absolute below 0.5λ during the elevation travel and with only gravity influence considered. The idea is to suppress if possible all alignment motors on the secondary mirror and make it a passive element. This would in consequence also increase the performance under windload and make the secondary unit practically maintenance-free.

3.5 Displaceability and Maintenance

The move of 50 to 60 telescopes should not last more than a month, i.e. two telescopes have to be displaced per day. It is proposed to have all stations equipped with the concrete tower as described under 3.1. The telescope structure is separated from the tower and transported to its new destination by a transporter as shown in Fig. 6.

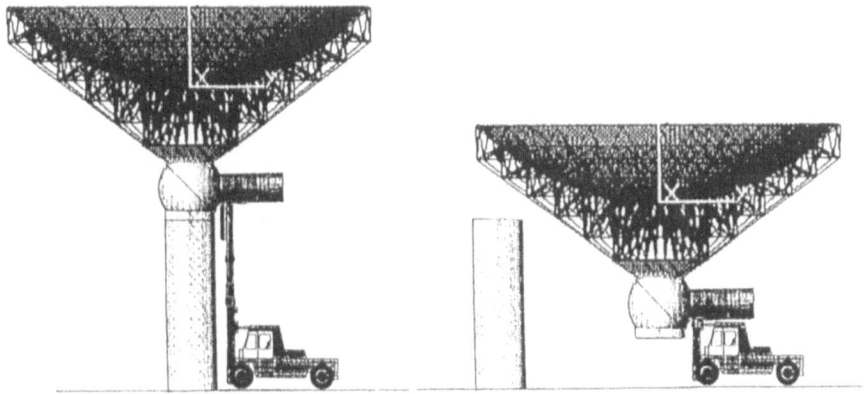

Fig. 6. Displacement of the Telescope

The time for reconfiguration is also the typical time for maintenance. To reduce the efforts for maintenance, the telescopes should be made of only a few but sophisticated items and the equipment be limited to an absolute minimum. All necessary devices and elements,however, should be of good quality with high long-term stability.

3.6 Costs

IRAM is now assembling its fifth antenna on the Plateau de Bure. The reflector diametre and the surface accuracy correspond more or less to the requirements of the LSA. Also the choice of the materials meets targets like thermal and long-term stability. This experience, the application of state of the art techniques and the advantages of a series production make it possible to consider costs of typically 12.5 MFF/telescope unit, including a warranty of 5 years for the industrial products.

The break-down of the costs could look like the following:

1. 15-m reflector unit   8.5 MFF
   - alu panels   2.5
   - secondary support   0.8
   - subreflector + support   0.7
   - alignment devices   0.5
   - nodes   0.4
   - CFRP struts   2.0
   - central support structure   0.3
   - deicing   0.4
   - cabling   0.4
   - assembly   0.5

2. mount   4.0 MFF
   - bearings   0.5
   - structures   1.0
   - drives and encoders   2.0
   - concrete tower and steel ring   0.5

3. unit price per 15-m telescope   **12.5 MFF**

4 Summary

The described slant axis telescope is based on the design and experience with the five IRAM antennae for mm-observations which operate successfully since 1986 on the Plateau de Bure in the French Alpes and the mostly identical SEST installed on La Silla in the Atacama Desert. During these 10 years, intensive studies have been carried out on the applied materials, long-term stability, maintenance requirements and astronomical performance, parameters which will influence considerably the costs and quality of the future LSA telescope.

References

Delannoy, J. (1985): Status of the IRAM 15-m Antennas and SEST in 1985, Proc. of the ESO-IRAM-ONSALA Workshop on (Sub-)Millimeter Astronomy, Aspenäs, Sweden

Fassaro, J.S., Porter, R.W. (1947): Photographic Giants of Palomar, Westernlore Press

Labeyrie, A.: private communication

Plathner, D., et al. (1987): VISIR Interferometer; Phase A Study on Telescopes of 1.5 to 2 Metres, Study prepared by IRAM under INSU contract

Millimetre Receiver Technology for a Large Array

James W. Lamb

IRAM, 300 rue de la Piscine, F-38406 St-Martin d'Hères, France

Abstract. The current state of receiver technology for millimetre wave receivers is reviewed with regard to the requirements of a large interferometric array. Optics, amplifiers, mixers, local oscillators and cryogenics are all considered. It is concluded that the technology exists today to construct receivers of high sensitivity, but technical developments should continue to improve reliability and useability.

1 Introduction

The possibility of making a large interferometer array rests on the ability to build large numbers of sensitive, reliable receivers at moderate cost. Current technology for receivers for single dish and small arrays is quite mature with sensitivities approaching the quantum limit (Carlstrom & Zmuidzinas 1996). Receivers today routinely use SIS mixers with niobium junctions as the first stages. Local oscillator (LO) sources are typically Gunn oscillators with frequency multipliers where necessary.

In such a short article it is impossible to review all aspects of the technology, and the References must be regarded as representative rather than exhaustive.

1.1 Requirements for Receivers

Some requirements for receivers are given below in Table I. For the purposes of this review these are indicative rather than definitive.

Table 1. Receiverr equirements for a large millimetre array

Parameter	Specification
Frequency range	40-350 GHz
Noise temperature	4hv/k (SSB)
IF Bandwidth	2 GHz
Polarisation	Dual-linear
Phase	Good stability, low noise
Reliability	"Good"
Cost	"Low"

A phased introduction of frequency bands should be envisaged, but ultimately one could consider contiguous coverage of the whole millimetre band. It is reasonable not to exclude atmospheric absorption bands which could be rather narrow at a good site. A 30 % bandwidth for each receiver would result in six bands from 65 to 400 GHz — dual-polarisation on 60 antennas would require 720 channels.

The importance of reliability can be seen by considering the result of failures in components. For example, assume each of N antennas has M backshort drive motors with an average time to failure of t_{mtbf} and that it takes a time t_{rep} to repair. For t_{mtbf} = 3 yr, t_{rep} = 1 day, N = 60, and M = 12 one expects a failed motor almost half of the time. Clearly the reliability of individual components should be extremely good.

1.2 General Receiver Scheme

Most receivers follow the block diagram given in Fig. 1. An optical system brings the signal from the antenna and perhaps performs other functions. At present few systems have input amplifiers and the first stage is usually a mixer. A phase-locked LO is required for the mixer to downconvert the signal to the intermediate frequency (IF) where it may be amplified and processed by later stages. Since all current low-noise receivers operate at cryogenic temperatures a refrigeration system capable of a temperature of 2-20 K, depending on the application, is required.

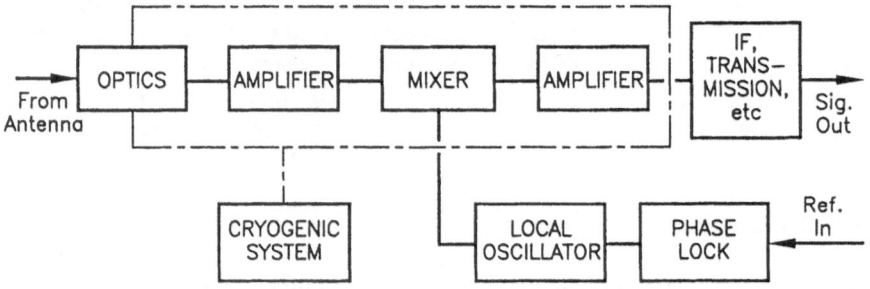

Fig. 1. General block diagram of a heterodyne, showing the sections discussed in the text.

2 Optics

The minimum function of the optics is to couple the signal from the antenna into the mixer. Typically a feed horn is used and one or more lenses or focusing mirrors may also be required. More complicated devices are often used, including: grids, quarter/half-wave plates, Fabry Perots, Martin-Puplett Interferometers, and prisms.

Although optical losses may be quite small in the millimetre band they are still significant in comparison with very low-noise receivers. Several loss mechanisms should be considered. *Diffraction* losses result from the truncation of the optical beam. Clearances should be sufficiently large to make this negligible, and large apertures are particularly required near images of the sky, while smaller ones can be accepted near images of the telescope aperture. Surface accuracies of ~ λ/100 are needed. *Ohmic* losses in reflectors are depend on conductivity, wavelength, and surface finish At 300 GHz the loss may 0.18 % for Al, and 0.13 % for Cu so some ohmic losses at room temperature are acceptable. *Dielectric* losses are proportional to frequency and depend on the material and preparation. Low-loss plastics attenuate by ~0.5 % mm^{-1}, so cold losses may be acceptable for lenses.

3 Mixers

Mixers may be classified according to the nonlinear device (SIS, SIN, Josephson junction, hot-electron bolometer); structure (waveguide, planar); and function (double sideband, image rejecting, image separating). Only superconductors are considered here as there are no other devices with competitive sensitivities at these frequencies.

3.1 Devices

SIS mixers are currently the most sensitive in the millimetre range, and representative state-of-the-art performance is shown in Fig. 2, taken from Karpov (1994), Payne et al. (1994), and Kooi et al. (1995). The double sideband noise approaches 2hv/k so that the SSB noise is close to the 4hv/k goal. SIN mixers have the potential benefit that there are no perturbing Josephson currents, but because the non-linearity is less than for SIS devices the performance is currently three times worse (Karpov et al. 1995b). Josephson mixers rely on pair tunneling and require much smaller area junctions than SIS mixers. Lack of a significant shunting capacitance means that many harmonics are significant and the noise which is down-converted by the shunt resistance results in relatively poor noise performance (Taur 1980). Furthermore, there are regimes where chaotic behaviour may dominate. Hot electron bolometers (Gershenzon et al. 1990; McGrath 1995) can have very short thermal time constants and therefore respond to the beat between an LO and signal to produce an IF output. As a relatively new device, the hot electron bolometer can be expected to improve. However, even theoretically, it is unlikely to be competitive with than the SIS mixer at millimetre wavelengths. Noise temperatures of 450 K have been obtained at 100 GHz, with IF's up to 2 GHz (expected to approach 5 GHz) (Okunev et al. 1995).

Niobium is the preferred superconductor for millimetre wavelengths: it is rugged, well understood and its moderate transition temperature yields good results at 4.5 K. Progress has been made with niobium nitride, but the best SIS results are not as good as for niobium (Karpov et al. 1995a). The higher transition temperature is not a significant advantage, though the larger gap voltage could may make the Josephson currents less important. "High-T$_c$" devices are still in early development.

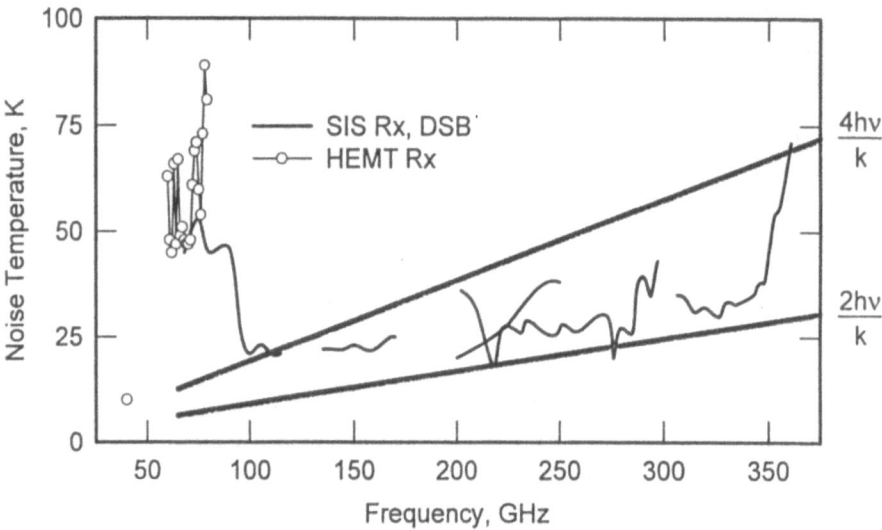

Fig. 2. Representative noise temperatures for SIS and HEMT receivers. See text for references.

3.2 Structures

The best results in the millimetre band have been achieved with waveguide mixers. Corrugated horns are almost ideal, having good beam shapes, low cross-polarisation, and good match over bandwidths of 40 % or more. As waveguide devices they have a cut-off frequenciy which helps to reject interference. Horns have successfully been made and tested for frequencies up to 3.1 THz (Ellison 1994).

Planar ("Quasioptical") structures have appeared in several forms, including the bow-tie, log-periodic, log-spiral, double dipole and double-slot. Fabrication of planar devices is simpler than waveguide devices, but they need accurate lenses with anti-reflection coatings. Low noise temperatures have been reported, particularly at higher frequencies, but often the beam couples poorly to a telescope, and there may be cross-polarisation. Good patterns are reported for double-slot antennas (Zmuidzinas & LeDuc 1992) but main beam efficiency is ~ 70 % (c.f., ~ 99 % for a corrugated horn).

3.3 Functionality

The simplest mixers are fixed tuned double-sideband devices with bandwidths of > 25 %. Due to the lack of mechanical tuning the reliability is high, and frequency changes are rapid. Image rejection requires a quasioptical or waveguide filter. (In the future a mixer with a very broad bandwidth could be envisaged which could be intrinsically single-sideband. Tuning would then be done within the IF band).

Image rejecting mixers with one or two backshorts are well established. Rejection may degrade as IF bandwidths are increased, but for an interferometer only modest rejection is required — line confusion is avoided by LO phase switching. Moving backshorts will have some reliability problems and there are also more likely to be problems with stability of the mixer.

An alternative to rejecting the image is to separate the signal from the image using two mixers. Some demonstrations of the concept have been made and 25 dB separaton has been achieved over narrow bandwidths, although at the expense of doubling the mixer noise (Akeson et al. 1993). The main advantages are when the atmospheric and spillover noise are dominant, which may not be true on a good site.

4 Local Oscillators

Powers of order 100 μW are required from LO sources. Although the power at the mixer is much lower there are significant losses for the phase-lock system, waveguides, and couplers. Injection without tuning implies that coupling of the order of -20 dB is required. To match receivers, bandwidths of ~ 30 % are required. For an interferometer LO's need to be phase-locked and have low phase and amplitude noise.

4.1 Solid State *vs* Superconducting Oscillators

Solid state oscillators have a well developed technology with high power, low noise and narrow intrinsic linewidths. They usually operate at ambient temperature, but loss into a dewar may be as low as 1 dB. There are mechanically or electrically tuned oscillators, and appropriate multiplier technology which is well developed. Supercond- ucting technology is less well established. Fundamental operation is possible over the millimetre band and the devices operate in the cryostat. Intrinsic linewidth is relatively large, and though they are electrically tuned suitable phase-locking has yet to be demonstrated.

4.2 Solid State Oscillators and Multipliers

Gunn oscillators, prevalent today, achieve fundamental operation to at least 170 GHz and second harmonic operation to 290 GHz with 300 μW output power (Eisele & Haddad 1994). Phase-locking is relatively simple and the oscillator phase and amplitude noise do not contribute significantly to the system sensitivity (Padin et al. 1988). Significant tuning bandwidths may only be obtained with mechanical tuners. IMPATT oscillators are electrically tunable but generally too noisy to use. Progress has been made with TUNNETT devices, having 70 mW at 105 GHz, low phase-noise, and electrical tuning, although not yet with large continuous tuning bandwidth (Eisele & Haddad 1995). Transistors oscillators may be made up to 215 GHz with ~ 1 μW (Rebeiz 1994) higher powers may be expected in the future. Electrically tuned low phase-noise YIG oscillators tune from 26 - 40 GHz (Mede et al. 1996).

In concert with oscillator development, improvements have been seen in multipliers. Advances in varactor technology (materials, air-bridges, parameter optimisation) have brought better efficiency and reliability (Crowe 1995). Some benefits can be obtained from cryogenic cooling including higher power handling and improved efficiency (Erickson et al. 1992). Improvements in waveguide circuit design have produced good fixed-tuned performance.

An electronically-tuned LO system for a Schottky mixer array has been built by Erickson et al. (1992). It uses a YIG tuned FET from 29 - 38 GHz, followed by a power amplifier and varactor tripler. Although the oscillator is electrically tuned a mechanically tuned filter is required to reject the amplifier noise. With lower noise amplifiers and smaller LO coupling for SIS receivers this may be avoided.

4.3 Superconducting Oscillators

Josephson junction (JJ) array and flux-flow oscillators (FFO) are both under development for LO sources. JJ's generate power by Josephson oscillations, while FFO's depend on the rapid transit of flux quanta along an extended Josephson junction. Although significant power can be generated, spectral linewidths are relatively large (~ 0.1 MHz). Both devices are rather complex and will require significant development. For JJ oscillators, arrays of devices are required to achieve required power levels. Effective synchronisation of the devices is complex and competes with the tuning bandwidth (Booi 1995). Investigations continue on FFO's to determine the constraints on linewidth and other properties (Henne et al. 1995)

5 HEMT Amplifier Technology

The maximum frequency of a HEMT (high electron mobility transistor) works is a function of material and geometry. Noise temperature is roughly proportional to frequency: 1 K GHz^{-1} is typical, 0.5 K GHz^{-1} achieved and 0.25 K GHz^{-1} predicted. The lowest noise has been obtained with discrete transistors. Noise temperatures of 50 K at 75 GHz have been reported for an amplifier at 20 K (Posieszalski 1993), comparable to the performance of SIS receivers at frequencies approaching 100 GHz (Fig. 2). Monolithic amplifiers are attractive for large quantities, and integrated amplifiers at 77 - 110 GHz (Wang et al. 1993), and 130 - 150 GHz (Wang et al. 1995) have been fabricated. Limits of the present 0.1 μm gate technology are being approached and major technological steps are needed to reach higher frequencies.

Although HEMT's may not replace SIS mixers as first stages in the near future, they may give wide bandwidths when used as IF amplifiers. Bradley (1995) has built an amplifier at 18 - 26 GHz with a noise temperature of ~ 7 K. Progress has been made in matching HEMT's to SIS mixers (Weinreb 1987; Padin et al. 1995).

6 Cryogenics

It seems almost certain that temperatures around 4 K will be required. Although hybrid dewars have been successfully used on small arrays (5 - 6 antennas, 2 channels, (Blondel et al. 1995)), liquid cryogens are impractical for a large interferometer array, and closed-cycle systems will be used. Joule-Thompson (JT) refrigerators have been in operation for many years with good reliability statistics. High capacities at 4 K are easily achieved with 2-3 W being common. Temperature stability of a couple of millikelvin can be obtained on timescales of up to several minutes or more. Such systems have been used on telescopes for several years for masers and, for a shorter period, SIS receivers (Woody et al. 1985; Payne et al. 1994)

More recently 2-stage or 3-stage Gifford-McMahon (GM) refrigerators have been developed using rare earth compounds for the heat exchanger. As with most GM refrigerators, the temperature is subject to variations over the displacer cycle, but it can be smoothed out by appropriate design. A 4-K GM refrigerator has been designed used for a millimeter receiver (Plambeck et al. 1992). No data are available yet on the long-term reliability or lifetime of these systems.

7 Conclusions

All the technology to build receivers for an array operating at millimetre wavelengths exists today. Sensitivities approach quantum limits, and frequency coverage for a single receiver may be ~ 30 %.

As a baseline model, the most likely design of receiver would have several channels in a 4 K dewar with a JT refrigerator. A simple optical path with a grid to separate the two polarisations and some mirrors to direct the beam into corrugated feed horns would be used. DSB waveguide SIS mixers would be followed by low noise wide-bandwidth (> 5 GHz) HEMT amplifiers. Solid state local oscillators at low frequencies (< 40 GHz) followed by fixed tuned multipliers and HEMT amplifiers could provide an electronically-tuned, low-noise, phase-locked source.

The challenge facing receiver builders is how to make the systems reliably and cheaply. A very significant engineering effort is required to take millimetre technology from the "custom" level to "production" volumes.

References

Akeson, R. L., Carlstrom, J. E., Woody, D. P., Kawamura, J., Kerr, A. K., Pan, S.-K, Wan, K (1993): Proc. Fourth Int. Symp. Space Terahertz Technology, LA, USA, 12-18

Blondel, J., Carter, M., Karpov, A. Lazareff, B. Mattiocco, F. (1995): Digest 20th Int. Conf. on Infrared and Millimeter Waves, Florida, USA, Dec. 1995

Booi, P. A. A. (1995): *High-Frequency Array Oscillators Based on Nb/Al-AlO$_x$/Nb Junctions*, Thesis, Universiteit Twente

Bradley, R. (1995): Private communication

Carlstrom, J. E., Zmudzinas, J. (1996): "Millimeter and Submillimeter Techniques," to appear in *Reviews of Radio Science 1993-1995* (OUP, Oxford)

Crowe, T. (1995): Third Int. Workshop on Terahertz Technology, Zermatt, Switzerland

Eisele, H. Haddad, G. I. (1994): Electron. Lett **30**, 1950-1951

Eisele, H. Haddad, G. I. (1995): IEEE Trans. Microwave Theor. Tech **43**, 210-212

Ellison, B. N., Oldfield, M. L., Matheson, D. N., Maddison, B. J., Mann, C. M. Smith, A. F. (1994): Internat. Seminar on Terahertz Electronics (Part II), Lille, France, June 13-14

Erickson, N. R., Goldsmith, P. F., Novak, G., Grosslein, R. M., Viscuso, P. J., Erickson, R. B., Preadmore, C. R. (1992): IEEE Trans. Microw. Theor. Tech. **40**, 1-11

Gershenzon, E. M., Golt'sman, G. N., Gogidze, I. G., Gusev, Y. P., Elantev, A. I., Karasik, B. S., Semenov, A. D. (1990): Superconductivity **3**, 1582-1597

Henne, P, Kohlstedt, H., Ustinov, A. V., (1995): Third Int. Workshop on Terahertz Technology, Zermatt, Switzerland

Karpov, Maier A., Blondel, J., Lazareff, B., Gundlach, K.-H., (1995a): Proc. Sixth Int. Symp. Space Terahertz Technology, Pasadena, USA, 344-354

Karpov, A., Karpov, Maier A., Blondel, J., Lazareff, B., Gundlach, K.-H., (1995b): Int. J. IR and Millimeter Waves **16**, 1299-1316

Karpov, A (1994): Proc European SIS Users Meeting, Köln, Germany

Kooi, J. W., Chan, M., Bumble, B., LeDuc, H. G., Schaffer, Phillips, T. G. (1995): Int J. IR and Millimeter Waves **16**

McGrath, W. R. (1995): Sixth Int. Sympos. Space Terahertz Technology, Pasadena, 216-227

Mede, F., Gleissner, J., Brennemann, A., Beyer, A. (1996): Microwave Eng. Europe, Jan, 29-33.

Okunev, O., Dzardranov A., Gol'tsman, G., Gershenzon, E.(1995): Proc. Sixth Int. Symp. Space Terahertz Technology, Pasadena, USA, 247-253

Padin, S., Woody, D., Scott, S. L. (1988): Radio Science **23**, 1067-1074

Padin, S. Woody, D. P., Stern, J. A., LeDuc, H. G., Blundell, R., Tong, C.-Y. E., Pospieszalski (1995): Proc. Sixth Int. Symp. Space Terahertz Technology, Pasadena, USA, 134-139

Payne, J. M., Lamb, J. W., Cochran J. G., Bailey, N. J. (1994): Proc. IEEE **82**, 811-823

Plambeck, R, Thatte, N., Sykes, P. (1992): Proc 7th Int. Cryocooler Conf., Santa Fe NM, Nov. 1992

Pospieszalski, M. W. (1993): Proc 23rd EuMC

Rebeiz, G. (1994): IEEE Antennas Propogat. Magazine **36**, 36-38.

Taur, Y. (1980): IEEE Trans. Electron Devices **ED-27**, 1921-1928

Wang, H., Lai, R., Chen, S. T., Berenz, J. (1993): IEEE Antennas and Propagat. **3**, 381-382.

Wang, H., Lai, R. Lo, D. C. W., Streit, D. C., Liu, P. H., Dia, R. M., Pospieszalski, M. W, (1995): IEEE Microw. and Guided Wave Lett. **5**, 150-152

Weinreb, S. (1987): IEEE Trans. Microw. Theor. Tech. **MTT-3**, 1067-1069

Woody, D, Miller, R. E., Wengler, M. J. (1985): IEEE Trans. Microw. Theor. and Tech. **MTT-32**, 9095

Zmuidzinas, J., LeDuc, H. G. (1992): Trans. Microw. Theor. Tech. **40**, 1797-1804

Correlator Developments
for (sub)Millimeter Telescopes

A. van Ardenne , A. Bos

Netherlands Foundation for Research in Astronomy, P.O.Box 2, 7990AA Dwingeloo,
The Netherlands

Abstract. The next generation correlators require an order of magnitude more processing power than the ones that are under construction today. Some of the key technological aspects are reviewed here, and the need for an R&D program is emphasised.

1 Introduction

The next generation correlators for the millimeter wave interferometers projected in the next decade, require at least an order of magnitude higher data handling capability when compared with todays instruments. The design experience gained with todays correlators and those under construction for lower frequency radiointerferometry, will undoubtedly constitute the starting point for their design. Straightforward extrapolations will however not provide the required performance for the novel instruments needed in (sub)millimeter astronomy. Hence other, yet fairly unexploited areas of technical R&D need to be addressed. These include high speed sampling techniques, the level of circuit integration and the possible need to apply optical techniques both external (in the case of digitized Intermediate Frequencies) as internal (inside the correlator) signal distribution.

Another concern which may impact the design, relates to the need to properly calibrating and correcting the atmospheric phase uncertainties for which more detailed input to the design is necessary. The correlator configuration and its supporting software, should allow for sufficient flexibility to cope with these effects.

2 Objectives

The next generation of (sub)millimeter radiotelescopes will generally consist of many-element arrays with the major characteristics summarized in Table 1.

The challenge for these large new arrays is to develop the correlator hard- and software capable to handle, control and process signals from a (say) 50-element interferometer for a multitude of astronomical needs. These include eg. a wideband/low resolution and smallerband/high resolution mode while making all four Stokes polarization parameters available in the calibrated and correlated output. This functional requirement is similar with those for existing albeit smaller interferometers at lower radio frequencies. In terms of processing

Table 1. Summary of existing and planned facilities

	European			U.S.			Japan
	IRAM	LSA	BIMA	OVRO	CMA	MMA	LMA
Telescopes	5(6)x15m	50x16m	9x6m	6(10)x10.4m	OVRO +BIMA	40x8m	50x10m
Coll. area (m^2)	880	10.000	255	510	800	2000	3900
Baseline (km)	0.5(1.5)	5-10(100)	<1	<1	1	3(10)	10
Nr of baselines	10	1225	36	15	105	780	1225
Frequ (GHz)	80-260	40-230	80-115	80-115	80-230	40-360	100-500
occasional	330-360	330-450	210-260	210-260	330-360	>360	600-900
Altitude (m)	2552	>3000	1043	1216	∼3000	>4000	>4000
Oper. $(year)$	current	>2005	current	current	∼2000	∼2005	2006

power ("bit operations per second"), there exists at least an order of magnitude difference due to the larger number of interferometers and the wider bandwidth.

The overall design should be configurable to cope with the different observing requirements i.e. wide bandwidth, choice of interferometers vis a vis baselines, spectral resolution eg. continuum vs. line and the bit correlation scheme. It should be flexible enough for functional extensions due to changing requirements (eg. extension of observing modes). These objectives are much related to the hard- and software architectures and of properly designed interfaces eg. to cope with the different data speeds and volumes. The primary set of different configuration modes are yet to be decided.

Furthermore, the design should be cost-effective, reliable and allow for ease of maintenance e.g. with regard to adequate monitoring and tests and ease of repair.

A starting point for target specifications of a potential correlator design, is given in Table 2. This is based on input from Downes (1995) and subsequent discussions.

Table 2: Target specifications assuming 50 telescopes. Configuration and calibration modes are not yet clear.

Total untuned (IF) bandwidth	$4 - 8$ GHz/pol/tel
Number of subbands (say: p)	TBD
Number of IF inputs	$p \times 50 \times 2$
Number of baselines	1225
Frequ resolution per polarisation	$100 - 50$ MHz (continuum mode)
	$500 - 25$ kHz (spectral mode) in 10 MHz
Number of complex points per subband	TBD

Total number of complex channels	$\approx 10^7$ (depending on architecture)
Polarisation	4 Stokes parameters (I,Q,U,V)
Configuration	TBD
Other functions	Phase switching (e.g. for sideband separation, crosstalk and offsets)
	Gating (e.g. beamswitching, blanking)
	Adding (compound interferometer)

It is unclear how to implement appropriate calibration schemes in order to cope with the variable atmospheric conditions. These variations occur from telescope to telescope (see Hills (1996)) as well as over any single telescope aperture, the radiofrequency equivalent of 'seeing' cells. Beamswitching and blanking, per band and/or per telescope are probably necessary as means to minimize these effects. In this respect, more advanced calibration schemes apart from the common ones i.e. amplitude and phase-closure and redundancy, for example by using multiple beams taking advantage of the near-field of the troposphere and even image-plane correlators, should be given some thought.

As also suggested in Escoffier (1995), it is worth considering a dedicated continuum correlator as this is the main driver for the highest speeds needed while requiring relatively simple operations on the data. In this case attention should be given to the so-called smearing effect due to the non-compensated delay across the band.

For spectral line observations, a correlating scheme with minimal loss should be considered. For example, a three bits per channel correlation results in only a few percent sensitivity loss as compared to the ideal (analog) case. This must be weighted against the expense resulting from the added complexity.

3 Architecture

The architectural design of the correlator hard- and software is affected by interrelated aspects which are increasingly less easy to separate from the overall system design. Some optimization on a high level is therefore required. As a separate issue on a high design level, experience with present day instrumentati on indicate that build-in flexibility is a matter to be considered seriously from the beginning.

Channelization of the total band vs. highspeed sampling and digitization eg. at the telescopes, is an important aspect to be considered for the hardware design. Channelization yield a robust yet very flexible design. For example, the design results in independently tunable frequency settings and filters per channel. The approach is rather straightforward with minimum design risks and requires the minimum correlation power. The need for many channels will allow a high level of functional integration probably "at the expense" of taking less advantage from the more rapid developments in the digital micro-electronics area including intelligent DSP's. A potential difficulty is the band to band alignment for wideband observations. The requirement for wideband Intermediate Frequencies are making it attractive that digitization is done at the individual telescopes. This has the immediate implication of the accurate distribution of the sampling clock. Incorpo ration of the fine-delay is to be considered in that case. This being solved then has the advantage that the signal distribution can be done relatively simple eg. through stable fibre-optic links. In the case of high speed sampling and apart from the required developments to result in reliable sampling and clock distribution schemes, all digital implementations may prove to be very attractive. An assessment has been done in the context of the MMA studies (see Escoffier (1995)).

The matter of sampling relates to the number of levels (1-bit versus multibit) correlation. From the point of view of sensitivity, the sensitivity loss of 57% (i.e. $\pi/2$) for a simpler 1-bit system is most pronounced for spectral observation. In case of a 2-bit system, this loss can already be reduced to a more acceptable level of less than 15%.

Another aspect deals with the interrelation between architectural design and the actual implementation. In actuality this is an interconnectivity matter defining the integration level versus modularity and flexibility.

Other factors related to the detailed design are to be found in the areas of Single-Side-Band versus Double-Side-Band IF-to-baseband conversion i.e. Real vs Complex correlation, the matter of using Fourier Transform techniques prior to correlation or the other way around (ie. XF vs. FX correlators, see Bos (1993)) and the question where and how to perform the delay compensation and fringe rotation to reduce the natural fringes.

With all these factors in mind, the development of a systematic approach at the system design level will prove to be an extremely useful optimization tool.

The supporting software mostly relates to Test and Configuration software the architectural design of which is connected to the hardware architectural design. Matters to be addressed relate to the level of integration, the embedded processing eg. FFT's or other algoritmes, control and test software interfaces, the interface to external, off-line software and by the appropriate datahandling. Todays design relies on software engineering approaches using a staged object-oriented design techniques with explicit classes for the userinterfaces and the mode of observation (see De Vos (1995)).

4 Technology assessment

The product of the maximum desirable bandwidth times the number of interferometers constituting the array i.e. $B.N(N-1)/2$, serves as a simple complexity measure of the correlator capacity. For the LSA, this product is about 5000 Giga(word) Operations/sec which should be compared with the 19.2 G Oper./sec of the new Westerbork correlator now under construction or the 5.4 G Oper./sec of the VLA correlator. This measure does not take into account the specific bitcorrelation scheme (ie. number of bits), the clockspeed nor the maximum number of real or complex channels given a certain instantaneous (IF)-bandwidth. The approximate correlator capacity for an XF-correlator with the performance of Table 2, indicates a capacity which is 1-2 orders of magnitude larger than the new correlator back-end presently under construction for Westerbork and the Smithonian Millimeter Array. To put this in perspective, the actual number of Gigabit operations per second for radioastronomical correlators over the last decades show a 1000 times capacity increase over the last 20 years. The future requirement in say, 2005-2010 is beyond the PetaOperations/Second but close to a straightforward extrapolation of these figures. Not too surprisingly, microelectronics developments in cost, power consumption and the number of transistors per chip ("Moore's Law") show similar developments over these decades. These are positive indications that our aim is ambitious but realistic.

Before making final choices, there are several areas of attention which need to be addressed:

Firstly, high speed A/D-converters are for example being developed for possible use in a new (multibeam) autocorrelator at the JCMT. For interferometry with different requirements, attention should in particular be given to the phase characteristics of these samplers. For its realization, developments in the military domain may be of use combined with a dedicated effort to suit our purpose. High speed sampling also have impact on processing elements after A/D conversion which need to be addressed as well including the engineering consequences. As a rule, it can be expected that sampling at lower speed not only reduces engineering risks due to the lower speeds involved but also of cost. The latter is due to a potentially higher level of integration in less demanding technology as well as a reduction of processing power in proportion to the number of filter channels.

Also, a high level of functional integration is likely to be necessary both for engineering and cost effectiveness. This approach should be balanced by the reduction in flexibility. Areas where integration may be particularly attractive are in the IF- and A/D-channels. The feasibility depends on their number and the required filtering but the design of IC's in these areas is not excluded and should be detailed further. Also, mixed mode designs in which both analog and digital functions are combined on a single IC are, helpful.

Most recent correlator designs already use IC's with some level of customization, i.e. 'customised intelligent' DSPs. These are presently characterized by 100k gates consisting of 1 Million transistors operating at up to 125 MHz clockspeed. Development trends in microelectronics will result in gate array designs 5-10 this size. Recent experience indicate that full custom designs in the all digital functions are probably difficult and should be replaced by gate array designs which industry supports. Low power, low voltage e.g. 2 Volt CMOS electronics both reduce the consumption of power locally while making the engineering challenge appreciable.

The further developments of supporting software and design, test and verification tools, enhance design capabilities and will be extremely helpful to realize the LSA's of modern times.

There are several areas where the role of optics in relation to the correlator, could be advantageous:

- Distribution of digitized "astro" signals from each telescope to the central

control area including clock distribution, depending on the architecture.
- Networks supporting monitoring and control functions.
- Signal(re)distribution inside the correlator.
- Optical processing functions. This is presumably the most speculative domain with a long lead time to actual application. Nevertheless, highly sophisticated functions are available in other applications.

Not mentioned is the use of optical delay lines as a posssibility. Although in principal possible, it seems that the use of dynamic storage in small memory buffers in digital logic, is easy and technologically consistent with other aspects of the design and hence is recommended.

Furthermore, it is not too wild a guess that the Object Oriented software technology now in development in the 'on'-line as well as in the 'off'-line packages under development, will have significant impact on the design of the software architectural approach in new correlators. The same applies with respect to an OSI- layered approach in the real time environment. This approach offers many advantages including reliability, supportability and flexibility.

There is a need to formalize and describe the impact of instrumental and atmospheric effects on each astrosignal as it passes through the telescope and other instruments including the correlator. This generic formalism is now being incorporated in the international AIPS++ effort coordinated through the NRAO in the US. Further studies and developments in these areas will prove to be useful to the the LSA effort.

5 Engineering challenges

Apart from the challenge to set up the appropriate level of R&D and to make this a manageable project, technical challenges remain both in the implementation and realization itself. In this context and apart from the cost issue which need to be addressed further, there are constraints related to:
- Structural design matters eg. power management and thermal control and time and clock distribution. Also, the level of integration will put constraints on manageable levels of crosstalk and EMI-shielding needs.
- Reliability and servicability/maintainability
- Monitoring and tests and finally the means and methods ie. the supporting development tools and engineer design methods excercising a system level design view.

6 Conclusions

Correlators for new millimeter arrays require an order of magnitude more processing power then those for existing applications and those under construction. With the results of a properly designed R&D-program partly based on todays collected experience, it seems feasible to design and build such an instrument within the forthcoming decade. Such R&D-program should primarily be aimed at given insight in the baseline architecture needed for an LSA-correlator, the desired technologies, the development effort and the costing envelope.

References

Downes, D. (ed) (Oct 1995): *LSA: Large Southern Array Project Report*

Hills, R.T. (1996): (This Workshop)

Escoffier, R. (1995): *A possible MMA correlator design* (Memo to the MMA system design group, May 5, 1995)

Bos, A. (1993): *The EVN/NFRA correlator: design considerations.* (NFRA ITR 202)

Vos, C.M. de (1993): *Towards a Telescope Management System for the WSRT.* (NFRA Report TMS1.v3)

Part 4

Workshop Summary

Concluding Remarks

L. Woltjer

Observatoire de Haute Provence, F–04780 Saint-Michel l'Observatoire

It is interesting to remember that 20 years ago there was still much discussion if there would be small enough structures to resolve with a 1 km interferometer or if large dishes were better suited to study the InterStellar Medium. This finally resulted in the creation of IRAM, a sort of forced marriage of a dish and an interferometer, which has been remarkably successful. By now there is no doubt that as Reinhard Genzel said "you can never have enough angular resolution", and since resolution without enough sensitivity to exploit it is not very useful, the need for large arrays becomes obvious.

The scientific case for such an array is overwhelming. From the nearest planets to the outer reaches of the universe it would have an enormous impact. During this conference we have seen that promising areas of research include the chemistry and dynamics of planetary and stellar atmospheres and of the ISM, star formation, the dynamics and evolution of galaxies, gravitational lensing, the formation of galaxies and cosmology. Surveys in the continuum, in CO and in C^+ will tell us much about evolution in the Universe up to perhaps a redshift of 10 – at least if such high redshift objects exist. Not only will a large array contribute to our understanding in all these areas in a quantitative way, there are also numerous aspects where such an array can lead to entirely new areas being opened up and thereby to unique and qualitative progress.

It is important to note that the array is a perfect counterpart to HST with comparable resolution, but unhindered by dust opacity. Also it is highly complementary to the VLT, as discussed earlier by James Lequeux. Of course, we should evaluate the impact of this array not in the context of today's science, but rather in the one to be foreseen for the period around 2010. But this gives even greater opportunities. HST will have explored the near infrared region by then, while the VLT with its adaptive optics will have attained high angular resolution in the IR windows accessible from the ground.

It appears to be a particularly fortuitous moment to propose an array for the mm region since adjacent communities are developing rapidly: ISO has received 1000 observing proposals, many from "optical" or "radio" astronomers, while the prospect of FIRST has led to the crystallization of an active submm community. Both of these communities will continue to struggle with poor angular resolution (10–30 arcsec for typical wavelengths of ISO and FIRST) for the foreseeable future, and consequently they will much benefit from better images at adjacent wavelengths.

The development of radio facilities follows rather closely that in the optical. The total collecting area of the world's optical telescopes has been doubling every 11 years for much of this century. That of European optical facilities came off to a slow start, but has accelerated over the last 30 years doubling on average every

7 years. These figures are very comparable to those given by Dennis Downes for cm facilities, while the development of mm facilities currently is slower. The great increase in optical collecting area has become possible by a substantial decrease in the cost per m^2 due to the introduction of new technology – in particular active optics. A similar development should be achievable in the construction of a mm array partly by the cost reduction due to the building of a large number of identical telescopes and undoubtedly also by the application of new technology. Of course, the replication advantages would be partly lost if the array would be composed of different types of telescopes (e.g. for mm and for submm) or if the construction time would be stretched out over too long a period.

Optimization of such an array will need much care. To be meaningful it has to be done at a fixed financial envelope and from a broad astronomical point of view. The optimization will involve the frequency range, the diameter of the unit telescopes, the required field of view, the movability of some or all of the telescopes, the altitude and the location of the site. It should also include the operating costs for a decade or so.

The maximum frequency to be achieved strongly affects both the unit telescope cost and the site water vapor requirements, and thereby possibly the necessary altitude. Clearly, at fixed cost an array that aims to function in the 0.3 mm window will be very much smaller than an array limited to 1 mm. The impact of telescope diameter and field of view should be less, but rather detailed technical studies will be needed. Movable telescope are obviously more expensive than fixed telescopes, while also operationally they should add further costs. However, for a proper coverage of the uv plane they have major advantages. Again, detailed studies are needed on the cost factors involved in making the telescopes all movable or only some of them.

Altitude is a major cost factor. Most of us function without too many problems at 4000 m. However, at 5000 m the situation appears to be radically different. This very much impacts the construction costs (installation on the site, building of the infrastructure). It probably even more affects the operating costs. Even at a site like La Silla, ESO has found it difficult and costly to attract qualified engineering personnel, and this would a forteriori be the case at a much higher and more remote place. Location is also important, but almost certainly the Chilean sites are to be preferred. Sites in Namibia probably are inferior to La Silla, while the east side of the Andes is wetter than the west side. Interannual variations may be rather large and daily ones even larger. In comparing a few selected sites it is therefore important to simultaneously measure atmospheric transmission in these, while in the final selection as long a series as possible should be obtained. At Paranal such measurements have been made during six years, and for much of the time comparable data were obtained at La Silla. During the ESO site selection process also numerous data on integrated atmospheric water vapor content and other parameters were obtained by Arne Ardeberg at higher sites further east. These sites were unsuitable for optical telescopes because of higher frequency of cirrus clouds over the Andes, but the extensive data obtained often nearly simultaneously at different locations well deserve to be integrated with the data being obtained now.

The final issue I should address is how to realize the project from a financial and political point of view.

First of all, it is necessary that such a project be supported by a broad astronomical community. This is not yet the case and much "missionary" work remains to be done: presentations at scientific meetings, etc.

A second and rather fundamental point is the relation of the project to ESO. A connection with ESO has three principal advantages: ESO has a good framework for its activities in Chile; even though recently some difficulties have occurred in the relationship between Chile and ESO, these should be regarded as temporary and undoubtedly will have been forgotten a decade from now. Secondly, ESO is engaged in a major project, the VLT, which will be completed very early in the next decade. While the continuing flow of funds to ESO for a subsequent project is not at all guaranteed, it is, nevertheless, clear that it is easier to redirect an existing flow of funds to another project than to start a new flow from scratch. Thirdly, ESO has the experience to deal with large projects, and it is not so simple to build this experience anew.

There are, of course, also problems. First of all, ESO has from a scientific and technical point of view only a limited connection to the mm community, though it should be added that the SEST experience has already improved this quite a bit. Secondly, it is not clear that all ESO member countries would have the same enthusiasm for such a project, while some countries not in ESO, like Spain or possibly the U.K., might wish to participate. The former problem could if necessary be solved by making the array a "special project" in the sense of the ESO Convention. This is a juridical and possibly also politically acceptable option, but it is not very attractive to split the ESO community and also administratively it is cumbersome. In this respect it would also be extremely regrettable if one or more European countries were to join the U.S. project, at least if this were to preclude their later participation in the European one. Unfortunately, this risk it not at all hypothetical.

The other problem could conceivably be solved by an ad hoc contractual arrangement; in the case of Spain, the problem would disappear if the current discussions about a possible ESO membership were to come to fruition.

Another possibility would be to create a joint ESO-IRAM venture. As an example of such an arrangement the European Coordinating Facility for the Space Telescope may be mentioned, which is sponsored conjointly between ESA and ESO. However, this would be more difficult in the case of IRAM because of its different legal framework. Nevertheless, if all partners wish so, there is generally a way to solve such problems.

The final question concerns a broader international cooperation with the U.S. and/or Japan. Considerable obstacles exist to the idea of building one joint array, in part because of the different time scales. Furthermore, industrial policy plays an important part in the funding process. It is not clear that all partners would agree to a 1/3 share, and I do not think that an arrangement like in GEMINI (2 × 8 m optical, 1/2 U.S.A., 1/2 rest of world) would be attractive for Europe.

However, it would be important to maximize compatibility between the three arrays so that medium baseline interferometry between them would be possible, yielding increased angular resolution.

More immediate than all this is to start work on a detailed technical study of the array, including some studies in industry. A sum of the order of 10 MECU would be needed and perhaps some of this could be found in the European Union. Such a project needs a name. How about LEMMA, for the Large European MilliMeter Array? A lemma in mathematics is an element in a proof, and the array undoubtedly will prove or disprove many hypotheses about the Universe. Should the array finally become really international instead of European only one letter has to be changed.

Springer-Verlag
and the Environment

We at Springer-Verlag firmly believe that an international science publisher has a special obligation to the environment, and our corporate policies consistently reflect this conviction.

We also expect our business partners – paper mills, printers, packaging manufacturers, etc. – to commit themselves to using environmentally friendly materials and production processes.

The paper in this book is made from low- or no-chlorine pulp and is acid free, in conformance with international standards for paper permanency.